CISM COURSES AND LECTURES

Series Editors:

The Rectors of CISM
Sandor Kaliszky - Budapest
Mahir Sayir - Zurich
Wilhelm Schneider - Wien

The Secretary General of CISM
Giovanni Bianchi - Milan

Executive Editor
Carlo Tasso - Udine

The series presents lecture notes, monographs, edited works and proceedings in the field of Mechanics, Engineering, Computer Science and Applied Mathematics.
Purpose of the series is to make known in the international scientific and technical community results obtained in some of the activities organized by CISM, the International Centre for Mechanical Sciences.

INTERNATIONAL CENTRE FOR MECHANICAL SCIENCES

COURSES AND LECTURES - No. 385

BEHAVIOUR OF GRANULAR MATERIALS

EDITED BY

BERNARD CAMBOU
CENTRAL SCHOOL OF LYON

 Springer-Verlag Wien GmbH

Le spese di stampa di questo volume sono in parte coperte da
contributi del Consiglio Nazionale delle Ricerche.

This volume contains 247 illustrations

SPIN 10682553

In order to make this volume available as economically and as
rapidly as possible the authors' typescripts have been
reproduced in their original forms. This method unfortunately
has its typographical limitations but it is hoped that they in no
way distract the reader.

ISBN 978-3-211-82920-2   ISBN 978-3-7091-2526-7 (eBook)
DOI 10.1007/978-3-7091-2526-7

PREFACE

Granular materials are of great importance in civil engineering or in manufacturing processes whether they be granular soils, powder for compaction, powder for sintering, powder for ceramics, agricultural products (grains) or raw materials for industries.

Due to their discontinuous nature, the behaviour of such materials is complex and their modelling is quite difficult. Many models have been proposed in the literature but some typical phenomena are not yet really taken into account (the role of fabric, strain localization in shear band, etc,...). On the other hand, new methods such as the micromechanical approach (homogenization techniques, Distinct Element Method) have been developed in the last few years providing new powerful tools for the modelling of granular media. Furthermore, modern technology stimulates these developments by requiring a more precise description of the behaviour of granular materials to be considered in numerical tools such as the Finite Element Method.

The goal of this book is to propose the state of the art on the behaviour of granular materials considered in a dense state and submitted to static or dynamic loadings (rapid flow is not considered). Our purpose is to present the three complementary approaches which are:

- experimental analyses which demonstrate the main features of the behaviour of granular materials;
- micromechanical analysis using homogenization techniques or numerical modelling (D.E.M.) which allow the understanding and the modelling of the behaviour of granular materials from the knowledge of local phenomena;
- phenomenological modelling which remains necessary to use numerical tools such as the F.E.M., allowing many kinds of boundary problems to be solved. Different kinds of constitutive laws will be presented as well as the modelling of coupling phenomena for undrained conditions. A particular

attention will be devoted to the analysis and modelling of shear banding in granular materials.

These three approaches are often presented in different papers or books. One of the purposes of this book is to show that these approaches are complementary and to contribute to a better understanding of the links between them.

This book has been written from the lecture notes proposed for the advanced school entitled: "Behaviour of granular materials", organised by the CISM (Udine) to which I would like to express my thanks for its invitation to coordinate this course. I also wish to express my appreciation to all the lecturers of this school for their kind cooperation and excellent work.

We hope that this book may serve as a good reference for scientists and engineers not only in geomechanics but also in wide disciplines in engineering dealing with granular materials.

Bernard Cambou

CONTENTS

Page

CONTENTS

EXPERIMENTAL BEHAVIOUR OF GRANULAR MATERIALS

P-Y. Hicher
Central School of Nantes, Nantes, France

1. Introduction

Granular materials are present in different branches of engineering such as Civil Engineering, Chemistry, Metallurgy, Pharmacy, Electrical Engineering. The shape, the size, the constituents of the grains can be very diverse, as can also be the mechanical loading to which they are submitted. We can for example study the stability of rockfill dams or the compaction of submicronic powders at very high stresses. These materials, in principle different from each other, have in fact common features due to their granular structure. In particular, their mechanical properties are strongly dependent on the mean stress (the first invariant of the stress tensor). When submitted to small mean stresses, the shear strength is also very small and the granular material can flow almost like a liquid. This is the case for example in silos. On the other hand, when the mean stress is high, the granular material will be able to bear high loading such as in the case of Civil Engineering infrastructures.

These materials contain voids and the void ratio (volume of voids/volume of solids) changes under monotonic or cyclic loading, which creates volumetric strains on the granular material as well as changes in its mechanical properties. These voids may be filled by fluids (liquids and gasses) at different pressures which will influence the mechanical behaviour of the mixture.

In the following chapters we will examine experimental data of the mechanical behaviour of diverse granular materials and analyse the results in order to propose a structured and comprehensive approach.

Most of the experimental results were obtained from French research laboratories, principally the Laboratoire de Mécanique : Sols, Structures et Matériaux de l'Ecole Centrale de Paris and the Laboratoire 3S de Grenoble [1], and partly from research work conducted within GRECO Géomatériaux [2] and ALERT Geomaterials.

2. Physical characteristics of granular material components

2.1 Geometry of the grains

The range of grain sizes can be very large, from ms to μms. An important parameter is the grain size distribution, which can be represented by a grading curve on a particle size distribution chart (Fig.2.1). If the slope is steep, one size predominates and the granular material is "poorly graded"; if the slope is flat , the material is "well graded". To characterise the size range we can use the coefficient of uniformity d_{60}/d_{10} provided that there is sufficient continuity in the grain size distribution curve to justify this.

Grain shapes differ considerably. Flatness, angularity, roughness are parameters which will affect the overall behaviour of granular materials when subjected to external loading (Fig.2.2).

2.1 Geometry of the arrangement

We will consider here only the isotropic aspect of the arrangement. (For the anisotropic aspect see Chapter 3). It can be characterised by the void ratio e, or porosity n or by the density γ_d or γ_d/γ_s (voids contain only air at atmospheric pressure).

$$e \text{ void ratio} = \frac{\text{volume of voids}}{\text{volume of solids}} = \frac{V_v}{V_s} = \frac{n}{1-n}$$

$$n \text{ porosity} = \frac{\text{volume of voids}}{\text{total volume}} = \frac{V_v}{V}$$

$$\gamma_d \text{ dry unit weight} = \frac{\text{dry weight}}{\text{total volume}} = \frac{W_s}{V} = \frac{\gamma_s}{1+e}$$

γ_s unit weight of grains

These parameters describe the state of packing in a particular granular material for a given loading history. This state of packing is strongly dependent on the parameters of the grain geometry characterised by grain size distribution and grain shapes. For example in the case of uniform spheres, the densest possible state corresponds to a specific volume $1 + e = 1.35$ and the loosest to $1 + e = 1.92$. These two extreme states are not easily measurable for a grain assembly with different shape and grain size. Therefore normalised tests were defined in order to derive two parameters called e_{max} and e_{min} which characterise a "loose" state and a "dense" state for a given material (ASTM ref.). These are not absolute limits and a state of packing can be found exceptionally outside these limits, but they are convenient numbers for characterising the potential volume change of a given granular material. These numbers are widely used in soil mechanics.

A given state of packing can therefore be compared to these loose and dense states for a given material. For this purpose we can define the relative density :

$$Dr = (e_{max} - e)/(e_{max} - e_{min})$$

Dr varies from 0 (loose state) to 1 (dense state). Dr has a more direct link with the mechanical properties of a granular material than does e (see Chapter 3).

2.3 Mechanical properties of the grains

Linear elasticity (characterised by Young's modulus E_g and Poisson's ratio υ_g) are often sufficient to describe the behaviour of grains (Chapter 3). In some cases, however, such as elevated stresses or low resistance grains, plastic deformations and maximum strength (ruptures of particles) have to be introduced because they are capable of modifying the overall behaviour of the granular assembly (Chapter 6).

The behaviour of contact between grains plays an important role in that it gives to the assembly a frictional behaviour characteristic of this type of materials (friction angle). In most of the following study, we will only consider a frictional contact (defined by a friction angle φ_μ) between the particles. We will not take into account the presence of "glue" (such as sintering, chemical or physical adhesion, surface forces...) which gives a cohesive resistance to the granular materials (cohesion).

2.4 Discontinuous to continuous medium

In the following chapters, we will focus our attention on mechanical test results interpreted by relating stresses and strains, presuming that the granular material under consideration is a continuous medium.

These parameters are obtained from the force and displacement of samples. The sample tested should be of sufficient size compared with the particles, so that the actual forces and displacements can be related with reasonable accuracy to the stress and strain in the fictitious continuum (Fig.2.3).

Stress continues to be defined as force per unit area (F/S) but the area cannot be allowed to tend to zero on account of the finite size of the particles. The question is therefore how large must S be, compared with the particle size in order for the stress definition to approximate that of a continuum? Gourves [1] measured the sum of the forces applied by cylinders on a plate, and showed that the scatter of the resultant force on the plate increased greatly as its size diminished. In order to have a coefficient of variation of the average stress less than 10%, a plate of at least 5 cm diameter for cylinders of 0.2-0.5 cm diameter must be used. This variation is smaller if the cylinders are replaced by three dimensional particles. In practice the sample dimensions in test apparatus are at least 10 times larger than the largest particles.

There is a further consideration in addition to the requirement that displacements should be measured over a length that is large compared with the particle size. This is that rotation of the particles and slippage between them make a contribution to the strain in addition to that from the compression of the particles. Consider two neighbouring points on either side of the point of contact of two particles. These two points do not in general remain close to each other but describe complex trajetories. Fictitious average points belonging to the fictitious continuous medium can be defined which remain adjacent so as to define a strain tensor. The problem presents itself differently for disordered particles compared with ordered spheres of equal sizes. In the latter case small zones appear in which there is no relative movement of particles. This can lead to specific behaviour such as periodic instabilities known as slip-stick.

The control of homogeneity in mechanical testing is very important because strains and stresses are usually measured at the boundary of the samples and therefore have to be considered reasonably homogeneous within the whole volume (Fig. 2.4).

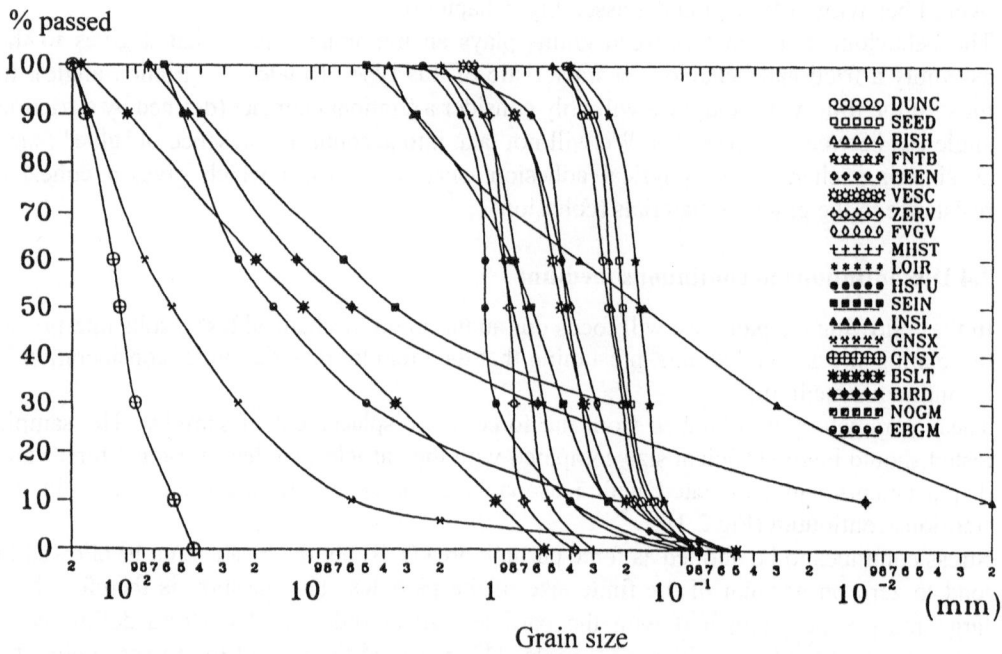

Fig. 2.1 : Examples of grain size distribution curves

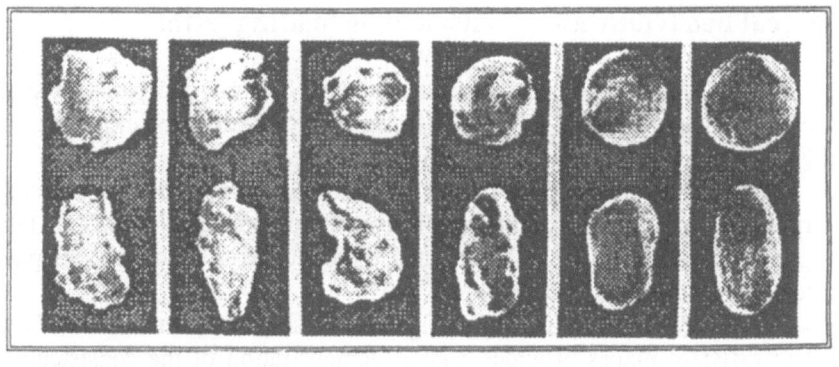

very angular   angular   sub angular   sub rounded   rounded   well rounded

Fig. 2.2 : Examples of grain angularity

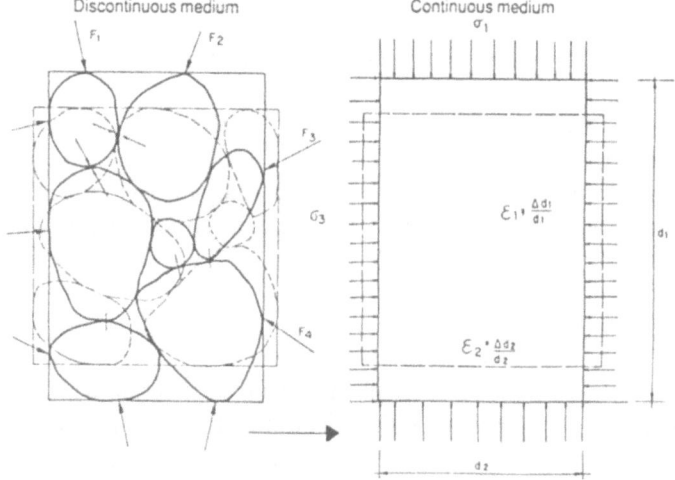

Fig. 2.3 : From discontinuous to continuous medium

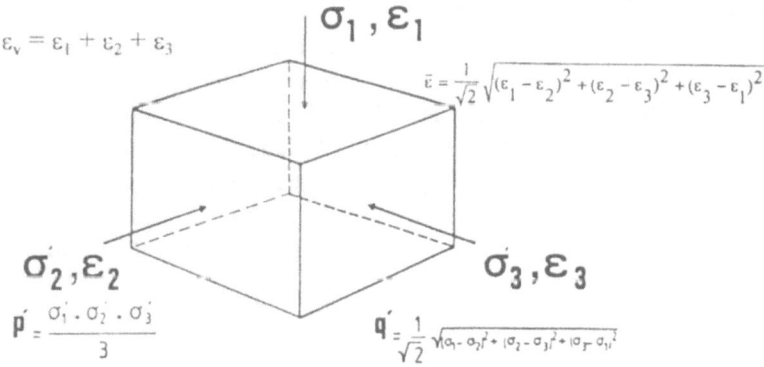

Fig. 2.4 : Stress and strain tensors and invariants

3. Mechanical behaviour along axisymetric loading paths

Mechanical properties can be expressed as the relation between the paths in stress and strain space as a function of time: $\sigma(t) = F(\varepsilon(t))$. In order to determine experimentally the functional F, each and every strain history must be followed closely.

In practise, simpler paths are imposed by the limitation of the apparatus, for example the axisymmetric path, $\sigma'_2 = \sigma'_3$ = constant, used in conventional triaxial tests. This test starts with isotropic compression ($\sigma'_1 = \sigma'_2 = \sigma'_3$) and thereafter axial loading takes place at a constant rate of axial strain.

A systematic representation of 5 or 6 linked figures that show different views of the paths projected onto different planes provides a good demonstration of the consistency of the results.

In this chapter we will present evidence of the properties of granular materials along simple paths such as:

1. Isotropic stress path
2. σ'_3/σ'_1 constant path
3. Conventional triaxial (compression with $\sigma'_2 = \sigma'_3$ = constant) path
4. Deviatoric stress path (p' = constant)
5. Deviatoric strain path (e = constant)
6. Oedometer path ($\varepsilon_2 = \varepsilon_3 \simeq 0$).

In Chapter 4 these results will be extended to 3D paths. We will first examine results on supposedly isotropic materials and then study the influence of structural anisotropy.

3.1 Compression tests

3.1.1 Isotropic compression

If we gradually increase the isotropic pressure p' on a granular material, we will observe an isotropic deformation (ε_i), if the sample is initially isotropic. It cannot be infinite, however, for it is limited by the total elimination of the voids; the void ratio e is in this case a more significant parameter than ε_i (Figs. 3.1 et 3.2).

At the beginning of loading, we observe a quasi elastic behaviour (high percentage of reversibility). The elasticity is non-linear as discussed in chapter 7 : the stiffness of the sample increases with the isotropic pressure.

Above an elastic limit pressure p'_{ic}, the decrease in the void ratio becomes especially non-retrievable (plastic strains occur). The value of p'_{ic} depends on the initial void ratio of the material; it increases when the initial density is higher (Fig. 3.3).

Isotropic behaviour is often represented on the (e, log p') plane. The quasi-elastic domain is represented by a straight line with a slope C_s for limited loading or unloading. Plastic behaviour can also be represented by a straight line with a slope C_c, called the volumetric compression index (Fig. 3.4). For elevated stresses, this slope tends to decrease in the absence of significant grain breakage.

$e = ei - Cc \log (p'/p'i)$

$E_{iso} = dp'/d\varepsilon_v = 2.3(1+e)p'/C_c$

The secant hydrostatic modulus is therefore proportional to the mean stress.

Values of C_c range between 0.1 and 0.2, depending on the geometry of the grains and the grain size distribution. Plastic deformations of the grains or significant grain ruptures lead to higher value of C_c (see Chapter 6)

3.1.2 One-dimensional compression

Powder compaction is often realised in a mould with rigid boundaries called oedometer. Under these conditions a given stress is applied in one direction (1) and the deformations in the perpendicular directions (2, 3) are equal to zero. The behaviour of the material can be represented by a relation $\varepsilon_1(\sigma_1)$ or $e(\sigma_1)$. We obtain similar behaviour as that observed during isotropic tests (Fig. 3.5).

We can identify two parts in this relationship: the first for loading up to a given σ'_{1c} and the second beyond this. Deformations are more nearly reversible in the first part than in the second (in which only a small part of the loading strains can be recovered). σ'_{1c} can therefore be considered as the limit of a quasi-elastic domain. Its value depends on the initial void ratio of the material. Preparation of specimens in the gravity field leads to initial void ratios (or relative density) which depend on the dimensions and the density of the particles.

In the plastic domain, if stresses $\sigma'_2 = \sigma'_3 = \sigma'_r$ are measured on the surface of the cylinder during the compression, one notices that

$$\sigma'_r/\sigma'_v = \sigma'_2/\sigma'_1 = \sigma'_3/\sigma'_1 = K_0$$

remains virtually constant as long as the grains deform elastically. Values of K_0 range between 0.3 and 0.7. This corresponds to q'/p' remaining constant because

$$(q'/p')_0 = 3(1 - K_0)/(1 + 2K_0)$$

Placed in the $(e, \log p')$ plane, the relation becomes straight with the same slope C_c as the one obtained in isotropic compression. The one-dimensional compression line is located in this plane under the isotropic line, which means that for the same mean stress, the compaction is more pronounced in oedometer condition : a deviatoric stress helps to densify a granular material, probably because the re-arrangement of the particles are facilitated (Fig.3.6).

When the stresses become high enough to provide plastic deformation of the particles, K_0 does not remain constant but increases with the density (Fig. 3.7).

During unloading, a quasi-elastic behaviour can be observed at the beginning of the stress reversal, corresponding to a straight line of slope C_s in the semi-log plane. If unloading continues to take place, the behaviour becomes less reversible and the relation is no longer linear. One must notice that the stress path during unloading does not coincide with that obtained during the loading stage. As a matter of fact, the value of K_0 increases during unloading and can acquire values higher than 1 (Fig. 3.7). The sample is therefore submitted to radial stresses higher than the axial one. The state of stress can reach the plastic criterion in extension and generate a failure plane or cracking perpendicular to the axis of loading.

3.1.3 Stress path with $\sigma'_3/\sigma'_1 = K$

A stress path at $\sigma'_3/\sigma'_1 = K$ or $\eta = q'/p' = 3(1 - K)/(1 + 2K)$ leads to similar behaviour as that observed along the isotropic stress path and plastic compressibility is also represented by a straight line of slope C_c on the (e, log p') plot (Figs. 3.8, 3.9).

A particular case is the oedometer test path which also represents a constant K stress path with $K = K_0$. A smaller value of K, corresponding to a higher value of η gives higher reduction in the void ratio for the same mean pressure. This result illustrates the effect in granular media of the deviatoric stress on volume changes, which will be discussed more widely in the following paragraph.

3.2 Conventional triaxial path ($\sigma'_2 = \sigma'_3$ = constant)

This test starts from a state of isotropic stress, i.e. $\sigma'_1 = \sigma'_2 = \sigma'_3$, which is represented by the point I on the space diagonal in stress path (Fig. 3.10). Compression at a constant strain rate follows. The cell pressure, $\sigma'_2 = \sigma'_3$ remains constant. Let us first consider specimen with low initial density (loose state) or small relative density (0<Dr<0.3). The test demonstrates a continuous increase of σ'_1 until it attains a maximum value unaffected by further changes in ε_1. This final state is named perfect plasticity.

The conventional representation of a triaxial test is given in Figure 3.11 by the two curves ε_1 - q' and ε_1 - ε_v. We can notice that the stress-strain relation is highly non linear. This non-linearity corresponds to a non reversible behaviour of the granular material which develops, as we will see in chapter 6, even for a small strain amplitude. At the same time a volume change (it corresponds here to an increase in density) develops which tends to a constant value as ε_1 increases and a state of perfect plasticity, characterised by no further volume change, is reached if the deformations remain homogeneous within the test sample.

The Coulomb hypothesis suggests that the envelope containing the perfect plasticity Mohr circles is a line in the Mohr plane, i.e.:

$$\tau = c' + \sigma'_n \tan \varphi'$$

c' is the cohesion and φ' is the angle of internal friction (friction angle).

For granular materials with no internal force (such as capillarity forces, see Chapter 4) and no "glue" between the grains, the cohesion c' is equal to zero.

Perfect plasticity may also be represented by the line q' = Mp' in the (q', p') plane where p' is the mean effective stress and q' is the deviatoric stress (Fig 3.10).

We use, instead of the classical second invariant

$$q' = \frac{1}{\sqrt{2}} \sqrt{\left(\sigma'_1 - \sigma'_2\right)^2 + \left(\sigma'_2 - \sigma'_3\right)^2 + \left(\sigma'_3 - \sigma'_1\right)^2}$$

which is proportional to the normal distance in principal stress space of the point representing the stress (σ'_1, σ'_2, σ'_3) from the space diagonal. If $\sigma'_2 = \sigma'_3$, then q' = σ'_1 - σ'_3. The stress path is represented by a line of slope 3:1 in the (q', p') plane. Note that in three-dimension, the relation q' = Mp' corresponds to the Drucker-Prager criterion (see Chapter 4).

Figure 3.12 presents the overall behaviour in five linked graphs which facilitate the analysis of test results. We prefer void ratio to volume change because it allows a better emphasis of the influence of the initial density as we shall see later on.

On the (e, log p') plot, perfect plasticity is represented by a line of slope C_c parallel to the isotropic plastic compression line.

For specimen whose initial void ratio corresponds to a point located on the isotropic plastic line, the relationship of q'/p' - ε_1 and Δe - ε_1 are mainly independent of the mean stress.

The (e, log p') plot shows straight lines which have the same gradient C_c for the isotropic, oedometric and perfect plasticity paths and more generally for any q'/p' = constant test.

Usually, due to the mode of preparation, the initial state of a granular material is not located on the isotropic plastic line, but on a point situated on the left side of this line in the (e, log p') plane on a line of slope C_s. In this case the volume change during triaxial testing will be different from the previous case. At the beginning of loading, the specimen remains contractant (decrease in void ratio), but at a given stress ratio, the void ratio starts to increase and the granular material becomes dilatant. This increase in void ratio even under compressive stresses is a typical behaviour for granular media called "dilatancy".

If we plot in the linked diagrams the results obtained on a same granular material at different initial void ratios and isotropic stresses, we can draw the following conclusions (Figs. 3.13, 3.14). The curve q' - ε_1 has a peak corresponding to the dilatancy of the sample which increases with the initial relative density of the sample. At fairly large homogeneous deformations, the same perfect plasticity is obtained at q' = Mp' and at the same void ratio e, for a given p', irrespective of the initial void ratio, i.e., "large deformations override the previous conditions".

In practise, it is difficult to obtain perfect plasticity because of strain localisation, especially in dense materials. Figure 3.15 illustrates the effect of strain localisation on the stress-strain response, the existence of a very pronouced peak in the curve q' - ε_1, and on the volume change, dilatancy is suddenly stopped and perfect plasticity is not reached in the (e, log p') plane. It is possible however to limit the effect of the localisation by reducing the slenderness of the specimen and by taking measures to minimise the friction between the rigid end plattens and the specimen (Fig. 3.16). If this is done, then much greater dilatancy is observed and the e - log p' curves approach the state of perfect plasticity. The stress-strain (q' - ε_1) curves continue to reduce after the peak, but in a smoother way.

Figures 3.17 to 3.21 present some triaxial test results on granular materials with different nature of grains. The same pattern is observed : influence of the mean stress, volume changes indicating contactancy or dilatancy according to the initial mean stress and relative density.

The volume change due to the application of a deviatiatoric stress affects the entire stress-strain relationship. Rowe [11] showed that the results of triaxial tests can be represented in a first approximation by a linear relationship between σ'_1/σ'_3 and (1 - $d\varepsilon_v/d\varepsilon_1$). This is independent of the initial voids ratio. If the slope of this line is expressed as $tg^2(\pi/4 + \varphi'/2)$, φ' then approximates the perfect plasticity friction angle φ'_{pp}. This conclusion is valid for granular materials having a relatively isotropic arrangement. Rowe's stress-dilatancy law can then be written as follows :

$\sigma'_1/\sigma'_3 = tg^2(\pi/4 + \varphi'_{pp}/2)(1 - d\varepsilon_v/d\varepsilon_1)$

The minimum, A (Fig. 3.22), on the curve of volumetric strain, ε_v, against axial strain ε_1, and the perfect plasticity B both correspond to $d\varepsilon_v/d\varepsilon_1 = 0$. Therefore from Rowe's relation, $\sigma'_1/\sigma'_3 = tg2(\pi/4 + \varphi'_{pp}/2) = K_{pp}$ at both A' and B'. Thus to obtain φ'_{pp}, it is sufficient to project A up to A' which should be at the deviatoric stress B' corresponding to perfect plasticity. The true point B is difficult to obtain because of the localisation associated with the large strains at B. The locus of point A' in the p' - q' plane is a straight line starting from the origin of the axes and separating contractant and dilatant behaviour. It is called "line of plane transformation" [12] or "characteristic line " [5].

Rowe's relation indicates that for a given ε_v - ε_1 curve, it corresponds one curve σ'_1/σ'_3 - ε_1 (Fig. 3.23). The stress-strain curve can be normalised by the value of the friction angle (or slope M) so that the behaviour of a granular medium can be represented by an assembly of curves q'/Mp' - ε_1, each of them corresponding to a ε_v - ε_1 relation which can be parametred by using $(d\varepsilon_v/d\varepsilon_1)max$, sometimes called the dilatancy angle. The total dilatancy can be expressed in the following way :

$e_{NC} - e_{OC} = (C_c - C_s)\log p'_{ic}/p'$

which gives a parameter by which different granular materials can be compared.

In the (q', p') plane the envelope of maximum strength is above the line of slope M for a dilatant material. This line of maximum resistance from the origin permits the definition of a peak friction angle φ' which decreases if the average pressure p' increases (Fig. 3.24). For a given pressure the peak friction angle φ' increases if the initial voids ratio decreases according to the approximation :

$e\ tg\ \varphi' = e_{pp}\ tg\ \varphi'_{pp} = constant = \Lambda$

The value of Λ depends on the particle geometry and rugosity. Typical values between 20 and 35° for φ'_{pp} were measured, depending on the type of particles. Rowe's relation gives φ' as a function of φ'_{pp} and the maximum dilatancy.

3.3 Extension tests

Extension tests can also be performed in the triaxial by decreasing the axial stress and by keeping the radial stresses (Fig. 3.25) constant. Qualitatively the results are the same than in compression showing also contractant and dilatant behaviour in relation with the initial void ratio. It is an interesting test to be conducted in order to verify the validity of the Mohr-Coulomb criterion: φ'_{pp} = constant (see Chapter 4). However the development of strain localisation in the early stages of testing, producing a "necking" phenomenon, renders the interpretation and such testing difficult and no clear response can be given, even if the majority of results produces higher friction angles in extension rather than in compression.

3.4 Deviatoric stress path (p' = constant)

The deviatoric stress path in principal stress space is orthogonal to the mean stress path (Fig. 3.26). Starting from an isotropic state of stress, one of the principal stresses is increased by keeping constant the mean pressure.

During compression a typical phenomenon of granular material is observed, namely that the deviator stress causes volume change. For granular materials whose initial state is located on the isotropic plastic line (slope C_c), this volume change is a reduction of the voids termed contractancy and characterised by the deviatoric compression index C_d. This is the slope of the line relating e to the logarithm to the base 10 of $(1 + \eta^2/M^2)$, i.e. :

$$e = e_0 - C_d \log (1 + \eta^2/M^2)$$

The former expression indicates that the variation of volume depends only on η.

If the strain exceeds a certain value, the stress tensor remains constant as does the void ratio. Once again this is the state of perfect plasticity.

If the initial state is such that the representative point is located below the perfect plastic line, dilatancy will occur during a deviatoric stress path in order for the point to join the perfect plasticity state.

Perfect plasticity in axisymetric compression is characterised by :

(a) a law of the Coulomb type : $q' = Mp'$

where M corresponds to the angle of friction of perfect plasticity : $\sin \varphi'_{pp} = 3M/(6 + M)$

(b) a straight line $e_{pp}(\log p')$ of gradient C_c passing approximately through the following points:

$p' = 0.1$ MPa for $e = e_{max}$, $p' = 5$ MPa for $e = e_{min}$

3.5 Strain control tests

Strain control triaxial tests can also be performed and it is interesting to consider among them the tests at constant $d\varepsilon_v/d\varepsilon_1$.

3.5.1 Isochoric test

The constant volume tests are tests where $d\varepsilon_v/d\varepsilon_1 = 0$. The corresponding stress paths depend on the sign of the volume change during conventional triaxial tests (Figs. 3.27, 3.28). If the material is contractant during these tests, a continuous decrease of the mean effective stress is observed because contraction is prevented. The deviatoric stress q' reaches a maximum and then decreases to a constant value, which can be equal or close to zero for very loose specimen (static liquefaction). If the material is dilatant during conventional testing, the mean pressure starts to decrease (initial contractancy) and then increases along a stress path at constant q'/p' until it reaches the perfect plastic state.

The perfect plastic state is the same as the one obtained during conventional tests in both planes (p', q') and (e, log p').

This kind of loading can in particular be applied on saturated granular materials where voids are filled with an incompressible fluid (see Chapter 5).

3.5.2 Constant $d\varepsilon_v/d\varepsilon_1$ tests

These tests lead to impose a constant contractancy or dilatancy to the specimen. Different cases can be obtained (Fig. 3.29). When the imposed contractancy is higher than the one obtained for conventional tests, the stress path reached a line of constant slope q'/p' with a continuous increase of p'. The slope decreases with an increase of $k = d\varepsilon_v/d\varepsilon_1$ and increases when the initial density increases. These results are in accordance with the fact observed previously that the plastic flow depends on the ratio $\eta = q'/p'$.

When dilatancy is imposed, the kinematics cannot be carried out to the end by the specimen and the stress path tends to return to the origin (p'= q'= 0) tangential to the failure surface. For loose specimen, a maximum value of q is first obtained before a decrease to zero, the ratio q'/p' increasing continuously. For dense specimen, the stress path can follow the failure surface with an increase of p' before a maximum stress is reached whereupon the stress path changes direction to return to the origin.

All these results illustrate the influence of the volumetric strain on the behaviour of granular materials.

3.6 Inherent and Induced Anisotropy

In general, granular materials are anisotropic; their response to a given stress depends on the orientation of that stress. This is due to the arrangement of the particles. If the distribution and shape of the particles is statistically independent of the orientation of the chosen axes, the material may be considered to be isotropic. Its mechanical properties are then isotropic and the mechanical law which govern its behaviour is also isotropic. This law can be defined independently of the axes chosen for defining the tensors.

In the classic case, there are two types of anisotropy:
- Initial anisotropy due to the mode of deposition ;
- Anisotropy induced by the history of irreversible deformations caused by the stresses which have been applied on the sample.

The initial anisotropy is due to the fact that it was created by the application of an anisotropic stress tensor during deposition in a gravitational field. If it were possible to reconstruct the complete history of a sample, it would not be necessary to distinguish between induced and initial anisotropy. It is, however, more convenient to start with the initial state.

3.6.1 Geometric anisotropy

If the particles have a spherical form, one can characterise the geometric anisotropy of the assembly by the orientation of the tangent planes at the contacts (Fig. 3.30). The distribution of these directions evolves with the deformation of the material. It produces a reorientation with a concentration of the tangent planes normal to the direction of the principal stress,

Biaxial tests carried out by Biarez & Wiendick [16], Konishi, Oda & Nemat-Nasser [17] on rolling cylinders provided evidence of the sliding and the rotation of different elements during the course of homogeneous straining of a continuous medium. If we represent the number of tangent planes in a given direction by a vector proportional in length to the number of planes per unit of angle, the extremities of these vectors form an ellipse. a and b, the lengths of the major and minor axes, represent the principal directions of anisotropy. Biarez & Wiendick proposed that the anisotropy be quantified by the ratio:

$A = (a - b)/(a + b)$

Preparation of a granular material in the gravitational field gives an initial value of A which is modified by subsequent deformations. These cause quantitative changes to A and/or change the orientation of anisotropy. In the latter case there is rotation of the principal strain directions. Tests performed on cylindrical rods by Lanier et al.[2] showed that numerous contacts were created or lost during an external loading. The maximum of lost contacts correspond to the direction of maximum extension for the kinematics of the equivalent continuous medium. This anisotropy in the structure of the granular material generates a mechanical anisotropy which is characterised by an increase of the stiffness in directions corresponding to an increase in the number of contacts and a stiffness decrease in directions corresponding to a lost of contacts.
Anisotropy reaches a maximum value as perfect plasticity is approached.

3.6.2 Mechanical anisotropy

The anisotropy of the arrangement of the particles in a granular material creates an anisotropy of the mechanical behaviour. This can be seen in various mechanical stress paths. Our study here is limited to an initial orthotropic anisotropy. Thus an isotropic compression creates plastic anisotropic strains. For a sample with orthotropy the strain along the axis of the orthotropy is less than that measured in the perpendicular direction. These plastic deformations modify the initial anisotropy, making the material more isotropic (Figs. 3.31, 3.32).
We can say that the initial stiffness is larger when the direction of the major principal stress coincides with the direction defined previously by the small axis of the ellipse. An example of axisymmetric loading in the direction of anisotropy and in a perpendicular direction shows a large decrease in stiffness and a larger contractancy in the second case (Fig. 3.33). The same observations can be made in the case of anisotropy created by an initial loading in compression or in extension (Fig. 3.34). For anisotropic specimen, an axisymmetric loading can lead to a three-dimensional state of strain as shown in Figure 3.35.
In general, irreversible deviatoric strain creates a quasi-elastic anisotropic region, which will not be centered around the hydrostatic axis. As a first approximation, one can imagine a cone in principal effective stress space with the apex at the stress origin and with the highest value corresponding to the maximum stress tensor.

3.7 Correlations for granular media

The study of the mechanical behaviour of a granular medium can be treated theoretically in two different ways: either by applying the method of mechanics of continuous media to a supposedly continuous granular sample or by working directly on the assembly of grain of the continuous medium. Combining these two approaches has produced a relationship between the mechanical properties of the discontinuous medium and the behaviour of the supposedly continuous medium. The relationship has established a methodology of classification and empirical correlations, which characterises particular aspects of the behaviour of granular media. Up to now, empirical correlations have successfully offered some links between physical parameters of the discontinuous medium and numerical parameters of simple constitutive equations modelling the behaviour of the equivalent continuous medium (elasticity, perfect plasticity).

3.7.1 From discontinuous medium to continuous medium

For the granular media to which the mechanics of continuous media are applied, it is advisable to consider the scale of the grains at the microscopic level and the imaginary continuous medium (CM) as composed of an assembly of grains, with particular boundary conditions for each. The response of such a point of the material to a given stress or strain rate is modelled using the rheological parameters of (CM) and should be dependent on the constitutive equations of the grains, on the grain-grain interactions, grain-fluid (liquid or gas) equations, and on both mechanical and geometrical boundary conditions of the grains and fluids [18].

Hence, the physical or identification parameters which explain the rheological parameters have to be considered and classified according to Figure 3.36.

Geometry of grains (size, shape, surface state) and the possible number of contacts (granulometric range) can be regarded as parameters of the interaction equations (DM). For convenience, we shall call the first two classes and this last one the nature of the discontinuous medium (DM). Since it is difficult to identify all these parameters representative of (DM), the nature of granular media can be synthesised by a particular arrangement of the grains in response to suitable normalised mechanical tests, i.e., maximum and minimum void ratios, proctor density (widely used in soil mechanics), etc.

The total assembly, as another indicator of the number of contacts, is characterised by the parameters of the grain arrangement. We shall call these parameters the compacity of the discontinuous medium. We therefore obtain the first basic equation for the classification and the connection of the parameters :

Nature (DM) + Compacity (DM) => Rheology (CM)

The previous equation shows that the rheology of the equivalent continuous medium is explained by joining together parameters of the discontinuous medium representing the nature of the grains and the arrangement of the grains.

We can also combine the first two terms of the equation by using the particular compacities e_{max}, e_{min}, representative of the nature of the grains together with parameters representative of the grain arrangement. We shall call these mixed parameters, the mechanical state of the discontinuous medium (DM). For example we can use relative density :

$Dr = (e_{max} - e_0)/(e_{max} - e_{min})$

with e_0 the actual void ratio of the granular medium.
The second basic equation can then be written as :

Mechanical State (DM) => Rheology (CM)

These classifications and connections are synthesised in Figure 3.36.

3.7.2 Examples of correlations for granular media

Empirical connections between physical parameters of the discontinuous medium and mechanical properties of the equivalent continuous medium are expressed in mathemetical relationships called correlations. The previous equations give a framework to obtain these correlations in a well-founded manner. In the following section we give some examples, assuming that the particle geometric description remains constant (no plastic deformation of the grains nor grain breakage).

The correlations can be expressed in the following manner:

(1) from within a class of parameters. Thus from Class I we can for example obtain the values of e_{max} and e_{min} in relation with grain angularity and grain size distribution ((Fig. 3.37).

(2) Between several classes by using the form of equations (1) and (3). For example:

$$130 \, (d_{10})^2 \, (e/e_{max}) \; = \; K \; (\text{permeability}) \; (\text{d cm, K cm/s})$$

$$e \, \text{tg} \, \varphi' = e_{pp} \, \text{tg} \, \varphi'_{pp} = \text{constant} = \Lambda$$

(3) Some parameters of the continuous medium are indepedent of the initial compacity. This is the case for example for the plastic compressibility (slope C_c in the plane (e, log p') in isotropic or oedometric loading) or the perfect plasticity parameters (φ'_{pp}, the position of the perfect plasticity line in the (e, log p') diagram).

We can write :

Nature (DM) => Rheology (CM)

Figure 3.38 gives the position of the perfect plasticity line as a function of e_{max} and e_{min} :

$$p' = 0.1 \text{ MPa for e} = e_{max}, \quad p' = 5 \text{ MPa for e} = e_{min}$$

Combining the correlations of Figure 3.37 and Figure 3.38, we obtain a chart for the position of the perfect plasticity line and a function of nature parameters of the grains (Fig. 3.39).

These results are very useful in analysing test data. Figures 3.13 and 3.14 illustrate the comparison between correlations and test results for the position of the perfect plasticity line. In many tests, however, strain localisation in dense granular material does not allow dilatancy to develop completely inside the specimen and the perfect plasticity line therefore is not reached (Fig. 3.15).

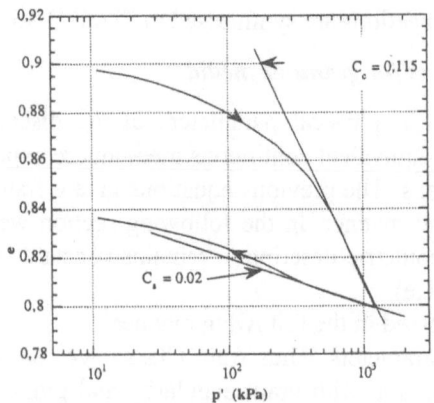

Fig. 3.1 : Isotropic loading on a loose sand [3]

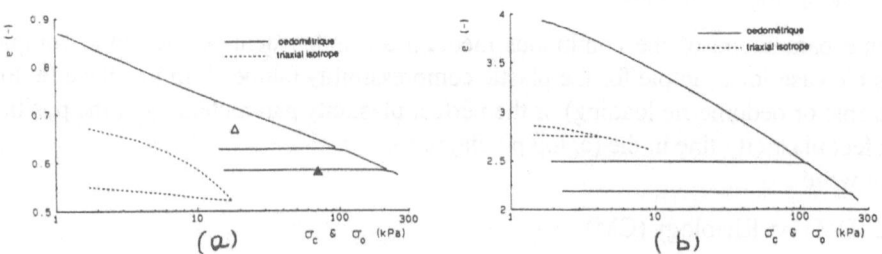

Fig. 3.2 : Isotropic and one-dimensional loading on a mineral (a) and an organic (b) powder [4]

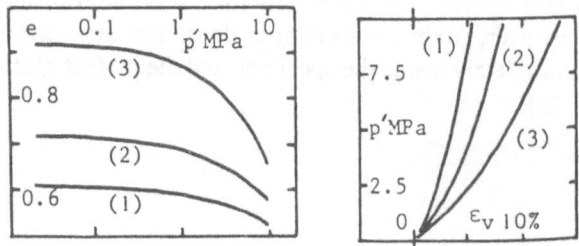

Fig. 3.3 : Isotropic loading on Fontainebleau sand, (1) Dr = 90%, (2) Dr = 62%, (3) Dr = 10% [5]

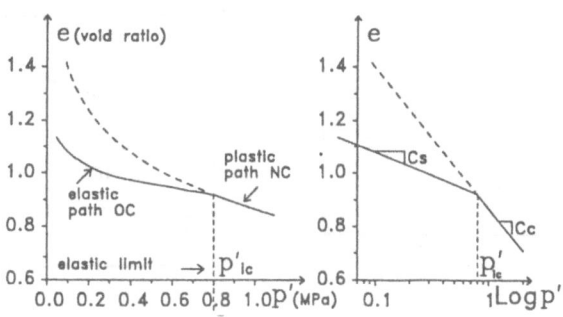

Fig. 3.4 : Representation of an isotropic loading path

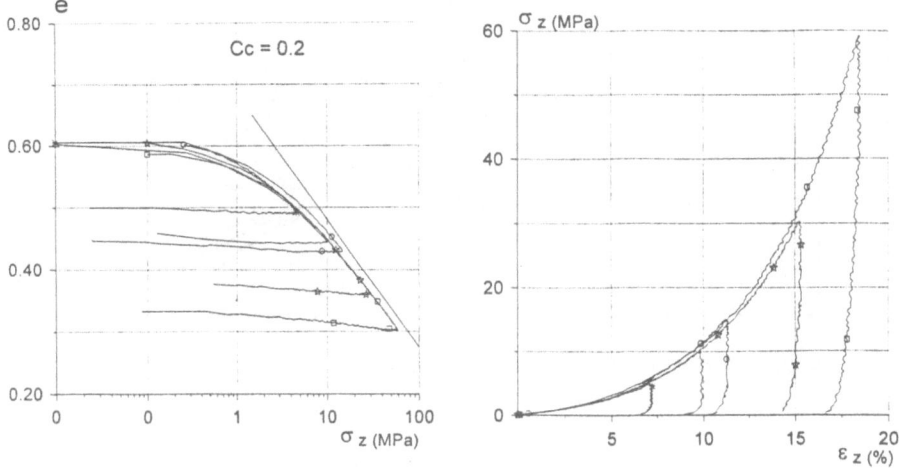

Fig. 3.5 : one-dimensional loading on crushed granite [6]

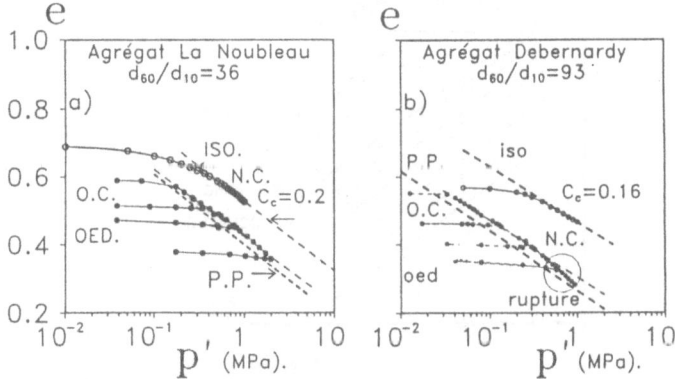

Fig. 3.6 : Isotropic and one-dimensional loading on gravels

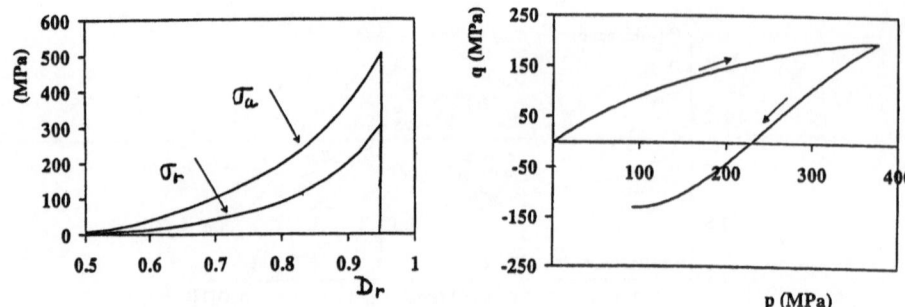

Fig. 3.7 : One-dimensional loading on copper powder [7]

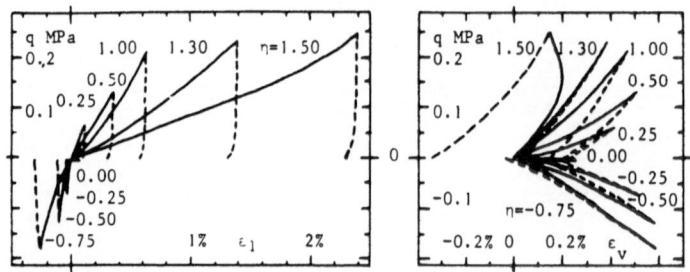

Fig. 3.8 : Radial loading on Fontainebleau sand [5]

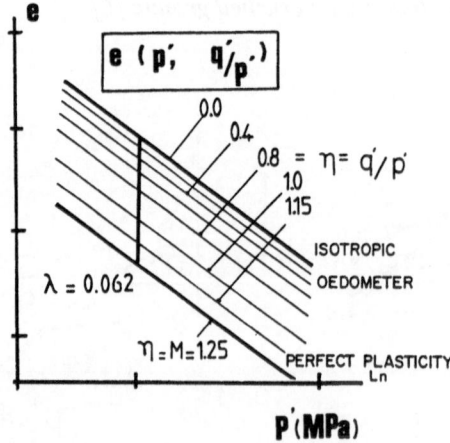

Fig. 3.9 : Constant q'/p' tests

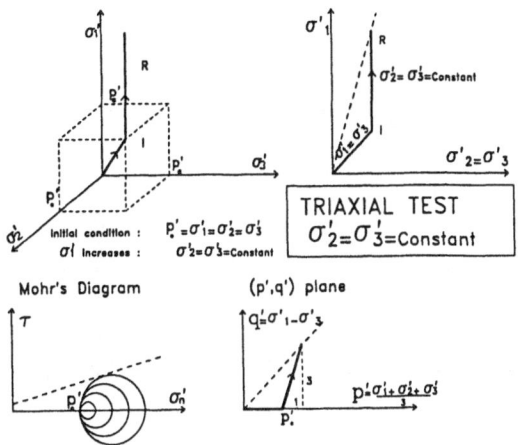

Fig. 3.10 : Triaxial test representation

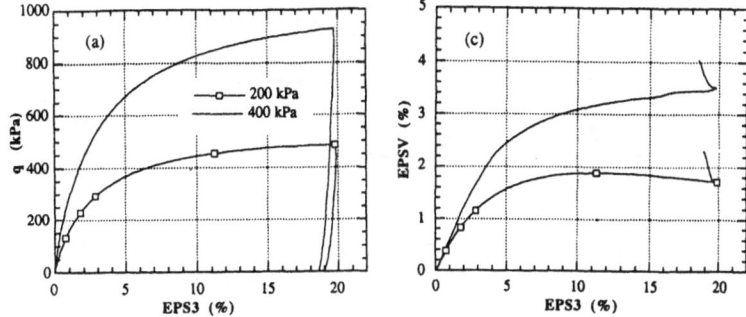

Fig. 3.11 : Triaxial tests on a loose sand [3]

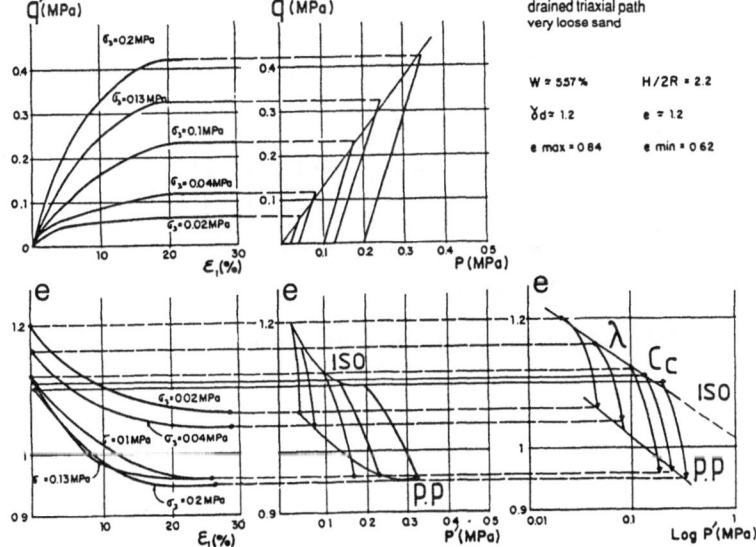

Fig. 3.12 : Stress-strain relationships in triaxial tests on loose sand

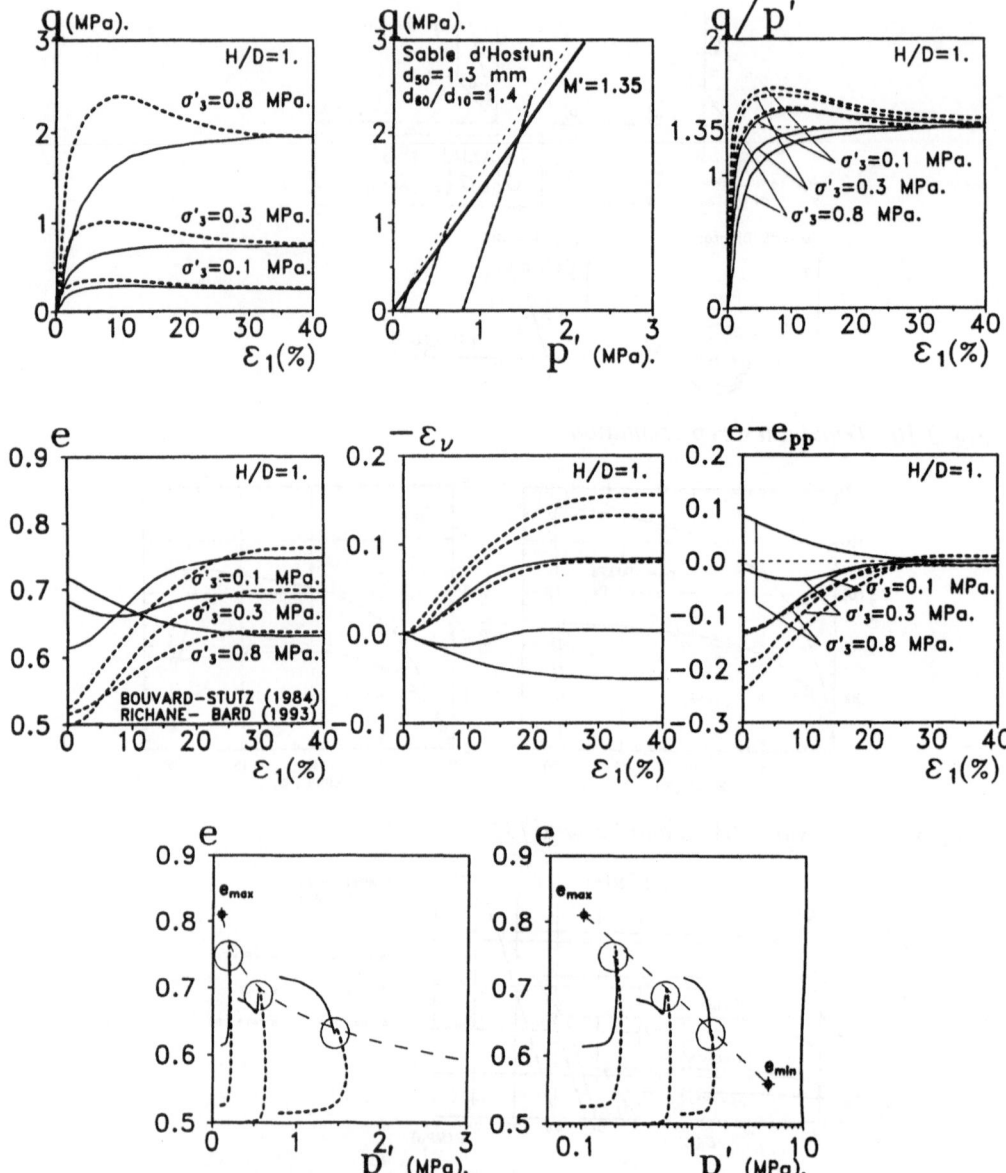

Fig. 3.13 : Stress-strain relationships in triaxial tests on sand at different densities and initial mean stresses

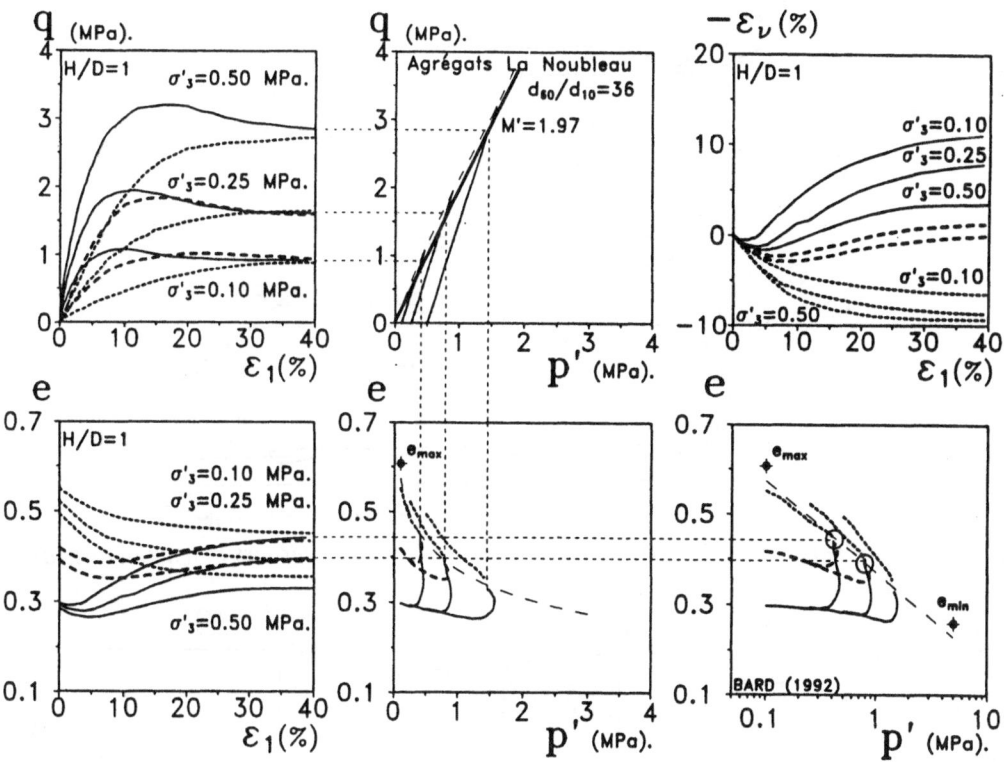

Fig. 3.14 : Stress-strain relationships in triaxial tests on gravels

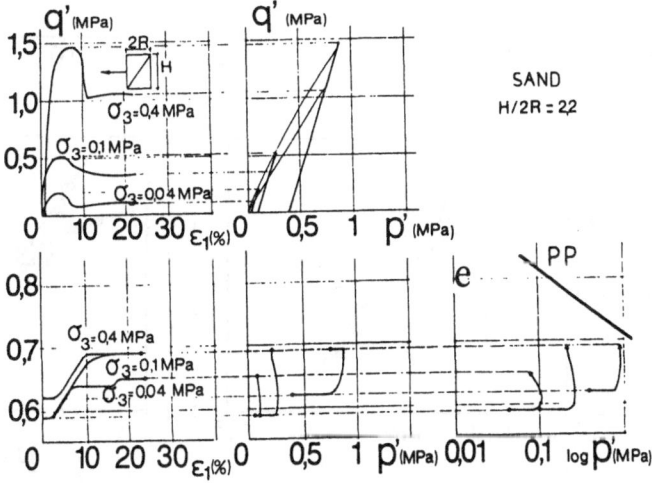

Fig. 3.15 : Strain localisation in triaxial tests on dense sand

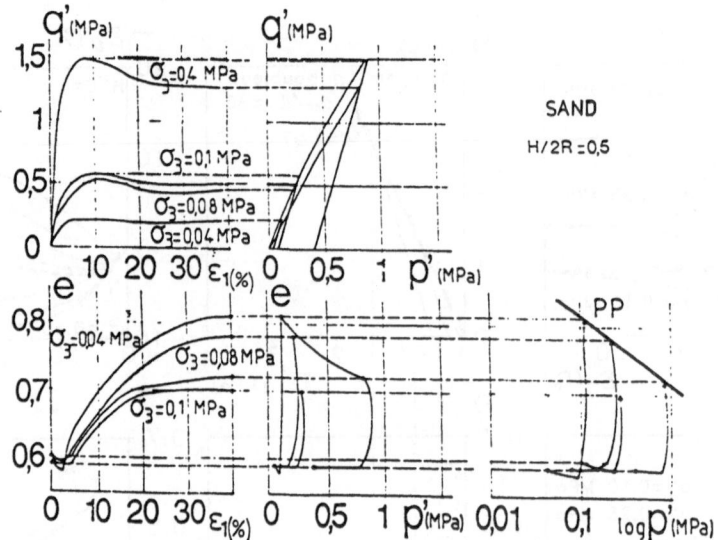

Fig. 3.16 : Influence of anti-frictional platen and reduced slenderness in triaxial tests on dense sand

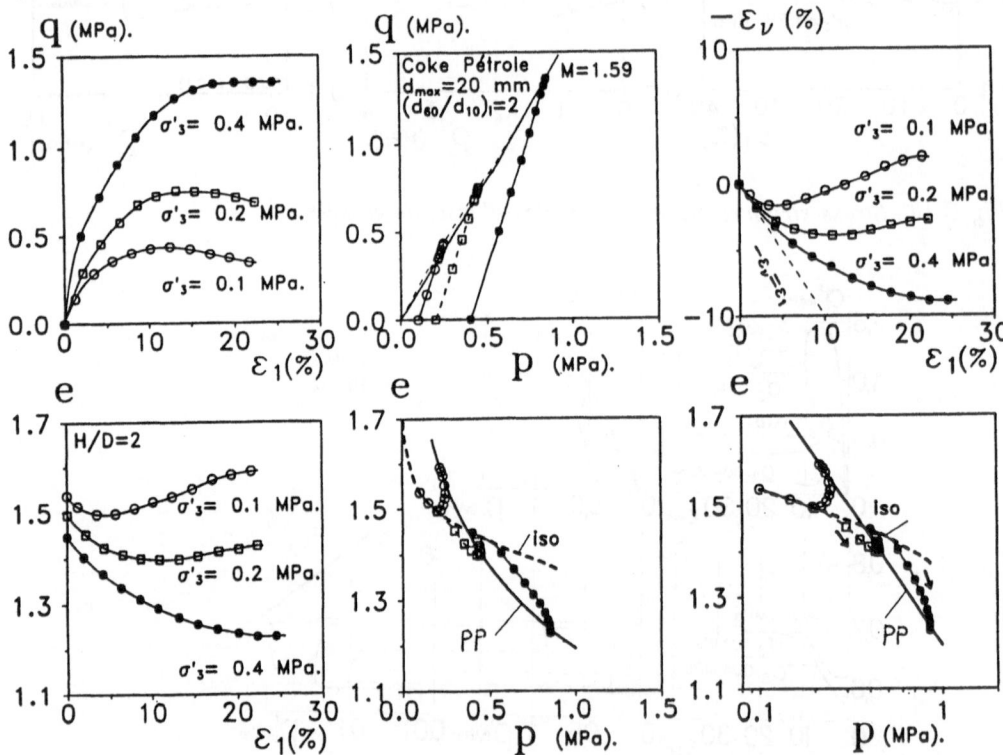

Fig. 3.17 : Triaxial tests on petroleum coke [8]

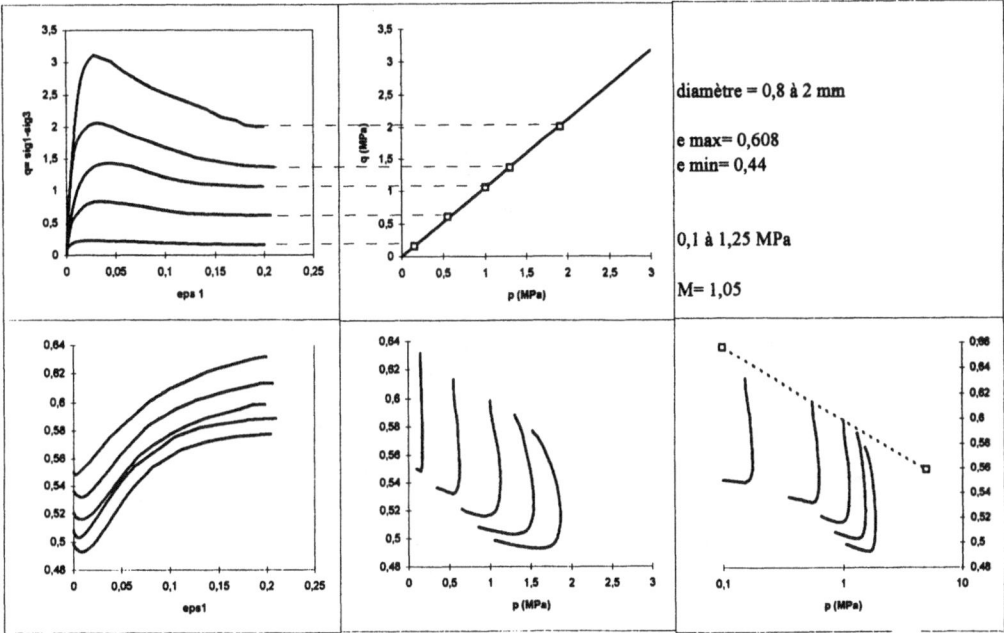

Fig. 3.18 : Triaxial tests on plastic balls [9]

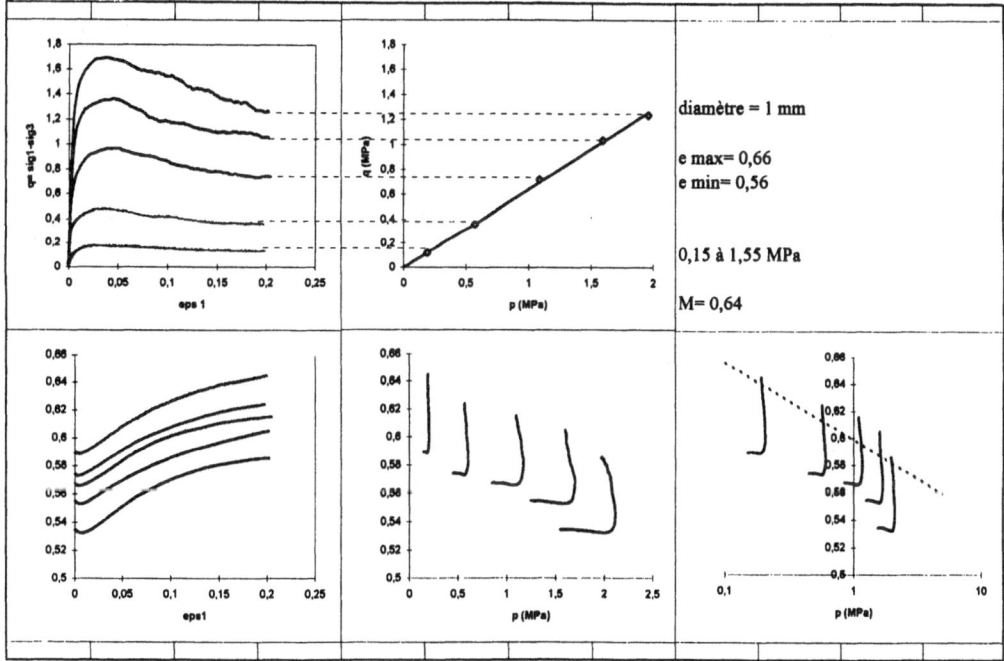

Fig. 3.19 : Triaxial tests on steel balls [9]

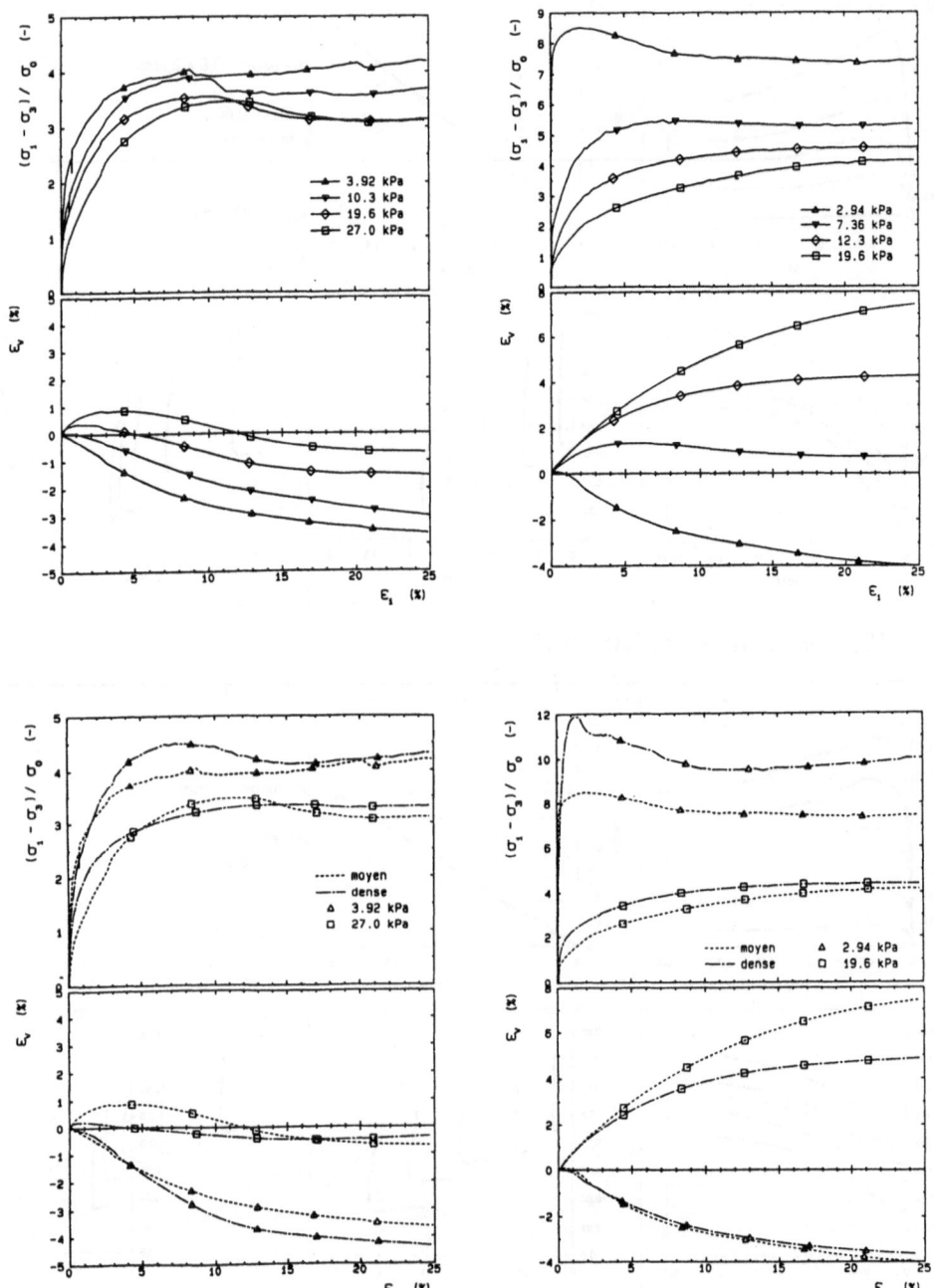

Fig. 3.20 : Triaxial tests on mineral and organic powders [4]

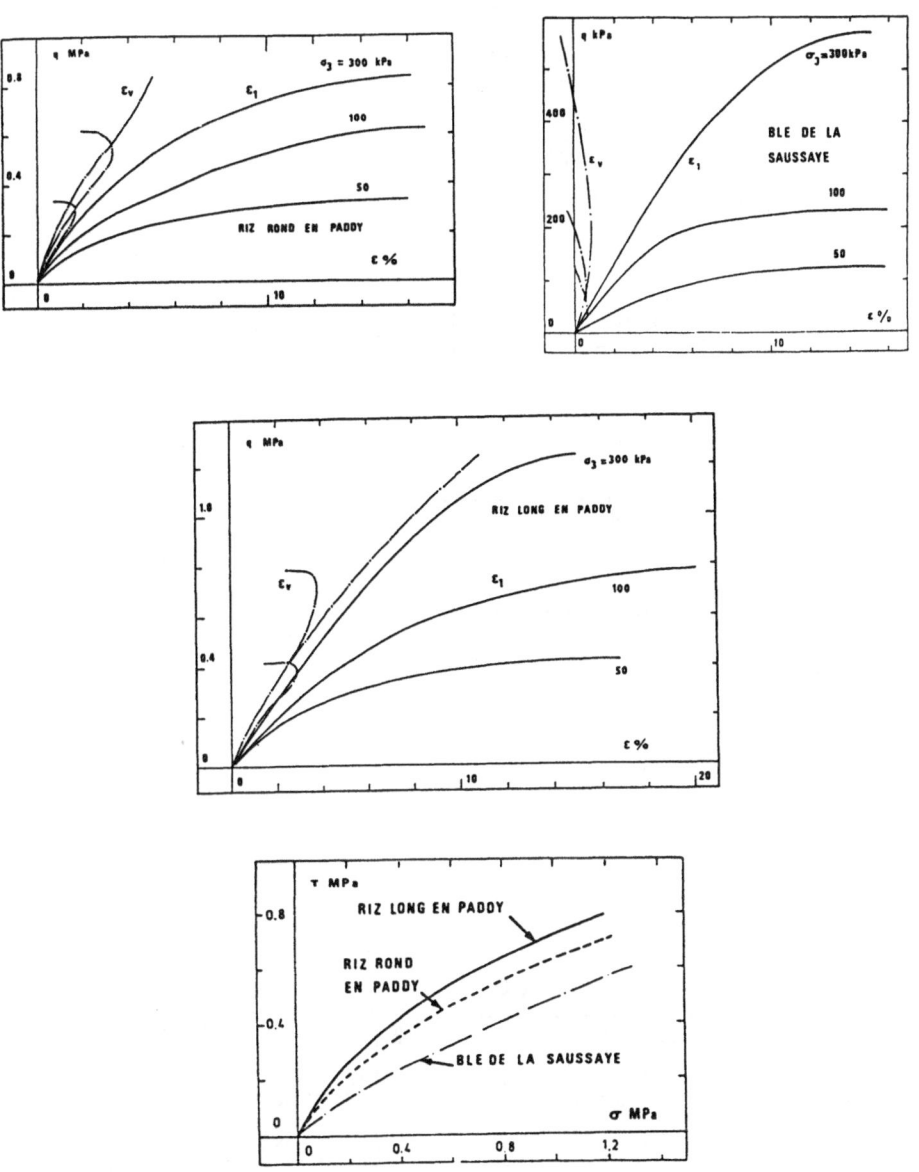

Fig. 3.21 : Triaxial tests on cereal grains [10]

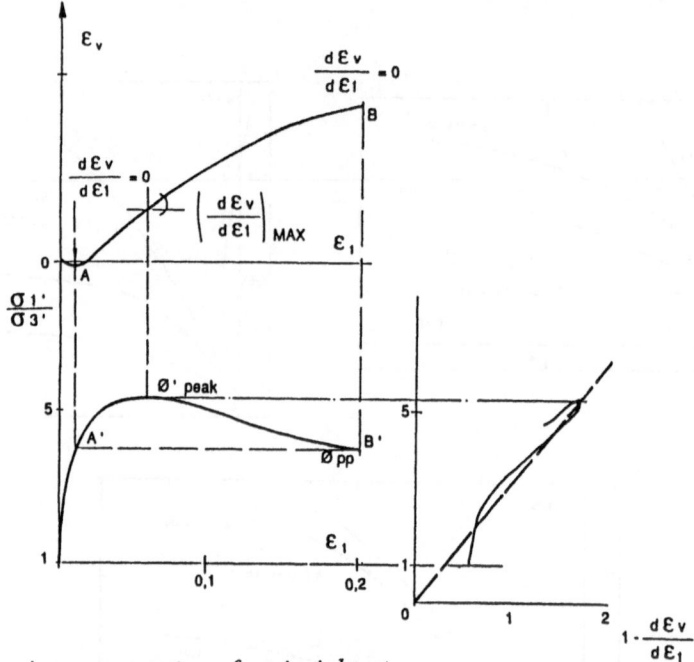

Fig. 3.22 : Rowe's representation of a triaxial test

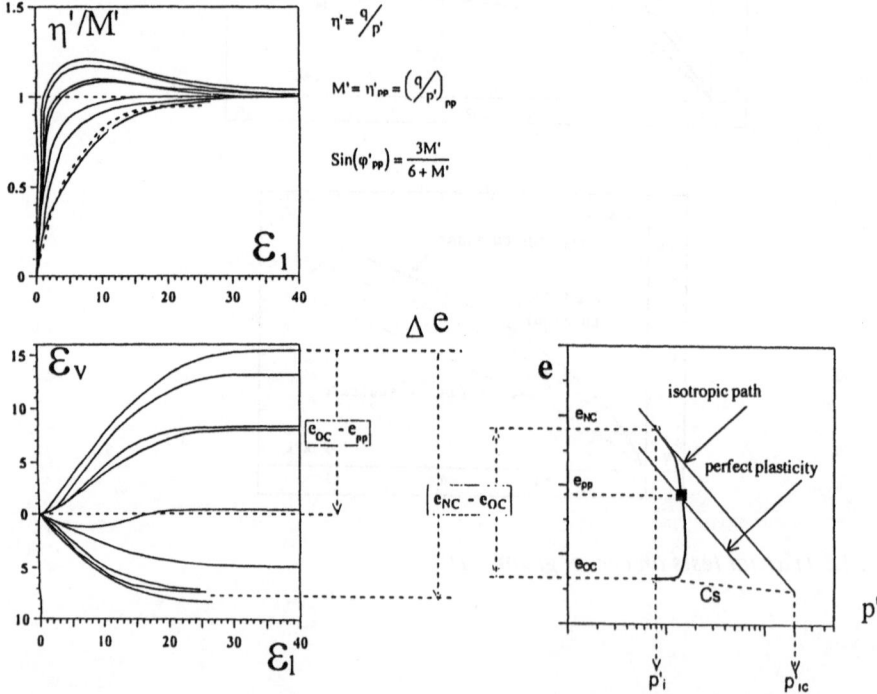

Fig. 3.23 : Influence of volume changes on stress-strain relationship in triaxial tests

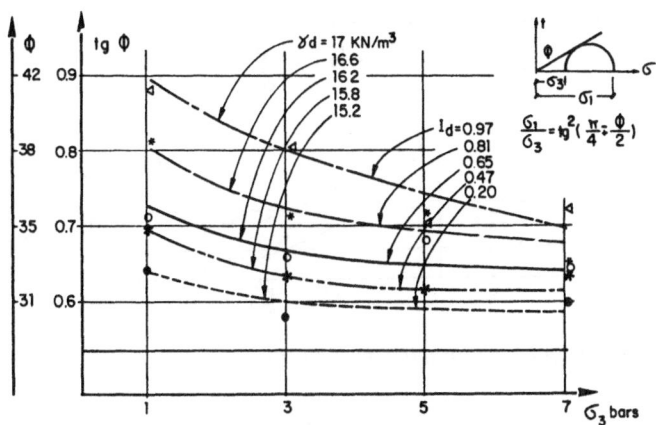

Fig. 3.24 : Evolution of peak friction angle with mean stress

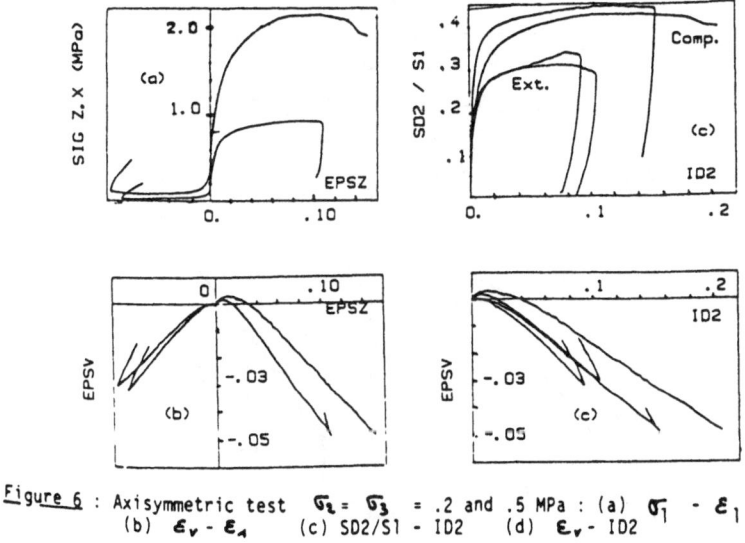

Figure 6 : Axisymmetric test $\sigma_2 = \sigma_3$ = .2 and .5 MPa : (a) σ_1 - ε_1
                (b) ε_v - ε_1 (c) SD2/S1 - ID2   (d) ε_v - ID2

Fig. 3.25 : Compression and extension triaxial tests on sand [13]

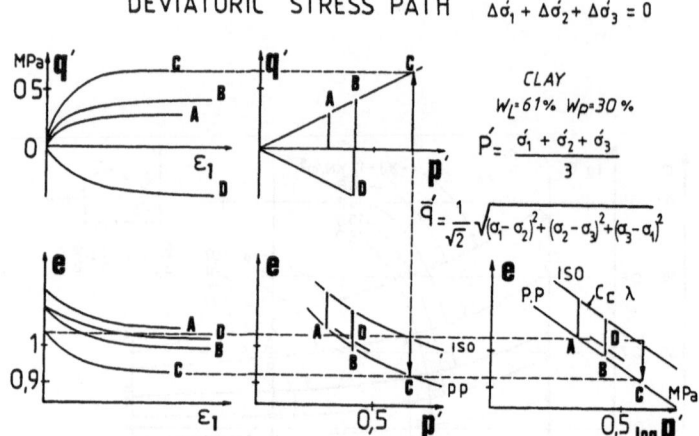

Fig. 3.26 : Representation of deviatoric stress paths

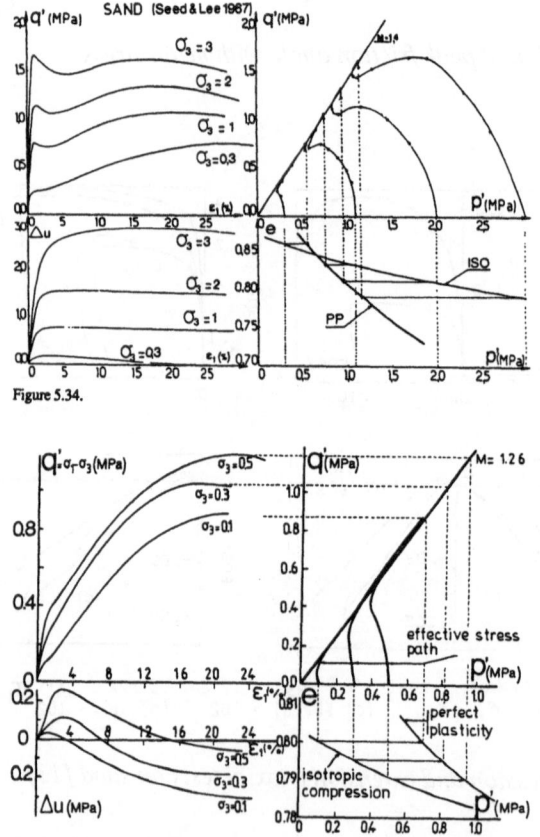

Fig. 3.27 : Isochoric tests on sand [14]

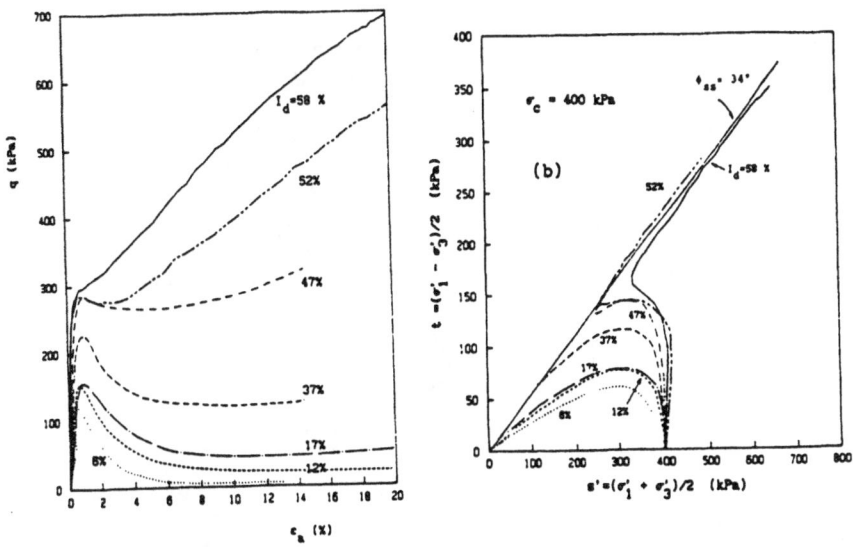

Fig. 3.28 : Influence of density in isochoric tests on sand [15]

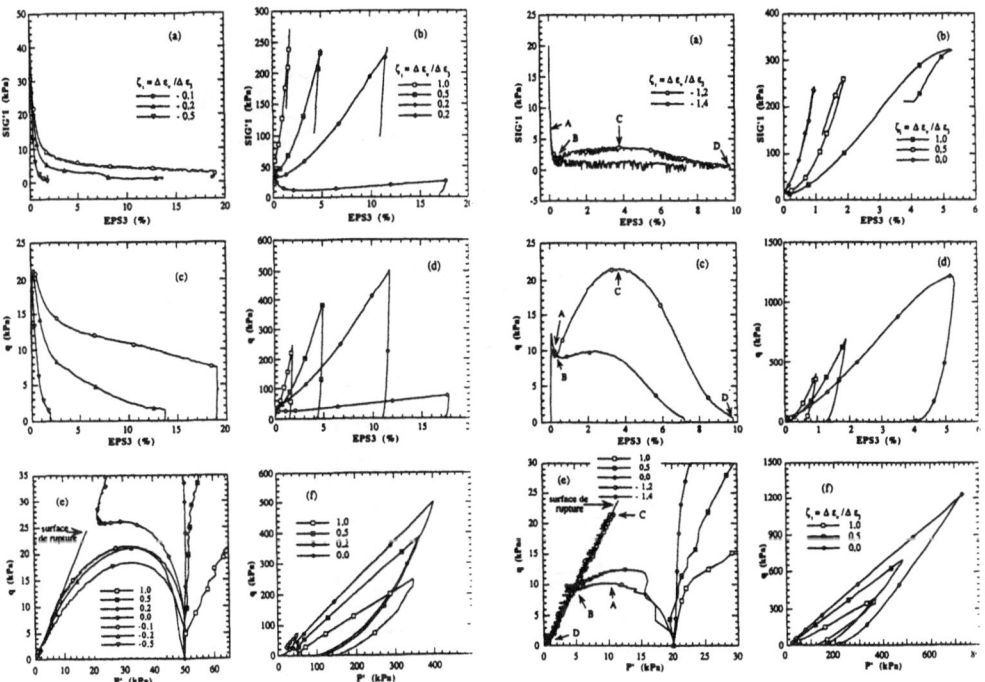

Fig. 3.29 : Triaxial tests at constant $d\varepsilon_v/d\varepsilon_1$ [3]

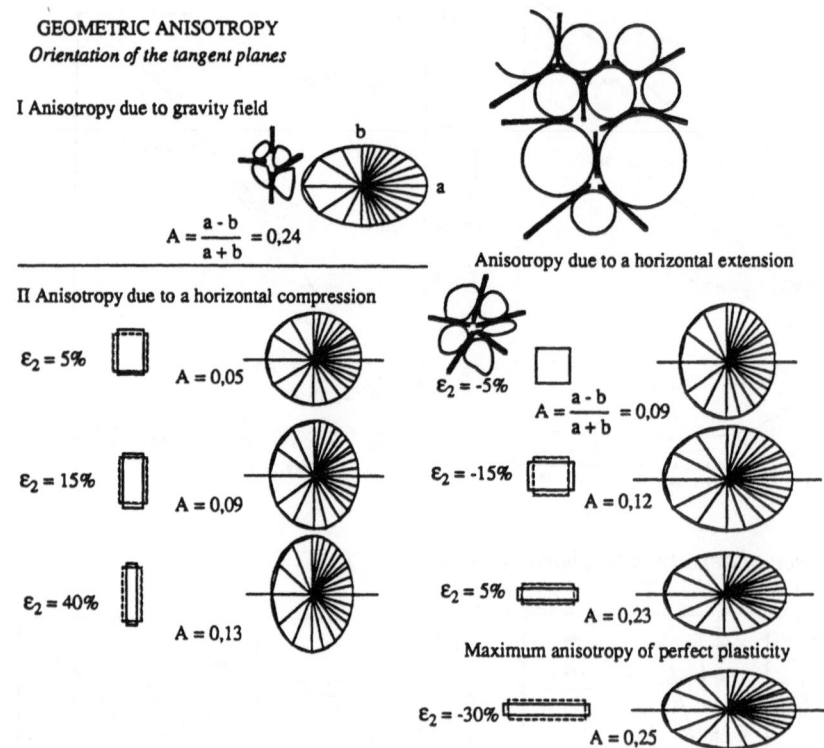

Fig. 3.30 : Description of geometric anisotropy

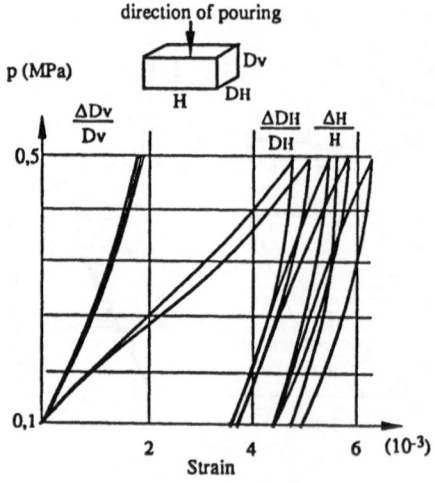

Fig. 3.31 : Cycles of isotropic loading after vertical pouring

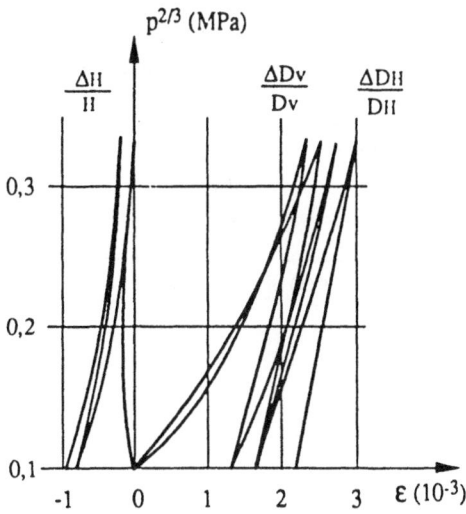

Fig. 3.32 : Cycles of isotropic loading after straining DH/H = 1.5% ; pouring in the direction D_v

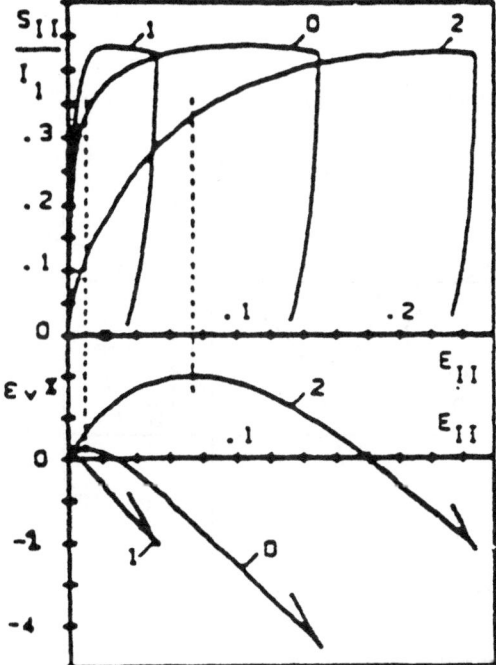

Fig. 3.33 : Induced anisotropy : comparison of three triaxial tests, (0) virgin sample, (1) reloading in the same direction, (2) reloading in a perpendicular direction [13]

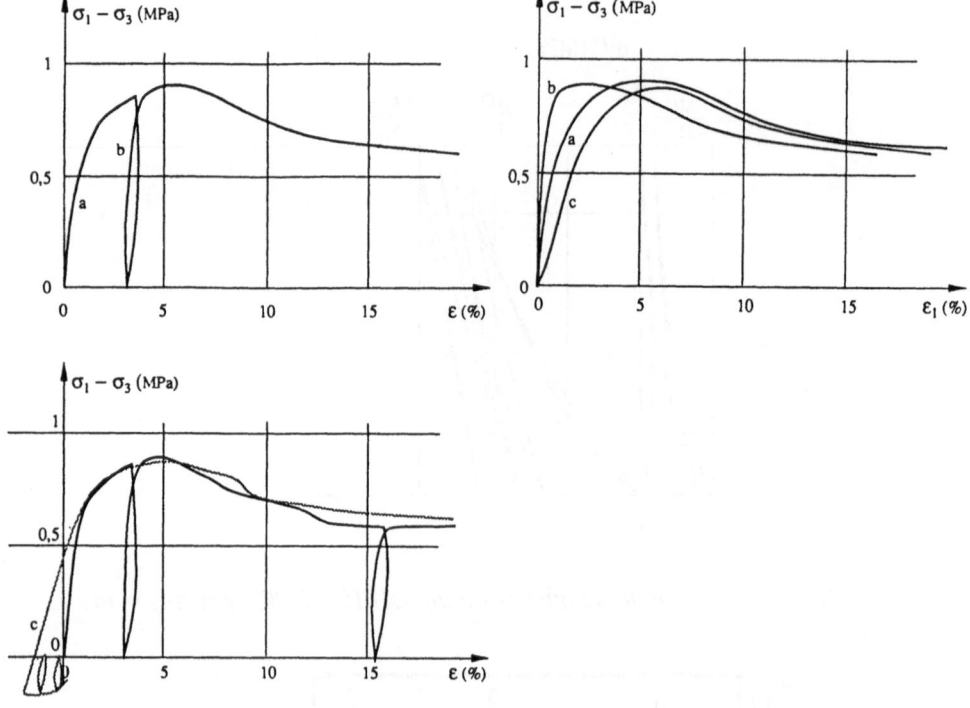

Fig. 3.34 : Influence of induced anisotropy on stress-strain relationship

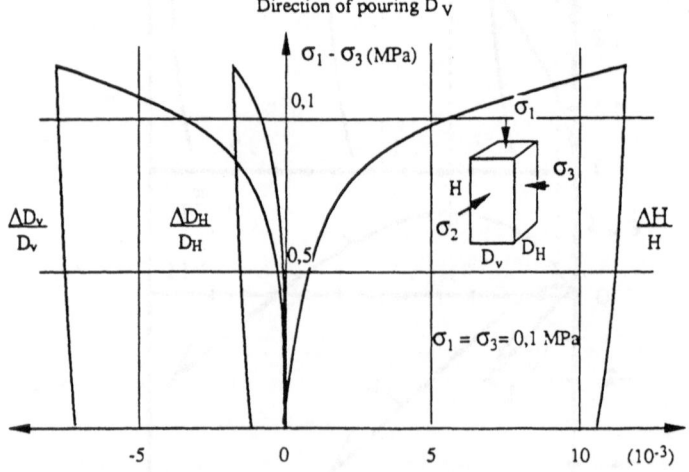

Fig. 3.35 : Axisymmetric loading on orthotropic material

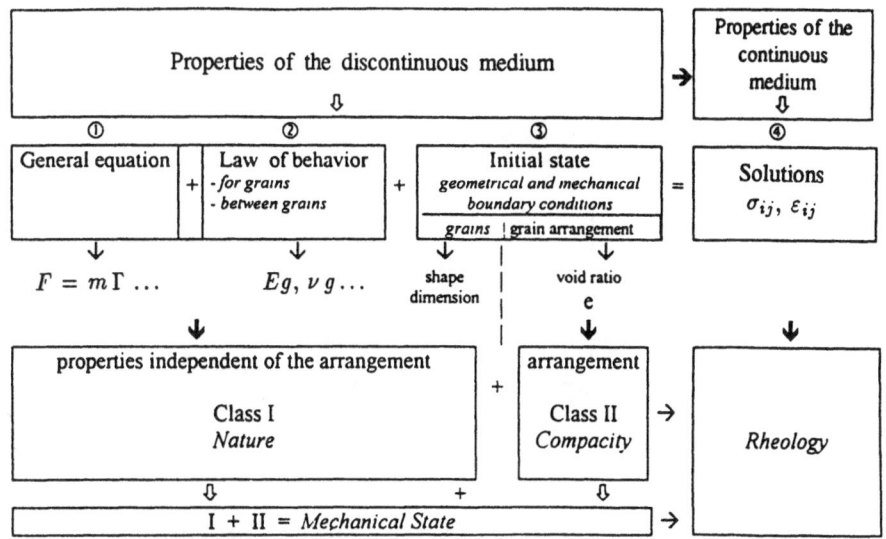

Fig. 3.36 : Framework of correlations in granular media

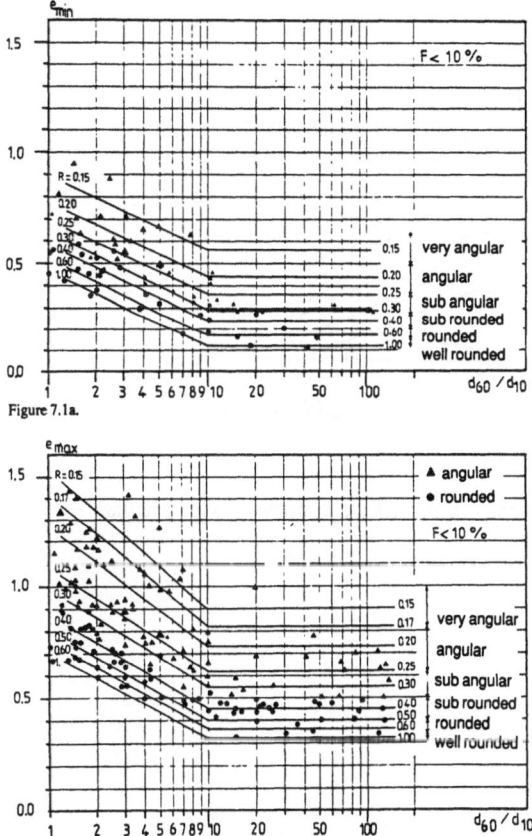

Fig. 3.37 : Correlations between e_{max}, e_{min} grain angularity and grain size distribution

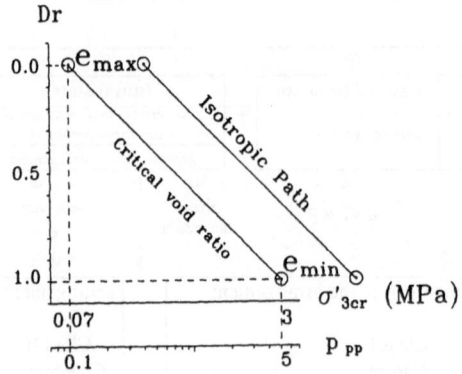

Fig. 3.38 : Position of perfect plasticity line in relation with e_{max} and e_{min}

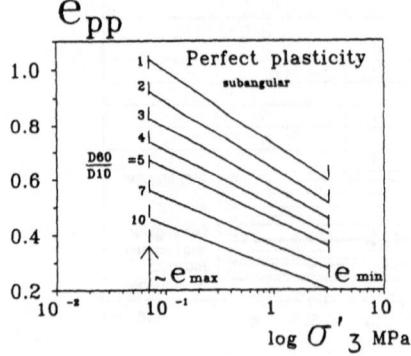

Fig. 3.39 : Perfect plasticity line as a function of nature parameters of grains

4. Three dimensional behaviour. Effect of the intermediate principal stress

Three dimensional tests were developed in the first place in order to study the effect of the intermediate stress σ'_2 on the strength of granular materials. If the Coulomb parameters c', φ' are obtained from axisymmetric compression tests, can they also be applied in the plane strain case for example?

Triaxial extension tests provide a preliminary approach to this question. The results, however, appear to be quite dispersed and sometimes contradictory. The problem involed with carrying out these tests are undoubtedly due to the rapid creation of kinematic discontinuities. A number of researchers have nevertheless shown angles of friction in extension greater than those obtained in compression. These results suggest therefore that the Mohr-Coulomb criteria (φ' = constant) is only an approximation for different values of the intermediate stress. Furthermore, plane strain tests have also produced angles of friction greater than those for axisymmetric compression.

More recently, the development of numerical models has made it necessary to investigate stress and strain paths other than axisymmetric ones. The construction of three dimensional testing apparatus allowed paths in principal stress and strain space to be followed in principle. These apparatus can be separated into four categories according to their system of loading :
- Apparatus which control strain (systems of 3 parts of rigid plates) (Fig. 4.1) ;
- Apparatus which control stress ;
- Combined equipment (Figs. 4.2, 4.3) ;
- Tests on hollow cylinders (Fig. 4.4).

The mode of operation of the different apparatus introduced a number of experimental problems related to the heterogeneity of the stresses and strains due to friction on the rigid end plattens, interference of the boundaries, and difficulties in measuring the strains on flexible boundaries. These difficulties have restricted the development of these apparatus which have therefore limited the number of available results.

For an isotropic material, a parameter b_σ may be defined which characterises the effect of the intermediate stress during monotonous radial loading. It is given by:

$$b_\sigma = (\sigma'_2 - \sigma'_3)/(\sigma'_1 - \sigma'_3)$$

4.1 Relationship between stress and strain

The sum of available results indicates in the general case that σ'_2 has little effect on the initial slope of the normalised deviatoric stress-strain curves q'/p' - ε. ε is the deviatoric strain (Fig.3.5). The weak role of σ'_2 on radial stress paths is illustrated in Figure 3.6 by the lines of equal deviatoric strain (isostrain lines) in the octahedral plane. These lines are nearly circular for small values of ε. For the remainder of the q'/p' - ε curves, there is a difference and the low values of b_σ are characterised by larger gradients and higher final values of q'/p'.

The variations of volume seem to be scarcely influenced by the value of b_σ, when the average pressure remains constant. The tests of Zitouni [20] on dense Hostun sand show an initial phase of weak contraction followed by dilation, the rate of which is practically independent of b_σ.

4.2 Development of principal strains

For linear stress paths (b_σ = constant), the linearity of the strain path in the deviatoric plane is observed when the strain becomes quite large before localisation (i.e., the development of shear discontinuities) occurs (Fig. 4.6).

There is, however, a phase difference between the stress and the strain paths. This is shown in Figure 4.7 by the cos $3\alpha_\sigma$ - cos $3\alpha_{d\varepsilon}$ graphs for two different densities of Hostun sand. (α is the direction of the path in the octahedral plane as indicated in Figure 4.7). The maximum deviation is greater for dense sand than for loose sand, but the corresponding values of b_σ are virtually the same ($b_\sigma = 0.3$).

The condition of plane strain (ε_2 =0) is reached, whatever the material, for values of b_σ near to 0.25.

4.3 Maximum Strength

Maximum strength (expressed by the ratio q'/p' or the angle of friction φ') varies with b_σ. The available data show a large dispersion. In general it can be said that :

- φ' depends on the intermediate stress. φ' clearly increases when b_σ varies from 0 to 0.5. Plane strain tests lead in the general case to angles of friction greater than those found for axisymmetric compression.

- When b_σ is raised above 0.5, the results are more contradictory. φ' may increase until b_σ = 1, but more commonly φ' reaches a maximum for b_σ between 0.5 and 0.7 and then decreases at an increasing rate until b_σ =1 (Fig. 4.8). For b_σ =1, the angle of friction has sometimes been found to equal the angle of friction for b_σ =0, although more frequently it is higher.

The lines on Figure 4.9 are deviatoric plane plots of several tests on glass assemblies [13]. Most of these lines end outside the Mohr Coulomb criterion (φ' = constant) and inside the Drucker-Prager criterion (q'/p' = constant) defined for a given value of the maximum strength obtained during axisymmetric compression. One must note that the Drucker-Prager criterion allows traction stress in the vicinity of extension (b_σ = 1) for friction angle larger than 38°, which is not possible for non-cohesive granular materials.

Different criteria have been proposed. Some of them are such that the angles of friction are different for b_σ =0 and 1. For example, Lade's criterion, widely used in elastoplastic models, is written in the form:

$$(I_1^3/I_3 - 27)\, I_1/p_a = \text{constant}$$

in which p_a is the atmospheric pressure.

It must be noted that an almost systematic type of kinematic discontinuity develops in the samples whatever the type of apparatus used. The deviatoric strain corresponding to the maximum strength decreases when b_σ increases, and this is indeed explained by the increased tendency for localisation to occur with higher b_σ values. Rupture occurs in a plane containing σ'_2. Under these conditions, the yield envelope is closer to a failure criterion (where homogeneity is lost) than to a criterion of perfect plasticity.

4.4 Three dimensional tests on anisotropic materials along the axes of orthotropy

Ochai & Lade [21] have investigated the three-dimensional behaviour of a dry sand prepared by deposition in a mould with the aid of a three-dimensional compression testing apparatus (Fig. 4.10). They have shown a stiffening of the material in the direction of the deposition. For the same value of b_σ and the same value of the deviator stress, the major principal compressive strain is smaller and the dilatancy larger when the major principal stress is aligned with the axis of orthotropy. More generally, the further the stress path in the deviatoric plane deviates from the axis of orthotropy, the more the material will "soften." For larger deformations, the effect of the initial anisotropy becomes less and less noticeable. The rupture envelope in the deviatoric plane is similar to the symmetric form obtained with an isotropic sample. Similarly, the directions of increment of plastic strain gradually move round until they approach those obtained in isotropic samples for the same stress path.

Comparable studies (supported by GRECO Geomaterials), carried out by Lanier [13] on a dense dry Hostun sand, support these conclusion. The initial anisotropy was in this case created by an axisymmetric loading up to a given value of the axial deformation ε_1. The following conclusions may be drawn from the results illustrated in the graphs (Fig. 4.11). The initial modulus decreases in a significant manner when the direction of the initial loading passes from the direction of the principal major stress to that of the intermediate stress and then to the minor stress. All the results have been reproduced on octaedral plane which show the evolution of the deviatoric strain along the different radial paths. The left hand side represents an initially isotropic material, that on the right hand, an initially anisotropic material. On the left hand side the curves of equal strain are practically circular for small values of ε. This implies that the gradients of the curves q'/p' - ε are independent of b_σ. On the right hand side this is not the case and there is a noticeable net decrease of the modulus when the direction of the path changes away from that of the initial loading.

The rupture surface is not affected by the initial anisotropy, which is progressively being erased by the subsequent loading.

If the iso-strain curves following triaxial compression and extension are compared, it can be seen that they keep the same shape but that they are shifted in the direction of the first load application. A kinematic hardening can be seen here. An initial compression of 3% creates a far weaker anisotropy than that obtained from compressions of 6% and 12% (which are virtually the same). One notices an increasing compaction when the direction of the principal major stress changes from the direction of the axis of anisotropy to a direction perpendicular to this axis.

The shape of the deviatoric strain path reveals a considerable change in the directions of the increments of strain during loading. If localisations are not produced, it can be stated that the final direction corresponds to that of an initially isotropic sample. There is a progressive modification of the anisotropy of the sample. In particular, axisymmetric stress paths ($b_\sigma = 0$ and 1) initially show non axisymmetric strain paths but they tend towards the paths of axisymmetric strain (i.e., to parallel stress and strain paths in the deviatoric plane) (Fig. 4.12).

4.5 Tests other than those on the orthotropic axes

4.5.1 Tests with the principal stress axis directions fixed

The inclination of the orthotropic axes of the sample with respect to the axes of the three dimensional triaxial test apparatus causes problems of nonhomogeneity. It is preferable to incline the principal stress axes. Two tests presently allow homogeneity to be maintained. These are : the torsion test on hollow cylinders and the directional shear cell developed by Wong & Arthur [22] and Sture et al. [23] which is a plane strain test with control over the normal and tangential stresses in the vertical plane (Figs. 4.13, 4.14). Tests carried out on sands with these apparatus have all shown a decrease in the gradient of q' - ε curves when the direction of the major principal stress deviates from the direction of anisotropy (the minor axis of ellipse) created by the precedent loading. This is accompanied by an increase in compaction (Fig. 4.15).

4.5.2 Influence of rotation of the principal stress axis

The influence of rotation can be observed from the tests carried out by Sture et al. [23] in the deviatoric plane. An initial loading with a fixed direction and different maximum values of σ'_1/σ'_3 (4 and 6 were used) created the anisotropy of the material (dry sand). A second loading with lower values of σ'_1/σ'_3 (2,3 and 4) was then applied, initially along the same axis but then with the direction of the principal stresses gradually changed from $\theta = 0°$ to $90°$ keeping σ'_1/σ'_3 constant (Fig. 4.16). In all cases an evolution of the deviatoric strain $\varepsilon_1 - \varepsilon_3$ occurred. This was accompanied by a variation of volume of which the amount and the sign depends on the ratio σ'_1/σ'_3. Contraction is larger for small values of σ'_1/σ'_3. This is shown for values of $\theta > 60°$. Tests with a fixed principal stress direction have also shown a net increase of contraction for $\theta > 60°$. The angle of $60°$ corresponds approximately to the orientation of the lines with no extension during the initial loading. An explanation could be that, during the initial loading, the number of inter-particle contacts increases in the zones of compression and decreases in the zones of extension. This was observed by Lanier [2] in a plane strain apparatus ($1\gamma 2\varepsilon$) with the use of a two-dimensional material (wooden rolls). During the rotation, a reorientation of the contacts takes place. This becomes more significant when the direction of σ'_1 is found to correspond to a line of extension (where there is a reduction in the modulus and an increase in compaction).

Kharchafi [24] carried out tests on hollow cylinders which are in accordance with the preceding results. A dense sample of Hostun sand, subjected first to pure torsion, shows a larger contraction and smaller rigidity during subsequent axisymmetric compression than an initially isotropic sample, whereas a sample subjected to a preliminary axisymmetric compression in torsion tests shows a smaller contraction and greater rigidity. In the first case, the principal major stress is in the direction of extension, in the second it is in the direction of compression (Fig. 4.17).

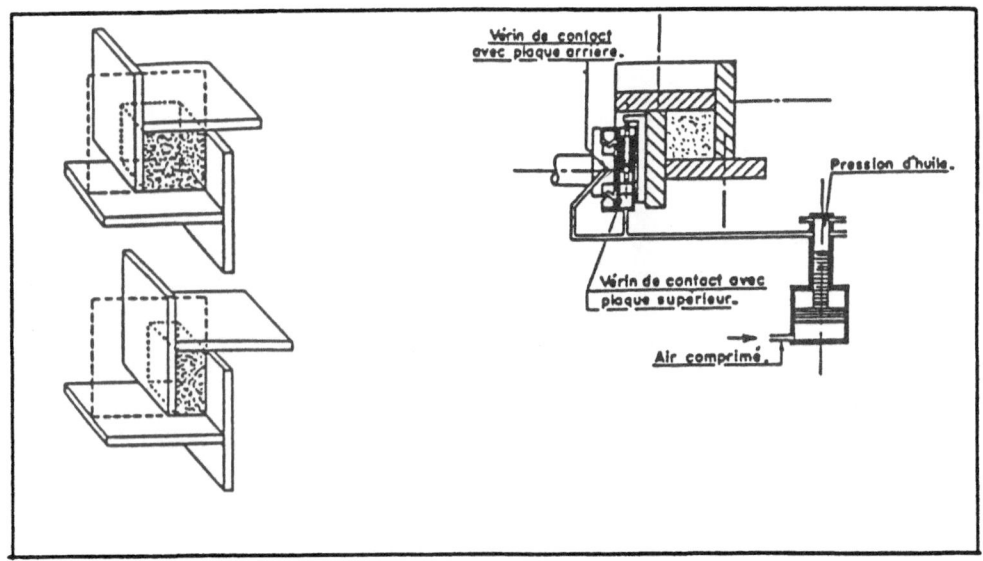

Fig. 4.1 : True triaxial apparatus with rigid boundaries [13]

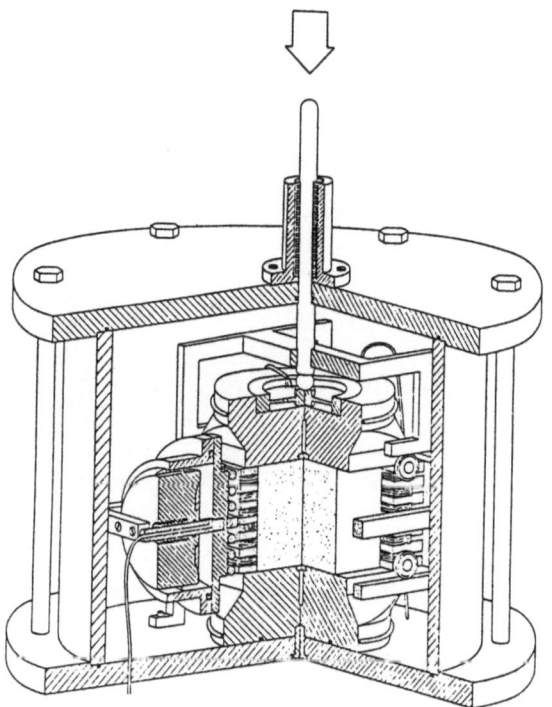

Fig. 4.2 : True triaxial apparatus with mixed boundaries [19]

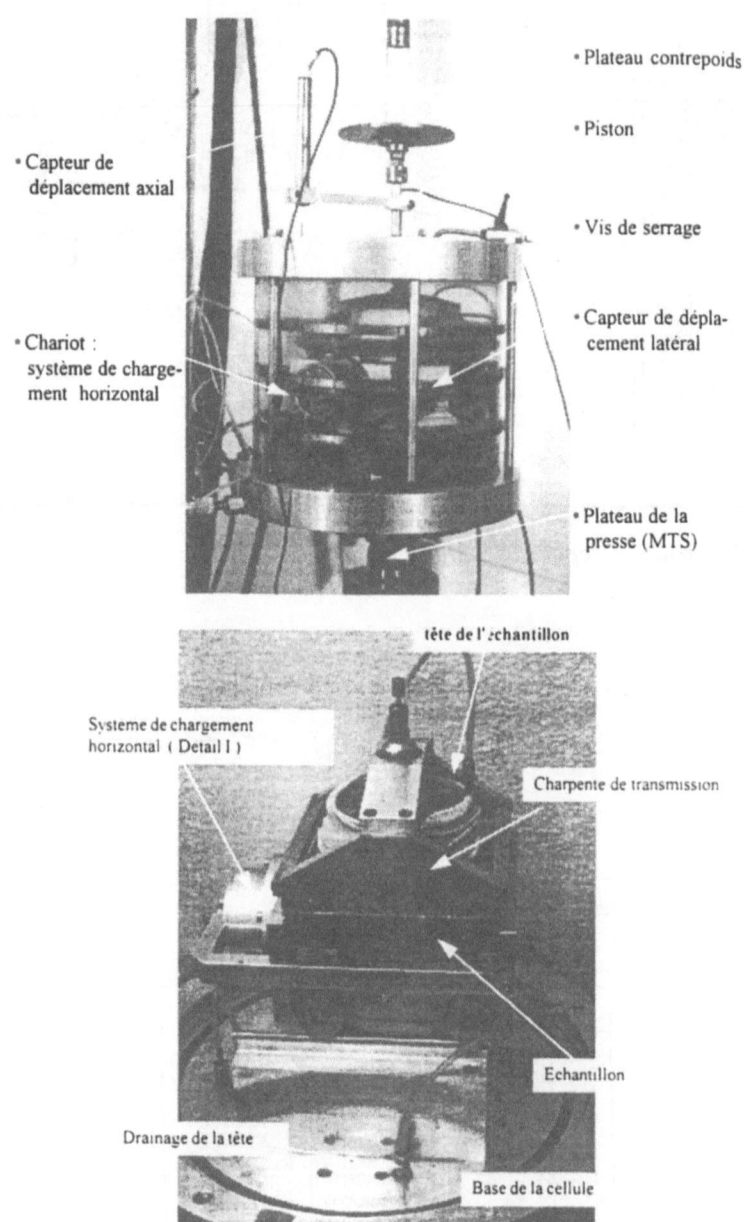

Fig. 4.3 : True triaxial apparatus with mixed boundaries (Hicher)

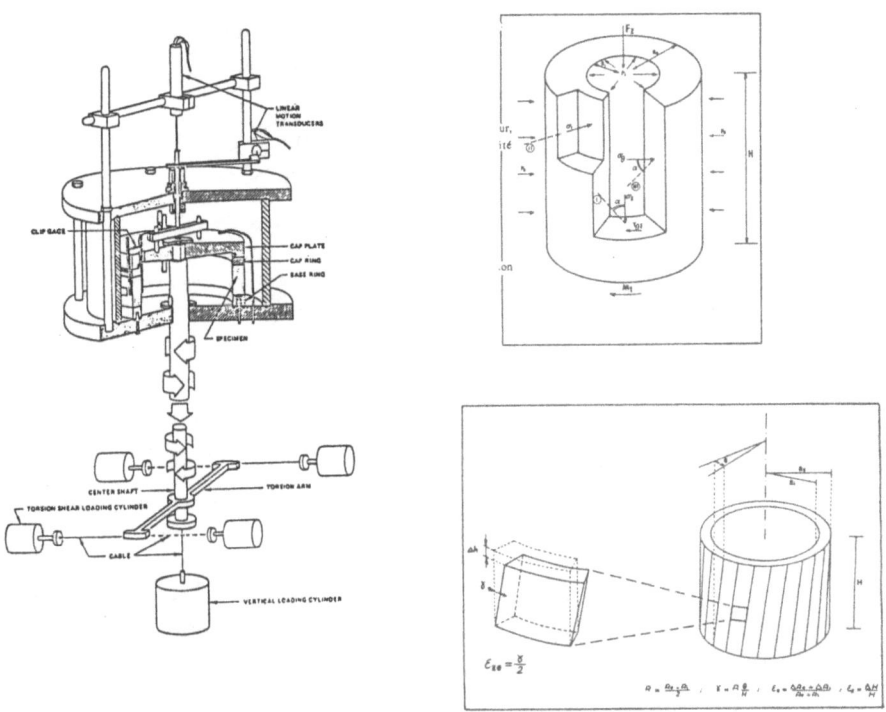

Fig. 4.4 : Torsion shear test on hollow cylinders

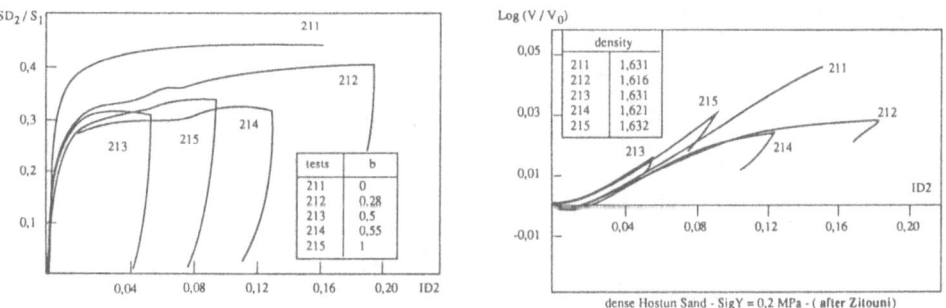

Fig. 4.5 : Stress-strain relationships in true triaxial testing [13]

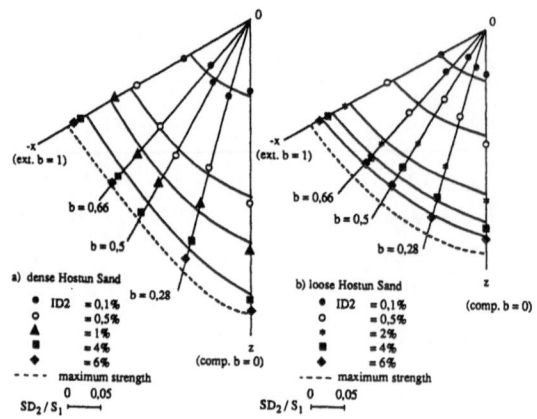

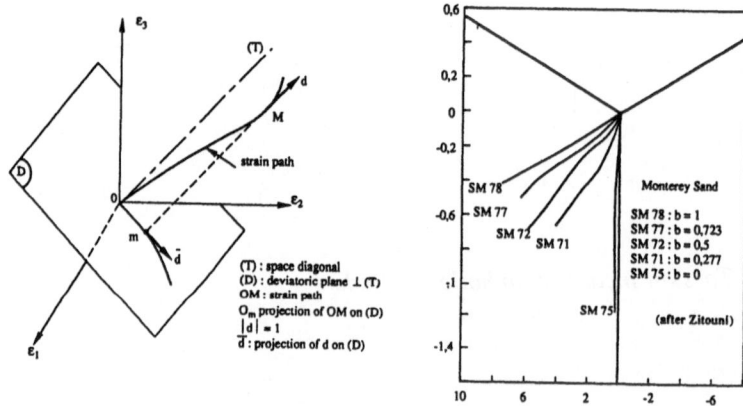

Fig. 4.6 : Evolution of strains during proportional stress paths (b_σ = constant) [20]

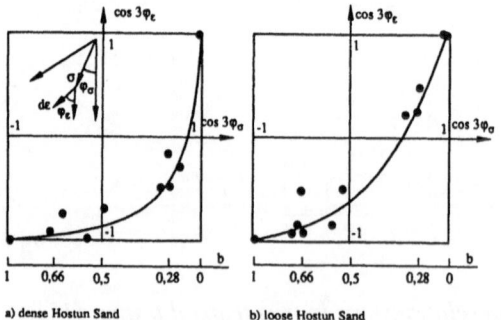

Fig. 4.7 : Comparison between stress and strain increment directions in deviatoric plane[20]

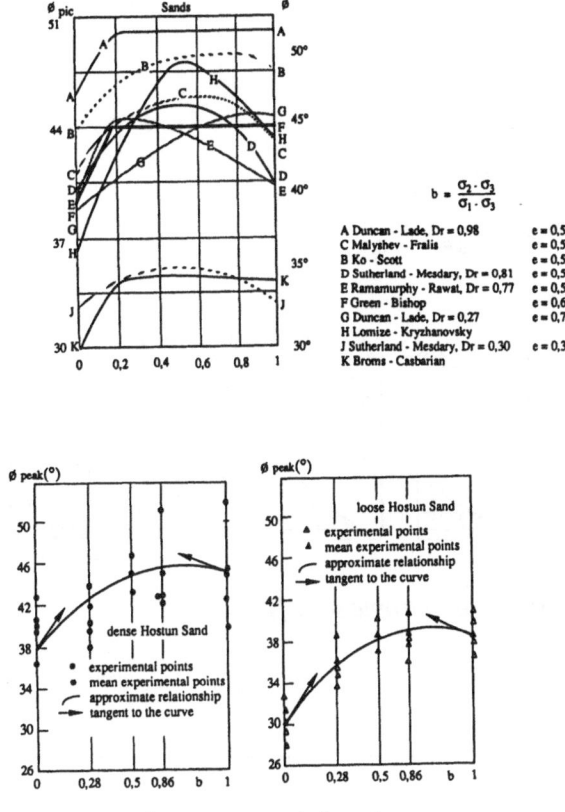

$$b = \frac{\sigma_2 - \sigma_3}{\sigma_1 - \sigma_3}$$

A Duncan - Lade, Dr = 0,98	e = 0,57
C Malyshev - Fralis	e = 0,51
B Ko - Scott	e = 0,52
D Sutherland - Mesdary, Dr = 0,81	e = 0,99
E Ramamurphy - Rawat, Dr = 0,77	e = 0,52
F Green - Bishop	e = 0,64
G Duncan - Lade, Dr = 0,27	e = 0,78
H Lomize - Kryzhanovsky	
J Sutherland - Mesdary, Dr = 0,30	e = 0,30
K Broms - Casbarian	

Fig. 4.8 : Maximum strength in true triaxial testing

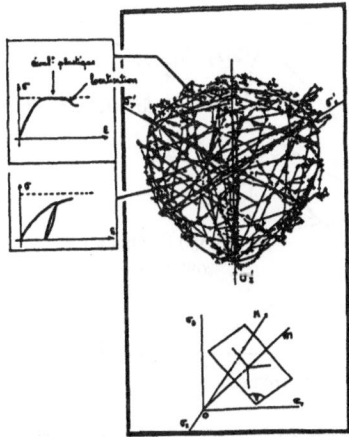

Fig. 4.9 : Allowed stress paths in a deviatoric plane [13]

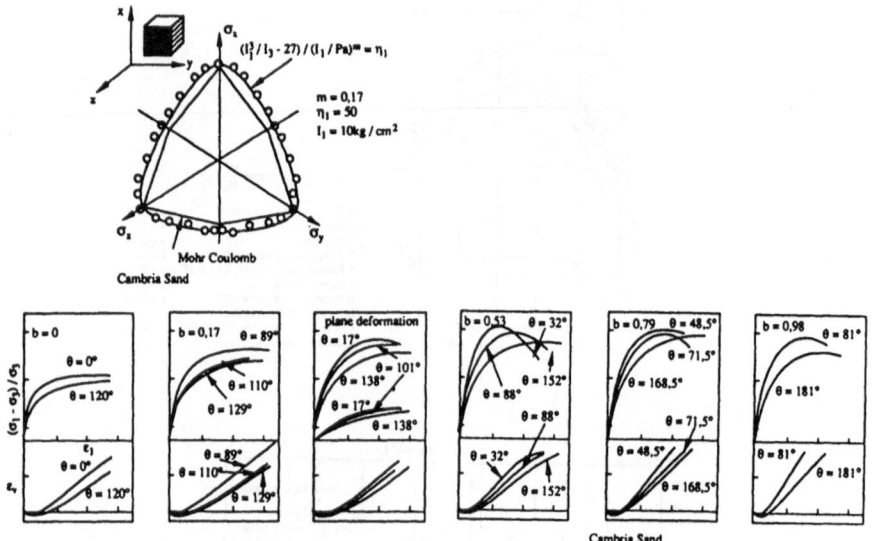

Fig. 4.10 : 3D tests on anisotropic sand [21]

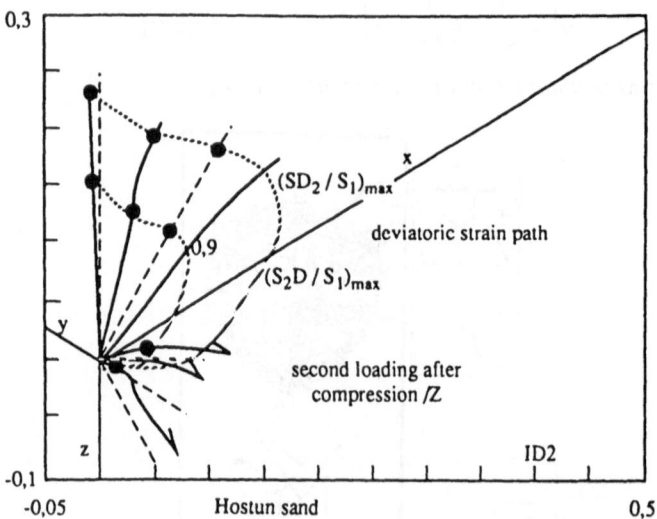

Fig. 4.12 : Strain response along radial stress paths in deviatoric plane [13]

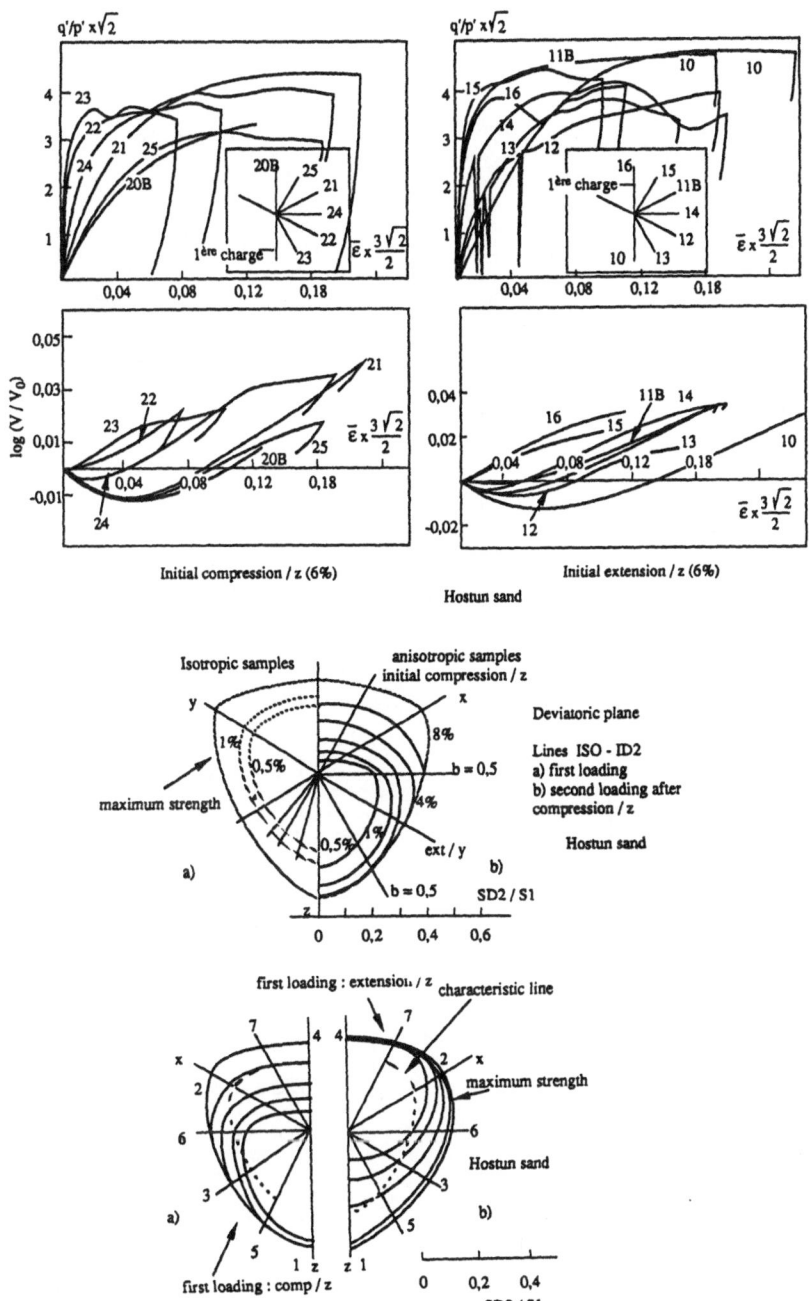

Fig. 4.11 : Influence of induced anisotropy on 3D stress-strain response of sand [13]

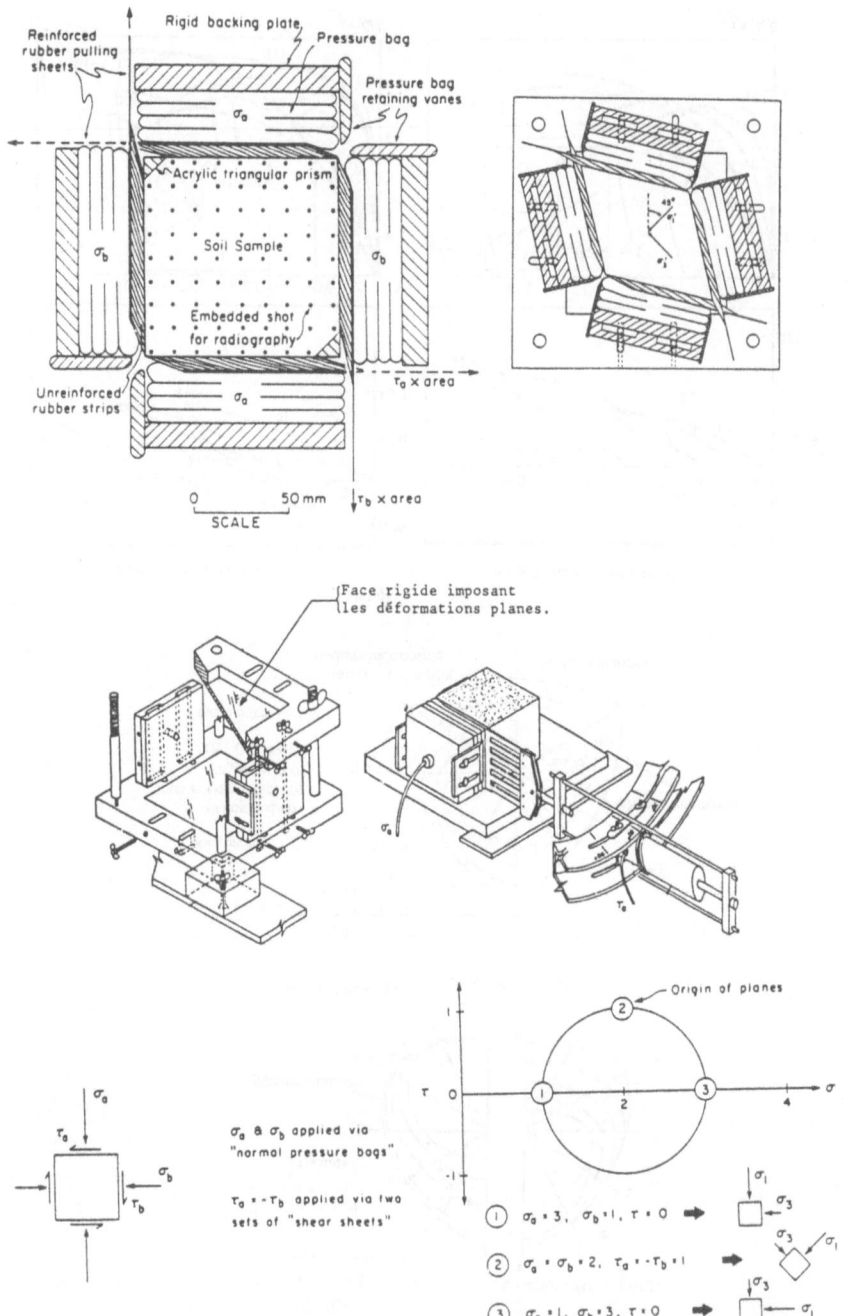

Fig. 4.13 : Directional shear cell [22]

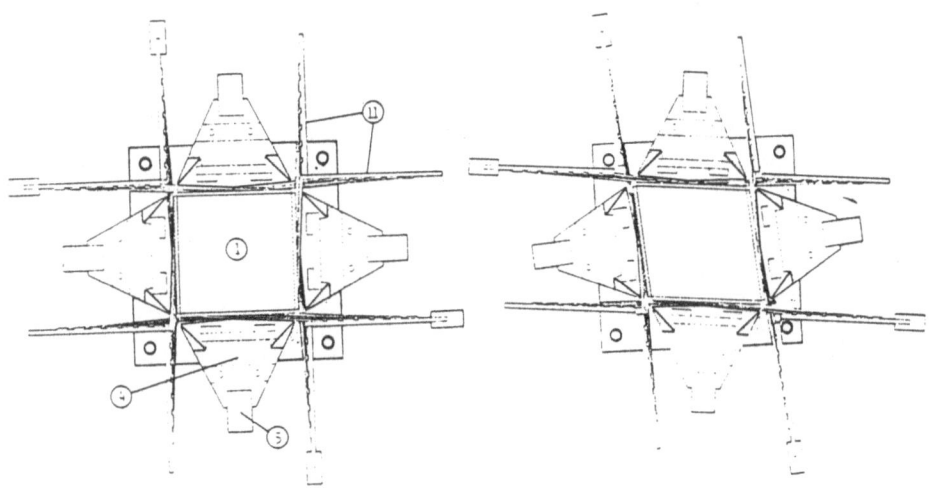

Fig. 4.14 : Plane strain shear apparatus [23]

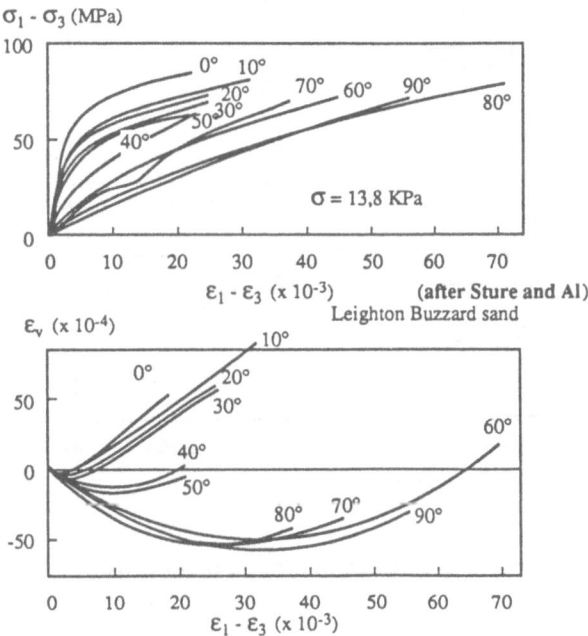

Fig. 4.15 : Influence of stress axis direction on anisotropic sand [23]

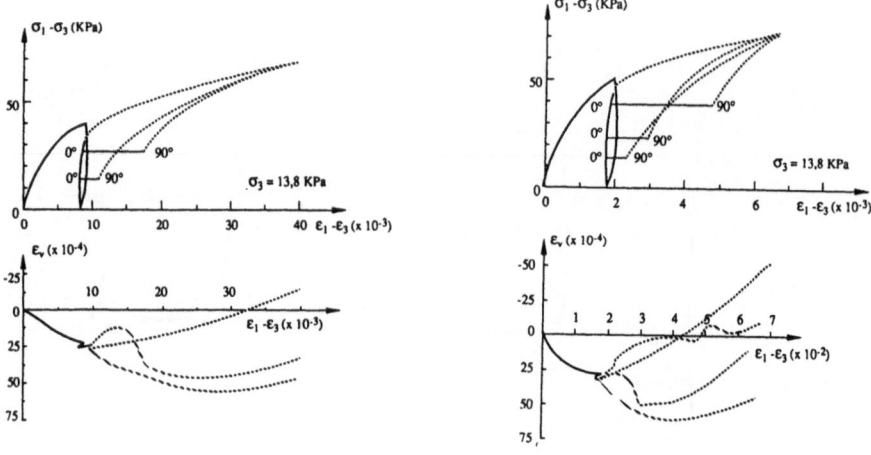

Fig. 4.16 : Influence of stress axis rotation on anisotropic sand [23]

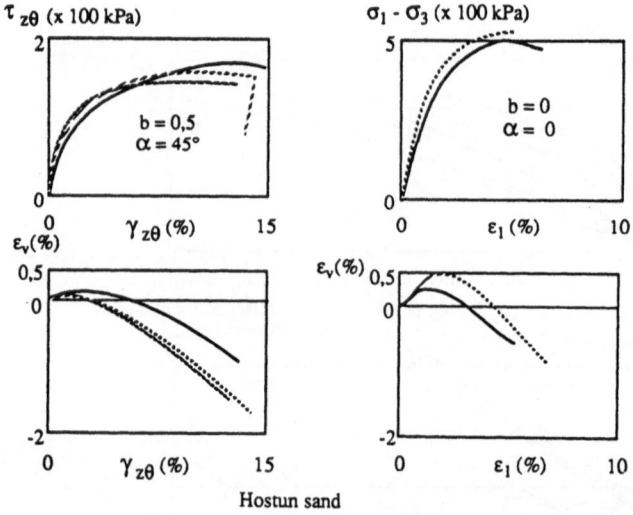

Hostun sand

Fig. 4.17 Influence of induced anisotropy in torsion shear tests on hollow cylinders [24]

5. Coupling phenomena in two or three-phases materials

In the previous chapters, we studied the behaviour of "dry" granular materials, i.e., where the voids were filled by air at atmospheric pressure. In different cases, the voids can be filled by fluids (liquids and gasses) at pressures different from the atmospheric pressure. In these conditions the granular medium becomes what can be called a "polyphasic" material and its mechanical behaviour will depend on the behaviour of the solid grains, the fluids, and the interactions at the interfaces.

As for the solid grains, the geometrical boundary conditions in the liquid and in the gas phases must be taken into account. This leads, according to Laplace's law, to different pressures in the liquid and in the gas, which depend on the principal radii of curvature of the menisci. The grain sizes, the degree of saturation (S_r) and the wettability of the solid particles by the liquid modify the principal radii of the menisci and, therefore, the value of the negative pressure in the pore liquid. This negative pore pressure induces adhesion forces between the grains, which develop in the continuous fictitious material what are called "effective stresses" or capillary cohesion.

Experimental results, therefore, can be very different according to the nature of the grains and of the pore fluids. Some examples are presented below but they are far from comprehensive for showing the possible behaviours of polyphasic materials.

5.1 Water saturated and non-saturated granular assemblies

5.1.1 Saturated samples

In saturated samples (all the voids are filled by water), we can apply Terzaghi's principle : the total stress tensor σ applied to the boundaries of a saturated sample is equal to the effective stress tensor σ' applied on the skeleton plus the pore water pressure $u\mathbf{I}$:

$$\sigma = \sigma' + u\mathbf{I}$$

The deformations of a saturated granular material are exclusively due to changes in the effective stresses. During a given test, only total stresses and pore pressure are measured. Effective stresses are therefore determined by using Terzaghi's postulate. Different stress and strain paths can be applied, in particular "drained" tests, where pore pressure is kept constant and "undrained" tests for which water is prevented from moving in or out of the samples. During drained tests, volume changes correspond to a change in water content : water moves outside of the sample for a contractant materal or inside the sample for a dilatant one. The results are identical to the ones obtained for dry samples (Chapters 3 and 4). Undrained tests correspond to isochoric tests (if we can neglect the compressibility of water and grains) and give also identical results when expressed in terms of effective stresses, validating Terzaghi's postulate.

5.1.2 Unsaturated Samples

Sand or glass balls are very wettable by water. Therefore capillary forces can easily develop in unsaturated samples. Terzaghi's expression of the effective stress must be modified to account for the fact that the contact area between particles and water is restricted to the area limited by the menisci. A simple expression has been proposed by Bishop and Blight [25] for effective stresses :

$$\sigma' = \sigma - (u_a + \chi(u_a - u_w))\mathbf{I}$$

where χ is an experimental parameter ranging from 0 to 1, which can be approximated by S_r, u_a is the air pressure and u_w is the water pressure. Often the air pressure corresponds to the atmospheric pressure and Bishop's expression becomes:

$$\sigma' = \sigma - S_r u_w \mathbf{I}$$

The variation of the negative pore pressure versus the water content is very much dependent on the pore distribution and sizes, i.e., the dimensions of the grains and the grain size distribution. Some results obtained by Taibi [26] are presented in Figure 5.1. Initially saturated samples are progressively dried (drying path) or initially dry samples are progressively watered (wetting path) and pore pressure is measured. Results on glass ball assemblies show the influence of the pore sizes; the smaller the pores are, the higher the negative pore pressure is when desaturation starts, corresponding to the point of air entry. For assemblies of the same diameter spheres, the desaturation occurs in the same amount of time in most of the pores which have the same diameter (Fig. 5.2). Laplace's law gives

$$-u_w(kPa) = 300/n \,.\, 1/d(\mu m)$$

where d is the diameter of the spheres and n a parameter depending on the assembly of voids. When the grains are of different sizes, the desaturation appears at first in the big pores and then continues into smaller pores when pore pressure continues to decrease. Therefore the slope of the relation w - u_w depends on the grain size distribution. Results on sand illustrate this influence (Fig 5.3). Superposition of drying and wetting paths indicate a slight hysteresis due to the topology of the voids.

Different results on granular assemblies demonstrate that Terzaghi's postulate applied to negative pore pressure, as long as the material remained saturated :

$$\sigma = \sigma' + u_w \mathbf{I}$$

u_w positive or negative with $S_r = 1$.

A negative pore pressure will increase the effective stresses inside the samples and therefore increase the mechanical properties of the granular material (stiffness and maximum strength). This will result in a measurement of an apparent cohesion when the results are expressed in terms of total stresses applied on the specimen. Figure 5.4 presents some results obtained by Taibi on glass ball assemblies by performing unconfined compression tests. According to Bishop's equation, the unconfined compression strength must be expected to be approximately proportional to the term Sr. Indeed, we observe an increase in the strength when the water content decreases, but after a maximum, a decrease in strength is observed for small water content (for $w = 0$, i.e., a dry material the unconfined strength is reduced to zero), but also the fact that strain localisation develops more rapidly in low water content samples. Permeability to water and air depends strongly on the degree of saturation (Fig. 5.5).

5.2 Behaviour of bituminous mixtures

Bituminous mixtures are polyphasic materials made of solid grains (sand and gravel), of a liquid phase (a bitumen) and of a gas phase (air). The difference with the previous material lies in the liquid phase : bitumen is much more viscous than water and the wettability of the grain is very reduced. So the capillarity forces do not act inside the sample, they are replaced by the mechanical properties of the bitumen (viscoelasticity).

An experimental study by Bard [8] on different bituminous mixtures illustrates the influence of different parameters linked to the fluid phases : viscosity of the bitumen, degree of saturation. Triaxial tests were performed at different confining pressures, different strain rates, different temperatures on bituminous mixtures, with different bitumen and different degrees of saturation (Fig. 5.6). The conclusions of the study can be summarised as follows :

For unsaturated samples ($S_r = 60\%$), the influence of the confining pressure indicates that a bituminous mixture behaves like a granular material : maximum strength increases with confining pressure (Fig. 5.7). The strength envelope is of the Mohr-Coulomb type with a term of cohesion, depending on the viscosity of the bitumen. When the viscous effect is small (small strain rate or high temperature), the behaviour of the bituminous mixture is close to the one of the granular assembly prepared at the same initial void ratio, due to the weak wettability of the grains by the bitumen (Fig 5.8). On the other hand, when the viscous properties are significant, the viscous behaviour overtakes the frictional one and the response becomes less sensitive to the mean stress with a high increase of the apparent cohesion.

The samples were prepared in a dense state (small void ratio = 0.2). They are therefore very dilatant and an increase of the void ratio is observed during the triaxial test, leading to a decrease of the degree of saturation. When the viscous properties of the bitumen are small, the liquid phase can adapt to the change in voids and the dilatancy of the bitumen mixture remains close to the one of the granular material. When the viscous properties increase, the bitumen cannot adapt to the new geometry and fissures appear in the samples, leading to a larger amount of dilatancy, due to low capillary forces (Fig. 5.9).

When the saturation in bitumen is very high (for this study $S_r = 90\%$), the behaviour of bitumen mixtures changes compared to the previous observations (Fig 5.10, 5.11). The initial stiffness appears to be less dependent on the confining pressure. There is practically no volume change, except for small confining pressure ($\sigma_3 < .2$ MPa) or elevated strain rate, where the material remains dilatant. The maximum strength envelope is no more a straight line but presents a decrease in slope. This decrease is more pronounced when the viscous effect is less marked, and the envelope tends to a horizontal line in the (p,q) plane indicating no more influence of the confining pressure. This behaviour is very similar to a granular material saturated with water and submitted to undrained loading. When the viscous effects are more pronounced, the bitumen cannot follow the movement of the grains and cracks appear inside the specimen leading to dilatancy and a bigger effect of the confining pressure.

5.3 Mechanical behaviour of carbon agglomerates

Industrial carbons and graphites are obtained by coking and then graphiting a polyphasic mixture consisting of a granular skeleton (cokes), an hydrocarbonic binder (pitch) and gas. The mechanical behaviour of these mixtures is affected by some important characteristics of the different phases : high porosity of coke grains, weak wettability of cokes by pitch, high viscosity of pitch. An extensive study by Bard [8] shows the influence of these characteristics on the overall behaviour of the carbon agglomerates.

One-dimensional compaction was performed at different temperatures and pitch content (p = pitch weight/coke weight). Some results are presented in Figures 5.12 and 5.13.The dry density (coke without pitch) remains approximately constant at a given compression test when the pitch content increases until saturation occurs. This behaviour is very different from what is observed in fine soils for example where a maximum dry density can be observed for a given water content. It can be explained by the weak wettability of coke by pitch.

Results at different temperatures show the influence of the pitch viscosity. Compression stresses have to be increased in order to obtain the same density. At a high temperature, the pitch is able to fill up both the external and internal (grain porosity) porosity. The point of saturation depends on the sum of these two porosities. When viscosity increases (low temperature) only the external viscosity becomes accessible for the pitch and an apparent saturation is reached earlier. This can be observed on the compressibility curves which show a decrease of the slope in the plane $e - \log\sigma_1$.

Triaxial tests on carbon agglomerates indicate a frictional cohesive behaviour (Fig. 5.14). The cohesion depends mainly on the viscous properties of the pitch (no capillary forces). However, at a given confining pressure, when the pitch content increases, the polyphasic material is nearer to a two-phase coke-pitch system and the effective stresses transmitted to the granular skeleton decrease (close to undrained conditions). Therefore, the apparent friction angle decreases. When saturation is completed, the frictional behaviour disappears (Fig 5.15). The results are similar in this case to those obtained on bituminous mixtures.

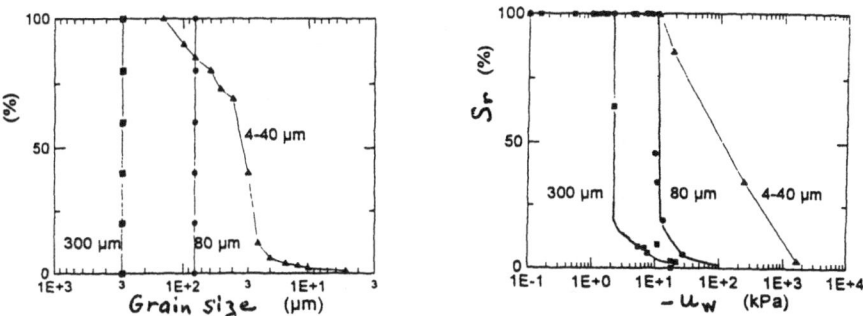

Fig. 5.1 : Drying paths in glass ball assemblies [26]

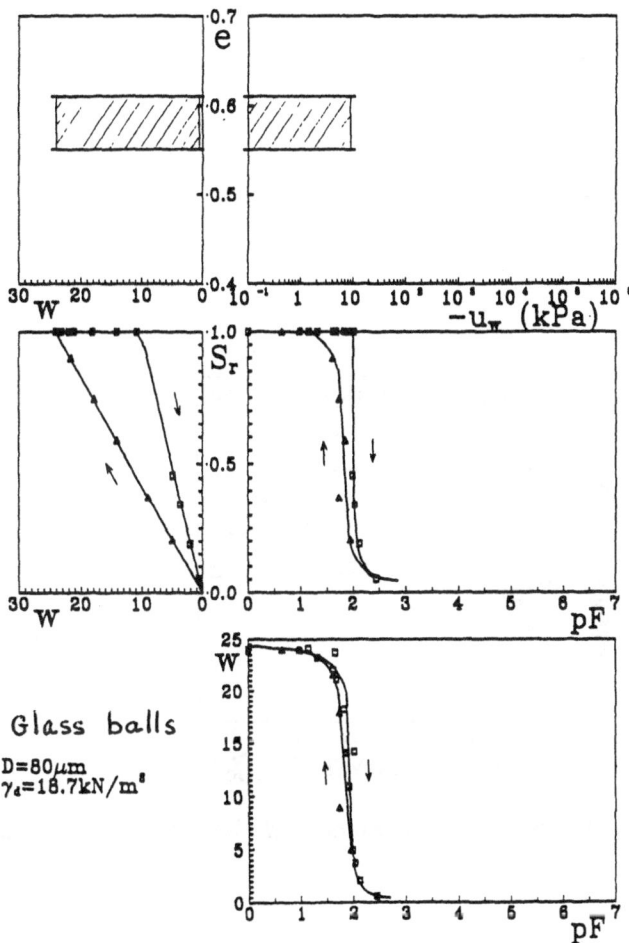

Fig. 5.2 : Wetting and drying paths in uniform glass ball assemblies [26]

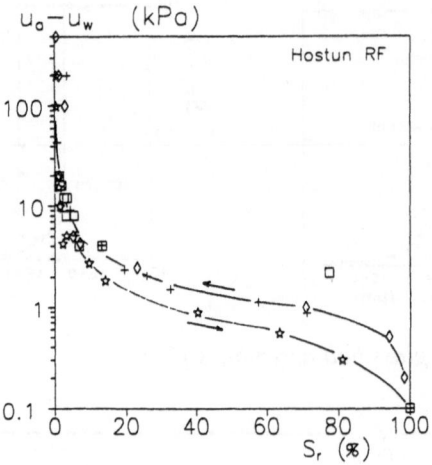

Fig. 5.3 : Wetting and drying paths in sand [26]

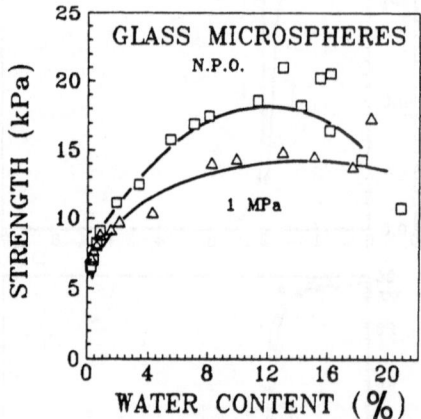

Fig. 5.4 : Unconfined compression tests on unsaturated glass ball assemblies [26]

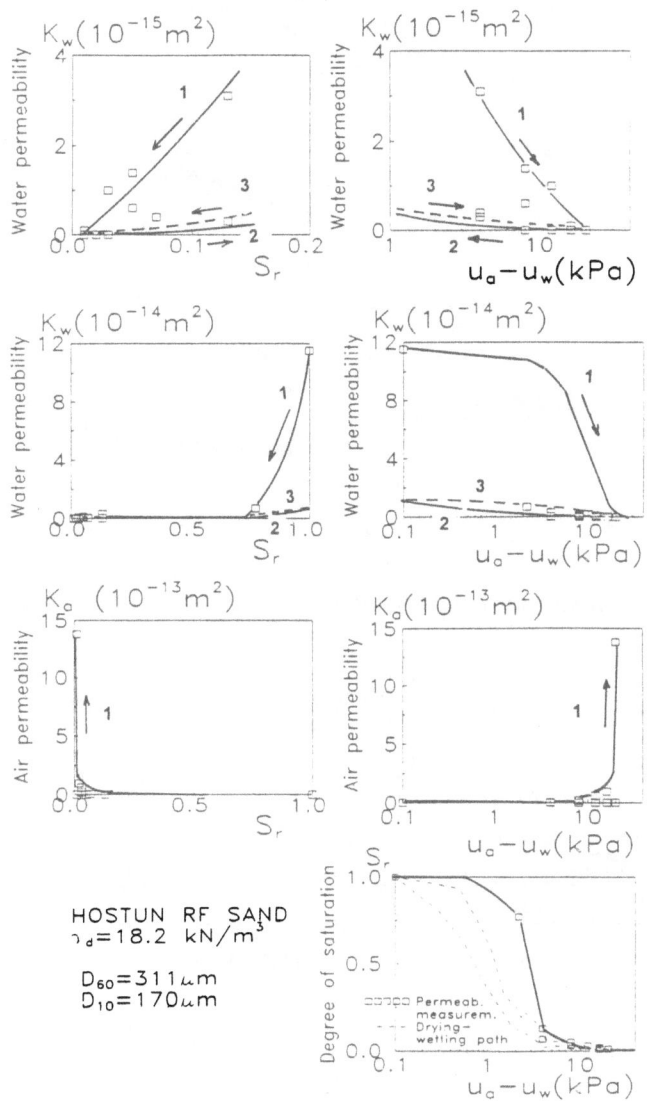

Fig. 5.5 : Air and water permeability of unsaturated sand [26]

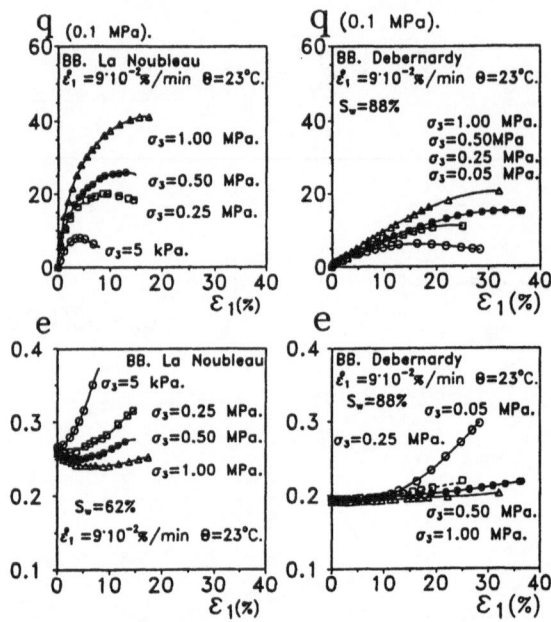

Fig. 5.6 : Influence of bitumen concentration on the behaviour of bitumen mixtures [8]

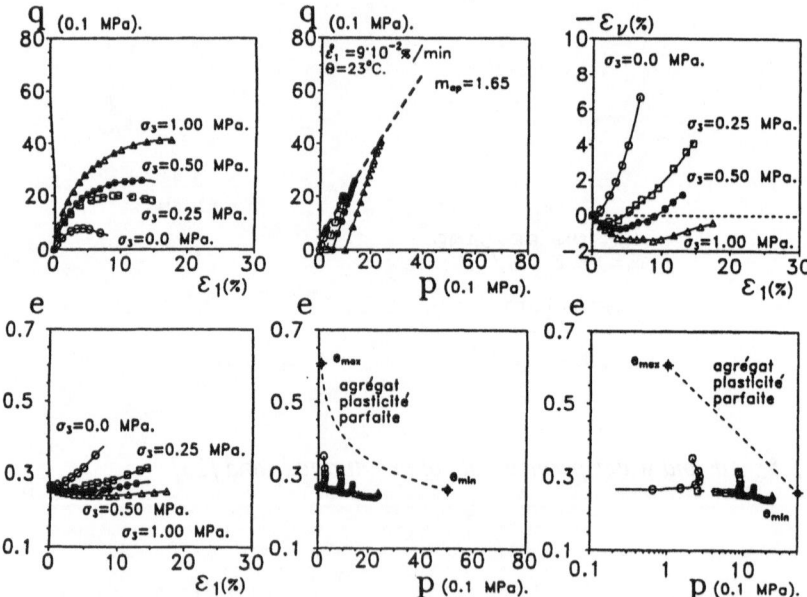

Fig. 5.7 : Stress-strain relationships in triaxial tests on unsaturated bitumen mixtures [8]

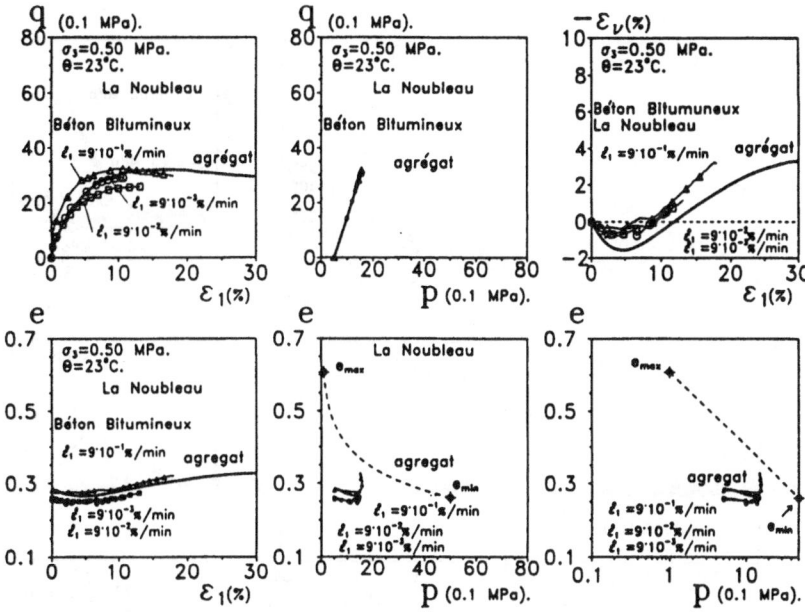

Fig. 5.8 : Influence of strain rate on unsaturated bitumen mixture at 23°C [8]

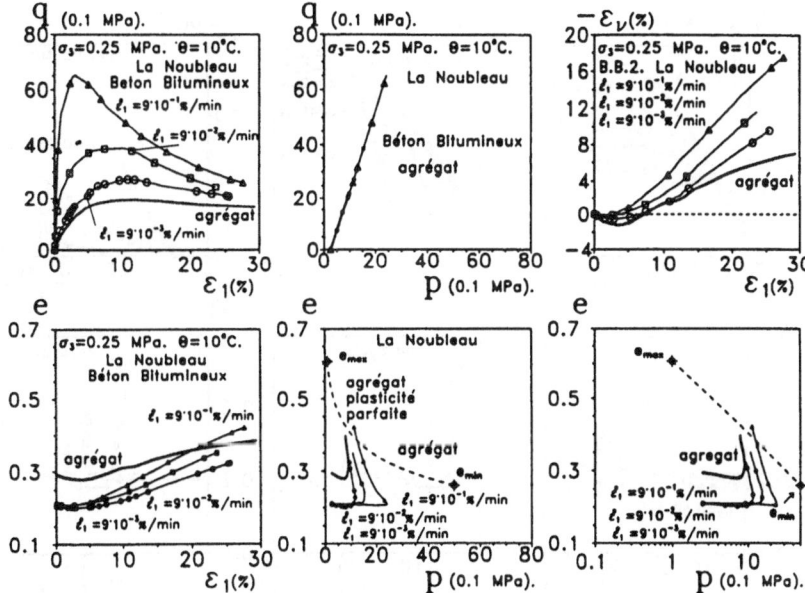

Fig. 5.9 : Influence of strain rate on unsaturated bitumen mixture at 10°C [8]

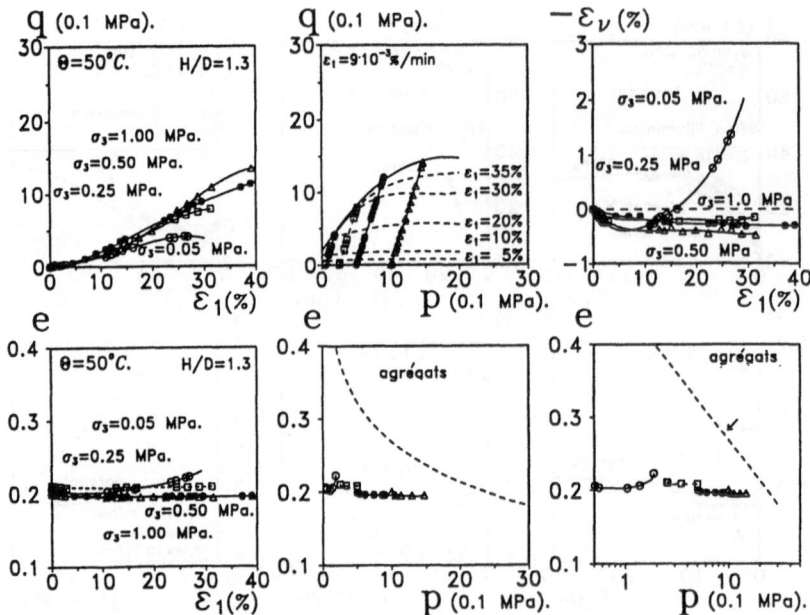

Fig. 5.10 : Stress-strain relationships in triaxial tests on saturated bitumen mixtures at 50°C [8]

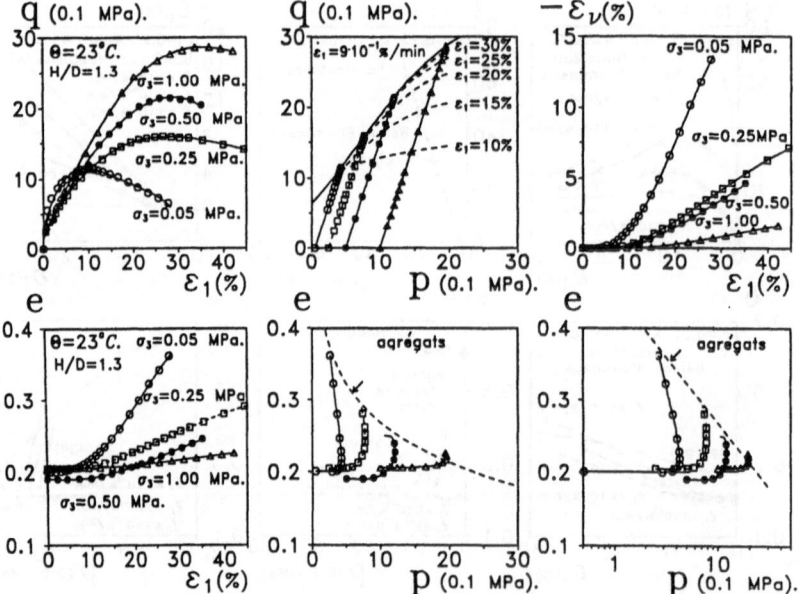

Fig. 5.11 : Stress-strain relationships in triaxial tests on saturated bitumen mixtures at 23°C [8]

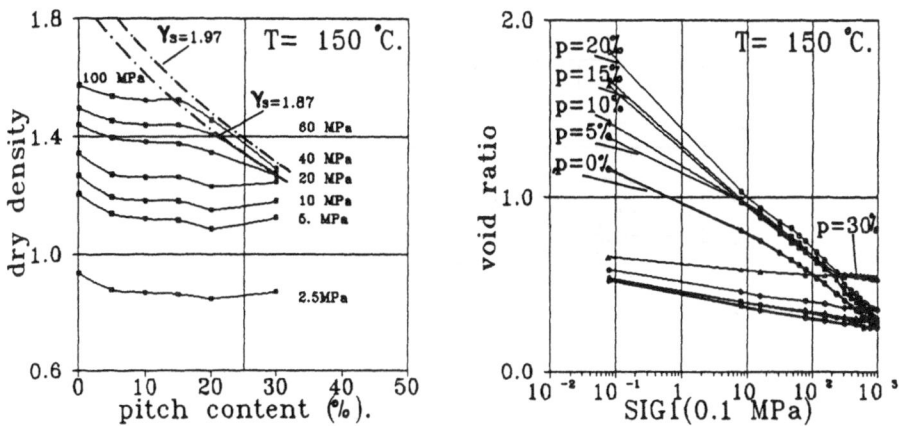

Fig. 5.12 : One-dimensional compression on fine-grained carbon agglomerates at 150°C [8]

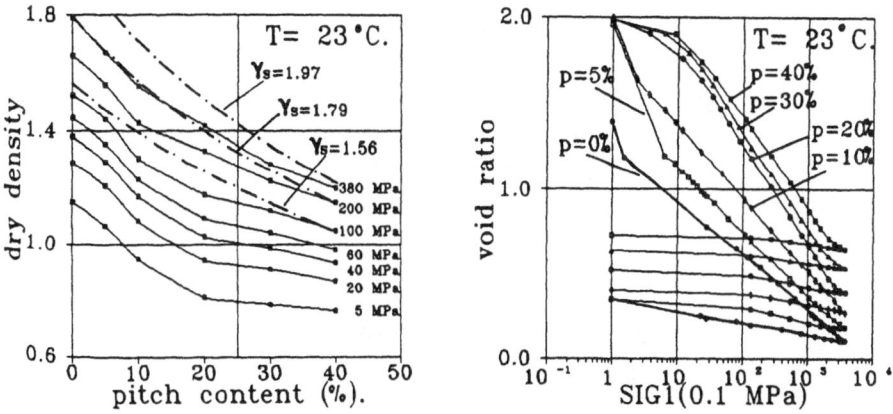

Fig. 5.13 : One-dimensional compression on fine-grained carbon agglomerates at 23°C [8]

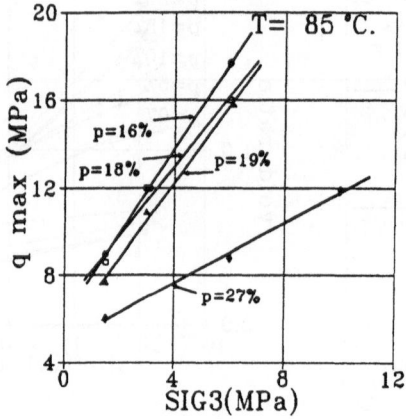

Fig. 5.14 : Maximum strength of carbon agglomerates as a function of confining pressure and pitch content [8]

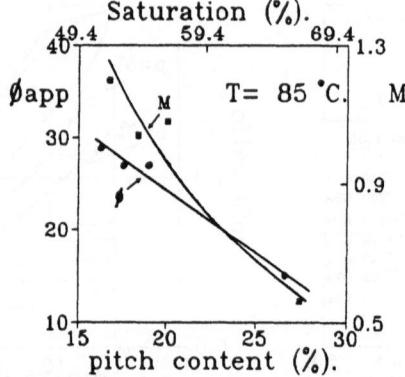

Fig. 5.15 : Evolution of apparent properties as a function of pitch content [8]

6. Rôle of inelastic deformations of particles

The inelastic deformations of individual particles caused by increase in stress or temperature leads to different laws from those applicable to media composed of elastic particles. Nevertheless, it can be interesting to examine how to interpret their behaviour. We cite here some examples where the role of plastic deformations with and without crushing of particles is seen.

6.1 Influence of grain ruptures

6.1.1 Parameters influencing grain breakage

Previous studies have demonstrated that grain ruptures are influenced by the applied stress level, which we will call a mechanical parameter and also by factors associated with the grains themselves, as mineralogy, shape, size, ..., which we will call nature parameters.

An interesting bibliographical study by Bard [8] has shown that the nature of the grains : mineralogy, shape, size, ..., plays a large role in the amount of grain breakage which is obtained under a given stress level. The strength of each individual grain is affected by its mineralogy as well as by its geological alteration (fissures, weak points...) In general, for grains of the same mineralogy, the amount of ruptures increases with grain size. This can be explained by the fact that the number of weak zones is bigger when the size increases, creating a decrease of its strength (scale effect) and also by the fact that the intensities of the forces at contact points increase due to a smaller number of these contact points. The angularity of the grains is another factor which influences grain breakage, this is due to a higher brittleness of the contacts that have a small curvature radius.

Hicher et al. [27] have studied the influence of the grain size distribution of granular assemblies made of crushed granite. The grains ranged from angular to very angular. Two materials were prepared, a well graded material (d_{60}/d_{10} = 10, d_{60} = 1 mm), called G1; a poorly graded material (d_{60}/d_{10} = 2, d_{60} = 1 mm), called G2. The grains could be considered as identical in mineralogy, size and shape, and therefore the grain size distribution appeared to be the sole different nature parameter between G1 and G2.

In Figure 6.1 we present the results of oedometer tests on G1 (relative densities Dr = 50% and 90%) and G2 (Dr = 90%) for axial stresses up to 60 MPa. Measurements of grain size distribution were performed after the tests and gave the results presented in Figure 6.2. We can clearly see a great evolution of the initial curve in the case of G2 all along the oedometric loading with d_{60}/d_{10} increasing from 2 to 10 at σ'_v = 60 MPa. In comparison, G1 evolution was much smaller. Drained triaxial tests on the same materials for mean stresses up to 15 MPa showed the same differences in the grain size distribution after testing. We can therefore say that a poorly graded material is more likely to demonstrate grain breakage than a well graded one.

6.1.2 Influence of mechanical parameters

It seems evident that, even if grain ruptures are not limited to elevated stresses, their number increases with the stress level. We can observe however that the stress path also plays an important role. For example, more ruptures can be achieved in a triaxial test than in an oedometer test for the same stress amplitude (Fig.6.3).

We have also studied the evolution of the grain size distribution at different strain amplitudes along triaxial loading. The results clearly show that, for identical stress levels, the amount of grain ruptures increase with the axial strain (Fig.6.3). This effect of strain amplitude can, at least partly, explain the differences obtained in triaxial and oedometer tests : in the oedometer test the increase of the assembly stiffness with stress limits the strain amplitudes at elevated stresses compared to the ones achieved in triaxial tests. There is therefore a combined effect of stress and strain on the amount of grain breakage in granular bodies.

6.1.3 Grain breakage influence on behaviour of granular materials

Oedometer test

If the sample is in the form of identical glass spheres, the fragile nature of this material leads to simultaneous crushing of the spheres. This creates a slope discontinuity in the (e, log σ_v) compression curve (Fig.6.4). In the case of a sand, however, this phenomenon is more progressive, due either to particle roughness or to particle size distribution (Fig.6.5). The consequences are mainly an increase of the compressibility along both isotropic and deviatoric stress paths. This can be seen for example in Figure 6.1 by measuring the maximum slope of the curves e-log σ'_v.

In the case of G1 the experimental values are between 0.15 and 0.20, according to the initial void ratio. Values of the same order of magnitude were obtained in different granular materials at small stresses without grain breakage. In comparison, the value obtained for G2 is much bigger : 0.38, and this difference reflects mainly the effect of the significant amount of grain ruptures produced by the material.

Similar results were obtained by Bard [8] on petroleum coke (Fig. 6.6). Ruptures of big grains were obtained for small vertical stresses and explained the high compressibility of the material. Due to internal grain porosity, this high compressibility is maintained even when $d_{60}/d_{10} > 10$. We observe at this stage a translation of the grain size distribution curves, with a constant d_{60}/d_{10}, towards smaller particle dimensions. Comparison of carbonaceous powders, with different initial grain sizes, demonstrates a similar behaviour at high strength. Grain breakage fabricates a new material which behaves the same way as initially well graded fine powders obtained by a previous crushing of the particles (Fig.6.7).

Triaxial test

The influence of particle crushing can be observed mainly on volume change curves. We can see that the volume change in G2 (Fig.6.8) is much higher than the one in G1 (Fig.6.9) at the end of a same test at $p'_0 = 15$ MPa and Dr = 90%.

The explanation is provided in a schematic way in Figure 6.10 where correlations are drawn between the position of the critical state line and nature parameters (e_{min}, e_{max}, d_{60}/d_{10}, grain angularity). One can see that an increase of d_{60}/d_{10} due to grain ruptures will produce a displacement of the critical state line towards smaller void ratios [29]. In order to reach the critical state during a triaxial test, a bigger volume change will be needed in case of significant grain breakage. This phenomenon will result also in a change of the stress-strain relationship and in particular in an increase of the axial strain corresponding to the maximum strength. There is therefore a significant change in the global mechanical behaviour of granular media due to grain breakage, which needs to be taken into account in constitutive modelling. (See in Figure 6.11 for an example of stress-strain relationship evolution with grain breakage on Hostun sand).

6.2 Influence of plastic deformation of particles

6.2.1 Oedometer tests

Figure 6.20 illustrates the influence of plastic deformation of individual particles on the stress-strain curves by the use of ductile spheres of lead or resin. The creation and the development of a courbure can be seen which continues until all the voids have been filled. The comparison between results obtained on lead and glass balls emphasizes the influence of individual grain behaviour (Fig 6.12). With a lead ball assembly, the evolution of the stress-strain curve is continuous with a slope increasing to a value of 0.69 in a plane (e-logσ_v), while for a glass ball assembly, the slope is initially smaller with a large increase that leads to a high compressibility when the balls start to rupture, then a decrease of the slope due to a change in grain size distribution leading to a large reorganisation of the grains. Figure 6.13 presents some results obtained by Mosbah [7] on copper powders. Photos 6.14 and 6.15 illustrate the plastic deformations of grains after compaction.

6.2.2 Triaxial tests

Triaxial tests were performed by Bard [8] on sintered iron fibers (d = 0.05 mm and L = 0.5 to 2 mm). Samples with high porosity (n = 0.6 to 0.8) were prepared in a mould and then submitted to axisymmetric loading. Typical results are presented in Figure 6.16. The high porosity allows large plastic deformations of the fibers to take place. As a result, the specimen developed a very high compressibility and large irreversible strains. Similar tests in sintered iron powders show very different stress-strain responses (Fig.6.17).
Triaxial tests on copper powders were performed by Mosbah [7]. Results in Figure 6.18 show the influence of the mean stress, indicating a transition between a frictional behaviour (granular material) for low stresses and a behaviour independent of the mean stress (dense metal) for high stresses. This is in accordance with the evolution of the relative density: for confining stresses higher than 250 MPa the density becomes close to 1, indicating that rearrangement and plastic deformation of the particles reduced the void ratio to a value close to 0.
Figure 6.19 summarises the influence of grain properties on the mechanical behaviour of a granular assembly along a one-dimensional loading path.

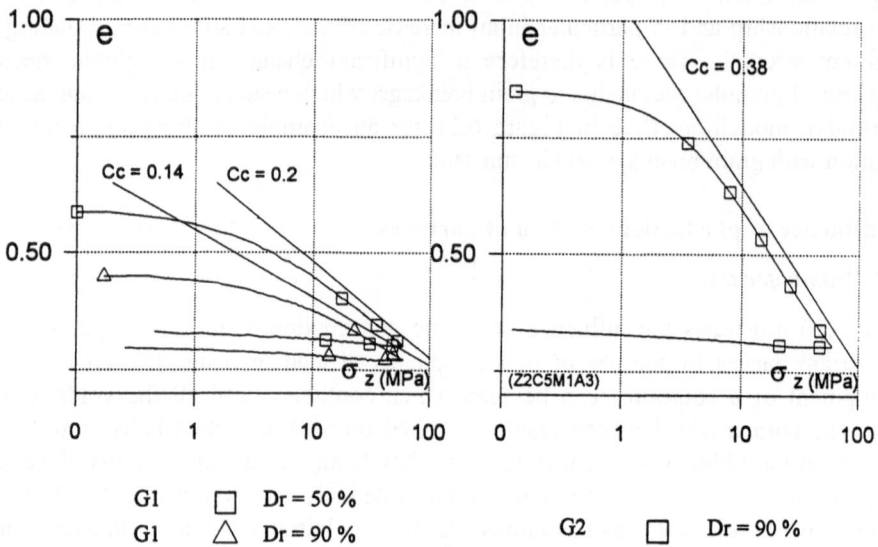

Fig. 6.1 : One-dimensional compression tests on crushed granite

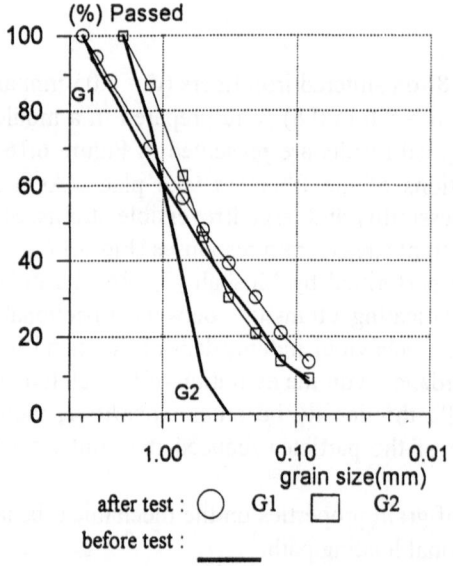

Fig. 6.2 : Grain size distribution curve for crushed granite before and after one-dimensional compression up to 60 MPa

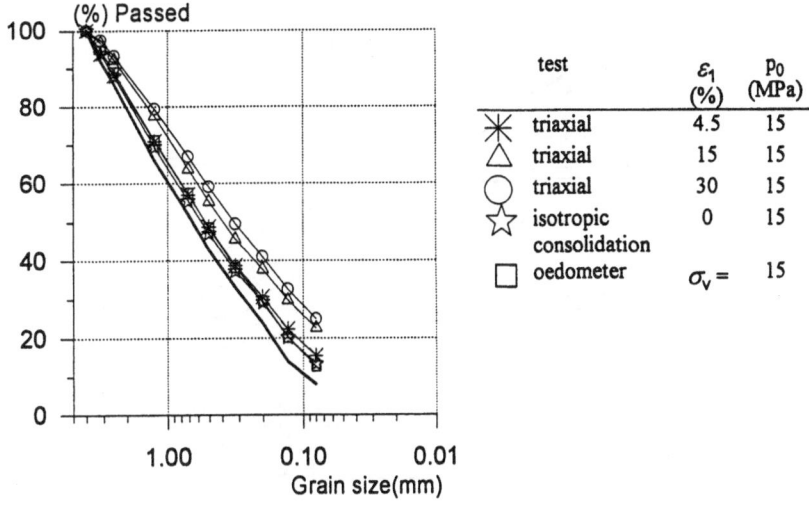

Fig. 6.3 : Influence of stress and strain paths on grain breakage of crushed granite

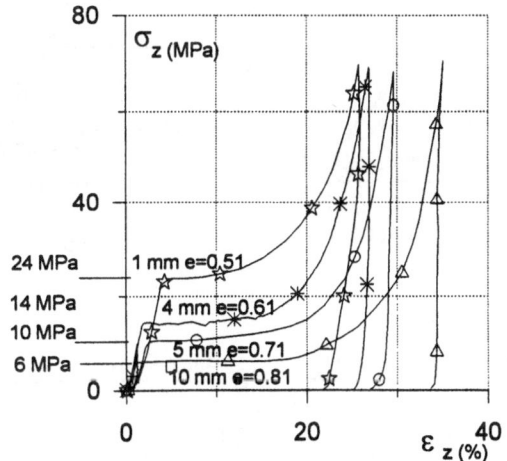

Fig. 6.4 : One-dimensional compression tests on glass balls

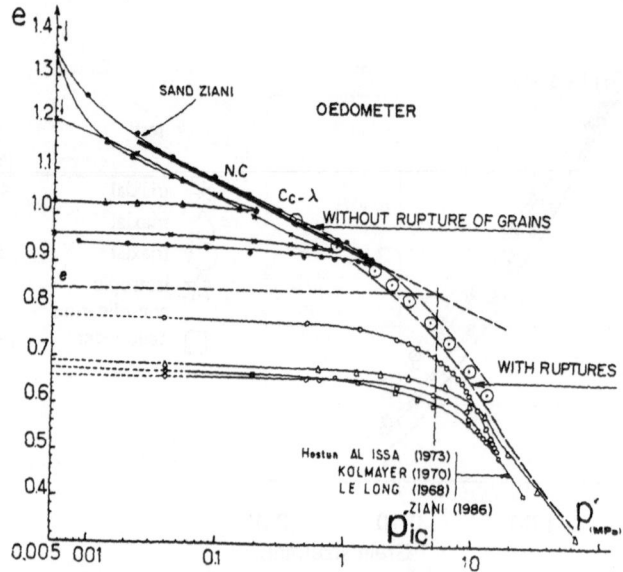

Fig. 6.5 : One-dimensional compression tests on sand

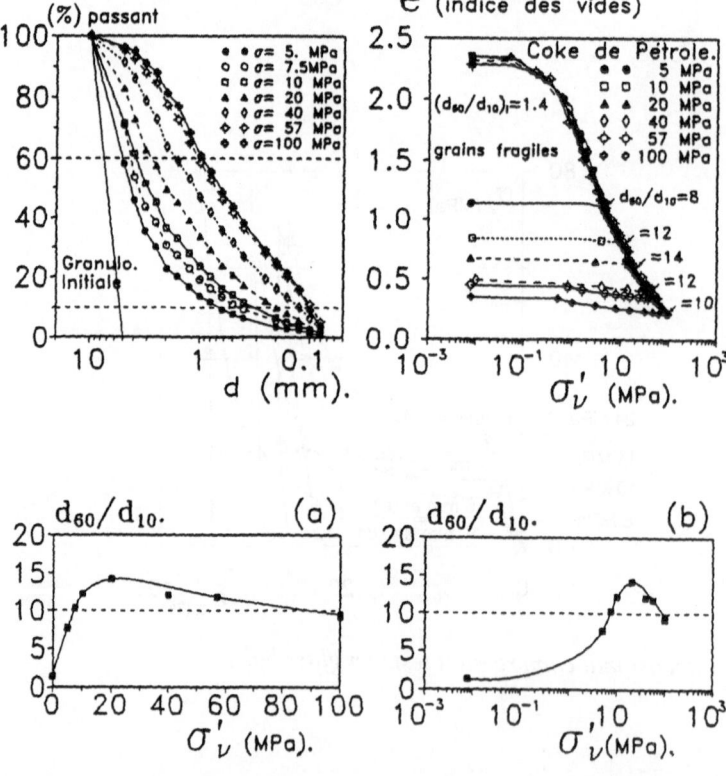

Fig. 6.6 : One-dimensional compression tests on petroleum coke [8]

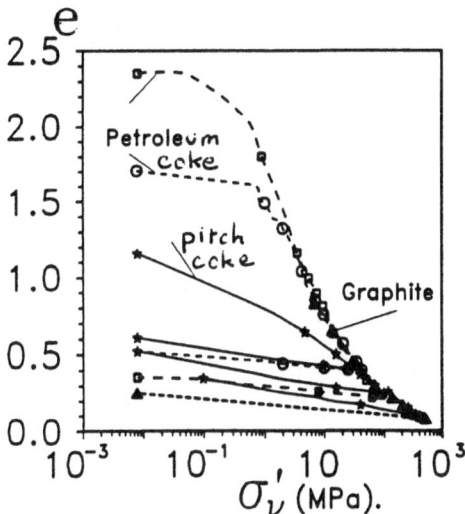

Fig. 6.7 : One-dimensional compression on carbonaceous powders [8]

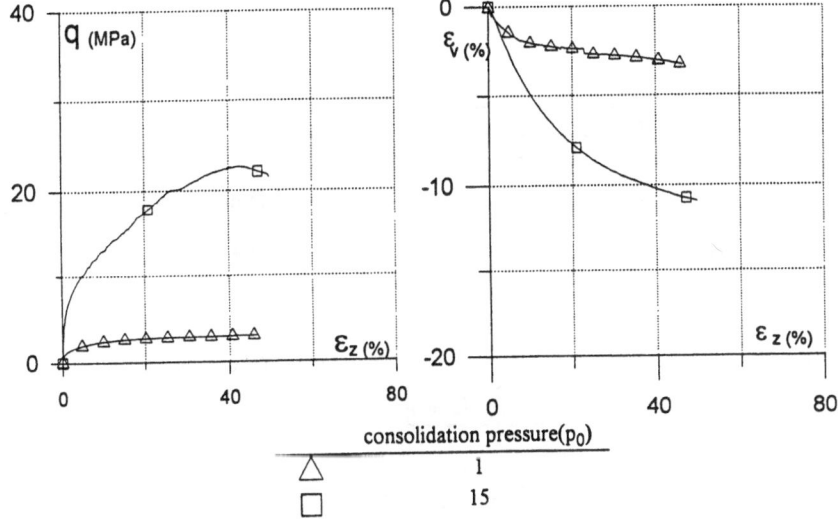

Fig. 6.8 : Triaxial tests on crushed granite (G2), Dr = 90%

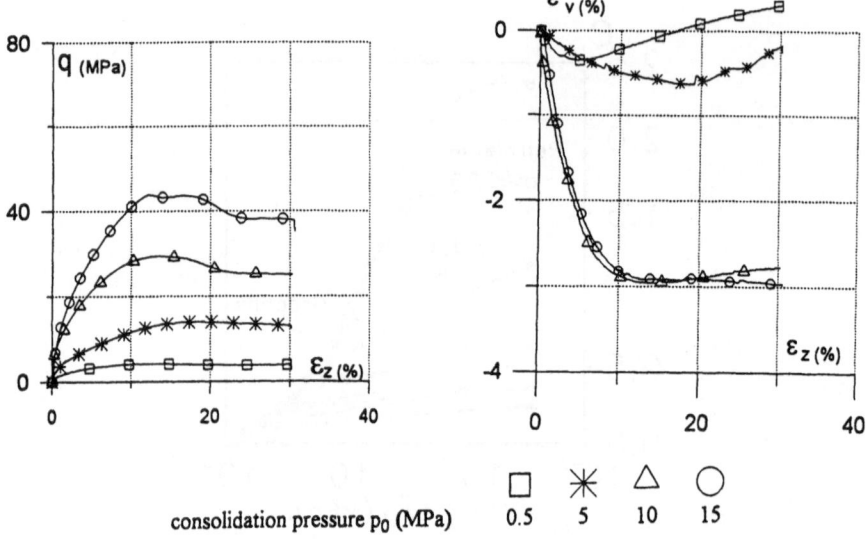

Fig. 6.9 : Triaxial tests on crushed granite (G1), Dr = 90%

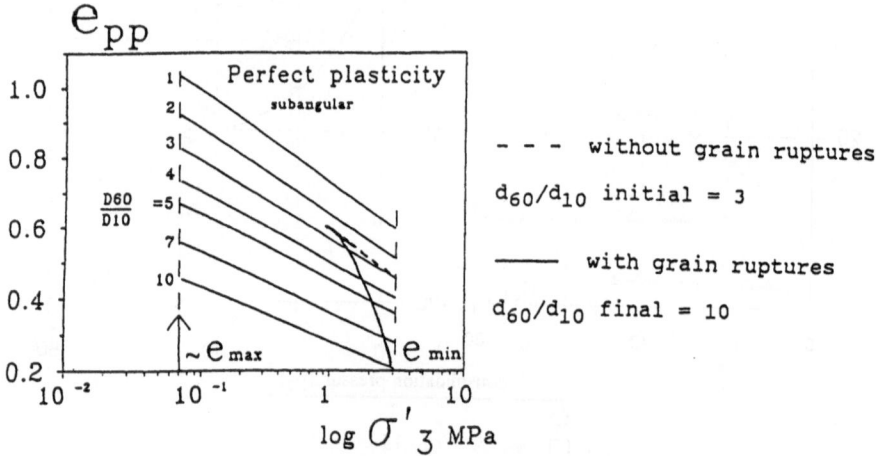

Fig. 6.10 : Chart of perfect plasticity line position as a function of grain size distribution

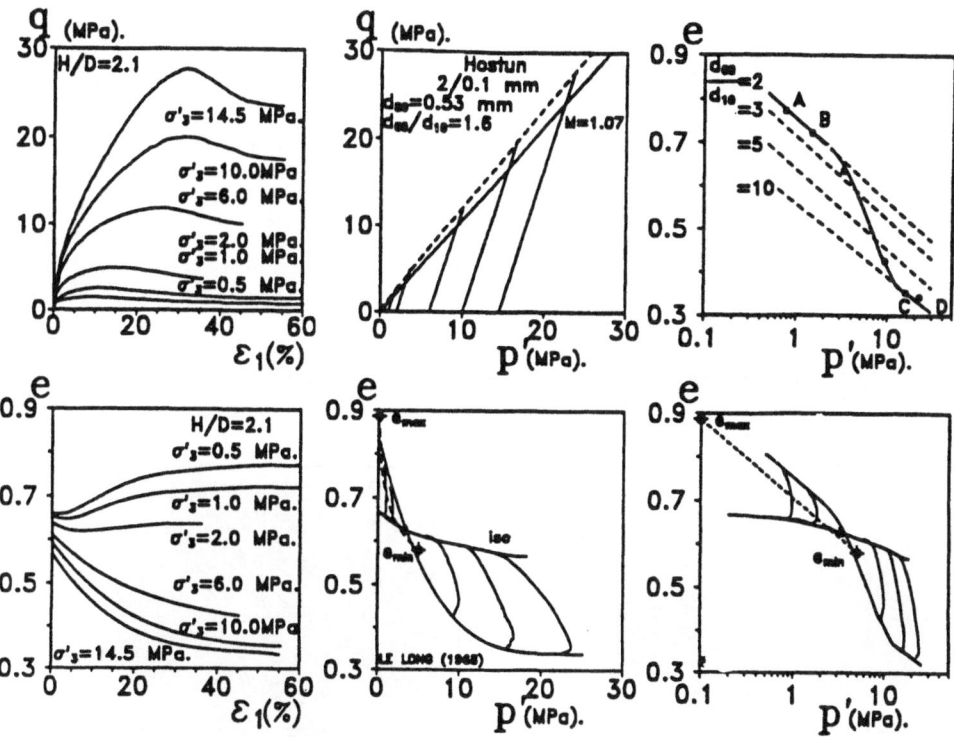

Fig. 6.11 : Stress-strain relationship in triaxial tests with grain breakage

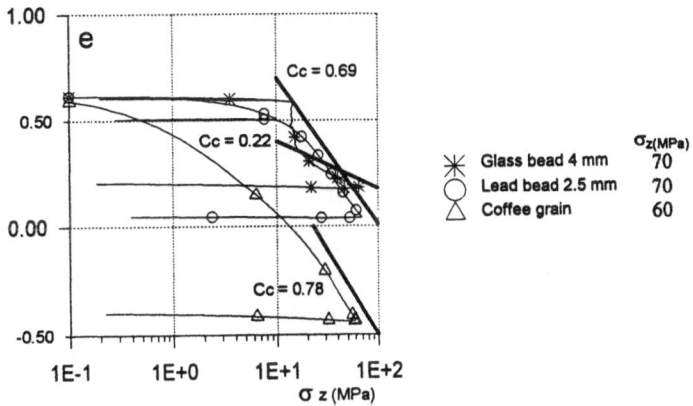

Fig. 6.12 : Comparison of one-dimensional compression tests on glass and lead balls

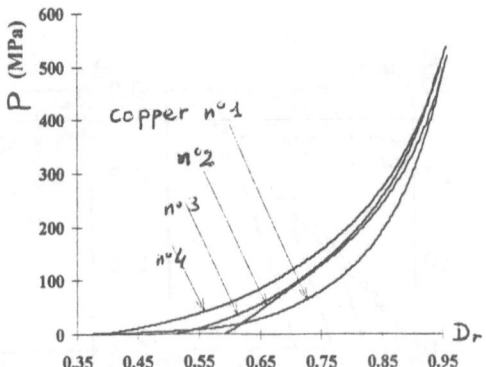

Fig. 6.13 : One-dimensional compression on copper powders [7]

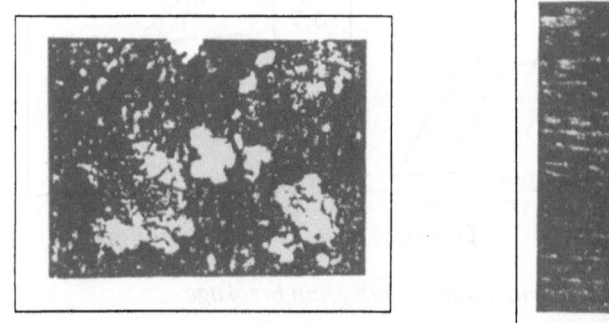
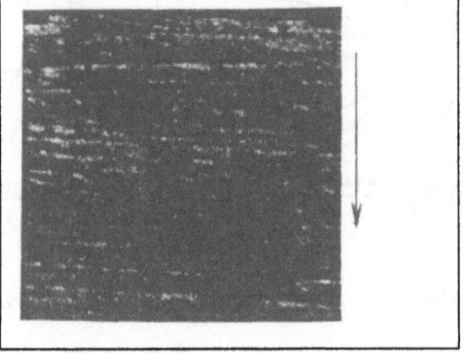

Fig. 6.14 : Evolution of grain shape during one-dimensional compression : copper powder[7]

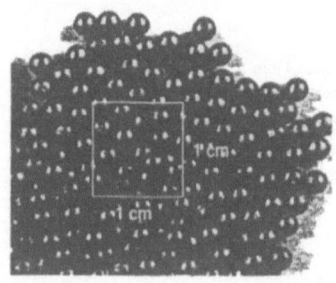

Fig. 6.15 : Evolution of grain shape during one-dimensional compression : lead balls

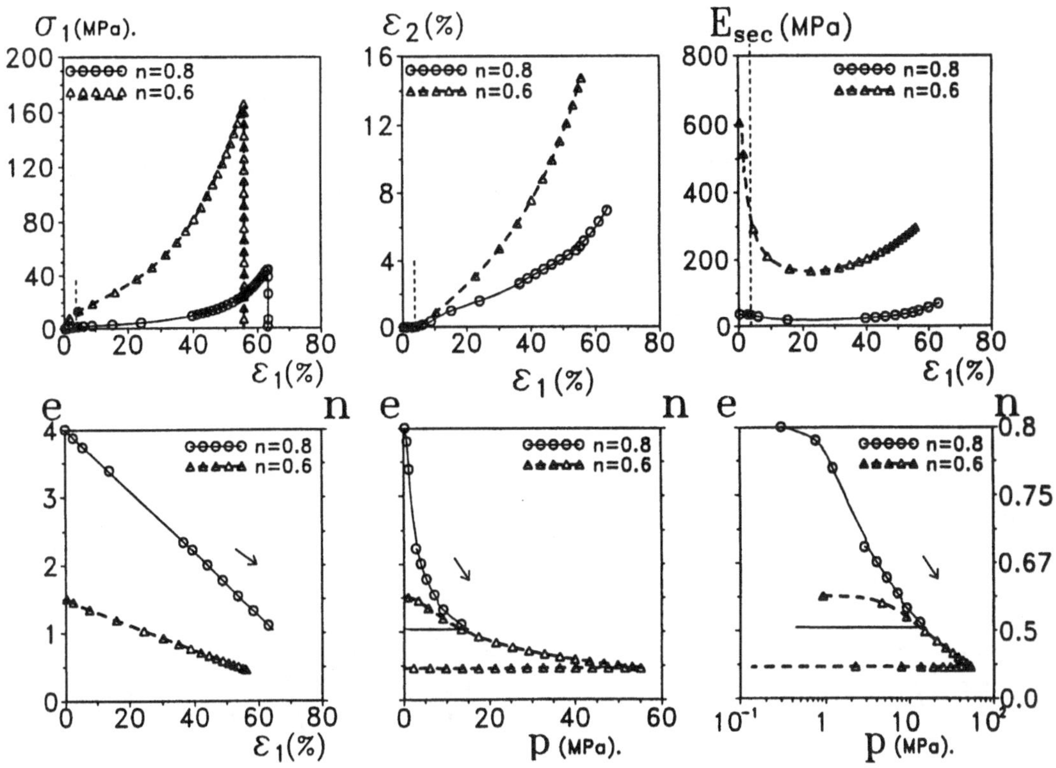

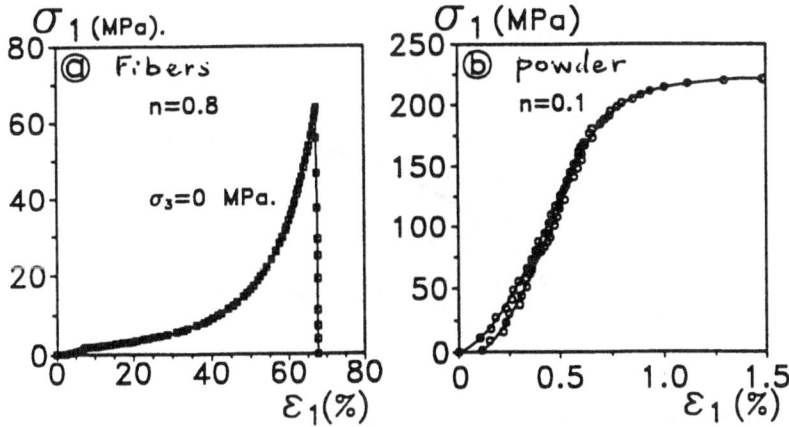

Fig. 6.16 : Unconfined compression tests on sintered iron fibers [8]

Fig. 6.17 : Influence of grain behaviour on stress-strain response in unconfined compression [8]

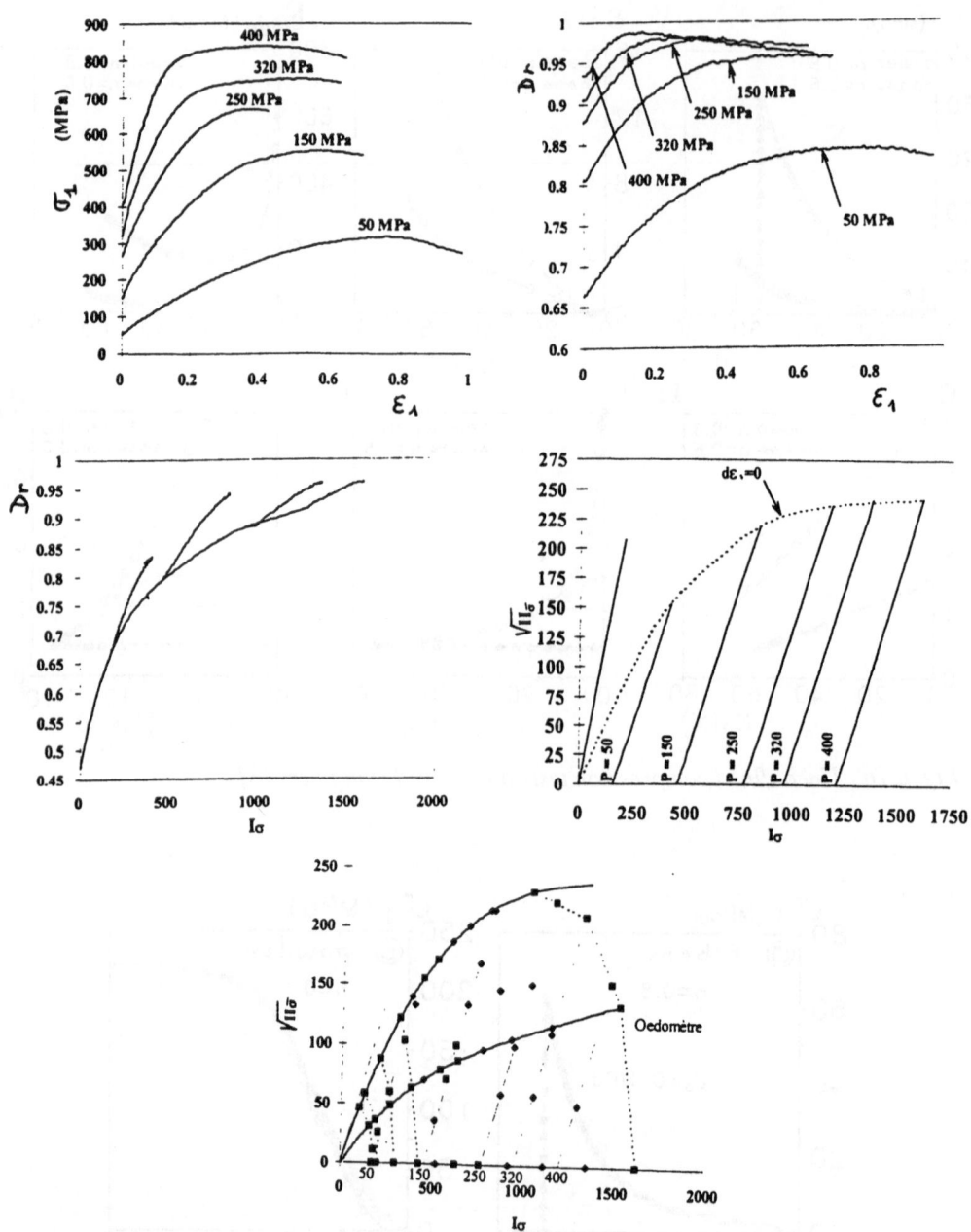

Fig. 6.18 : Triaxial tests on copper powder at elevated stresses [7]

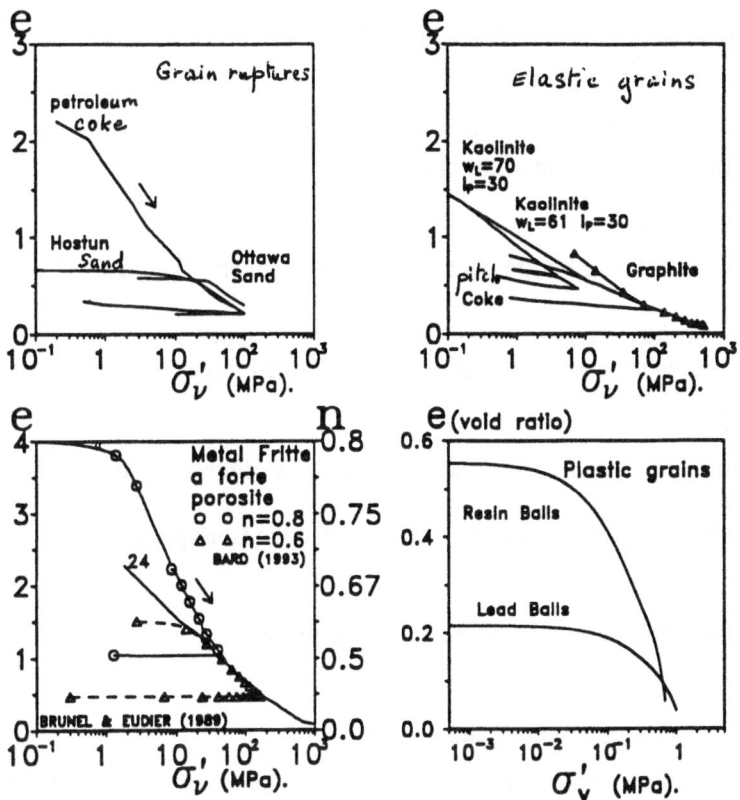

Fig. 6.19 : Influence of grain behaviour on stress-strain relationship in one-dimensional compression

7. Cyclic and dynamic loading

7.1 Characterisation of non-linear behaviour

7.1.1 Non linear elasticity

If we consider two elastic spheres in contact, subjected to a normal force N, we can write for the relative displacement of the two centers, w, the following equation :

$$w^{3/2} = \{3(1-\upsilon_g^2)/E_g4\sqrt{R}\}N$$

with E_g and υ_g : linear elastic coefficients of the spheres, and 2R : diameter of the two spheres.

If we now consider a regular assembly of the same spheres, we can extend equation (1), assuming that the spheres are subjected only to normal forces :

$$w^{3/2} = 3/4.R^{3/2}\{(1-\upsilon_g^2)/E_g\}G(e)p'$$

with p' : isotropic pressure applied at the boundaries of the assembly and G(e) : function of the assembly void ratio.

The isotropic strain $\varepsilon_i = w/R$ can then be related to the isotropic stress p' :

$$p'/\varepsilon_i = 3/2.\xi^{2/3}p'^{1/3} = E_i$$

The isotropic modulus E_i can then be written as follows :

$$E_i = 3/2\alpha p'^n$$

with : $\alpha = \{4E_g/3(1-\upsilon_g^2)G(e)\}^{2/3}$ and n = 1/3

Thus it follows that the assembly of linear elastic spheres under isotropic stress undergoes reversible but non linear strains.

This study, originated by Hertz, was extended by Mindlin to take into account the supplementary influence of a tangential force T at the contact of two spheres. The application of the elastic theory has, in this case, produced evidence of infinite tangential stress upon the circumference of the surface contact, which has prompted Mindlin to introduce Coulomb's friction law along the surface of the contact zone. Under these conditions, plastic dissipation appears along with non-reversible strains. As long as T < fN, these strains are localised on an annular surface. When T becomes equal to fN, this zone extends to the entire zone of contact and, consequently, relative displacement of the two spheres takes place, creating irreversible deformation of the assembly.

From the above analysis, we can derive that an even very small deviatoric stress applied on a granular assembly leads to irreversible strains. This is due to the fact that the distribution of the orientation of the contact forces created by an isotropic stress tensor can vary inside the friction cone between -φ and +φ. An additional deviatoric stress will produce in some contact points an orientation of the force equal to φ, leading to the relative displacement of some grains. After this first loading, any additionnal stress of intensity equal or lower to the first one, applied in the same direction will lead to smaller irreversible strains because the first cycle will have eliminated the most unstable contacts : this is the hardening phenomenon usually observed in granular media subjected to repeated loading (see below). All these considerations have been confirmed in recent years by several studies which, based on micromechanics computations, use elastic spheres with Coulomb's friction at contact.

We can thus posit the following conclusions concerning the behaviour of granular media :

i) the search of an elastic behaviour has to be made for very small strains ;

ii) the law of elasticity is non linear, with a dependancy of the modulus upon the mean effective stress ;

iii) if the structure of the granular medium is isotropic, the elastic domain will also be isotropic, but any irreversible strain will create an anisotropic structure leading to an anisotropic elastic domain and an anisotropic elastic law.

We will give some experimental evidence of the mechanical behavior of granular media, subjected to very small strains in order to investigate their elastic properties.

Experimental procedures

The standard triaxial apparatus is not satisfactory for studying material behaviour for strains less than 10^{-2}. We developed for this purpose a triaxial apparatus with the sensors placed inside the cell in direct contact with the sample (Fig.7.1) [30].

Reversible behaviour along a given stress path

Compression and extension triaxial tests have been performed starting from an isotropic state of stress on different granular materials (glass balls, sands, gravels). In order to investigate reversible behaviour, each loading was followed by an unloading at different levels of stress or strain. Figures 7.2, 7.3 and 7.5 present some typical results. They show a reversible behavior for all these materials along a compression-extension stress path around the isotropic initial state, as long as the strain amplitudes are lower than 1 to 3 10^{-5}

For a given value of the horizontal stress σ'_3, the relation between $\sigma'_1 - \sigma'_3$ and ε_1 is linear and reversible and gives the value of the Young's Modulus E. The relation between ε_1 and ε_3 is also linear and gives the value of Poisson's ratio υ (Fig. 7.4, 7.6). The strain amplitude for which the relation $(\sigma'_1 - \sigma'_3)$-ε_1 ceases to be linear and reversible was found to be between 10^{-5} and 3 10^{-5} for all the tests. It seems to be lower when the grain size distribution is more uniform and when the initial isotropic stress is smaller.

Figure 7.2 presents an example of cyclic stress-strain curves at small strains. The first cycles appear to be closed, whithout any hysteresis. Dynamic tests, however, conducted in a longitudinal resonant column show a small amount of damping (about 1%), in accordance with the work of Mindlin. In the case of our tests, 1% of damping represents the limit of accuracy in our measurements.

The main important conclusion is then the following : all the results show the existence of a reversible behavior when the samples are subjected to compression and extension triaxial loading. The maximum reversible strain amplitudes depend upon the nature of the granular medium and are generally located around 10^{-5}.

For larger deformations, the secant modulus continuously decreases with the increase of strain amplitude. Simultaneously the hysteresis loop becomes larger and damping ratio increases. A more and more significant part of the deformation is plastic (non recoverable). In looking back to Hertz' study one can assume, for a granular material, that Young's Modulus depends upon the mean effective stress, the void ratio, and the nature of the grains.

Influence of mean effective stress on Young's Modulus

Results of triaxial tests at different initial isotropic stresses show that the Young's Modulus increases with the mean effective stress (Fig. 7.3 and 7.5). Generalising the previous equation (4) we can write :

$$E = ap'^n$$

From our test results on dry granular media (glass balls, sands, gravels), we found that the value of n is close to .5 for all materials when the void ratio is kept constant (Fig. 7.7). This observation is in accordance with those of other studies, which used different experimental techniques.

Influence of void ratio

Parameter a in the above relation depends on the void ratio. As a first approximation, a general relation can be obtained by plotting $E/p'^{.5}$ as a function of e (Fig. 7.8). This gives :

$$E/p'^{.5} = 450/e \qquad \text{(E and p' in MPa)}$$

Poisson's ratio

In Figures 7.4 and 7.6 we have plotted the relation between axial and radial strains obtained on a dry sand and a glass ball assembly. We can see that this relation is linear over a range of strain amplitude from 10^{-6} to 10^{-4}. The slope of this straight line gives the value of Poisson's Ratio. In this case we obtained $\upsilon = .22$.
For all the tested granular materials, values of υ were located in a narrow range from .18 to .23. For any given material, υ was found independant of initial mean effective stress, in agreement with Hertz and Mindlin's studies. Void ratio has also little influence on Poisson's Ratio. It appears to depend slightly on the grain size distribution : it increases from $\upsilon = .18$ for a poorly graded Hostun sand and up to $\upsilon = .23$ for a well graded Hostun sand.

Dynamic testing

The elastic parameters can also be obtained by measuring elastic wave propagation in the specimen by the travel time method or by using a resonant column device. The resonant column is widely used in soil dynamics. A soil column is excited to vibrate in one of its natural modes either longitudinally or torsionally (Fig.7.9). When the frequency of resonance is known, the wave velocity can be determined and the elastic parameters are derived. Figure 7.10 presents some results obtained by Hardin and Richard [31] which show the influence of void ratio and mean stress on the shear modulus. These results are similar to those discussed above. Using a longitudinal resonant column, Boelle [32] obtained results on sand which could be compared to the ones obtained in cyclic triaxial tests (static conditions). No significant difference was found for this material which does not exhibit any viscous effect.The slight differences observed in Figure 7.11 can be more likely explained by a small difference in anisotropy due to the mode of preparation, which was not taken into account in the interpretation of resonant tests.

7.1.2 Non-linear plastic behaviour

Outside of the elastic domain, non-reversible strains take place and, as a consequence, the stress-strain relationship at a given mean stress is no longer linear. The secant modulus is thus a decreasing function of strain amplitude. Figure 7.12 presents some results on Hostun Sand. One can see that the evolution of the secant modulus E with mean stress can still be written as $E_s = a_s p^n$ with n increasing when the strain amplitude increases. Values of secant modulus for strain amplitude around 10^{-2} represent only 20% of the Young's Modulus. Similar results have been obtained for example by Drnevich et al. [33] using a resonant column. For large strain amplitudes, they used hollow cylindrical specimen instead of solid specimen in order to maintain a high degree of strain homogeneity.

Damping ratio can be determined in a resonant column (longitudinal or torsional) from either a free-vibration or a forced-vibration test. In a forced-vibration test, the sample is excited by varying the frequencies. A resonant curve is obtained from which the damping coefficient is derived. In a free-vibration test, the excitation force is suppressed when resonance is obtained and a measure of the logarithmic decrement of successive maximum amplitudes gives a damping ratio. Results at very small amplitudes show a small amount of damping, around 1% for $\varepsilon = 10^{-5}$. When the strain amplitude increases, the damping ratio also increases; typical values for sands are presented in Figure 7.13. This increase in the damping ratio is mainly due to a non-reversible behaviour of granular materials when subjected to strain amplitudes higher than 10^{-5}. Little influence of strain rate was observed. The damping in granular materials is due mainly to plastic behaviour rather than to viscous behaviour.

Cyclic triaxial loadings are also used in order to derive damping ratio values. The damping ratio is usually defined by the ratio $D = A_1/\pi A$ (Fig.7.14) where A_1 is the area of the loop (dissipated energy) and A the area of the triangle ACD (two times the elastic energy). Typical results are presented in Figure 7.15. A simple linear relationship was obtained by Charif and Hicher [35] in the case of dry Hostun Sand between the ratio E/E_{max} (secant modulus versus Young's) and D (Fig. 7. 16).

7.2 Isotropic and oedometric cyclic stress paths

A succession of isotropic loading and unloadings creates a progressive compaction of granular materials. For a given number of cycles the compaction increases with the stress amplitude. (Figs. 7.17,7.18). The cycles become more and more reversible but a complete stabilisation can be achieved only for very high numbers of cycles.

7.3 Cyclic triaxial paths

7.3.1 One-way cyclic tests

One-way tests in which q is cycled between zero and q_c show a significant plastic strain during the first cycle. The cycles retain practically the same shape as they move along the ε_1 axis (Fig.7.19).

The plastic strain during the first loading is important. It is clearly less so for the following cycles even though the plastic deformations continue. If after one or several cycles up to the same stress amplitude, the maximum stress amplitude is increased, a 'bend' in the stress-strain curve appears. There is a clear reduction in the gradient when the stress passes the maximum reached in preceding cycles. However, the following cycles are more regular and are straighter at the top of the cycle.

There is therefore a 'memory phenomenon' which can be characterised by the maximum value of the cyclic stress amplitude. This becomes a memory parameter analogous to p'_{ic} which we have used to define a loading surface. Its increase is associated with significant plastic deformations. Inside the surface there is also plastic strain but of lower amplitude.

Volume changes depend on the position of the stress path with respect to the contractant domain (in general $q'/p' < M$) defined by monotonic tests. If the stress path is situated entirely within this region, the cycles will produce a progressive compaction even for initially dense materials. If it crosses the boundary of the region during each cycle, there will be a period of compaction and a period of dilation. If the average level of the cycle lies within the contractant domain, cyclic loading produces a compaction; otherwise it produces a dilation [5].

7.3.2 Two-way cyclic triaxial tests

In this type of test each cycle, consisting of alternating axisymmetric compression and extension, involves a sudden 90° change in the directions of the major and minor principal stresses as they interchange. The results showing the influence of a rotation of the principal stress (chapter 3) reveal that considerable rotation tends to cause compaction when the stress is close to the isotropic state and therefore well inside the contractant domain. The results of two-way cyclic tests confirm this analysis and progressive compaction is always observed irrespective of the amplitude and the average level of the cycles. Figure 7.20 shows the results obtained on sand which can be compared with the previous graphs. For the same number of cycles at the same amplitude, the final density is always higher for two-way tests. The shape of the stress-strain cycles may alter. There is no longer a simple translation of the loop for each cycle along the ε_1 axis, but there is a modification of its slope. Tests with fixed strain amplitudes lead to analogous results. Due to the increasing density of the material with each cycle, the value of the maximum stresses in compression or in tension increases with each cycle. The variation in volume show clearly a region of contraction bounded (in compression as for extension) by a limiting value of q'/p' beyond which dilation occurs (Fig.7.21).

7.4. Influence of rotating principal axes

Wong & Arthur [37], using the directional shear, carried out tests on loose sand, keeping the ratio σ'_1/σ'_3 constant and cyclically turning the axis of σ'_1 and σ'_3 through a constant amplitude of rotation (θ). Figure 7.22 shows the variations of volume obtained. For an amplitude of $\theta = 30°$, there is no significant variation ; for $\theta = 55°$ and $70°$ they obtained considerable contraction during the cycles.

Ishihara & Tohwata [12] carried out tests on hollow cylinders with p' and σ'_1 - σ'_3 both constant and with cyclic rotation of the direction of the major principal stress between -45° and +45° with respect to the initial orthotropic axes. Here also progressive compaction of the sand sample, accompanied by a straigthening of the cycle occurred (Fig.7.23).

Tests were carried out by Joer [38] on a two-dimensional material (PVC rods) in a plane strain device called "$1\gamma2\varepsilon$" (Fig.7.24). Tests with constant principal stresses and continuous rotation of the principal axes were performed (Fig. 7.25). The cyclic effect of the rotation induced a compaction of the specimen. Along the circular stress path, incremental strain vectors were super-imposed. It is clear that the principal incremental strain axes do not coincide with the principal stress axes.

A cyclic rotation of principal directions could therefore produce significant variations of volume in the material which would be accompanied by a modification of the stress-strain relationships during the cycles.

7.5 Isochoric triaxial tests

In the isochoric test, variations of mean stress will depend on the tendency of the soil to contract or dilate.

We have seen in the preceding paragraphs that cyclic tests generally lead to an increase in the density of granular materials. In the isochoric test, this will translate into a decrease in mean stress. The decrease could in certain cases be sufficiently large to cause complete loss of stress : this is the phenomenon of liquefaction (Figs.7.26, 7.27).

Figure 7.29 shows the behaviour of a granular material during a liquefaction test. This process can be broken down into three stages :

(a) Small cyclic deformation, constant decrease of mean stress ;

(b) The stress path reaches the intrinsic line. The mean stress continues to decrease, the cyclic strain accelerates and the cycle develops a 'stepped' like shape ;

(c) The stress path becomes stable. It passes the point of zero effective stress twice per cycle. The mean stress varies periodically. Strains become large and continue to increase.

There exists therefore a region called liquefaction in which deformations are associated with near zero stress. This explains the long 'flat' portion of the q'-ε_1 curves either side of the origin during which the material has very little rigidity. Beyond a certain strain however, depending on the shape of the cycle, the mean stress increases and the material stiffens. If monotonic loading is continued beyond the value of the cyclic stress amplitude, the same strengths are obtained after liquefaction as before.

Liquefaction only occurs with two-way cyclic tests. One way tests also produce decreases in mean stress, but these tend to stabilize before liquefaction occurs.

Many factors influence whether a granular material will or will not liquify :

    - The relative density (Dr): the higher Dr the more cycles are required for liquefaction to occur ;
        - The mean stress ;
        - The cyclic stress amplitude ;
        - The particle size distribution : a uniform particle size facilitates liquefaction.
The effect of principal axis rotation can also be observed in isochoric tests. Figure 7.29 presents results obtained by Joer [38] in the 1γ2ε device. Continuous rotations of principal strain axes were imposed at constant volume. In the deviatoric stress plane (τ_{xy}, (σ'_x - σ'_y)/2) the path takes the form of a spiral towards the origin. The mean effective stress (σ'_x + σ'_y)/2 increases at the beginning of the tests, then decreases continuously to zero. The state of stress tends to move to the origin and a liquefaction state occurs due to the cyclic rotation of the principal strain axes.

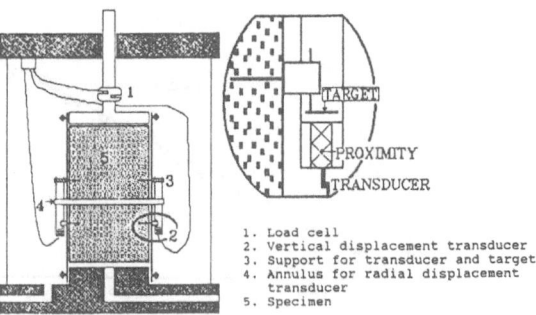

Fig. 7.1 : Triaxial apparatus for small strain amplitudes

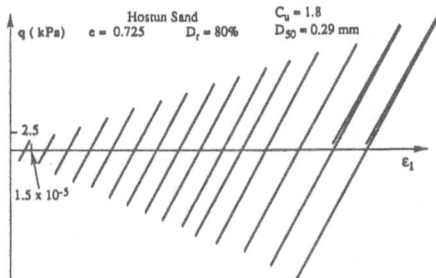

Fig. 7.2 : Stress-strain relationship in cyclic loading at small strains in sand

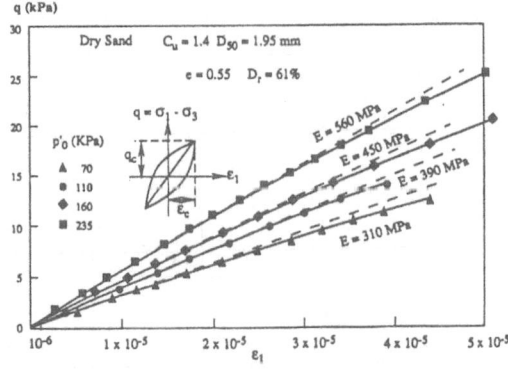

Fig. 7.3 : Stress-strain relationship at small strains in triaxial tests on sand for different mean stresses

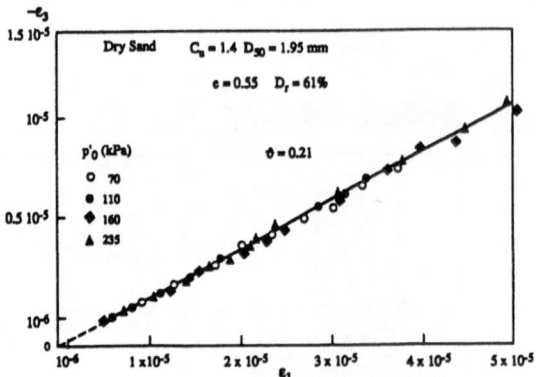

Fig. 7.4 : Relationship between axial and radial strains in sand

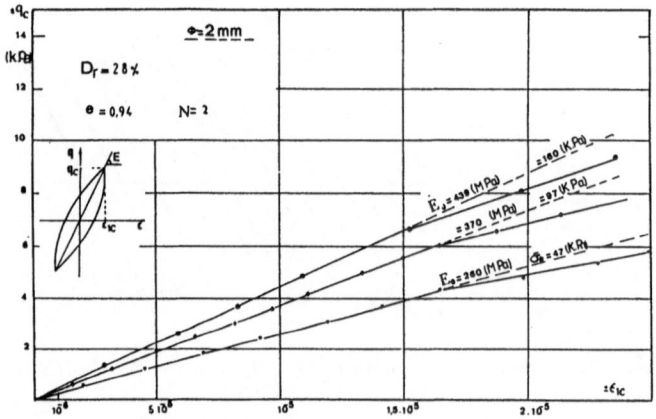

Fig. 7.5 : Stress-strain relationship at small strains in triaxial tests on glass ball assemblies for different mean stresses

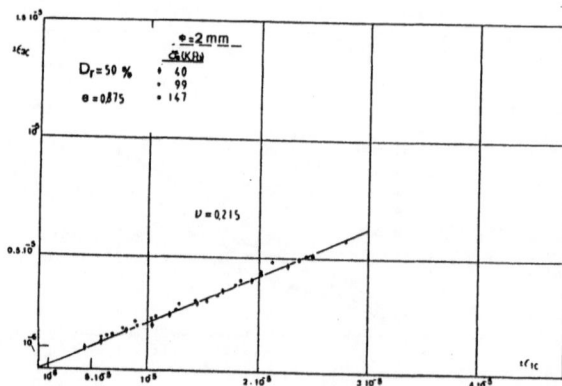

Fig. 7.6 : Relationship between axial and radial strains in glass ball assemblies

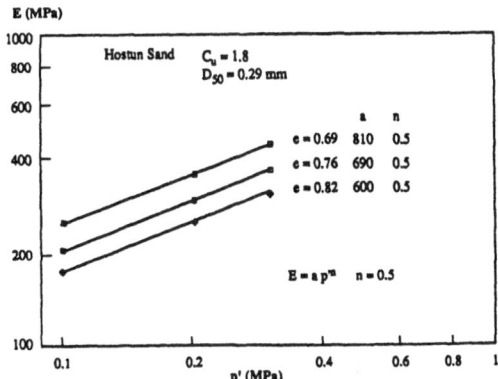

Fig. 7.7 : Relation between Young's Modulus and mean stress in sand for different void ratios

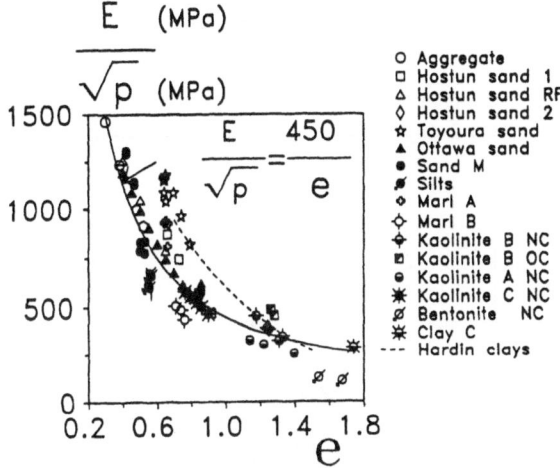

Fig. 7.8 : Relation between Young's Modulus and void ratio for sands and clays

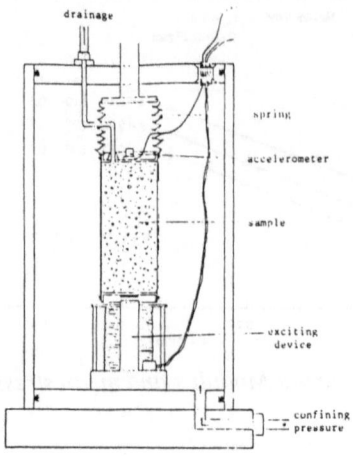

Fig. 7.9 : Apparatus device for the measurement of Young's modulus and Poisson's ratio from longitudinal resonant frequencies [32]

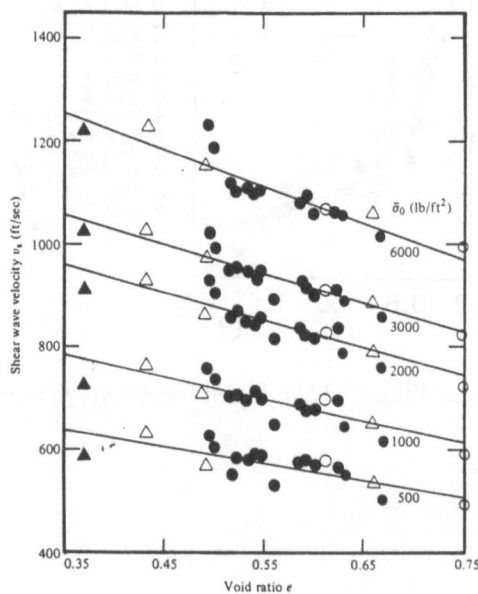

Fig. 7.10 : Variation of shear wave velocity with mean stress for Ottawa Sand [31]

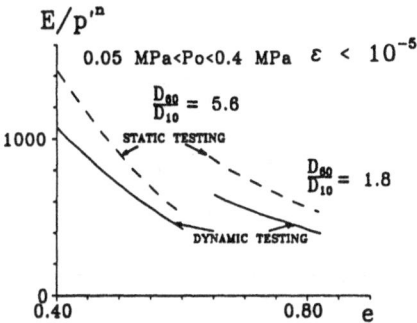

Fig. 7.11 : Comparison between moduli measured in dynamic and static testing in sand

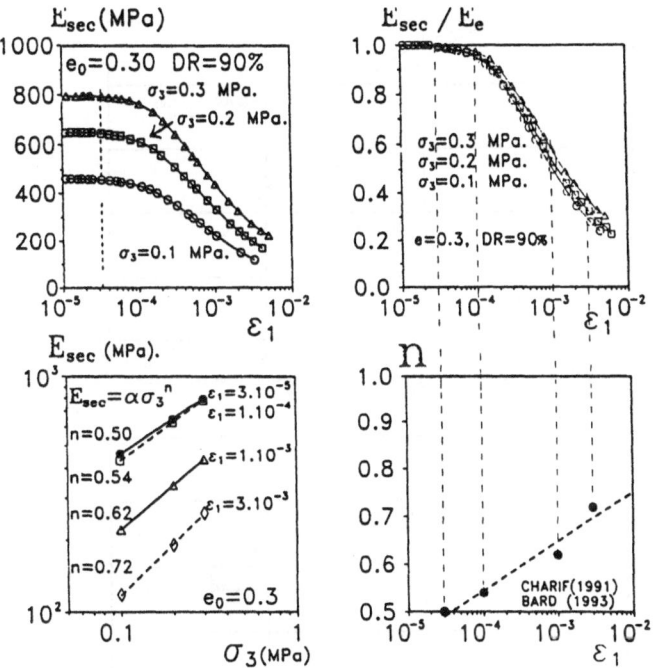

Fig. 7.12 : Non-linear behaviour of sand

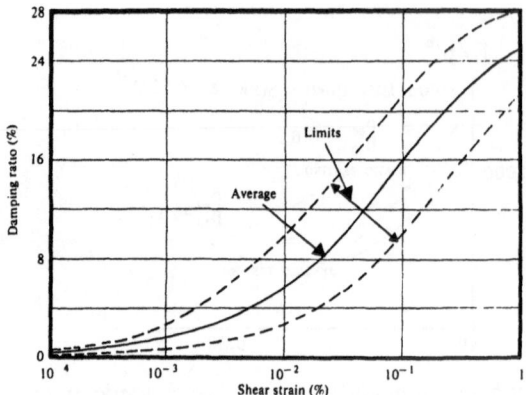

Fig. 7.13 : Damping ratio for sand [34]

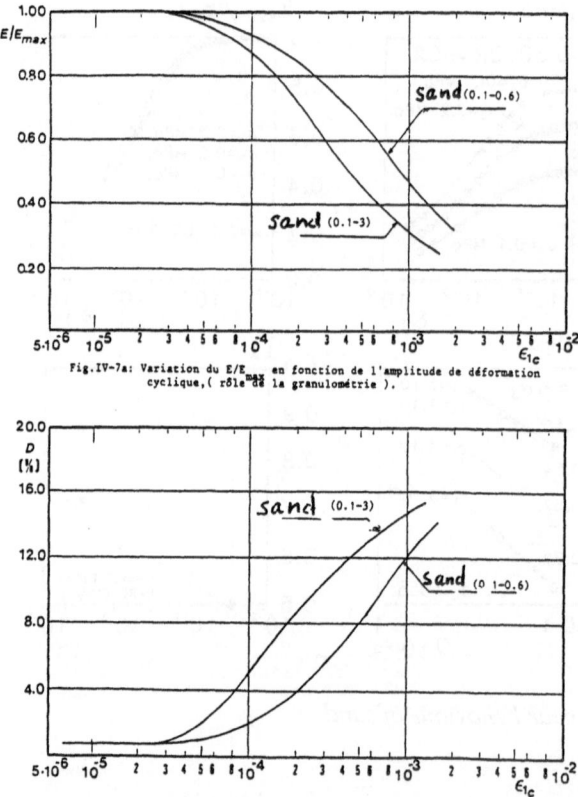

Fig. 7.15 : Measurement of secant modulus and damping ratio in triaxial tests

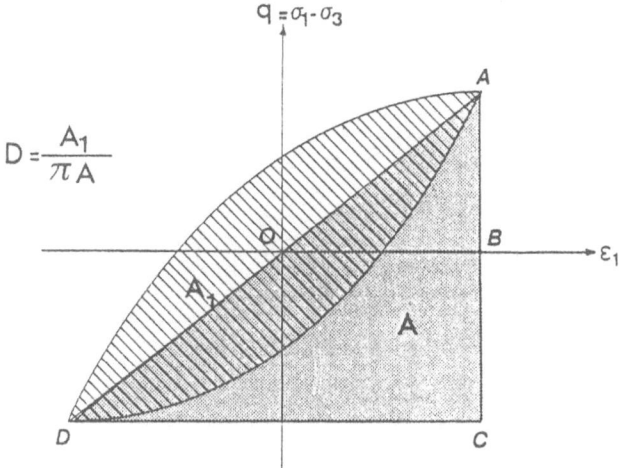

Fig. 7.14 : Definition of damping ratio in cyclic triaxial test

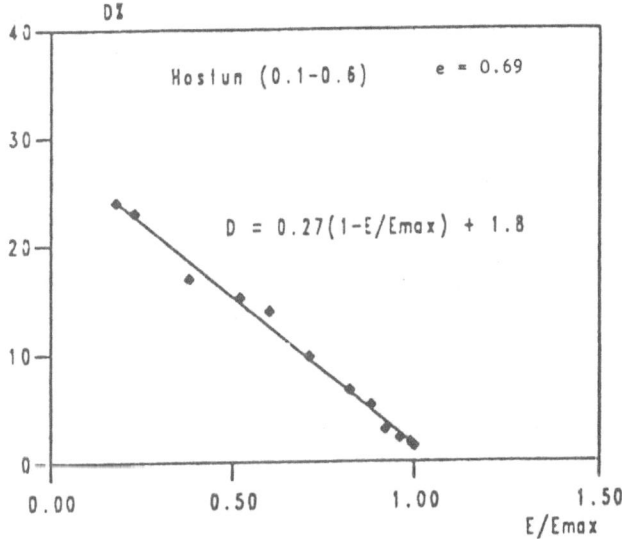

Fig. 7.16 : Relation between secant modulus and damping ratio in triaxial tests

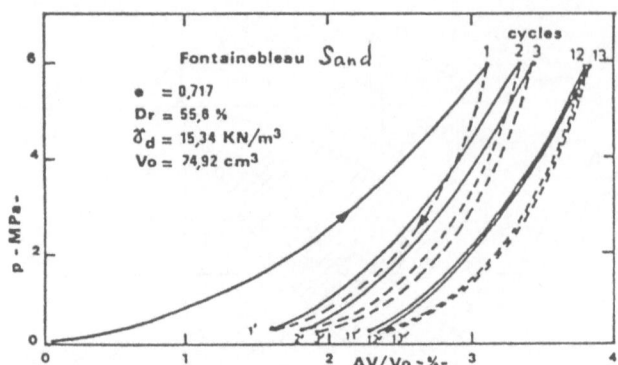

Fig. 7.17 : Cyclic isotropic tests on sand [5]

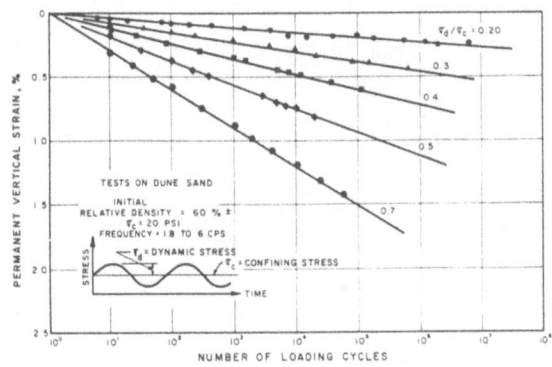

Fig. 7.18 : One-dimensional cyclic compression on sand [36]

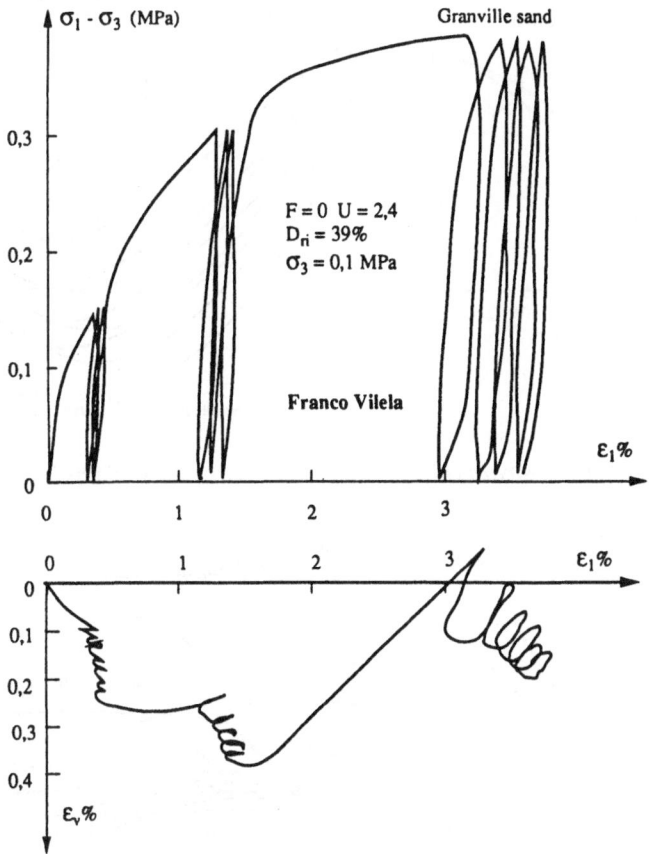

Fig. 7.19 : One-way triaxial test on sand

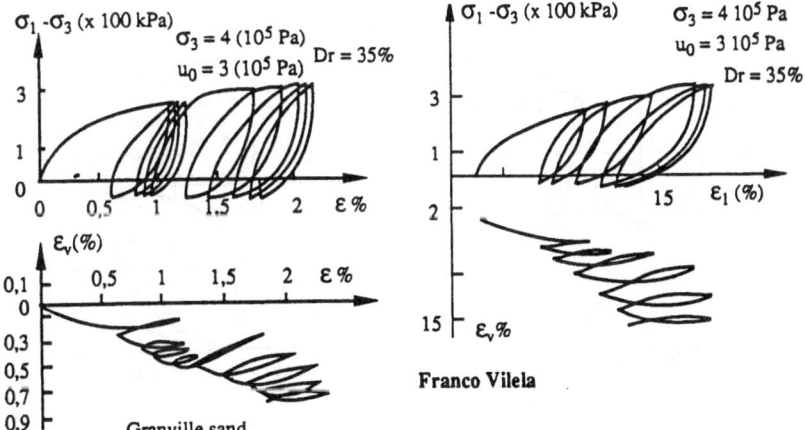

Fig. 7.20 : Two-ways triaxial tests on sand

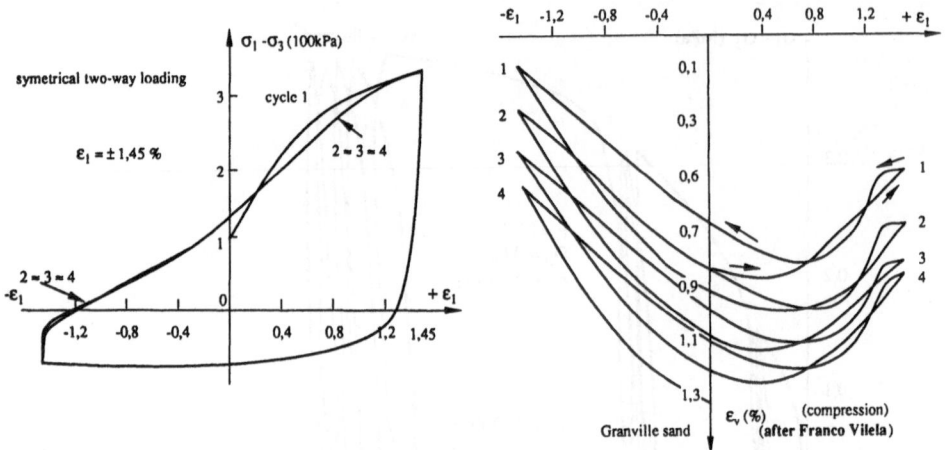

Fig. 7.21 : Evolution of stress-strain relationship during two-ways triaxial test on sand

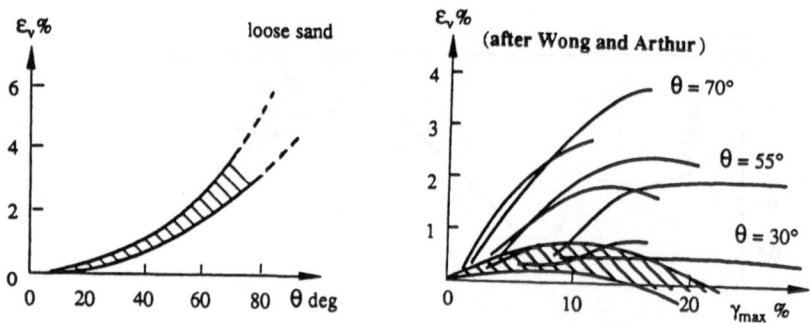

Fig. 7.22 : Influence of stress axis cyclic rotation on strain response in sand [37]

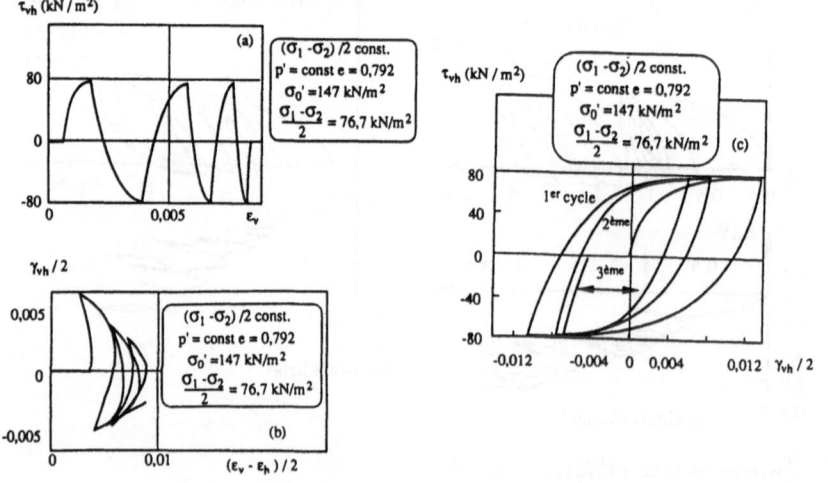

Fig. 7.23 : Cyclic rotation of principal stress direction in hollow cylinder test on sand [12]

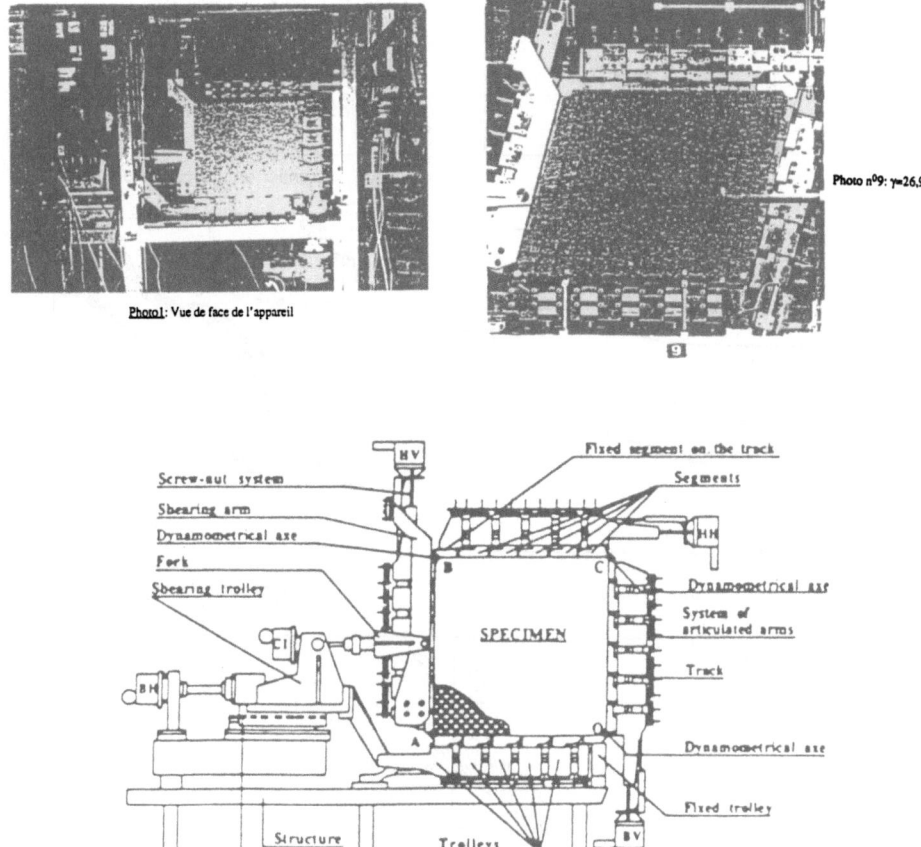

Photo1: Vue de face de l'appareil

Photo n°9: γ=26,9%

Fig. 7.24 : 1γ2ε apparatus [38]

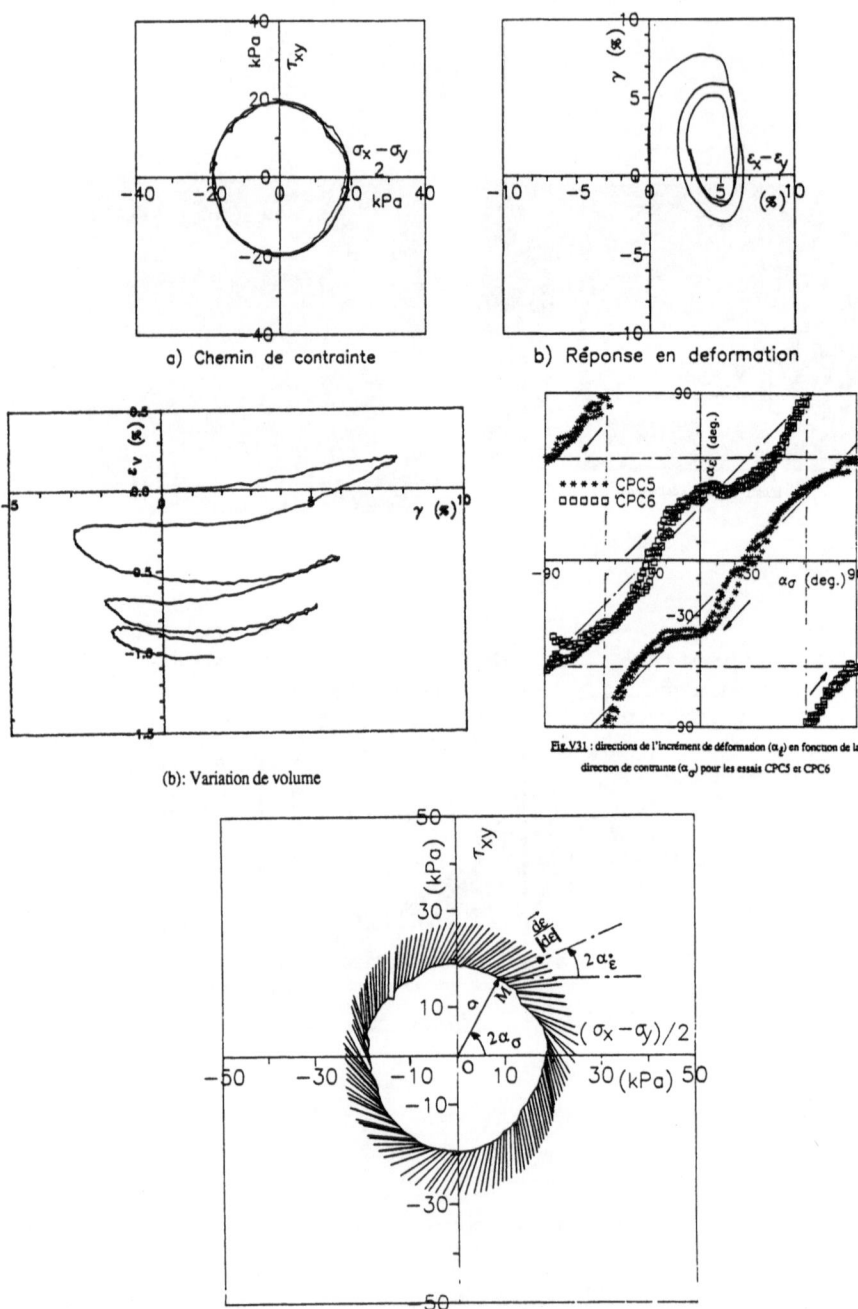

a) Chemin de contrainte

b) Réponse en deformation

(b): Variation de volume

Fig. V31 : directions de l'incrément de déformation ($\alpha_{\dot{\varepsilon}}$) en fonction de la direction de contrainte (α_σ) pour les essais CPC5 et CPC6

Fig. 7.25 : Cyclic compaction due to rotation of principal stress axes in 2D condition [38]

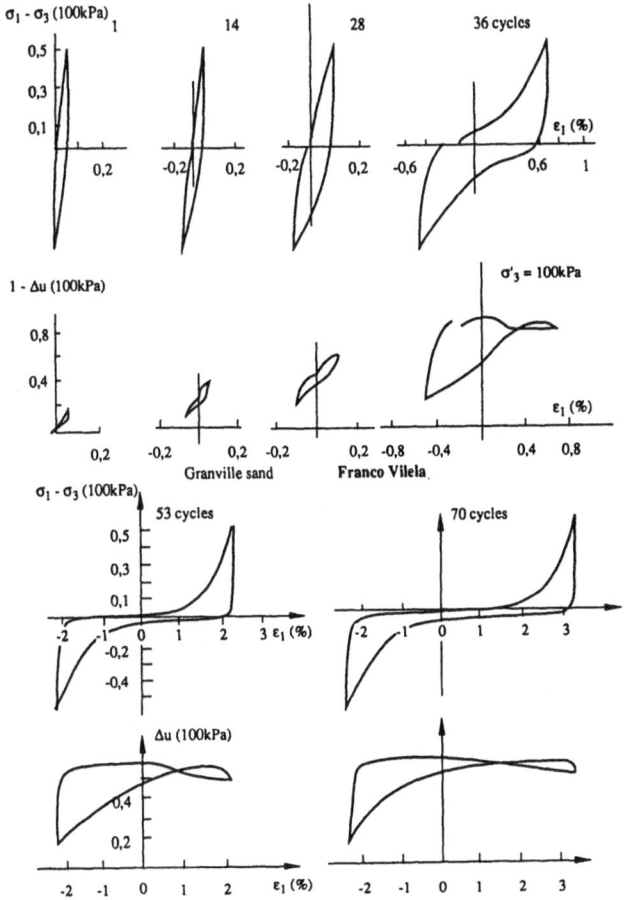

Fig. 7.26 : Isochoric cyclic triaxial test on sand

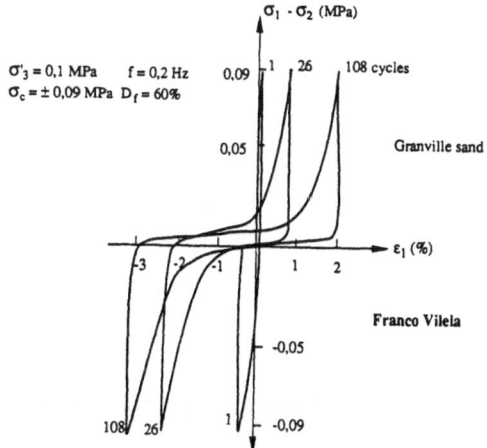

Fig. 7.27 : Evolution of stress-strain response in isochoric cyclic triaxial test on sand

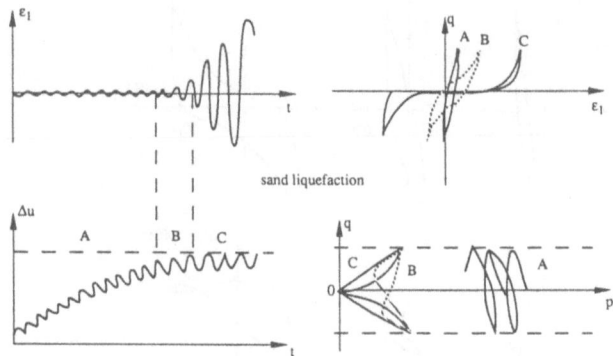

Fig. 7.28 : Liquefaction phenomenon during isochoric triaxial test on sand

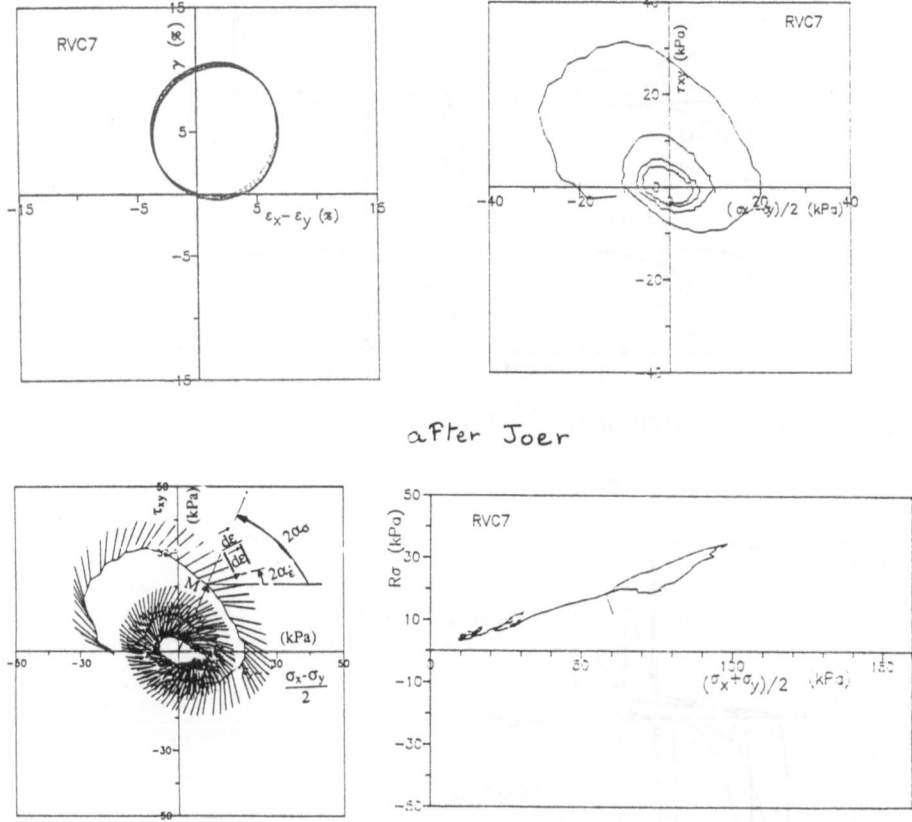

after Joer

Fig. 7.29 : Liquefaction due to rotation of principal strain axes in 2D condition [38]

Conclusion

We presented the main features of experimental behaviour of granular materials tested along various stress and strain paths. Having compared results on different types of grains, such as sands and gravels, metal and organic powders, cereal grains, we can at first say that all these materials have a common behaviour when tested in the same conditions.

Two parameters have a major importance in such materials, the void ratio of the assembly and the mean stress. If we consider tests in which the grains can be considered as elastic (small grain deformations), the notion of perfect plastic state allows us to define contractant or dilatant material according to their initial void ratio and state of stress. The volume changes during a deviatoric stress path, which is a typical behaviour of granular media, affect the stress-strain relationship : higher stiffness and maximum strength for initial dense states. The maximum strength is a function of the mean stress, corresponding to a frictional behaviour of the granular materials. Non-linear elasticity is also a consequence of the effect of the mean stress on the mechanical properties.

Inelastic deformations or ruptures of the grains, caused by increases in stresses, lead to variations in the mechanical behaviour of granular media. They mainly affect the volume changes and as a consequence the stress-strain relationships along isotropic and deviatoric stress paths. When the void ratio becomes very small one can observe a transition in the behaviour from a frictional one to one independant of the mean stress (with no plastic volume change).

Three-dimensional behaviour is very much affected by the initial and/or induced anisotropy of the specimen. As a consequence, rotations of principal stress and strain axes can induce large plastic deformations. The anisotropy affects also the behaviour during cyclic loadings. In particular, cyclic compaction is usually observed whose amplitude depends on the cyclic parameters. This tendancy to compact cyclically can lead, in isochoric condition, to a particular phenomenon called liquefaction, which corresponds to a loss of mechanical properties due to a drastic reduction of the mean stress.

We have also demonstrated the influence of the fluid phases inside the voids. The mechanical behaviour of the polyphasic materials depends on the behaviour of the solid grains, of the fluids and of the interactions at the interfaces.

Références

1. Biarez, J. & Hicher, P.Y. (1994) Elementary Mechanics of Soil Behaviour. Saturated remoulded Soils, Ed. Balkema

2. Darve, F., Hicher, P.Y. & Reynouard, J.M. (1995) Les géomatériaux : théories, expériences et modèles, Ed. Hermès

3. Al Mahmoud, M. (1997) Etude en laboratoire du comportement des sables sous faibles contraintes, thèse de Doctorat, Univ. Des SC. et Techn. de Lille

4. Lancelot, L. (1990) Etude expérimentale et modélisation du comportement de poudres de l'industrie chimique, thèse de Doctorat, Univ. Des Sc. et des Techn. de Lille

5. Luong, M.P. (1980) Stress-strain aspects of cohesionless soils under cyclic and transient loading, Int. Symp. on Soils under cyclic and transient loading, Swansea, pp. 353-376

6. Kim, M.S. (1995) Etude expérimentale du comportement mécanique des matériaux granulaires sous fortes contraintes, thèse de Doctorat, Ecole Centrale de Paris

7. Mosbah, P. (1995) Etude expérimentale et modélisation du comportement des poudres métalliques au cours du compactage en matrice fermée, thèse de Doctorat, Univ. Joseph Fourier, Grenoble I

8. Bard, E. (1993) Comportement des matériaux granulaires secs et à liants hydrocarbonés, thèse de Doctorat, Ecole Centrale de Paris

9. Duchesne, A. (1997) Comportement de matériaux granulaires secs au triaxial, rapport de recherches, Ecole Centrale de Paris

10. Luong, M.P. (1989) Rhéologie des grains agro-alimentaires ensilés, Construction métallique, n° 2

11. Rowe, P.W. (1962) The stress dilatancy relations for static equilibrium of an assembly of particles in contact, Proc. Royal Society, London, Series A, Vol. 269, pp. 500-527

12. Ishihara, K. & Towhata, I. (1983) Cyclic behaviour of sand during rotation of principal axes, Mechanics of Granular Materials, Ed. Elsevier, pp. 55-73

13. Lanier, J. (1994) Experimental behaviour of soils, 6th european automn school on constitutive equations for geomaterials, ALERT Geomaterials

14. Lee, K. & Seed, .B. (1967) Undrained strength characteristics of sand, Proc. ASCE, Vol. 93, No. SM6, pp. 333-360

15. Canou, J., El Hachem, M. & Kattan, A. (1990) Propriétés de liquéfaction statique d'un sable lâche, 25° Colloque du Groupe Français de Rhéologie, Grenoble

16. Biarez, J. & Wiendick, K. (1963) La comparaison qualitative entre l'anisotropie mécanique et l'anisotropie de structure des milieux pulvérulents, CRAS, 256, pp. 1217-1220

17. Konishi, J., Oda, M. & Nemat-Nasser, S. (1983) Induced anisotropy in assemblies of oval cross-sectional rods in biaxial compression, Mechanics of Granulal Materials, Ed. Elsevier, pp. 31-40

18. Hicher, P.Y. & Rahma, A. (1994) Micro-macro correlations in granular media. Application to the modelling of sands, Eur. J. of Mech., A : Solids 13, n° 6, pp. 763-781

19. Lade, P.V. & Duncan, J.M. (1973) Cubical triaxial test on cohesionless soil, ASCE, SM 10, pp. 793-811

20. Zitouni, Z. (1988) Comportement tridimensionnel des sables, thèse de Doctorat, Univ. Joseph Fourier, Grenoble I

21. Ochai, H. & Lade P.V. (1983) Three-dimensional behavior of sand with anisotropic fabric, J. of Geotechnical Eng., Vol. 109,No 10, pp. 1313-1328

22. Wong, R.K.S. & Arthur, J.R.F. (1985) Induced and inherent anisotropy in sand, Geotechnique 35, No 4, pp. 471-481

23. Sture, S., Budiman, J.S., Ontuna, A.K. & Ko, H.Y. (1987) Directional shear cell experiments on a dry cohesionless soil, Geotechnical Testing J. 10, No 2, pp. 71-79

24. Kharchafi, M. (1988) Contribution à l'étude du comportement des matériaux granulaires sous sollicitations rotationnelles, thèse de Doctorat, Ecole Centrale de Paris

25. Bishop, A.W. & Bligth, G.E. (1963) Some aspects of effective stress in saturated and unsaturated soils, Geotechnique, No 3, pp. 177-197

26. Taibi, S. (1994) Comportement mécanique et hydraulique des sols soumis à une pression interstitielle négative. Etude expérimentale et modélisation, thèse de Doctorat, Ecole Centrale de Paris

27. Hicher, P.Y., Kim M.S. & Rahma, A. (1995) Experimental evidence and modelling of grain breakage influence on mechanical behaviour of granular media, Int. Workshop on homogeneisation, theory of migration and granular bodies, Gdansk, pp. 125-133

28. Le Long (1968) Contribution à l'étude des propriétés mécaniques des sols sous fortes pressions, Thèse de Docteur Ingénieur, Univ. De Grenoble

29. Biarez, J. & Hicher, P.Y. (1997) Influence de la granulométrie et de son évolution par ruptures de grains sur le comportement mécanique de matériaux granulaires, Revue Française de Génie Civil, Vol. 1, n° 4

30. Hicher, P.Y. (1996) Elastic properties of soils, J. Geot. Eng., ASCE, Vol. 122, No 8, pp. 641-648

31. Hardin, B.O. & Richard, F.E. (1963) Elastic wave velocities in granular soils, J. Soil Mech. And Found. Div., ASCE 89, SM1, pp. 33-65

32. Boelle, J.L. (1983) Mesure en régime dynamique des propriétés mécaniques des sols aux faibles déformations, thèse de Docteur Ingénieur, Ecole Centrale de Paris

33. Drnevich, V.P., Hardin, B.O. & Shippy, D.J. (1977) Modulus and damping of soils by the resonant column, Dyn. Geot. Testing, ASTM STP 654, ASTM, Philadelphia, pp. 91-125

34. Seed, H.B. & Idriss, I.M. (1970) Soil moduli and damping factors for dynamic responses analysis, report No EERC 75-29, Eathquake Eng. Research Center, Univ. of California, Berkeley

35. Charif, K. & Hicher, P.Y. (1991) Influence of anisotropy on elastic and cyclic properties of sand, 5th Int. Conf. On Soil Dyn. and Earthquake Eng., Karlsruhe

36. D'Appolonia, E.P. (1970) Dynamic Loading, J ; Soil Mech. And Found. Div., ASCE 96, SM1, pp. 4ç-72

37. Wong, R.K.S. & Arthur, J.R.F. (1986) Sand sheared by stresses with cyclic variations in direction, Geotechnique 36, No 2, pp. 215-226

38. Joer, H.A. (1991) 1γ2ε : une nouvelle machine de cisaillement pour l'étude du comportement des milieux granulaires, thèse de Doctorat, Univ. Joseph Fourier, Grenoble I

INTRODUCTION TO COMPUTATIONAL GRANULAR MECHANICS

J.P. Bardet
University of Southern California, Los Angeles, CA, USA

1. Introduction

The first discrete modeling of soils can be traced to Hertz [90] who formulated a contact law between spheres, and Reynolds [167] who proposed a dilatancy theory. Dantu [55] and Schneebli [180] idealized real soils as assemblies of rigid rods, and noticed some striking similarities between the mechanical responses of these mechanical analogs and real soils. Duffy and Mindlin [70], Deresiewicz [59, 60], and Thurston and Deresiewicz [204] examined the response of soil models made of spheres. Biarez [21] used glass beads and duralumium rods to examine the elastic and limit response of soils, and applied his observations to analyze practical problems in geotechnical engineering. These pioneer works were later followed by photoelastic investigations [e.g., 67, 68] to visualize stresses within granular media.

The discrete modeling of soils benefited substantially from the development of computers in the 1970's. The computational discrete modeling of soils can be attributed to Cundall, who developed the computer code BALL [52]. At this time, computers had very limited capabilities, comparable to those of today's hand-held calculators. They were slow, and had limited memory and storage capacity. Yet, Cundall developed BALL, a program which many researchers still use. The program is fully documented in a two-volume report to the US National Science Foundation [52]. Since 1978, many researchers have adapted the original version of BALL to solve specific problems. Researchers have realized the power of computer simulations to understand the mechanics of granular materials. The major advantage of computations over real experiments is the generation of abundant information on particle displacements, contact forces and other physical quantities, which can be processed rapidly to comprehend the physics of granular assemblages.

This chapter reviews the basics of the discrete modeling of granular media, with the intent of providing readers with some understanding of granular material behavior based on the first principles in mechanics and computational methods. Three particular aspects of computational granular mechanics will be covered: (1) physical modeling of granular media, (2) numerical methods for computational granular mechanics, and (3) transitions from discrete to continuous media. Before going into these subjects, we will first try to answer the following questions. What exactly is discrete modeling? What are the applications of discrete modeling? Has discrete modeling any advantages over continuum mechanics? What are the limitations of discrete modeling? What are the main computational tools in discrete modeling?

1.1 Examples of discrete modeling in soil mechanics

Discrete modeling can best be described by considering two particular examples: one is relevant to the study of constitutive behavior, and the other to the failure of foundations.

1.1.1 Example 1

As shown in Fig. 1, the sample was constructed with 1848 cylindrical rods of diameter 4, 6 and 8 mm. There are 1040 particles of 4-mm diameter, 532 of 6-mm diameter, and 266 of 8-mm diameter. The sample slenderness ratio is 2.43. The rods were made of identical transparent acrylic material, which has an average density of 1.30 g/cm^3, and were cut to a 10-cm length. Their front ends were half painted in black to visualize their rotation and displacement simultaneously. Figures 1 and 2 show the experimental setup. The sample is enclosed in a 0.15-mm thick transparent latex membrane with two circular loading platens at the top and bottom. The membrane is clamped to the platens with compression rings. During the sample fabrication, the membrane is stretched on a rectangular mold, and the rods are manually positioned inside the mold with their axis parallel to one another. During the test, the specimen is subjected to a vacuum inside the membrane, which is equivalent to applying an external constant confining pressure equal to 95 kPa. The axial compression is applied by raising the lower platen at the constant rate of 5.6 mm/min, while the upper platen remains fixed. Figure 2 shows a side view of the experimental setup. The photograph of Fig.1 was obtained by using a 35-mm camera positioned approximately one meter in front of the sample, with its aim perpendicular to the front of the sample. The particles were lighted from behind by using a light box. The sample was tested by increasing the axial strain while keeping the confining pressure constant. The stress-strain curve is shown in Fig. 3. The axial strain ε is

$$\varepsilon = \Delta h/h_0 \tag{1.1}$$

where Δh is the vertical displacement of the lower platen, and h_o is the initial sample height. The axial stress σ is

$$\sigma = F/A_0 \tag{1.2}$$

where F is the measured axial load, and A_o is the initial cross-sectional area. The Initial Young's modulus was 300 MPa, and the peak friction angle was 18.4 deg.

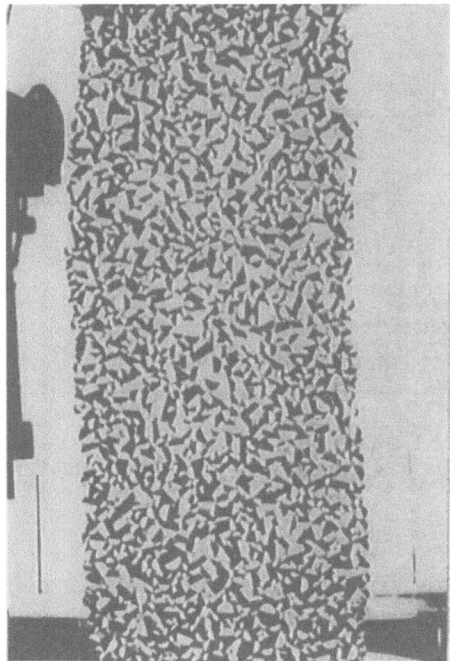

Fig. 1.   An assembly of cylindrical rods subjected to axial compression.

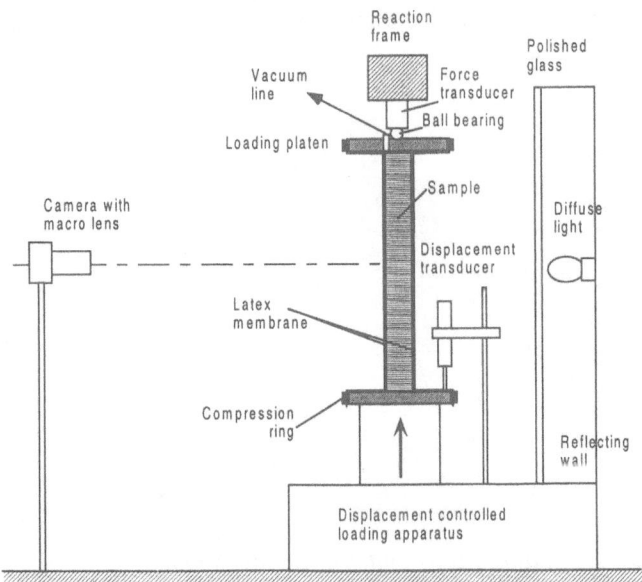

Fig. 2.   Side view of experimental setup for axial compression of sample in Fig. 1.

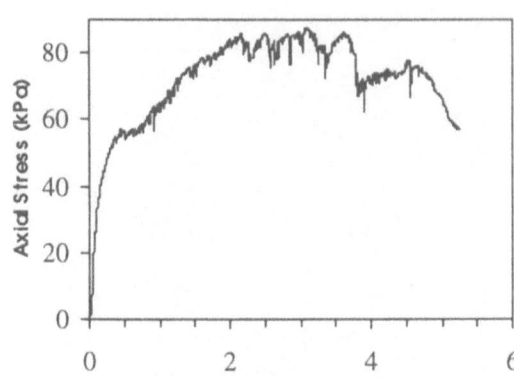

Fig. 3. Stress-strain response of sample shown in Fig. 1 (the confining pressure is 95 kPa).

1.1.2 Example 2

Figure 4 shows an early example of the application of discrete modeling for understanding the failure mechanism under shallow footing [120]. This photograph was taken by using a slow shutter speed, which blurs the moving particles. It shows the pattern of motion at failure within a stack of duralumium rods loaded by a rigid punch. The rods are 150-mm long and are of two shapes and sizes (round, 3 mm and 6 mm diameter; and hexagonal 4.8 and 7,9 mm across flats) to simulate the interlocking which occurs in actual soils. Additional examples of discrete models for understanding the failures of foundations and retaining walls can be found in [21].

Fig. 4.   Failure zone under a shallow foundation [120].

1.2 Applications of discrete modeling in engineering and applied sciences

One may attempt to survey the applications of discrete modeling in engineering and applied sciences by consulting the proceedings of the following conferences:
(1) the 1st International Conference on Powders and Grains [22];
(2) the 2nd International Conference on Powders and Grains [203];
(3) the 3rd International Conference on Powders and Grains [19];
(4) the 1st US Conference on Discrete Element Methods [145]; and
(5) the 2nd International Conference on Powders and Grains [228].
Discrete modeling is used for many applications in engineering and applied sciences. A review of the literature on discrete modeling, which is by no means exhaustive, indicates that past work can be sorted in the following general categories:

- *Experimental studies of granular media*: [e.g., 21, 55, 67, 68, 128, 146, 152-159, 189, 192, 194].
- *Studies of contact between particles*: [e.g., 2, 3, 11, 20, 30, 59, 60, 66, 74, 75, 76, 84, 90, 102, 105-110, 121, 123, 132-134, 167, 168, 195, 219, 226, 229].
- *Numerical techniques for discrete modeling*: [e.g., 9, 14, 25, 26, 34, 49, 50, 53, 80, 83, 88, 89, 103, 104, 114, 119, 128, 136-141, 144, 151, 160, 162, 206-209, 211, 215, 216, 220-224, 227, 230, 231].
- *Elastic behavior of granular media*: [e.g., 17, 33, 40, 69, 96, 97, 105].
- *Failure of granular media*: [e.g., 31, 193, 199].
- *Stress-strain behavior of granular media*: [e.g., 4, 18, 31, 36, 35, 39, 41, 43, 54, 56, 60, 65, 70, 98, 113, 116-118, 125, 127, 130, 148, 149, 150, 168, 172-174, 175, 197, 199-204, 225].
- *Higher-order continuum theories:*
  - *Second order gradient continuum*: [e.g., 37].
  - *Micropolar continuum*: [e.g., 10, 11, 38, 64, 111, 188].
- *Strain localization and shear bands*: [e.g., 8, 10, 50, 61-63, 142, 143, 146, 155, 158].
- *Powders and sintering processes*: [e.g., 1, 86, 94, 124, 196].
- *Suspension*: [e.g., 81, 82].
- *Granular Flow*: [e.g., 76, 29, 179].
- *Fluid and solid mixtures, including fluidized beds*: [e.g., 112, 190, 191, 214].
- *Rock mechanics*: [e.g., 13, 47-49, 182-184].
- *Physics of granular media*: [e.g., 44, 115, 126, 128, 129, 166].

1.3 Discrete and continuous modeling

Continuum mechanics is a powerful approach to solve scientific and engineering problems. It is based on mathematical assumptions, which are well described in the continuum mechanics literature [e.g., 73, 212]. It formulates engineering problems as mathematical boundary value problems (BVP). The main components are the governing equilibrium equations (i.e., partial differential equations translating basic physical balances, such as stress equilibrium); the boundary conditions (prescribed values of the unknown quantities or their derivatives on external surfaces); and the constitutive equations (generalized relation between stress and strain, or their respective rates). In the past 20 years, the continuum approach has been implemented using finite elements and finite differences, and has successfully been applied to solve engineering problems [e.g., 232].

One of the major assumptions of continuum mechanics is that the material properties can be scaled from small laboratory samples to large material masses using constitutive relations. Researchers have now produced a myriad of constitutive relations for various types of engineering materials. Many constitutive models are clever fittings of experimental results in the laboratory. Unfortunately, most models are not based on physics.

Through the use of bifurcation theory, the continuum approach was discovered to exhibit problems, especially associated with the loss of uniqueness of boundary value problems. The problems mainly result from local material instability and limitations of constitutive equations [e.g., 217]. These findings questioned some applications of continuum mechanics to soil mechanics, and pointed out the need for seeking a fundamental understanding of material behavior. Discrete modeling can be of great help to continuum mechanics, for not only developing constitutive models based on physics, but also for understanding the physical origins of material instability, and the limitations of continuum mechanics.

1.4 Limitations of discrete modeling in soil mechanics

Discrete modeling has obvious limitations in soil mechanics. The sheer number of individual soil particles, especially those with smaller diameters, within soil masses of practical interest to engineering, prohibits the simulation of their overall response even with the most advanced computers. For instance, in a cubic centimeter of soil, there may be as many as 5×10^{12} clay particles, when those are assumed to be identical square platelets (1-μm x 1-μm x 0.1-μm), and the void ratio is equal to one. Therefore, the largest computers yet built would not be sufficient to handle 1 cm^3 of fine-grained soils, which is quite irrelevant for engineering purposes. The number of soil particles in a volume decreases roughly with the cube of the particle size. In a cubic centimeter, there may be as many as 1,000 particles of coarse sand, when those are assumed rounded with 1-mm diameter and the void ratio is still equal to one. Yet again, the most powerful computers of today, with numerous parallel processors, would have extreme difficulty handling 1 m^3 of coarse sand, which corresponds to 1×10^9 particles.

In view of the excessively large number of particles in soils, it seems more feasible to combine the advantages of discrete modeling and continuum mechanics to solve engineering problems. In summary, discrete and continuum mechanics should be perceived as complementary, not adversary, tools in soil mechanics to understand the mechanics and physics of soil behavior.

1.5 Numerical methods for discrete modeling

The main numerical tool of discrete modeling is the *discrete element method* [53]. Cundall [51] proposed that this appellation apply to a computer program only if it (a) allows finite displacements and rotations of discrete bodies, including detachment, and (b) recognizes new contacts automatically as the calculation progresses. To my knowledge, there are eight main classes of numerical methods corresponding to this definition.

1. *Distinct element methods (DEM).* These methods use explicit and time-marching algorithms to solve the equation of motion. Bodies may be rigid or deformable, but contacts are always deformable. Examples of DEM codes are TRUBAL [50]; UDEC [48]; 3DEC [89]; DIBS [220]; 3DSHEAR [224]; and JP2 [7]. In granular statics, DEM calculates the equilibrium states of particle systems by using dynamic transitions, the convergence of which are generally accelerated and optimized by introducing an artificial viscous damping. Such optimizations include density scaling [46] and adaptative dynamic relaxation [9].

2. *Modal methods.* These methods, which are similar to DEM in the case of rigid bodies, use modal superposition for deformable bodies, [e.g., 227]. A representative code is CICE [93].

3. *Discontinuous deformation analysis (DDA).* In DDA [182], contacts are rigid, and bodies may be rigid or deformable. The condition of no-interpenetration is achieved by an iteration scheme; the body deformability comes from superposition of strain modes.

4. *Momentum-exchange methods.* The contacts and bodies are both rigid: momentum is exchanged between two contacting bodies during an instantaneous collision. Frictional sliding can be represented [e.g., 29, 88].

5. *Multibody Dynamics methods (MDM).* Moreau [136] treats the problem of non-penetrability by using *Convex Analysis*. The unilateral mechanical constraints of frictional non-penetrability are mathematically formulated in [27, 58, 136-141, 164]. MDM was implemented in [103] using an implicit algorithm, and was applied by several investigators to examine various aspects of particulate behavior [e.g., 56]. MDM considers only purely rigid bodies, and ignores the deformability of individual grains and contacts.

6. *Structural Mechanics methods (SMM).* These methods derive from the numerical techniques used in computational mechanics, especially finite element methods for continuum plasticity and contact mechanics [e.g., 25, 26, 114, 232]. The equilibrium equations for the system of particles are solved quasi-statically, and not dynamically, which eliminates the spurious oscillations of dynamic relaxation. SMM obtain the incremental transition between equilibrium states by relying on a tangential stiffness matrix, which is determined from the local contact stiffness between particles. Unfortunately, this tangential operator consumes a large amount of computer memory, rendering it inapplicable to large numbers of particles, and often becomes singular and causes numerical problems. However, SMM guarantees a strict convergence, when the physical problem has a solution, and is capable of detecting bifurcation points. This approach, which benefits from the progress in finite element methods for structural and continuum mechanics [e.g., 15, 232], reveals that there are many similarities between the numerical techniques in finite and discrete element methods.

7. *Mean Field method* [e.g., 14, 231]. The transitions between static states are calculated by imposing a mean field of displacement and rotation to the particles, and by restoring a new equilibrium configuration by means of incremental motions or *"fluctuations"* of each particle about the mean. The fluctuating motions of individual particles are determined statically by a global stiffness matrix, which is determined from the local contact stiffness between particles, as in structural mechanics. The problems arising from the

matrix singularity are overcome by solving the linear equations with the well-known relaxation method [187].

8. *Energy minimization method* [e.g., 42]. The transition between static states is obtained by incorporating the geometric constraints of non-penetrability, Mohr-Coulomb friction criterion for relative sliding, and a minimization function that translates the energy dissipated at the contacts due to internal friction. The interparticle forces are calculated at each time step by using an explicit finite difference scheme based on linear programming technique.

Following the classification in [51], Fig. 5 summarizes the attributes, advantages and shortcomings of the methods listed above, which include:
- contact and body stiffness
- number and shapes of bodies
- capabilities of fracturing individual particles
- packing density
- amplitude of displacement and strain, and
- static and dynamic capabilities.

A shown in Fig. 5, the method performances are grouped in three categories, ranging from good to not applicable. At the present, it is difficult to conclude on the superiority of a particular method, due to a lack of in-depth comparative studies. In this chapter, we will not cover all these methods, but will only introduce the basics of discrete element methods.

2. Physical modeling of granular media

2.1 Geometry of grains

As described in soil mechanics textbooks [e.g., 6, 120], soil grains have irregular and various shapes including spheres, ellipsoids, platelets, cylinders, and tubes, when they are observed with the naked eye and microscopes. Their wide range of grain sizes vary from colloids (<1 μm) to boulders (>100 cm). The diversity of grain shape, size, distribution, and structures are one of the major causes of the multiplicity of soil behavior observed in the laboratory and the field.

For the sake of simplification, hereafter we idealize soil grains as two-dimensional rods. This convenient assumption obviously departs from reality. However, it is sustainable as long as it provides us with some useful hints about material response. Some researchers have already moved to 3D geometry, and are getting new insights into material behavior [e.g., 202]. 2D models are educational for a first hands-on experience of granular mechanics. Many 2D concepts can readily be extended to 3D modeling.

2.2 Particles

As shown in Fig. 6, the set B represents a generic assembly of N rigid particles with nonlinear interaction at contacts, i.e., $B = \{1,...,N\}$. The particles, which are subjected to external forces and moments excluding body forces, belong to the set $B_E = \{N_I,...,N\}$, and the remaining ones belong to the set [   ]. The sets B_I and B_E are complementary, i.e., $B = B_I \cup B_E$.

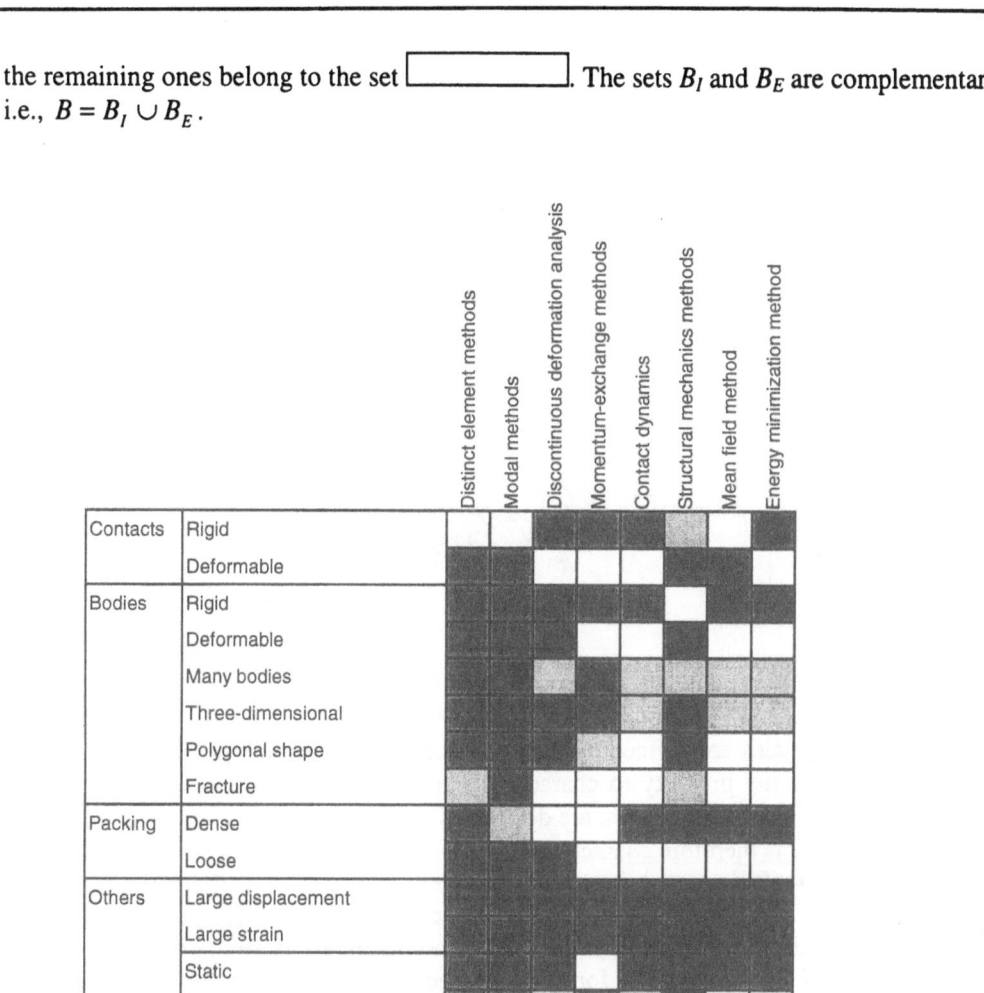

Fig. 5. *Attributes of the various classes of discrete element methods (adapted from [51]).*

The particles are assumed rigid (i.e., they cannot deform). The assumption of rigid particles will be discussed later. The conservation of mass implies that B is constant, i.e., no rigid particle should be lost. Some particles of B_I may become part of B_E forces/displacement. However B_I and B_E can vary, provided that they remain complementary.

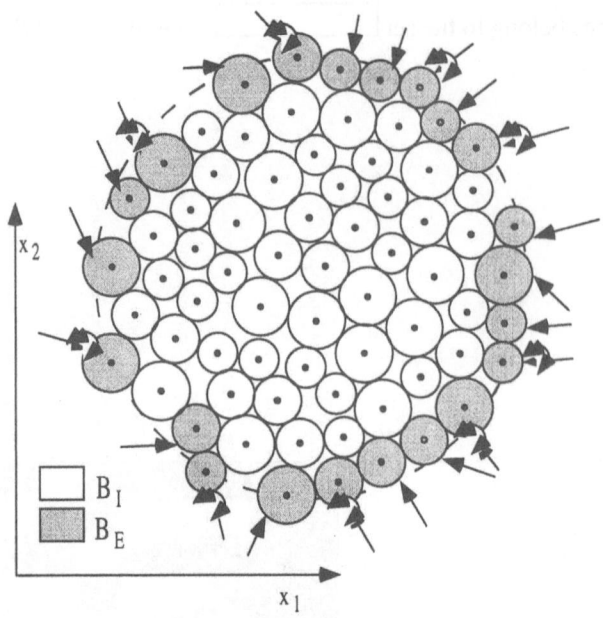

Fig. 6.   An assembly of discrete particles.

Cylindrical particles are defined by their radius and mass per unit area. Elliptical particles have an additional property to characterize their aspect ratio. The positions of rigid
particles, irrespective of their shapes, are defined by the center position and their rotation θ.
The motion of set B is therefore characterized by $3N$ independent variables.

The kinematics of the granular media is completely defined by a finite number of degrees of freedom. There is no need to use continuous interpolation functions to represent
the kinematics, unlike in finite elements. Indeed, a continuous interpolation between discrete points would be inappropriate for granular media because the displacement is discontinuous at the particle contacts.

2.3 Contacts

Contact mechanics [e.g., 108] is a vast subject, which is beyond the scope of this chapter. We will only introduce the basic concepts required in DEM. With the exception of
body forces which are applied at the particle contacts, the external and internal forces acting on particles are assumed to be applied at points and not to be distributed on surfaces.
There are M contact points belonging to the contact set C, i.e., $C = \{1, ..., M\}$. These contacts can be subdivided between the set I, $I = \{1, ..., M_I\}$, which corresponds to contact
between particles of B, and the set E, $I = \{M_{I+1}, ..., M\}$, which describes the points of application of external concentrated actions. The sets I and E are complementary, i.e.,
$C = I \cup E = \{1, ..., M\}$. By definition, the sets I_a and E_a represent the contact points of I
and E on particle a, respectively. The subsets I_a and E_a have the following properties:

$$I = \bigcup_{a \in B} I_a \quad \text{and} \quad E = \bigcup_{a \in B} E_a \tag{2.1}$$

and

$$E_a \cap E_b = \varnothing \quad \text{and} \quad I_a \cap I_b = \{c\} \quad \text{for} \; \forall \; a \neq b \in B \tag{2.2}$$

Equation 2 implies that two different sets I_a have at most one point in common. The number of contacts may vary during a deformation process, i. e., I varies depending on the state of B.

2.3.1 Contact detection

In the case of spherical or cylindrical particles, there is a contact between particles a and b when the following criterion is met:

$$\left\| \mathbf{x}_a - \mathbf{x}_b \right\|^2 = \left(x_i^a - x_i^b \right)^2 \leq R_a^2 + R_b^2 \tag{2.3}$$

where x_i^a and x_i^v are the center position of particles a and b, and R_a and R_b their radii, respectively. The contact detection criterion becomes more complicated for elliptical [e.g., 206] and polygonal [e.g., 48, 49] particles.

For large numbers of particles, the detection of contacts becomes a serious computational issue in granular mechanics. The most naive method for detecting contacts consists of applying Eq. 3 N-1 times for each ball of an assembly of N particles. There are therefore $N\text{x(N-1)}$ searches, a task which may rapidly consume a large fraction of the calculation time. For instance, the computer would make 10^6 searches for 10^3 particles, but 10^{12} searches for 10^6 particles, which is an excessively large number of calculations. Several detection algorithms have been proposed. The most efficient ones are $N \log N$ [e.g., 144]. We will present one of the most commonly used algorithms [e.g., 53].

As shown in Fig. 7, the space is divided in a uniform grid forming square boxes. The box size is larger than the diameter of the largest particles, so that no particle covers more than four boxes at once. Assuming that there is a maximum number of M particles per box, the maximum number of searches is $4NM$ as a single particle may cover 4 boxes at once. This search technique requires associating boxes and particles, and updating this association when particles move across the grid.

This simple search technique becomes less efficient when M increases, which is the case when the particles have a wide range of sizes. In these cases, one may consider alternate contact algorithms.

2.3.2 Contact geometry

The contact geometry of rigid cylindrical particles is defined as shown in Fig. 8. In theory, for rigid particles which do not overlap, the contact is reduced to the point of tangency between two particles (Fig. 8a).

$$\mathbf{x}_C = \frac{R_A}{R_A + R_B} \mathbf{x}_A + \frac{R_B}{R_A + R_B} \mathbf{x}_B \tag{2.4}$$

For particles which overlap slightly, the contact geometry is no longer a point but becomes a surface. However, this contact area will be reduced to a point (Fig. 8b) defined as follows:

$$\mathbf{x}_C = \left(\frac{1}{2} - \frac{R_A - R_B}{AB} \right) \mathbf{x}_A + \left(\frac{1}{2} + \frac{R_A - R_B}{AB} \right) \mathbf{x}_B \qquad (2.5)$$

For rigid cylindrical particles, the contact point becomes the tangency point (i.e., Eqs. 4 and 5 coincide). In the case of elliptical or polygonal particles, the contact point is more difficult to define as described for elliptical [206] and polygonal particles [49].

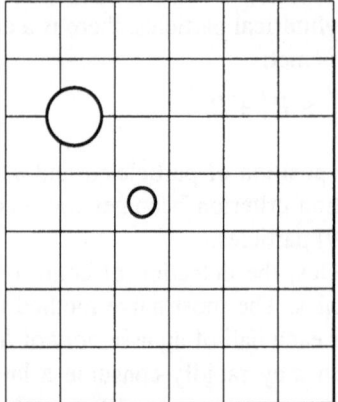

Fig. 7.   Uniform grid used for detection of contact between two particles.

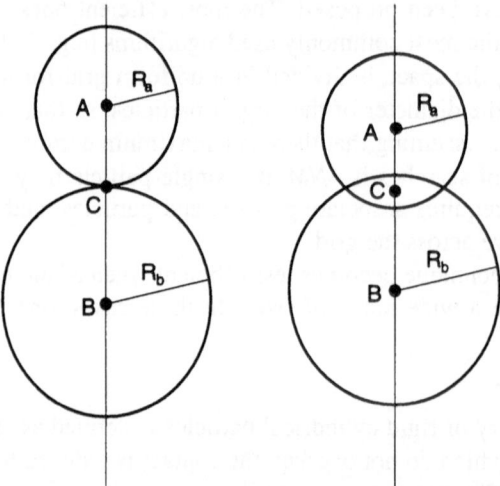

Fig. 8.   Contact point between two cylindrical particles without and with overlap.

2.3.3 Contact kinematics

The kinematics of the contacts characterizes the relative motion between the particles. As shown in Fig. 9, the relative displacement of particles a and b at contact point c is:

$$\Delta u_i^c = u_i^b - u_i^a + e_{ijk} \left(\omega_j^b r_k^{bc} - \omega_j^a r_k^{ac} \right) \qquad (2.6)$$

where u_i^a is the displacement vector of particle center, ω_i^a the rotation of particles located at x_i^a, and

$$r_i^{ac} = x_i^c - x_i^a, \text{ and } \quad r_i^{bc} = x_i^c - x_i^b \tag{2.7}$$

The relative rotation of particles a and b:

$$\Delta\omega_i^c = \omega_i^b - \omega_i^a \tag{2.8}$$

In the case of cylindrical particles,

$$r_i^{ac} = R_a n_i^c, \; r_i^{bc} = -R_b n_i^c, \; \omega_i^a = \omega_a n_3^c, \text{ and } \omega_i^b = \omega_b n_3^c \tag{2.9}$$

where $\mathbf{n}^c$ is the unit vector normal to the contact area:

$$n_i^c = \left(x_i^c - x_i^a\right)/R_a = -\left(x_i^c - x_i^b\right)/R_b \tag{2.10}$$

Equation 6 becomes:

$$
\begin{aligned}
\Delta u_1^c &= u_1^b - u_1^a - n_2^c\left(\omega_a R_a + \omega_b R_b\right) \\
\Delta u_2^c &= u_2^b - u_2^a + n_1^c\left(\omega_a R_a + \omega_b R_b\right)
\end{aligned}
\tag{2.11}
$$

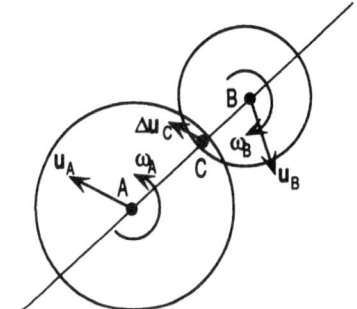

Fig. 9.   *Relative displacement between two particles.*

Equation 6 describes exactly the relative displacement between two particles which do not overlap (i.e., $AB = R_a + R_b$). However, Eq. 6 becomes approximate in case of overlap (i.e., $AB < R_a + R_b$). Fortunately, this overlap is usually very small in most cases, and is thought to have no significant consequences.

2.3.4 Contact actions

The actions at contact c are represented by contact forces f_i^{ac} and contact moments m_i^{ac}. Both forces and moments are applied at the contact points. In many instances, the contact moments are neglected, and the contact actions are reduced to forces alone. This assumption may apply to small contact areas between cylindrical and spherical particles, which cannot transmit significant moments due to their small size. However, this assumption may become questionable for particles of arbitrary shapes and large normal contact forces.

2.3.5 Contact relations

Various types of contact relations between particles were recently reviewed in [134], including contacts between smooth, spherical, non-spherical, cylindrical, and non-cylindrical elastic particles with friction and surface adhesion, rough elastic particles, and viscous bridge. Hereafter, we will only review the normal stiffness between spherical and cylindrical particles, and some findings on contacts with friction.

2.3.6 Elastic contact between smooth particles

The distribution of contact pressure proposed by Hertz [90] is:

$$p(r) = p_0 \sqrt{1 - (r/a)^2} \tag{2.12}$$

where p_0 is the maximum contact pressure, a is the radius of the circular contact area, and r is the polar coordinate. The total load P is related to the contact pressure through:

$$P = \int_0^a p(r) 2\pi r \, dr = \frac{2}{3} p_0 \pi a^2 \tag{2.13}$$

Therefore the maximum pressure p_0 is 3/2 times the mean pressure p_m. The radius of the contact area is:

$$a = \pi p_0 R / 2E^* = \sqrt[3]{\frac{3PR}{4E^*}} \tag{2.14}$$

where:

$$\frac{1}{E^*} = \frac{1-v_1^2}{E_1} + \frac{1-v_2^2}{E_2} \quad \text{and} \quad \frac{1}{R} = \frac{1}{R_1} + \frac{1}{R_2} \tag{2.15}$$

In the case of identical elastic properties, $E^* = E/2(1-v^2)$. The mutual approach of distant points in the two solids is:

$$\delta = \pi a p_0 / 2E^* = \frac{a^2}{R} = \sqrt[3]{\frac{9P^2}{16RE^{*2}}} \quad \text{or} \quad P = \frac{4}{3} E^* R^2 (\delta/R)^{3/2} \tag{2.16}$$

The secant normal stiffness k_n is therefore load-dependent:

$$k_n = \frac{P}{\delta} = \frac{4}{3} E^* \sqrt{R} \sqrt{\delta} = \sqrt[3]{\frac{16}{9} E^* PR} \tag{2.17}$$

In the case of cylindrical solids, the half-width of the contact area is:

$$a = \sqrt{\frac{4PR}{\pi E^*}} \tag{2.18}$$

where P is now a force per unit length (i.e. $[P] = F L^{-1}$). The maximum contact pressure is:

$$p_0 = \frac{2P}{\pi a} = \frac{4}{\pi} p_m = \sqrt{\frac{PE^*}{\pi R}} \tag{2.19}$$

The relative displacement δ is given in [108]:

$$\delta = \frac{P}{\pi E^*}\left(Ln\left(4\pi R E^*/P\right)-1\right) \tag{2.20}$$

The secant normal stiffness k_n is therefore load-dependent:

$$k_n = \frac{P}{\delta} = \frac{\pi E^*}{Ln\left(4\pi R E^*/P\right)-1} \tag{2.21}$$

Figure 10 shows the variation of normal load P with normal relative displacement δ in the case of spherical and cylindrical contact. Figure 11 shows the corresponding secant normal stiffness k_n. The relative displacement δ is normalized with the particle radius R. The normal load P is normalized with RE* for cylindrical particles and RE*^{2} for spherical particles. In the two-dimensional case (i.e., cylinder), [P]= FL^{-1}, and in the three-dimensional case (i.e., sphere) [P]= F.

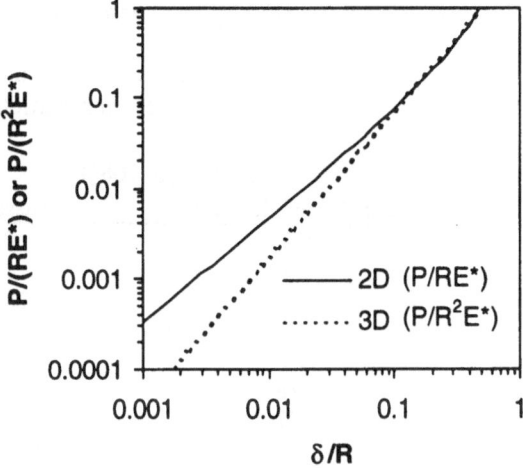

Fig. 10. *Variation of normal load with relative displacement δ/R for cylindrical (2D) and spherical (3D) particles*

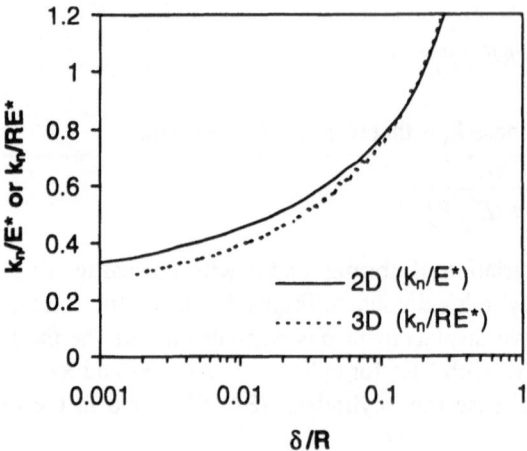

Fig.·11. Variation of normal secant stiffness corresponding to Fig. 10.

Table 1 lists typical values of elastic properties for various particle materials. These values are useful to define realistic values of contact stiffness between particles.

Table 1. Elastic properties for particle materials (after [6, 108]).

Material	Young's modulus (Gpa)	Poisson ratio
Perpex	3	0.38
Glass	55	0.25
Steel	200	0.28 - 0.29
Aluminium	55 - 76	0.34 - 0.36
Duraluminium	74	0.32
Cast Iron	113	0.25
Tungsten Carbide	732	0.22
Amphibolite	93 - 121	0.28 - 0.30
Anhydrite	68	0.30
Diabase	87 - 117	0.27 - 0.30
Diorite	75 - 108	0.26 - 0.29
Dolomite	110 - 121	0.30
Dunite	149 - 183	0.26 - 0.28
Feldspathic Gneiss	83 - 118	0.15 - 0.20
Gabbro	89 - 127	0.27 - 0.31
Granite	73 - 86	0.23 - 0.27
Limestone	87 - 108	0.27 - 0.30
Marble	87 - 108	0.27 - 0.30
Mica Schist	79 - 101	0.15 -0.20
Obsidian	65 - 80	0.12 - 0.18
Oligoclasite	80 - 85	0.29
Quartzite	82 - 97	0.12 - 0.15
Rock salt	35	0.25
Slate	79 - 112	0.15 - 0.20
Ice	7.1	0.36

2.3.7 Deformable and rigid particles

As previously mentioned, soil particles were assumed to be rigid and the contact deformable. However, in reality, soil particles are not rigid; they deform when they are subjected to contact forces. The assumption of rigid particles and deformable contacts is acceptable as long as the contact deformations represent the particle deformations. This condition may be met for particles with simple geometry (e.g., spheres and cylinders) undergoing elastic deformations. However, this condition may not be met for complex deformation patterns and inelastic deformation, and distributed contact loads [e. g., 51]. The case of deformable particles is not considered hereafter. One may refer to [48, 182] to account for elastic deformation of particles.

We will illustrate the fact that the contact deformation may represent the particle deformation in some simple cases. This will be demonstrated by considering the compression of the cylinder of Fig. 12, which is subjected to diametrically opposed concentrated forces. The stress distribution in the cylinder is given in [205]. The stresses at point A are:

$$\sigma_x = \frac{P}{\pi}\left(\frac{1}{R} - \frac{2(a_1^2 + 2z^2)}{a_1^2\sqrt{a_1^2 + z^2}} + \frac{4z}{a_1^2} \right) \text{ and } \sigma_z = \frac{P}{\pi}\left(\frac{1}{R} - \frac{2}{2R-z} - \frac{2}{\sqrt{a_1^2 + z^2}} \right) \quad (2.22)$$

In plane strain, the vertical strain is:

$$\varepsilon_z = \frac{1-v^2}{E}\left(\sigma_z - \frac{v}{1-v}\sigma_x \right) \quad (2.23)$$

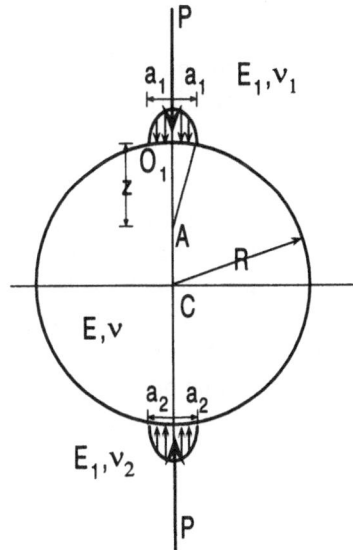

Fig. 12. *Compression of a cylinder due to diametrically opposed concentrated loads.*

The compression of the upper half of the cylinder O_1C is found by integrating ε_z from $z = 0$ to $z = R$, where $a \ll R$, to give:

$$\delta_1 = \int_0^R \varepsilon_z dz = \frac{P(1-v_1^2)}{\pi E}\left(2Ln(4R/a_1)-1\right)$$
(2.24)

A similar expression is obtained for the compression of the lower half of the cylinder so that the total compression of the diameter through the mid-points of the contact areas O_1O_2 is:

$$\delta = \frac{2P(1-v^2)}{\pi E}\left(Ln(4R/a_1)+Ln(4R/a_2)-1\right)$$
(2.25)

Equation 25 can be used to derive Eq. 20. As clearly indicated by Eqs. 22 to 25, the contact stiffness originates from the particle deformation. In this particular case, the contact deformation accounts for the particle deformation.

2.3.8 Frictional contact

In soil mechanics, the frictional characteristics between two particles is commonly thought to be a basic material property, which can easily be determined from laboratory experiments and tables of physical constants. However, this common belief is unfortunately unfounded. The determination of friction between two particles is still a complicated problem as described in [185]. As shown in Table 2, the values of friction angle ϕ_μ vary not only with mineral type but also on the contact cleanliness, water content, and level of normal load.

2.4 Contact models

Several relations were proposed for relating the contact actions and contact kinematics. We will review only two basic ones: the elastic-perfectly plastic relations with and without rotational stiffness and friction.

2.4.1 Elastic-perfectly plastic contact

Figure 13 represents the force-displacement relationship between two cylindrical particles which is used to simulate the intergranular behavior. The contact relationship is activated when two disks overlap. The contact geometry between two disks is characterized by a contact point, and a contact direction which passes through the centers of the particles in contact. N and S denote the projections of the contact force that are respectively tangential and normal to the contact direction. The value of the contact force at time $t+\Delta t$ is calculated from its value at time t by using the following relation

$$\begin{cases} N(t+\Delta t) = N(t)+k_n\Delta n \\ S(t+\Delta t) = S(t)+k_s\Delta s \end{cases}$$
(2.26)

where k_n and k_s are the tangential normal and tangential stiffness, respectively, and Δn and Δs are the normal and tangential components of the relative displacement of the contact between time t and $t+\Delta t$. In the case of linear elastic contacts, k_n and k_s are constant. In the case of Hertz contact, k_n and k_s vary with the contact load N and S. In general k_n and k_s are assumed to be equal. More realistic expressions are reported in [134].

Table 2. Interparticle friction angle ϕ_μ measured from laboratory tests (after [134]).

Material	Type of test	Conditions	Value ϕ_μ	Reference
Biotite	Along cleavage faces	Dry	14.6-17.2	[95]
		Saturated	7.4	—
Calcite	Block on block	Dry	8	—
		Water-saturated	34.2	—
Chlorite	Along cleavage faces	Dry	19.3-27.9	—
		Saturated	12.4	—
Feldspar	Block on block	Dry	6.8	—
		Water-saturated	37.6	—
	Direct shear box, free particles on flat surface		36	[71]
	Free particle		37	[122]
	Particle-plane	Saturated	28.9	[165]
Feldspar (micro-Feldspar (microline)	Direct shear box, fixed parti-	Dry	6	[95]
		Water-saturated	37	—
Glass	Free particles on plate	Dry	10-12	[85]
		Water-saturated	17	—
	Three balls on glass plate	Dry, Low load	9	—
		Water-saturated and low load	19	—
		After Redrying	16	—
Glass ballotini	Ball on ball	Dry, low load	2	[186]
		Dry, high load	4	—
		Flooded, low load	28-40	—
		Flooded, high load	38-40	—
		Dry, low load	3	—
		Dry, high load	7	—
	Direct shear box, free parti-	Water-saturated	17	[175]
		Dry, tested in dry nitrogen	9	[210]
		Acetone cleaned, tested in dry nitro-	16	—
		Trichloroethylene, acetone, detergent	21	—
		Water-saturated	15	—
		Cleaned with soap, water and acetone	15	—
	Particles fixed with wax after initial sliding		14-15	—
Muscovite	Along cleavage faces	Dry	16.7-23.3	[95]
		Saturated	13	—
Phlogopite	Along cleavage faces	Dry	14-17.2	—
		Saturated	8.5	—
Phosphor-bronze	Ball on ball	Water-saturated	21	[161]

Table 2 (continued)

Material	Type of test	Conditions	Value ϕ_μ	Reference
Quartz	Block on block	Dependent on surface condition	0-45	[28]
		Dry	7.4	[95]
		Water-saturated	24.2	—
	Block over particle set in	Dry	6	[213]
		Moist	24.25	—
		Water-saturated	24.25	—
	Direct shear box fixed parti-	Dry	6	—
		Moist and water- saturated	25	—
		Dry	11	[163]
		Water-saturated	33	—
		High load	19	—
		Low load	29	—
		Air dried	22	[23]
		Atmospheric	28	[28]
		Water-saturated	23-30	[175]
		High vacuum	38	[28]
		Water-saturated	26	[210]
		Moist and water-saturated	28	—
		Dry, tested in dry nitrogen	15	—
	Particle-particle	Saturated	26	[165]
	Particle-plane	Saturated	22.2	—
		Dry	7.4	—
	Particles on polished block	Water-saturated	22-31	[175]
	Three fixed particles over	Water-saturated	21-27	[87]
Quartz (clean)	Direct shear box, fixed parti-	Dry	6	[95]
		Water-saturated	23	—
		Amylaraine	31	—
		Carbontetrachloride	11	—
Quartz (milky)		Dry	9	—
		Water-saturated	27	—
Quartz (rose)		Dry	7	—
		Water-saturated	24	—
Steel	Ball on ball	Dry, polished and cleaned with car-	7	[85]
	Direct shear box, free parti-	Air	7	[175]
		Water-saturated	9	—
	Free particles on plate	Water-saturated, dry and cleaned with	8.8	[85]
	Light load apparatus	Dry, polished	9	—
		Dry, polished and cleaned with car-	9.5-12	—
	Rods on rods	Dry, cleaned with carbontetrachloride	9-14	—
Steel balls	Friction apparatus	Dry	16-32	[186]
Zircon	Direct shear box, free parti-	Water-saturated	23	[71]

The contact force of Fig. 13 obeys the Coulomb friction law:

$$S(t) \leq \tan\phi_\mu N(t) + c \tag{2.27}$$

where ϕ_μ and c are the friction angle and cohesion between two disks, respectively.

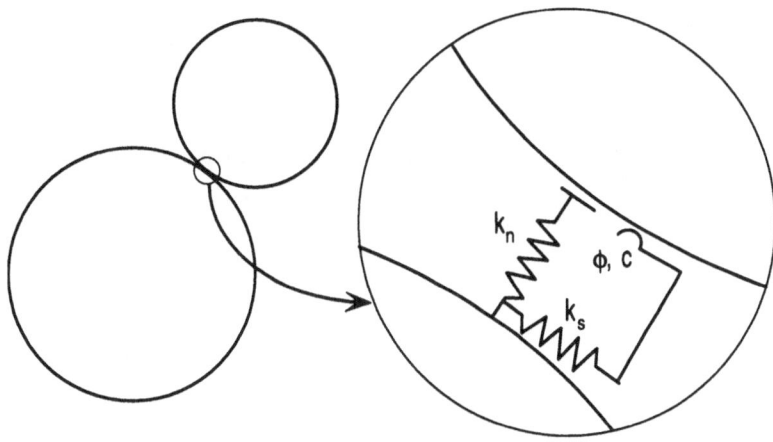

Fig. 13. Elastic perfectly-plastic model for contact without rotational stiffness and friction.

2.4.2 Linear-elastic perfectly plastic contact with rolling stiffness and friction

In idealized granular materials, rolling friction is included at the particle contacts by generalizing the elastic perfectly-plastic contact relation. Such a generalized model is schematized in Fig. 14, including its normal, tangential, and rotational stiffnesses, and rolling and sliding frictions. At the particle contacts, the increments of normal force, shear force, and moment are:

$$\Delta F_n = k_n \Delta n, \quad \Delta F_s = k_s \Delta s, \quad \text{and} \quad \Delta M = R k_\theta \Delta \theta \tag{2.28}$$

where k_n, k_s, and k_θ are the normal, tangential, and rotational stiffness of grain contacts, respectively. R is the average particle radius, Δn and Δs are the normal and tangential relative contact displacement, and $\Delta \theta$ is the relative rotation of disks, which is the rotation of the contact surface. Eq. 28 holds provided that the normal and tangential forces, and the moment at contact, obey the following generalized Coulomb friction law:

$$b = -R_2 |\Delta \theta_2 - \beta| \, sign \, \{ (\Delta \theta_1 - \beta)(\Delta \theta_2 - \beta) \} \tag{2.29}$$

where ϕ_μ is the sliding friction angle, and β the rolling friction angle. The rotational stiffness k_θ is based on an analytical expression derived from the two-dimensional theory of elastic contact [12]. k_θ increases proportionally to the cylinder radius and the normal load, which is in agreement with experimental results on hard rubber cylinders [12].

2.5 Governing equations of statics

The equilibrium of forces on particle a is:

$$\sum_{c \in I_a} f_i^{ac} + \sum_{e \in E_a} f_i^{ae} = 0 \quad i = 1,2,3 \tag{2.30}$$

where f_i^{ac} is the internal force at contact c, and f_i^{ae} the external force at point e. The equilibrium of moments about center of particle a is:

$$\sum_{c\in I_a}\left(m_i^{ac}+e_{ijk}r_j^{ac}f_k^{ac}\right)+\sum_{e\in E_a}\left(m_i^{ae}+e_{ijk}r_j^{ae}f_k^{ae}\right)=0 \quad i=1,2,3 \tag{2.31}$$

where m_i^{ac} is the internal moment at contact point c; m_i^{ae} the external moment at point e; e_{ijk} the alternating tensor; $r_j^{ac}=x_j^c-x_j^a$, $r_{j.}^{av}=x_j^v-x_j^a$; x_j^a and x_j^o the center coordinates of particles a and b; and x_j^c the position of contact point c.

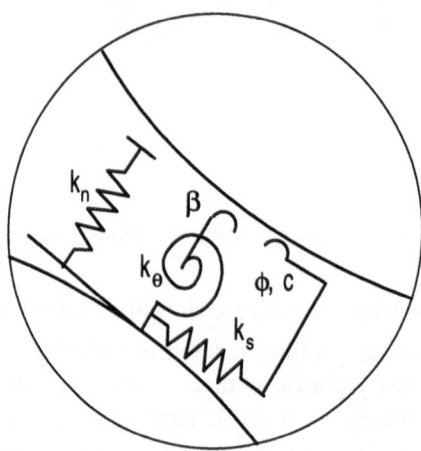

Fig. 14. Representation of generalized contact with rotational stiffness and rolling friction.

2.6 Boundary conditions

There are several types of boundary conditions, which can be specified on an assembly of particles, including (1) prescribed displacement, velocity, acceleration, and force boundary, (2) periodic boundary, (3) rigid boundary, and (4) flexible boundary.

2.6.1 Prescribed force/displacement

As in the boundary value problems of continuum mechanics, the boundary conditions in granular mechanics can be either prescribed displacement/ velocity/ acceleration, or prescribed force. External forces and moments can be applied to any point of particles.
The motion of a group of particles can conveniently be prescribed by a cluster. Clustered particles are subsets of particles that move as a rigid body and are useful to represent rigid objects. As shown in Fig. 15, the particles may overlap, and move as a solid object. The motion of the ath particle in a cluster is defined by its translation u_i^a and rotation $\Delta\theta_a$ as follows:

$$u_i^a=U_i+e_{ijk}\Delta\theta\,(X_i-x_i^a)\quad\text{and}\quad\Delta\theta_a=\Delta\theta \tag{2.32}$$

where U_i is the translation vector and $\Delta\theta$ is the rotation of a reference point for the cluster.

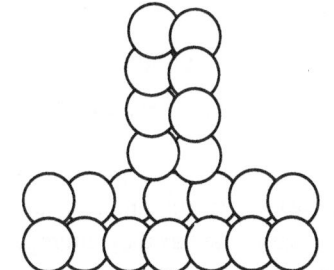

Fig. 15.   Rigid cluster of particles.

2.6.2 Periodic boundaries

Periodic boundaries are illustrated in Fig. 16. The particles leaving segment AB are re-introduced on segment CD. The periodic boundary conditions can be used in both horizontal and vertical directions, therefore filling the complete space with particles. Periodic boundary conditions have extensively been used for determining average continuum quantities (e.g., stress and strain) free of the heterogeneities caused by rigid boundaries. However, periodic boundaries introduce kinematic constraints in some circumstances. These conditions are met when the deformation patterns have a length scale that is a sub-period of the box. One of these adverse effects is observed for shear bands as explained in [8].

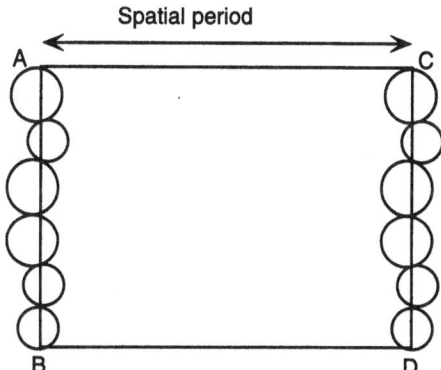

Fig. 16.   Periodic boundaries.

2.6.3 Rigid walls

Rigid walls simulate the loading platens which transmit loads to laboratory samples. They are usually made of a single segment, but can also be made of several connected segments for generating various polygonal shapes, such as superficial footings. Displacement, force, or moment can be specified for rigid walls.

2.6.4 Flexible membranes

Flexible membranes were introduced in [8] to simulate the flexible membranes used in the triaxial test. Forces and moments are externally applied to the boundary particles by specifying the prescribed stress tensor. The forces distributed on the boundary are calcu-

lated from the unit vectors normal to the boundary segments and the prescribed stress tensors. For instance in Fig. 17, the force F_i^v applied to the particle center O is:

$$F_i^O = BC\sigma_{ij}n_j^{BC} (OC+CB/2)/OA + CD\sigma_{ij}n_j^{CD} + DE\sigma_{ij}n_j^{DE} (OD+DE/2)/OF \qquad (2.33)$$

where σ_{ij} is the prescribed stress tensor and n^{BC}, n^{CD}, and n^{DE} are the unit vectors normal to
segments BC, CD and DE. The flexible membrane can sustain large deformation. Particles
are inserted into the flexible boundary chain as they attempt to force their way between two
particles of the flexible boundary.

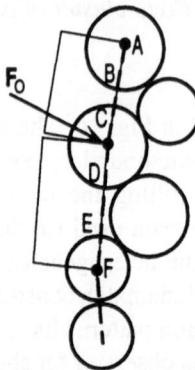

Fig. 17. Flexible membrane for stress-controlled boundaries.

3. Computational methods for granular mechanics

This section introduces the numerical methods for solving the equilibrium equations of an assembly of particles. The main emphasis is to present the basic techniques and to unveil potential problems in the use of computer programs in granular mechanics. We will first review the Newmark algorithms, which are numerical techniques in finite elements and differences commonly used to solve hyperbolic problems. Following this review, we will introduce the concept of dynamic relaxation - one of the main numerical techniques in computational granular mechanics - and the automatic determination of its parameters.

3.1 Review of Newmark integration schemes

The equation of motion is written in a general form as:

$$\mathbf{Ma} + \mathbf{N}(\mathbf{v}, \mathbf{d}) = \mathbf{F}^{ext} \tag{3.1}$$

where $\mathbf{M}$ is the generalized mass matrix, $\mathbf{a}$ the generalized acceleration vector, $\mathbf{v}$ the generalized velocity vector, $\mathbf{d}$ the generalized displacement vector, $\mathbf{N}$ the nonlinear term, and $\mathbf{F}^{ext}$ the vector of externally applied force. The term $\mathbf{N}$ arises from various types of nonlinearity, including nonlinear elasticity and strain-hardening elastoplasticity. The term $\mathbf{N}$ is also referred to as generalized internal force vector. In computational mechanics, $\mathbf{N}$ is usually assumed differentiable. The corresponding tangent stiffness $\mathbf{K}_T$ and tangent damping $\mathbf{C}_T$ are defined as:

$$\mathbf{K}_T = \partial \mathbf{N}/\partial \mathbf{d} \quad \text{and} \quad \mathbf{C}_T = \partial \mathbf{N}/\partial \mathbf{v} \tag{3.2}$$

The initial conditions complete the discrete equation of motion and form an *initial value problem*:

$$\mathbf{d}(0) = \mathbf{d}_0 \quad \text{and} \quad \dot{\mathbf{d}}(0) = \mathbf{v}_0 \tag{3.3}$$

This initial value problem can be solved by time-discretization of Eq. 1. There are several time-discretization algorithms which can be used for this purpose. Some of them are described in [100, 101, 232]. In this section, we will only consider the implicit and explicit versions of the Newmark algorithm.

3.1.1 Implicit Newmark algorithms

The basic equation of the implicit Newmark algorithm is as follows

$$\mathbf{Ma}_{n+1} + \mathbf{N}(\mathbf{d}_{n+1}, \mathbf{v}_{n+1}) - \mathbf{F}^{ext}_{n+1} \tag{3.4}$$

$$\mathbf{d}_{n+1} = \tilde{\mathbf{d}}_{n+1} + \Delta t^2 \beta \, \mathbf{a}_{n+1} \tag{3.5}$$

$$\mathbf{v}_{n+1} = \tilde{\mathbf{v}}_{n+1} + \Delta t \gamma \, \mathbf{a}_{n+1} \tag{3.6}$$

$$\tilde{\mathbf{d}}_{n+1} = \mathbf{d}_n + \Delta t \mathbf{v}_n + \frac{\Delta t^2}{2}(1 - 2\beta)\mathbf{a}_n \tag{3.7}$$

$$\tilde{\mathbf{v}}_{n+1} = \mathbf{v}_n + \Delta t(1-\gamma)\mathbf{a}_n \tag{3.8}$$

where Δt is the time step, $\mathbf{F}_n^{ext} = \mathbf{F}^{ext}(t_n)$, $\mathbf{d}_n = \mathbf{d}(t_n)$, $\mathbf{v}_n = \mathbf{d}(t_n)$, $\mathbf{a}_n = \ddot{\mathbf{d}}(t_n)$, β and γ are constants of time integration, $\tilde{\mathbf{d}}_{n+1}$ and $\tilde{\mathbf{v}}_{n+1}$ are predictors, and $\mathbf{d}_{n+1}$ and $\mathbf{v}_{n+1}$ are correctors.
Tables 3 and 4 summarize the complete implementation procedure of the a- and d- form for the implicit Newmark algorithm. In the d-form, the unknown is the generalized displacement, whereas in the a-form, it is the generalized acceleration. The d-form, which defines an effective stiffness matrix K*, is common used. The a-form, which defines an effective mass M* instead, offers the advantage of optionally selecting the time integration parameter β equal to zero, therefore giving both implicit and explicit formulations at once.

The effective stiffness K* and mass M* are full matrices because the tangential stiffness matrix K_T, and the tangential damping matrix C_T are full matrices in some instances. The linear system of equations must be solved at each time step, which is lengthy but should not cause problems as long as K* or M* is not singular. However, numerical problems are to be expected when the effective matrices K* or M* become singular. The system of equations has multiple solutions, which are referred to as bifurcations. The origins of these bifurcations may be physical or numerical. Most analysts prefer to avoid these cases, and very often resort to adding constraints to stabilize the numerical systems. These constraints are however not always justified physically; they may change the nature of the system under consideration.

Table 3. Implicit Newmark algorithm (d-form).

(1) Predictor phase ($i=0$) where i is the iteration counter. $$\mathbf{d}_{n+1}^{(i)} = \tilde{\mathbf{d}}_{n+1} \quad \mathbf{v}_{n+1}^{(i)} = \tilde{\mathbf{v}}_{n+1} \quad \mathbf{a}_{n+1}^{(i)} = 0$$
(2) The residual or "out-of-balance" force is calculated $$\Delta\mathbf{F} = \mathbf{F}_{n+1}^{ext} - \mathbf{M}\mathbf{a}_{n+1}^{(i)} - \mathbf{N}(\mathbf{d}_{n+1}^{(i)}, \mathbf{v}_{n+1}^{(i)})$$
(3) The linear matrix system is solved for $\Delta\mathbf{d}$: $\mathbf{K}^{*}\Delta\mathbf{d} = \Delta\mathbf{F}$ where the effective stiffness matrix $\mathbf{K}^{*}$ is $\quad \mathbf{K}^{*} = \dfrac{1}{\Delta t^2 \beta}\mathbf{M} + \dfrac{\gamma}{\Delta t\beta}\mathbf{C}_T + \mathbf{K}_T$
(4) Corrector phase $$\mathbf{d}_{n+1}^{(i+1)} = \mathbf{d}_{n+1}^{(i)} + \Delta\mathbf{d}$$ $$\mathbf{a}_{n+1}^{(i+1)} = \left(\mathbf{d}_{n+1}^{(i+1)} - \tilde{\mathbf{d}}_{n+1}\right)\big/(\beta\Delta t^2)$$ $$\mathbf{v}_{n+1}^{(i+1)} = \tilde{\mathbf{v}}_{n+1} + \Delta t\gamma\mathbf{a}_{n+1}^{(i+1)}$$
(5) additional iterations? If yes, $i \leftarrow i+1$, go to (2). If no, continue
(6) Update solution: $\mathbf{d}_{n+1} = \mathbf{d}_{n+1}^{(i)} \quad \mathbf{v}_{n+1} = \mathbf{v}_{n+1}^{(i)} \quad \mathbf{a}_{n+1} = \mathbf{a}_{n+1}^{(i)}$
(7) Execute next time step: $n \leftarrow n+1$, go to (1)

Table 4. Implicit Newmark algorithm (a-form).

(1) Predictor phase ($i=0$) where i is the iteration counter. $\quad \mathbf{d}_{n+1}^{(i)} = \tilde{\mathbf{d}}_{n+1} \quad \mathbf{v}_{n+1}^{(i)} = \tilde{\mathbf{v}}_{n+1} \quad \mathbf{a}_{n+1}^{(i)} = 0$
(2) The residual or "out-of-balance" force is calculated $\quad \Delta \mathbf{F} = \mathbf{F}_{n+1}^{ext} - \mathbf{M}\mathbf{a}_{n+1}^{(i)} - \mathbf{N}(\mathbf{d}_{n+1}^{(i)}, \mathbf{v}_{n+1}^{(i)})$
(3) The linear matrix system is solved for $\Delta \mathbf{a}$: $\quad \mathbf{M}^{*}\Delta \mathbf{a} = \Delta \mathbf{F}$ where the effective mass matrix $\mathbf{M}^{*}$ is $\mathbf{M}^{*} = \mathbf{M} + \gamma \Delta t \mathbf{C}_{T} + \Delta t^{2} \beta \mathbf{K}_{T}$
(4) Corrector phase $\quad \mathbf{a}_{n+1}^{(i+1)} = \mathbf{a}_{n+1}^{(i)} + \Delta \mathbf{a}$ $\quad \mathbf{v}_{n+1}^{(i+1)} = \tilde{\mathbf{v}}_{n+1} + \gamma \Delta t \mathbf{a}_{n+1}^{(i+1)}$ $\quad \mathbf{d}_{n+1}^{(i+1)} = \tilde{\mathbf{d}}_{n+1} + \beta \Delta t^{2} \mathbf{a}_{n+1}^{(i+1)}$
(5) additional iterations? If yes, $i \leftarrow i+1$, go to (2). If no, continue
(6) Update solution: $\mathbf{d}_{n+1} = \mathbf{d}_{n+1}^{(i)} \quad \mathbf{v}_{n+1} = \mathbf{v}_{n+1}^{(i)} \quad \mathbf{a}_{n+1} = \mathbf{a}_{n+1}^{(i)}$
(7) Execute next time step: $n \leftarrow n+1$, go to (1)

3.1.2 Explicit Newmark algorithm

The explicit Newmark algorithm is used to decouple the linear system of equation, and avoids the solution of a linear system, which may become singular. The explicit Newmark algorithm is very similar to its implicit counterpart. The only difference is as follows:

$$\mathbf{M}\mathbf{a}_{n+1} + \mathbf{N}(\tilde{\mathbf{d}}_{n+1}, \tilde{\mathbf{v}}_{n+1}) = \mathbf{F}_{n+1}^{ext} \tag{3.9}$$

The effective stiffness matrix is therefore proportional to the mass matrix:

$$\mathbf{K}^{*} = \frac{1}{\Delta t^{2}\beta}\mathbf{M} \tag{3.10}$$

When the mass matrix is diagonal, the solution of Eq. 10 is trivial. There is only one pass through the iterative phase:

$$\mathbf{d}_{n+1} = \mathbf{d}_{n+1}^{(1)} \quad \mathbf{v}_{n+1} = \mathbf{v}_{n+1}^{(1)} \quad \mathbf{a}_{n+1} = \mathbf{a}_{n+1}^{(1)} \tag{3.11}$$

Table 5 summarizes the implementation procedure for the explicit Newmark algorithm. The implementations of the implicit and explicit algorithms are very similar. They are often implemented together. There is even the option of selecting which degree of freedom will be treated explicitly and implicitly. This mixed technique is called the implicit/explicit integration scheme [99].

Table 5. Explicit Newmark algorithm (a-form, $\beta=0$).

---

(1) The predictors are calculated

$$\tilde{d}_{n+1} = d_n + \Delta t v_n + \frac{\Delta t^2}{2} a_n$$

$$\tilde{v}_{n+1} = v_n + \Delta t (1-\gamma) a_n$$

(2) The residual or "out-of-balance" force is calculated

$$\Delta F = F_{n+1}^{ext} - N(\tilde{d}_{n+1}, \tilde{v}_{n+1})$$

(3) The linear matrix system is solved for Δa: $\mathbf{M} a_{n+1} = \Delta F$

(4) Update solution

$$v_{n+1} = \tilde{v}_{n+1} + \gamma \Delta t a_{n+1}$$

$$d_{n+1} = \tilde{d}_{n+1}$$

(5) Execute next time step: $n \leftarrow n+1$, go to (1)

The advantage of explicit formulation over its implicit counterpart is to use the diagonal mass and damping matrix to decouple the equations. In other words, there is no need for calculating a stiffness matrix, and solving a fully coupled system of linear equations. The disadvantage of the explicit scheme is that it is unconditionally stable. The concept of stability and convergence will now be described briefly.

3.1.3 Stability and convergence analysis

Due to the modal decomposition of multiple-degree freedom systems, the stability and convergence of Newmark algorithm can be studied in the case of the single degree of freedom (SDOF) model:

$$\ddot{d} + 2\xi\omega\dot{d} + \omega^2 d = F$$

$$d(0) = d_0 \tag{3.12}$$

$$\dot{d}(0) = v_0$$

The SDOF model can be recast as follows:

$$y_{n+1} = Ay_n + L_n \quad \text{and} \quad y_n = \begin{Bmatrix} d_n \\ v_n \end{Bmatrix} \tag{3.13}$$

where A is the amplification matrix and L_n is the load. The details of the stability analysis can be found in [100].

The local truncation error T is defined as:

$$y(t_{n+1}) = Ay(t_n) + L_n + \Delta t.T(t_n) \quad \text{and} \quad y(t_n) = \begin{Bmatrix} d(t_n) \\ v(t_n) \end{Bmatrix} \tag{3.14}$$

If the amplitude of T satisfies the following relation for all $t > 0$:

$$\|T(t)\| \leq c\Delta t^k \tag{3.15}$$

where c is a constant independent of Δt, and $k>0$, the algorithm is called *consistent*; k is called the *order of accuracy* or *rate of convergence*. Stability plus consistency implies convergence. The error at time t_n is defined as:

$$e(t_n) = y_n - y(t_n) \tag{3.16}$$

It can be shown that the error is:

$$e(t_n) = \mathbf{A}^n e(0) - \Delta t \sum_{i=0}^{n-1} \mathbf{A}^i \mathbf{T}(t_{n-1-i}) \tag{3.17}$$

The stability is unconditional when

$$\gamma \geq 1/2 \quad \text{and} \quad \beta \geq (\gamma + 1/2)^2 / 4 \tag{3.18}$$

and is conditional when:

$$\gamma \geq 1/2 \quad \text{and} \quad \Omega = \omega \Delta t \prec \Omega_{crit} \tag{3.19}$$

where Ω_{crit} is in the damped case ($\xi > 0$):

$$\Omega_{crit} = \frac{\xi (\gamma - 1/2)/2 + \sqrt{(\gamma + 1/2)^2/4 - \beta + \xi^2 (\beta - \gamma/2)}}{(\gamma + 1/2)^2 / 4 - \beta} \tag{3.20}$$

and in the undamped case ($\xi = 0$):

$$\Omega_{crit} = \frac{1}{\sqrt{(\gamma + 1/2)^2 / 4 - \beta}} \tag{3.21}$$

Note that Ω_{crit} increases with damping ratio ξ. Thus the undamped value of Ω_{crit} serves as a conservative estimate when ξ is unknown. Table 6 lists the properties of a few well-known Newmark algorithms.

The concept of stability is not to be confused with that of convergence. A numerically unstable system will not converge toward the exact solution. However, a numerically stable system may not converge toward the right solution. This case is demonstrated in Example 1.

Table 6. Properties of a few Newmark algorithms.

Method	Type	β	γ	Stability condition	Order of accuracy
Average acceleration	Implicit	1/4	1/2	Unconditional	2
Linear acceleration	Implicit	1/16	1/2	$\Omega_{crit} = 2\sqrt{3}$	2
Fox-Goodwin	Implicit	1/12	1/2	$\Omega_{crit} = \sqrt{6}$	2
Central difference	Explicit	0	1/2	$\Omega_{crit} = 2$	2

3.1.3.1 Example 1

A single degree of freedom (SDOF) system is shown in Fig. 18. It has for mass m, stiffness constant k, and damping coefficient c. The specific values are: $m = 1$, $c = 2\omega\xi$, $k =$

ω^2, and $\omega = 2\pi$ ($T=1$). The SDOF is initially at a state of rest ($v_0 = 0$ and $x_0 = 0$) and becomes subjected to a constant force $F_0 = 40$ at $t = 0$.

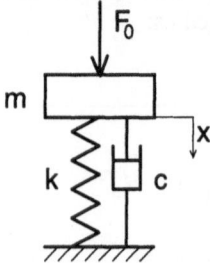

Fig. 18. Single degree of freedom (SDOF) system subjected to an initial load F_0.

When $\xi < 1$, the general closed-form analytical solution for the SDOF response is:

$$x(t) = \frac{F_0}{km} + e^{-\xi\omega t}\left[\frac{v_0 + \xi\omega(u_0 - \frac{F_0}{km})}{\omega\sqrt{1-\xi^2}}\sin\left(\omega\sqrt{1-\xi^2}t\right) + \left(u_0 - \frac{F_0}{km}\right)\cos\left(\omega\sqrt{1-\xi^2}t\right)\right] \quad (3.22)$$

When $m = 1$, $v_0 = 0$ and $x_0 = 0$, the response becomes:

$$x(t) = \frac{F_0}{\omega^2}\left(1 - \left[\xi/\sqrt{1-\xi^2}\,\sin\left(\omega\sqrt{1-\xi^2}t\right) + \cos\left(\omega\sqrt{1-\xi^2}t\right)\right]e^{-\xi\omega t}\right) \quad (3.23)$$

The implicit Newmark algorithm may be used to calculate the time history of the response of this SDOF in two particular cases: undamped ($\xi = 0$) and damped ($\xi = 0.1$).

Figure 19 lists the formulas of an Excel spreadsheet that follows the implementation steps of Table 4. The procedure applies to both implicit and explicit algorithms. Figures 20 and 21 show the corresponding numerical results for the time step $\Delta t = 0.31$ and the Newmark constants $\beta = 0$ and $\gamma = 0.5$.

The algorithm is a central-difference type. It is explicit: $\Omega_{crit} = 2$. The time step $\Delta t = 0.31$ is less than the critical time step $\Delta t_{crit} = \Omega_{crit}/\omega = 1/\pi \approx 0.32$. When Δt is larger than Δt_{crit}, the numerical integration diverges completely. As shown in Figs. 20 and 21 , when Δt is close to Δt_{crit} (e.g., 0.31), the numerical solution is stable but inaccurate for both damped and undamped SDOF. When Δt is smaller than to Δt_{crit}, the numerical solution coincides practically with the theoretical solution (not reported in Figs. 20 and 21).

	A	B	C	D	E	F
1	T = 1		ω = =2*PI()/T		Δt = 0.31	
2	ξ = 0		$\omega\Delta t$ = =w*Dt		γ = 0.5	
3	d_0 = 0		m* = =1+g*Dt*2*x*w+k		β = 0	
4	v_0 = 0				Ω_{crit} = =(x*(g-0.5)+SQRT(g...
5	F_0 = 40					
6						
7	t	a	v	d	vp	dp
8	0	=F0-2*x*w*v0-w^2*d0	=v0	=d0		
9	=A8+Dt	= (F0-2*x*w*E9-w^2*F9)/m	=E9+g*Dt*B9	=F9+b*Dt^2*B9	=C8+(1-g)*Dt*B8	=D8+Dt*C8+Dt^2*(0.5...
10	=A9+Dt	= (F0-2*x*w*E10-w^2*F10)/...	=E10+g*Dt*B10	=F10+b*Dt^2*B1(	=C9+(1-g)*Dt*B9	=D9+Dt*C9+Dt^2*(0.5...

Fig. 19. Formulas of Excel spreadsheet for implementation of implicit/explicit Newmark algorithm for a SDOF.

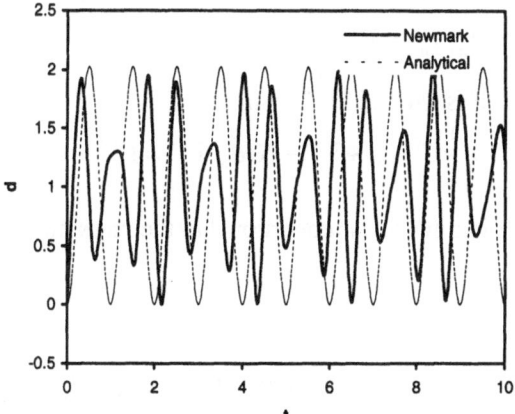

Fig. 20. Example of poor convergence but stability for explicit Newmark integration for an undamped SDOF ($\xi =0$, $\omega=2\pi$, $F_0=40$, $\Delta t=0.31$, $\beta = 0$, $\gamma = 0.5$).

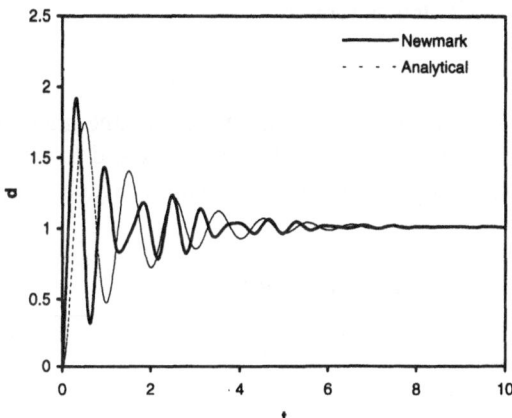

Fig. 21. Example of poor convergence but stability for Newmark integration for a damped SDOF ($\xi =0.1$, $\omega=2\pi$, $F_0=40$, $\Delta t=0.31$, $\beta = 0$, $\gamma = 0.5$).

3.1.4 Central difference time integration ($\beta = 0$, $\gamma = 1/2$)

The following central difference expressions are used for the acceleration at time t_n:

$$\mathbf{a}_n = \frac{1}{\Delta t}(\tilde{\mathbf{v}}_{n+1} - \tilde{\mathbf{v}}_n) \quad \tilde{\mathbf{v}}_{n+1} = \frac{1}{\Delta t}(\mathbf{d}_n - \mathbf{d}_{n-1}) \tag{3.24}$$

where Δt is the time interval between times t_n and t_{n+n} (i. e., $\Delta t = t_{n+1}\text{-}t_n$). The velocity $\mathbf{v}_n$ at time t_n is:

$$\mathbf{v}_n = \frac{1}{2}(\tilde{\mathbf{v}}_{n+1} + \tilde{\mathbf{v}}_n) \tag{3.25}$$

The equation of motion at time t_n is:

$$\mathbf{M}\mathbf{a}_n + \mathbf{C}\mathbf{v}_n + \mathbf{N}(\tilde{\mathbf{d}}_n, \tilde{\mathbf{v}}_n) = \mathbf{F}_n^{ext} \tag{3.26}$$

and after substitution of Eqs. 24 and 25, becomes:

$$\frac{1}{\Delta t}\mathbf{M}(\tilde{\mathbf{v}}_{n+1} - \tilde{\mathbf{v}}_n) + \frac{1}{2}\mathbf{C}(\tilde{\mathbf{v}}_{n+1} + \tilde{\mathbf{v}}_n) + \mathbf{N}(\tilde{\mathbf{d}}_n, \tilde{\mathbf{v}}_n) = \mathbf{F}_n^{ext} \tag{3.27}$$

Assuming that the damping matrix is proportional to the mass matrix,

$$\mathbf{C} = \alpha\mathbf{M} \tag{3.28}$$

then Eq. 27 becomes:

$$\tilde{\mathbf{v}}_{n+1}(1 + \alpha\Delta t/2) = \tilde{\mathbf{v}}_n(1 - \alpha\Delta t/2) + \Delta t\mathbf{M}^{-1}\left(\mathbf{F}_n^{ext} - \mathbf{N}(\tilde{\mathbf{d}}_n, \tilde{\mathbf{v}}_n)\right) \tag{3.29}$$

or equivalently:

$$\tilde{\mathbf{v}}_{n+1} = \frac{1 - \alpha\Delta t/2}{1 + \alpha\Delta t/2}\tilde{\mathbf{v}}_n + \frac{\Delta t}{1 + \alpha\Delta t/2}\mathbf{M}^{-1}\left(\mathbf{F}_n^{ext} - \mathbf{N}(\tilde{\mathbf{d}}_n, \tilde{\mathbf{v}}_n)\right) \tag{3.30}$$

The displacements are calculated as follows:

$$\mathbf{d}_{n+1} = \mathbf{d}_n + \tilde{\mathbf{v}}_{n+1}\Delta t \tag{3.31}$$

Table 7 summarizes the steps of a central difference time integration scheme. The steps are simpler and less numerous than those of Tables 3, 4 and 5. Only the predicted velocity and the displacement are needed during the calculation, which makes this method computationally efficient.

Table 7. Central difference synchronous algorithm.

(1) Initialize $n=0$, $\tilde{\mathbf{v}}_1 = \mathbf{v}_0$, and $\tilde{\mathbf{d}}_1 = \mathbf{d}_0$
(2) Calculate the residual force $\Delta\mathbf{F} = \mathbf{F}_n^{ext} - \mathbf{N}(\tilde{\mathbf{d}}_n, \tilde{\mathbf{v}}_n)$
(3) Calculate the predicted velocity $\tilde{\mathbf{v}}_{n+1} = \dfrac{1 - \alpha\Delta t/2}{1 + \alpha\Delta t/2}\tilde{\mathbf{v}}_n + \dfrac{\Delta t}{1 + \alpha\Delta t/2}\mathbf{M}^{-1}\Delta\mathbf{F}$
(4) Update solution $\mathbf{d}_{n+1} = \mathbf{d}_n + \tilde{\mathbf{v}}_{n+1}\Delta t$
(5) Execute next time step: $n \leftarrow n+1$, go to (2)

Equation 30 is an explicit central difference algorithm of the synchronous type [101]. An asynchronous system can also be devised as follows:

$$\mathbf{M}\mathbf{a}_n + \mathbf{C}\tilde{\mathbf{v}}_n + \mathbf{N}(\tilde{\mathbf{d}}_n, \tilde{\mathbf{v}}_n) = \mathbf{F}_n^{ext} \tag{3.32}$$

This asynchronous system implies that

$$\mathbf{M}\frac{1}{\Delta t}(\tilde{\mathbf{v}}_{n+1} - \tilde{\mathbf{v}}_n) + \mathbf{C}\tilde{\mathbf{v}}_n + \mathbf{N}(\tilde{\mathbf{d}}_n, \tilde{\mathbf{v}}_n) = \mathbf{F}_n^{ext} \tag{3.33}$$

and

$$\tilde{\mathbf{v}}_{n+1} = \tilde{\mathbf{v}}_n(1 - \alpha\Delta t/2) + \Delta t\mathbf{M}^{-1}\left(\mathbf{F}_n^{ext} - \mathbf{N}(\tilde{\mathbf{d}}_n, \tilde{\mathbf{v}}_n)\right) \tag{3.34}$$

The asynchronous and synchronous formulations coincide in the absence of damping.

3.1.4.1 Example 2

The problem of Example 1 can be solved by using the synchronous and asynchronous explicit algorithms of Table 7. Figures 22 and 23 list all the Excel spreadsheet formulas to implement these two algorithms for an SDOF.

As in Example 1, $\Delta t_{crit} = 0.32$. The synchronous and asynchronous algorithms give the same results for time step Δt smaller than 0.05. As shown in Fig. 24, for $\Delta t = 0.2$, the synchronous algorithm converges more rapidly than its asynchronous counterpart, which is not surprising because asynchronicity modifies the damping force.

	A	B	C	D
1	T = 1		$d_0 = 0$	
2	ω = =2*PI()/T		$v_0 = 0$	
3	α = =2*w*x		$F_0 = 40$	
4	ξ = 0.1			e = =(1-a*Dt/2)/(1+a*Dt/2)
5	Δt = 0.2			f = =Dt/(1+a*Dt/2)
6	Δt_{crit} = =2/w			
7	t	d	vp	
8	0	=d0	=v0	
9	=A8+Dt	=B8+Dt*C9	=e*C8+f*(F0-w^2*B8)	
10	=A9+Dt	=B9+Dt*C10	=e*C9+f*(F0-w^2*B9)	
11	=A10+Dt	=B10+Dt*C11	=e*C10+f*(F0-w^2*B10)	

Fig. 22. *Formulas for the implementation of synchronous explicit algorithm for an SDOF in Excel spreadsheet.*

	A	B	C	D
1	T = 1		$d_0 = 0$	
2	ω = =2*PI()/T		$v_0 = 0$	
3	α = =2*w*x		$F_0 = 40$	
4	ξ = 0.1			e = =(1-a*Dt/2)
5	Δt = 0.2			f = =Dt
6	Δt_{crit} = =2/w			
7	t	d	vp	
8	0	=d0	=v0	
9	=A8+Dt	=B8+Dt*C9	=e*C8+f*(F0-w^2*B8)	
10	=A9+Dt	=B9+Dt*C10	=e*C9+f*(F0-w^2*B9)	
11	=A10+Dt	=B10+Dt*C11	=e*C10+f*(F0-w^2*B10)	

Fig. 23. *Implementation of asynchronous explicit algorithm for an SDOF using an Excel spreadsheet.*

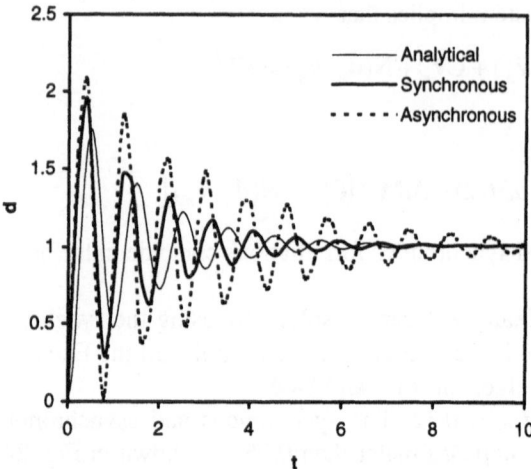

Fig. 24. *Comparison of numerical results for asynchronous and synchronous explicit algorithms for an SDOF (ξ =0.1, ω=2π, F_0=40, Δt=0.2).*

3.1.5 Stability of explicit schemes

The critical time step corresponding to the stability limit for the undamped case is:

$$\Delta t_c = 2/\omega_{max}$$ (3.35)

One of the advantages of the asynchronous over synchronous formulation is that there already exists a stability limit for the damped case. The critical time step corresponding to the stability limit for the damped case is:

$$\Delta t_c = \frac{2}{\omega_{max}}(\sqrt{1+\xi^2} - \xi)$$ (3.36)

where the damping ratio is in the case of Rayleigh damping:

$$\xi = \frac{a}{2\omega}+\frac{b\omega}{2} \quad C = aM + bK$$ (3.37)

In the synchronous algorithm, the damping, which is proportional to mass, is on the left hand side of the equations, and that which is proportional to damping, is on the right hand side. This partition is similar to that of implicit-explicit integration schemes [99].

The advantages of explicit integration are as follows: it requires no computation of a stiffness matrix, it saves storage and computer time, and it is very easy to program even of case of complex physical phenomena and constitutive equations. However, its main shortcoming is conditional stability of the time step. It provides no information on bifurcation points.

3.1.6 Dynamic relaxation

Dynamic relaxation, which is conveniently abbreviated as DR, consists of finding a static solution by using a dynamic transient analysis method. Reviewed in [215], dynamic relaxation has been applied to various types of problems in nonlinear structural analysis. The method originates from the second order Richardson method developed in [77].

DR uses the central difference algorithm previously described, especially the central difference synchronous algorithm. Table 8 summarizes the DR algorithm, which differs from that of Table 7 by the choice of mass matrix and damping coefficient. DR can be applied to solve nonlinear problems as illustrated in Example 3.

Table 8. Dynamic Relaxation (DR) algorithm.

(1) Initialize

$n=0$, $\tilde{\mathbf{v}}_1 = \mathbf{v}_0$, and $\tilde{\mathbf{d}}_1 = \mathbf{d}_0$

(2) Define mass matrix $\mathbf{M}$ and damping coefficient α

(3) Calculate the residual force

$$\Delta \mathbf{F} = \mathbf{F}_n^{ext} - \mathbf{N}(\tilde{\mathbf{d}}_n, \tilde{\mathbf{v}}_n)$$

(4) Calculate the predicted velocity

$$\tilde{\mathbf{v}}_{n+1} = \frac{1-\alpha\Delta t/2}{1+\alpha\Delta t/2}\tilde{\mathbf{v}}_n + \frac{\Delta t}{1+\alpha\Delta t/2}\mathbf{M}^{-1}\Delta \mathbf{F}$$

(5) Update solution

$$\mathbf{d}_{n+1} = \mathbf{d}_n + \tilde{\mathbf{v}}_{n+1}\Delta t$$

(6) Execute next time step: $n \leftarrow n+1$, go to (3)

3.1.6.1 Example 3

The SDOF of Example 1 has a nonlinear spring; the spring force is a power function of displacement:

$$N(d,v) = k_0 d^m \tag{3.38}$$

When the applied force is constant and equal to F_0, there is a closed-form analytical solution for the final displacement (static case):

$$d^* = \sqrt[m]{F_0/k_0} \tag{3.39}$$

This solution pertains to the linear case (i.e., $m=1$). The determination of optimal DR parameters is not trivial in the nonlinear case. The error is defined as follows:

$$E_n = (d_n - d^*)/d^* \tag{3.40}$$

Figure 25 lists all the formulas used to implement the DR algorithm for nonlinear SDOF. The damping coefficients α and Δt have arbitrarily been selected ($\alpha = 2$ and $\Delta t = 0.1$). Figure 26 shows the convergence of the solution toward the static solution with time, and Fig. 27 shows the decrease of error with the number of time steps. As shown in Fig. 27, the DR solution converges toward the numerical solution in about 200 steps. At this point, it reaches the computer precision.

	A	B	C	D	E
1	α = 2			e = =(1-a*Dt/2)/(1+a*Dt/2)	
2	Δt = 0.1			f = =Dt/(1+a*Dt/2)	
3	d₀ = 0			k = 10	
4	v₀ = 0			n = 4	
5	F₀ = 40		Exact solution = =(F0/k)^(1/n)		
6	t	d	vp	Error	Step
7	0	=d0	=v0	=ABS((B7-df)/df)	1
8	=A7+Dt	=B7+Dt*C8	=e*C7+f*(F0-k*B7^n)	=ABS((B8-df)/df)	2
9	=A8+Dt	=B8+Dt*C9	=e*C8+f*(F0-k*B8^n)	=ABS((B9-df)/df)	3
10	=A9+Dt	=B9+Dt*C10	=e*C9+f*(F0-k*B9^n)	=ABS((B10-df)/df)	4

Fig. 25. Formulas for the implementation of DR algorithm for a nonlinear SDOF using Excel spreadsheet.

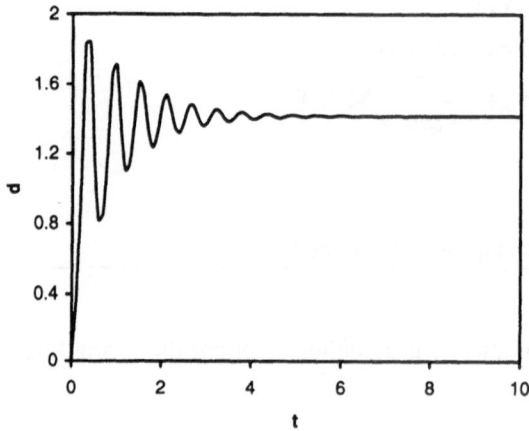

Fig. 26. Convergence of DR solution toward static solution.

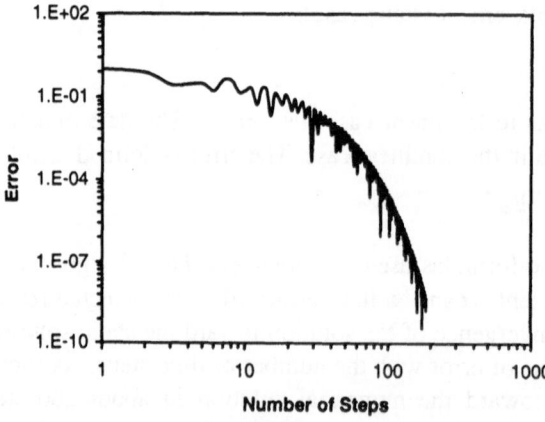

Fig. 27. Decrease of error E_n with number of time steps.

3.1.7 Adaptative Dynamic relaxation

Papadrakakis [160] and Underwood [215] optimized the DR parameters by developing adaptative dynamic relaxation techniques, hereafter denoted ADR. This technique is used in [7].

3.1.8 Convergence and stability of DR integration scheme

By definition, the error e_n at time t_n is the difference between the calculated value d_n at time t_n and the exact solution d^*, which satisfies $N(d^*) = f_n$:

$$e_n = d_n - d^* \tag{3.41}$$

Using Eqs. 31 and 41, one obtains:

$$\tilde{v}_{n+1} = (d_{n+1} - d_n)/\Delta t = (e_{n+1} - e_n)/\Delta t \tag{3.42}$$

and after substitution of Eq. 30:

$$e_{n+1} - e_n = \beta(e_n - e_{n-1}) + \kappa M^{-1}\Delta F \tag{3.43}$$

where

$$\beta = \frac{2-\alpha\Delta t}{2+\alpha\Delta t} \text{ and } \kappa = \frac{2\Delta t^2}{2+\alpha\Delta t} = (\beta+1)\Delta t^2/2 \tag{3.44}$$

Equation 43 can be rewritten:

$$M^{-1}\Delta F = M^{-1}(F_n^{ext} - N(\tilde{d}_n, \tilde{v}_n)) \tag{3.45}$$

$$F_n^{ext} = N(d^*) \tag{3.46}$$

By using Taylor expansion, one obtains:

$$N(d_n) - N(d^*) \approx \frac{\partial N}{\partial d}(d_n - d^*) = \frac{\partial N}{\partial d}e_n \tag{3.47}$$

The matrix A is defined as:

$$Ae_n = M^{-1}\frac{\partial N}{\partial d}e_n \tag{3.48}$$

Finally, Eq. 43 can be rewritten:

$$e_{n+1} = (1(1+\beta) - \kappa A)e_n - \beta e_{n-1} \tag{3.49}$$

For linear problems, A becomes:

$$A - M^{-1}K \tag{3.50}$$

where K is the stifness matrix.
In order to examine the convergence, we look for a solution of Eq. 49 which obeys the following relation:

$$\mathbf{e}_{n+1} = \lambda_{DR}\, \mathbf{e}_{n+1} \tag{3.51}$$

By substituting Eq. 51 into Eq. 49, λ_{DR} is the solution to the following quadratic equation:

$$\left(\mathbf{A} - \frac{\lambda_{DR}^2 - (\beta+1)\lambda_{DR} + \beta}{\kappa\lambda_{DR}}\mathbf{1}\right)\mathbf{e}_{n-1} = 0 \tag{3.52}$$

Equation 52 is an eigenvalue problem, which can be written as follows:

$$(\mathbf{A} - \lambda\mathbf{1})\mathbf{e}_{n-1} = 0 \tag{3.53}$$

The coefficient λ_{DR} and the ith eigenvalue λ_i of $\mathbf{A}$ are related through:

$$\lambda_{DR}^2 - (1+\beta - \kappa\lambda_i)\lambda_{DR} + \beta = 0 \tag{3.54}$$

Equation 54 is a quadratic equation in λ_{DR} having for discriminant:

$$\Delta = (\beta + 1 - \kappa\lambda_i)^2 - 4\beta \tag{3.55}$$

or equivalently:

$$\Delta = 4\left(\lambda_i^2\Delta t^4 - 4\lambda_i\Delta t^2 + \alpha^2\Delta t^2\right)/(2+\alpha\Delta t)^2 \tag{3.56}$$

In the case $\Delta<0$, there are 2 conjugate complex roots:

$$\lambda_{DR} = \left(\beta + 1 - \kappa\lambda_i \pm \sqrt{(\beta+1-\kappa\lambda_i)^2 - 4\beta}\right)/2 \tag{3.57}$$

which have the same amplitude independent of λ_i

$$|\lambda_{DR}| = \sqrt{\beta} = \sqrt{\frac{2-\alpha\Delta t}{2+\alpha\Delta t}} \tag{3.58}$$

In the case $\Delta=0$, there are two real double roots:

$$\lambda_{DR} = (\beta + 1 - \kappa\lambda_i)/2 = \frac{2 - \lambda_i\Delta t^2}{2+\alpha\Delta t} \tag{3.59}$$

which corresponds to:

$$\alpha^2\Delta t^2 = 4\lambda_i\Delta t^2 - \lambda_i^2\Delta t^4 \tag{3.60}$$

In the case $\Delta>0$, there are two real roots, the maximum absolute value of which is:

$$|\lambda_{DR}| = \left(|\beta + 1 - \kappa\lambda_i| + \sqrt{(\beta+1-\kappa\lambda_i)^2 - 4\beta}\right)/2 \tag{3.61}$$

or equivalently:

$$|\lambda_{DR}| = \frac{\left|2 - \lambda_i\Delta t^2\right| + \sqrt{\lambda_i^2\Delta t^4 - 4\lambda_i\Delta t^2 + \alpha^2\Delta t^2}}{2+\alpha\Delta t} \tag{3.62}$$

Figure 28 illustrates the variation of $|\lambda_{DR}|$ as a function of $\alpha\Delta t$ for various values of $\lambda_i\Delta t^2$. The most rapid convergence is obtained for the smallest possible $|\lambda_{DR}| < 1$. This minimum is obtained when $\Delta = 0$.

Figure 29 illustrates the variation of $|\lambda_{DR}|$ as a function of $\lambda_i \Delta t^2$ for various values of $\alpha \Delta t$. The minimum value of $|\lambda_{DR}|$, which corresponds to the fastest convergence, is obtained for one eigenvalue when $\alpha \Delta t = 2$, and for a range of eigenvalues when $\alpha \Delta t < 2$. The convergence is obtained provided that $\lambda_i \Delta t^2 < 4$.

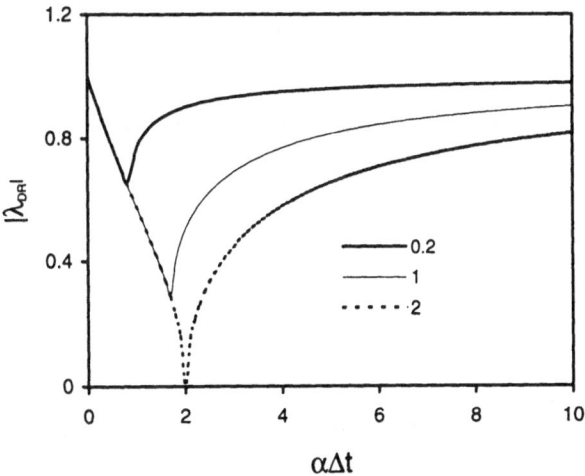

Fig. 28. Variation of $|\lambda_{DR}|$ versus $\alpha \Delta t$ for three different values of $\lambda_i \Delta t^2$.

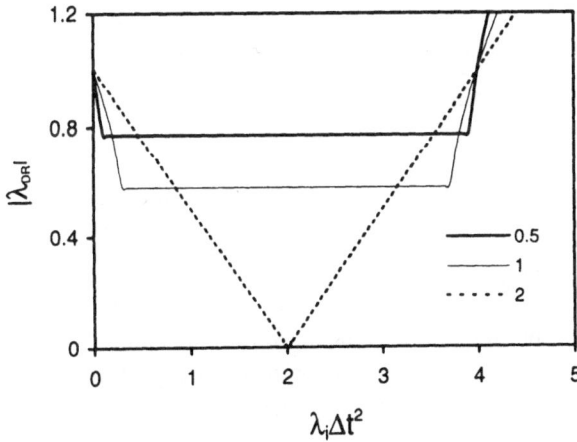

Fig. 29. Variation of $|\lambda_{DR}|$ versus $\lambda_i \Delta t^2$ for three different values of $\alpha \Delta t$.

The convergence should be obtained for all eigenvalues λ_i ranging from λ_{max} to λ_{min}, which are the maximum and minimum eigenvalues of $\mathbf{A}$.

Papadrakakis [160] proposed to select the optimum algorithm parameters so that both λ_{max} and λ_{min} satisfy $\Delta = 0$:

$$\begin{cases} \alpha^2 \Delta t^2 = 4\lambda_{max} \Delta t^2 - \lambda_{max}^2 \Delta t^4 \\ \alpha^2 \Delta t^2 = 4\lambda_{min} \Delta t^2 - \lambda_{min}^2 \Delta t^4 \end{cases} \tag{3.63}$$

The optimum values of Δt and α are therefore:

$$\Delta t = \frac{2}{\sqrt{\lambda_{max}} + \lambda_{min}} \quad \text{and} \quad \alpha = 2\sqrt{\frac{\lambda_{max}\lambda_{min}}{\lambda_{max} + \lambda_{min}}} \tag{3.64}$$

In terms of the periods T_{max} and T_{min} corresponding to λ_{max} and λ_{min}

$$T_{min} = 2\pi / \sqrt{\lambda_{max}} \quad T_{max} = 2\pi / \sqrt{\lambda_{min}} \tag{3.65}$$

the optimum convergence is obtained when:

$$\Delta t = \frac{T_{min} T_{max}}{\pi \sqrt{T_{min}^2 + T_{max}^2}} \quad \text{and} \quad \alpha = \frac{4\pi}{\sqrt{T_{min}^2 + T_{max}^2}} \tag{3.66}$$

In the particular case of $T_{max} \gg T_{min}$, Eq. 66 becomes:

$$\Delta t = T_{min} / \pi = 2/\omega_{max} \quad \text{and} \quad \alpha = 4\pi / T_{max} = 2\omega_{min} \tag{3.67}$$

In the particular case of SDOF, $T_{max} = T_{min} = T$, Eq. 67 becomes:

$$\Delta t = \frac{T}{\pi\sqrt{2}} = \sqrt{2}/\omega \quad \text{and} \quad \alpha = \frac{4\pi}{T\sqrt{2}} = \sqrt{2}\omega \tag{3.68}$$

which implies that $\alpha\Delta t = 2$.
The critical damping ratio for each mode varies as follows:

$$\xi_i = \frac{\alpha}{2\omega_i} = \frac{1}{\omega_i}\sqrt{\frac{\lambda_{max}\lambda_{min}}{\lambda_{max} + \lambda_{min}}} = \frac{1}{\omega_i}\frac{\omega_{max}\omega_{min}}{\sqrt{\omega_{max}^2 + \omega_{min}^2}} \tag{3.69}$$

The minimum and maximum values of critical damping ratio are:

$$\xi_{max} = \frac{\omega_{max}}{\sqrt{\omega_{max}^2 + \omega_{min}^2}} \quad \text{and} \quad \xi_{min} = \frac{\omega_{min}}{\sqrt{\omega_{max}^2 + \omega_{min}^2}} \tag{3.70}$$

In summary, the properties of the ADR algorithm depend on the spectral bandwidth of the dynamic system. The minimum period controls the algorithm stability, while the maximum period controls the optimal rate of convergence.

3.1.8.1 Example 4

The problem of Example 3 can be solved by using ADR. Figure 30 lists the formulas used in an Excel spreadsheet to implement the ADR algorithm. The parameters are determined by using Eq. 66. Figure 31 shows the convergence of the solution toward the static solution with time, and Fig. 32 shows the decrease of error with the number of time steps. As shown in Fig. 32, the ADR solution converges toward the numerical solution in about 8 steps, which is much faster than the DR solution of Fig. 27.

	A	B	C	D
1	$d_0 = 0$		$k = 10$	
2	$v_0 = 0$		$n = 4$	
3	$F_0 = 40$		$\alpha\Delta t/2 = 1$	
4			Solution = =(F0/k)^(1/n)	
5	t	d	vp	Δt
6	0	=d0	=v0	0.3
7	=A6+D6	=B6+C7*D6	=D6*(F0-k*B6^n)/2	=SQRT(2/(k*n*ABS(B7)^(n-1)))
8	=A7+D7	=B7+C8*D7	=(1-aDT)/(1+aDT)*C7+D7/(1+aDT)*(F0-k*B7^n)	=SQRT(2/(k*n*ABS(B8)^(n-1)))
9	=A8+D8	=B8+C9*D8	=(1-aDT)/(1+aDT)*C8+D8/(1+aDT)*(F0-k*B8^n)	=SQRT(2/(k*n*ABS(B9)^(n-1)))

Fig. 30. Application of ADR to linear SDOF.

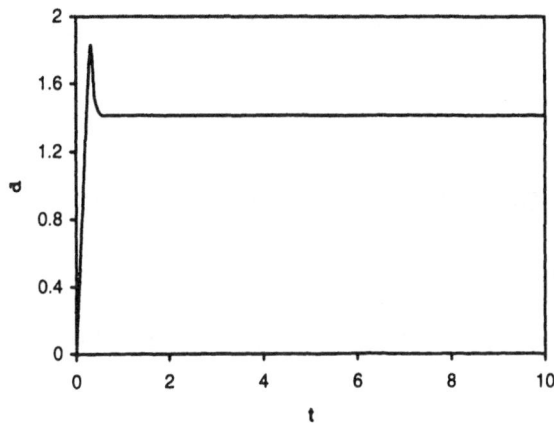

Fig. 31. Convergence of ADR solution toward static solution.

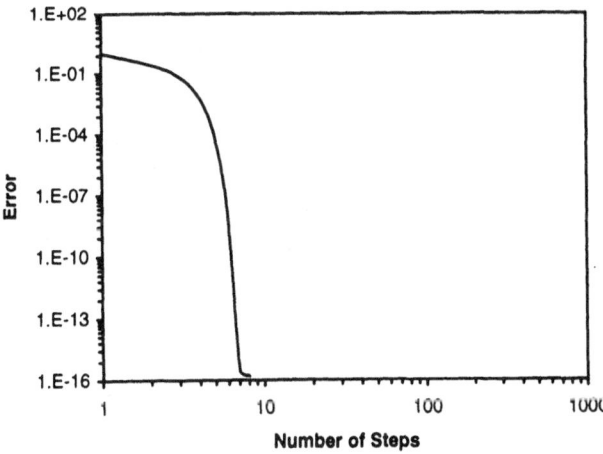

Fig. 32. Decrease of error E_n with number of time steps.

3.1.9 Automatic adjustment of DR parameters

The DR parameters can be adjusted as suggested by Papadrakakis. An upper bound for the maximum eigenvalue may be determined from the Gershgorin bound theorem, which states:

$$|\lambda_{max}| < \max_i \sum_{j=1}^{N} |A_{ij}| \tag{3.71}$$

$$\mathbf{d}_{n+1} - \mathbf{d}_n = \lambda_{DR}(\mathbf{d}_n - \mathbf{d}_{n-1}) \tag{3.72}$$

$$\lambda_{DR} = \|\mathbf{d}_{n+1} - \mathbf{d}_n\|/\|\mathbf{d}_n - \mathbf{d}_{n-1}\| \tag{3.73}$$

When λ_{DR} has converged to an almost constant value, it means that the dominant eigenvalue is

$$\lambda_{min} = -\frac{\lambda_{DR}^2 - (\beta+1)\lambda_{DR} + \beta}{\kappa\lambda_{DR}} = -\frac{(2+\alpha\Delta t)\lambda_{DR}^2 - 4\lambda_{DR} + (2-\alpha\Delta t)}{2\Delta t^2 \lambda_{DR}} \tag{3.74}$$

Papadrakakis [160] suggests defining the optimum convergence rate by using the value of the parameter χ:

$$\chi = -Ln\left[\left(\sqrt{\lambda_{max}} - \sqrt{\lambda_{min}}\right)/\left(\sqrt{\lambda_{max}} + \sqrt{\lambda_{min}}\right)\right] \tag{3.75}$$

The current rate of convergence of the iteration process is defined by $-Ln(\lambda_{DR})$. Therefore the ratio of current to optimal convergence rate is:

$$Q = -\frac{Ln(\lambda_{DR})}{\chi} \tag{3.76}$$

4. Transitions from discrete to continuous media

The passage from discrete to continuous quantities is a necessary step for interpreting the behavior of particulate material in terms of continuum mechanics. Hereafter, we will examine some basic transitions from discrete to continuous quantities, including stress, displacement gradient, finite and infinitesimal strains, void ratio, volumetric strain, macrorotation, interpolated displacement, and fabric tensor. We will also examine the rolling-sliding at particle contacts, the corresponding energy dissipation, and the fabric tensor.

4.1 Calculation of stress

4.1.1 Background

Many researchers have worked on stresses in granular media [e.g., 10, 18, 38, 43, 52, 81, 111, 131, 171, 177]. The definition of stress in granular media remains a controversial topic. Some researchers [e.g., 24, 38, 111, 143] claim that the Cauchy stress tensor is not symmetric in granular media, and that there are couple stresses, and micropolar effects which are important to understand material instability such as shear bands. Others [e.g., 43, 52] state that these micropolar effects are absent or can be neglected for all practical purposes. In this chapter, we will not be able to resolve these issues related to stress in granular material, but will limit our presentation to two basic definitions of stress in idealized granular materials.

4.1.2 Definition of granular media

As previously defined , a granular medium is made of N particles in Volume V. N_I particles are free of external forces or moments, and N_E particles have external forces or moments. The particles are defined by the following set and subsets:

$$B_I = \{1,...,N_I\}, B_E = \{N_I,...,N\}, \quad \text{and} \quad B = B_I \cup B_E = \{1,...,N\} \tag{4.1}$$

The contacts are characterized by M points. There are characterized by the set $C = \{M$ points$\}$ and the two subsets $I = \{$points between two particles of $B\} = \{1, ..., M_I\}$ and $E = \{$points with external actions$\} = \{M_{I+I}, ..., M\}$. The subsets I and E are complementary, i.e.:

$$C = I \cup E = \{1,..., M\} \tag{4.2}$$

Similar subsets can be defined for each particle a, e.g., $I_a = \{$points of I on particle $a\}$, and $E_a = \{$points of E on particle $a\}$. These sets have the following properties

$$I = \bigcup_{a \in B} I_a \quad \text{and} \quad E = \bigcup_{a \in B} E_a \tag{4.3}$$

and

$$E_a \cap E_b = \varnothing \quad \text{and} \quad I_a \cap I_b = \{c\} \text{ for } \forall \; a \neq b \in B \tag{4.4}$$

The equilibrium of forces on particle a, and that of moments about center of particle a imply that:

$$\sum_{c\in I_a} f_i^{ac} + \sum_{e\in E_a} f_i^{ae} = 0 \quad i = 1,2,3 \tag{4.5}$$

$$\sum_{c\in I_a} e_{ijk} r_j^{ac} f_k^{ac} + \sum_{e\in E_a} e_{ijk} r_j^{ae} f_k^{ae} = 0 \quad i = 1,2,3 \tag{4.6}$$

where f_i^{ac} = internal force at contact c, and f_i^{ue} = external force at point e, and e_{ijk} = alternating tensor, x_j^a, x_j^b and x_j^c = coordinates of points a, b, and c, and $r_j^{ac} = x_j^c - x_j^a$, and $r_j^{ae} = x_j^e - x_j^a$ $j = 1,2,3$

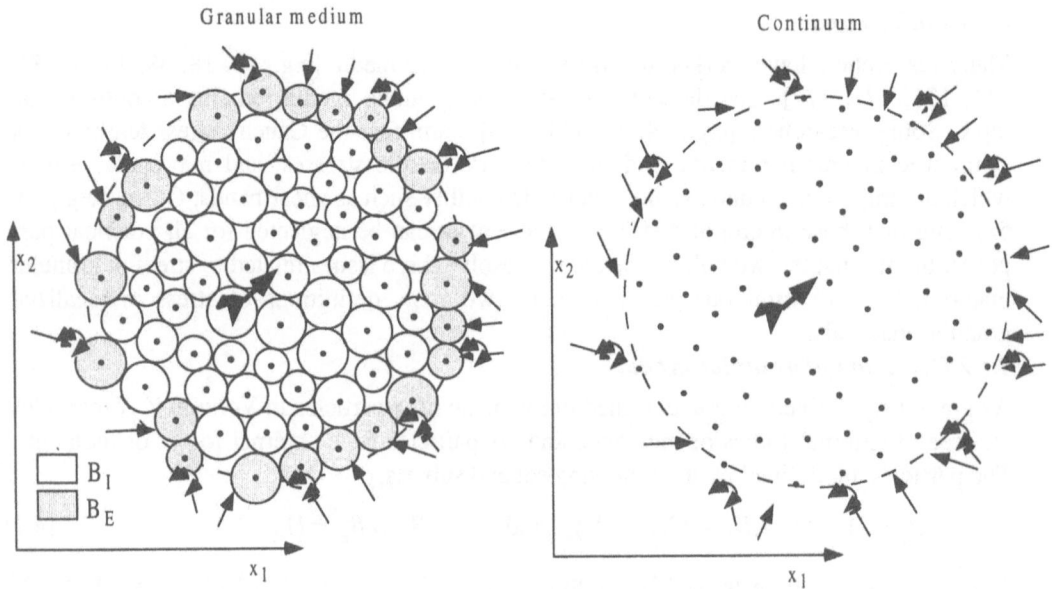

Fig. 33. *Granular medium, and its equivalent continuum.*

4.1.3 Stress and virtual work in continuum
By definition, the Cauchy stress σ_{ij} is related to the traction vector T_i through:

$$T_i = \sigma_{ji} n_j \tag{4.7}$$

where n_i is the unit normal vector of the surface on which T_i is applied. In the absence of body force and couple stresses, the equilibrium equations are:

$$\sigma_{ij,i} = 0 \text{ and } \sigma_{ij} = \sigma_{ji} \tag{4.8}$$

We will invoke the principle of virtual work, which states that the virtual work δW_E of external force should be equal to the virtual work δW_I of internal forces:

$$\delta W_E + \delta W_I = 0 \tag{4.9}$$

The internal and external virtual work are [78, 79]:

$$\delta W_E = \int_S T_i \delta u_i \, dS, \quad \text{and} \quad \delta W_I = \int_V \sigma_{ij} \left(\delta u_{i,j} - \delta \omega_{ij} \right) dV \tag{4.10}$$

where $\delta \omega_{ij}$ is the particle rotation:

$$\delta \omega_{ij} = -e_{ijk} \delta \omega_k \tag{4.11}$$

The virtual strain and virtual rigid rotation are:

$$\delta \varepsilon_{ij} = \delta u_{(i,j)} = \frac{1}{2} \left(\frac{\partial \delta u_i}{\partial x_j} + \frac{\partial \delta u_j}{\partial x_i} \right) \qquad \delta w_{ij} = \delta u_{[i,j]} = \frac{1}{2} \left(\frac{\partial \delta u_i}{\partial x_j} - \frac{\partial \delta u_j}{\partial x_i} \right) \tag{4.12}$$

The continuous field is generated by the displacement of the particle centers and by the particle rigid rotation. The translation and rotation components of the virtual field are $\delta u_i(x_i)$ = displacement and $\delta \omega_i(x_i)$ = rotation. These fields can be approximated by using a Taylor expansion in a small neighborhood as follows:

$$\delta u_i = a_i + b_{ij} x_j + ..., \quad \text{and} \quad \delta \omega_i = \alpha_i + ... \tag{4.13}$$

where a_i, b_{ij}, and α_i are arbitrary constant coefficients. The strain corresponding to the variational field is:

$$\delta u_{i,j} + e_{ijk} \delta \omega_k = b_{ij} + e_{ijk} \alpha_k + ... \tag{4.14}$$

The virtual work can be expressed as:

$$\delta W_I = \frac{1}{V} \int_V \sigma_{ij} \left(b_{ij} + e_{ijk} \alpha_k \right) dV + ... \tag{4.15}$$

or equivalently:

$$\delta W_I = b_{ij} \frac{1}{V} \int_V \sigma_{ij} dV + \alpha_j \frac{1}{V} \int_V \sigma_{ik} e_{ikj} dV + ... \tag{4.16}$$

Equation 16 can be written:

$$\delta W_I = b_{ij} \overline{\sigma}_{ij} + \alpha_j e_{ikj} \overline{\sigma}_{ik} + ... \tag{4.17}$$

and is useful to introduce the average Cauchy stress tensor in volume V as follows:

$$\overline{\sigma}_{ij} = \frac{1}{V} \int_V \sigma_{ij} dV \quad i, j = 1, 2, 3 \tag{4.18}$$

The symmetry of σ_{ij} can be examined by considering the second term of Eq. 17:

$$\frac{1}{V} \int_V \sigma_{ik} e_{ikj} dV = \frac{1}{V} e_{ikj} \int_V \sigma_{ik} dV = e_{ikj} \overline{\sigma}_{ik} = \overline{\sigma}_{ij} - \overline{\sigma}_{ji}, \quad i \ne j = 1, 2, 3 \tag{4.19}$$

where $e_{ikj} \overline{\sigma}_{ik} = 0$ for symmetric stress (i.e., $\overline{\sigma}_{ij} = \overline{\sigma}_{ji}$). When $j=3$, Eq. 19 becomes:

$$e_{ik3} \overline{\sigma}_{ik} = e_{123} \overline{\sigma}_{12} + e_{213} \overline{\sigma}_{21} = \overline{\sigma}_{12} - \overline{\sigma}_{21} \tag{4.20}$$

4.1.4 Kinematics and virtual work in granular medium

As illustrated in Fig. 33, the virtual displacement and rotation are assumed identical in the granular medium and the continuum at the grain centers:

$$\delta u_i^a = \delta u_i(x_j^a), \quad \text{and} \quad \delta \omega_i^a = \delta \omega_i(x_j^a) \quad i,j = 1,2,3 \; \forall a \in B \tag{4.21}$$

where δu_i^a is the displacement vector of particle center, and $\delta \omega_i^{\cdots}$ is the rotation of particles located at $x_j^{\cdots}$. The relative displacement of particles a and b at contact c is:

$$\Delta \delta u_i^c = \delta u_i^b - \delta u_i^a + e_{ijk}\left(\delta \omega_j^b r_k^{bc} - \delta \omega_j^a r_k^{ac}\right) \tag{4.22}$$

where:

$$r_i^{ac} = x_i^c - x_i^a, \quad \text{and} \quad r_i^{bc} = x_i^c - x_i^b \tag{4.23}$$

Equation 22 can be rewritten:

$$\Delta \delta u_i^c = \delta u_i^b - \delta u_i^a + e_{ijk}\left(\delta \omega_j^b \left(x_k^c - x_k^b\right) - \delta \omega_j^a \left(x_k^c - x_k^a\right)\right) \tag{4.24}$$

The relative rotation of particles a and b is:

$$\Delta \delta \omega_i^c = \delta \omega_i^b - \delta \omega_i^a \tag{4.25}$$

The internal and external works are:

$$\delta W_I^D = \sum_{c \in I} f_i^c \Delta \delta u_i^c, \quad \text{and} \quad \delta W_E^D = \sum_{e \in E} f_i^e \delta u_i^e \tag{4.26}$$

By using the Taylor expansion of Eq. 13, the relative displacement and relative rotation at contact point c becomes:

$$\Delta \delta u_i^c = b_{ij}\left(x_j^b - x_j^a\right) + \alpha_j e_{ijk}\left(x_k^a - x_k^b\right) + \dots \quad \text{and} \quad \Delta \delta \omega_i^c = 0 + \dots \tag{4.27}$$

Therefore the internal work is:

$$\delta W_I^D = b_{ij} \sum_{c \in I} f_i^c \left(x_j^b - x_j^a\right) + \alpha_j \sum_{c \in I} f_i^c e_{ijk}\left(x_k^a - x_k^b\right) \tag{4.28}$$

4.1.5 Definition of average stresses

The average stress is defined by stating that the virtual work in the continuum is equal to the virtual work in the granular medium for equivalent virtual displacement, i.e.,

$$\delta W_I^D = \delta W_I \tag{4.29}$$

Therefore, for arbitrary values of b_{ij},, the average Cauchy stress is:

$$\overline{\sigma}_{ij} = \frac{1}{V} \sum_{c \in I} f_i^c \left(x_j^b - x_j^a\right) \quad i,j = 1,2,3 \tag{4.30}$$

Equation 30 can also be written as follows:

$$\overline{\sigma}_{ij} = \frac{1}{V} \sum_{c \in I} f_i^c l_j^c \quad i,j = 1,2,3 \tag{4.31}$$

where:

$$l_j^c = x_j^b - x_j^a \quad j = 1,2,3 \tag{4.32}$$

Equation 31 is identical to those obtained in [43, 81, 171]. In the case of spherical (or cylindrical) particles, Eq. 31 becomes:

$$\bar{\sigma}_{ij} = \frac{1}{V} \sum_{c \in I} f_i^c (R_a + R_b) n_j^c \quad i,j = 1,2,3 \tag{4.33}$$

where n_i^c is the unit normal vector of contact c, and $l_j^c = (R_a + R_b) n_j^c \quad j = 1,2,3$. Equation 33 can also be rewritten:

$$\bar{\sigma}_{ij} = \frac{1}{V} \sum_{a \in B} \sum_{c \in I_a} f_i^{ac} R_a n_j^{ac} \quad i,j = 1,2,3 \tag{4.34}$$

after invoking the following result:

$$f_i^{ca} n_j^{ca} = (-f_i^{cb})(-n_j^{cb}) = f_i^{cb} n_j^{cb} \tag{4.35}$$

The symmetry of the average Cauchy stress can be examined within our simplified derivation. For arbitrary values of α_{ij},:

$$e_{ikj} \bar{\sigma}_{ik} = \frac{1}{V} \sum_{c \in I} e_{ijk} f_i^c \left(x_k^a - x_k^b \right) \quad j = 1,2,3 \tag{4.36}$$

After summing the equilibrium equations of moments on all the particles, one obtains:

$$\sum_{a \in B} \sum_{c \in I_a} \left(e_{ijk} r_j^{ac} f_k^{ac} \right) + \sum_{a \in B} \sum_{e \in E_a} \left(e_{ijk} r_j^{ae} f_k^{ae} \right) = 0 \quad i = 1,2,3 \tag{4.37}$$

which is equivalent to:

$$\sum_{c \in I} \left(e_{ijk} (r_j^{ac} f_k^{ac} + r_j^{bc} f_k^{bc}) \right) + \sum_{e \in E} \left(e_{ijk} r_j^e f_k^e \right) = 0 \quad i = 1,2,3 \tag{4.38}$$

Due to the fact that the forces at contact c between particles a and b have opposite sign:

$$f_i^{ac} = -f_i^{bc} \quad i = 1,2,3 \tag{4.39}$$

one obtains:

$$\sum_{c \in I} \left(e_{ijk} f_k^c (r_j^{ac} - r_j^{bc}) \right) + \sum_{e \in E} \left(e_{ijk} r_j^e f_k^e \right) = 0 \quad i = 1,2,3 \tag{4.40}$$

The distance between the particle center is also:

$$r_j^{ac} - r_j^{bc} = x_j^a - x_j^b \quad j = 1,2,3 \tag{4.41}$$

Therefore:

$$\sum_{c \in I} \left(e_{ijk} f_k^c (x_j^a - x_j^b) \right) + \sum_{e \in E} \left(e_{ijk} r_j^e f_k^e \right) - 0 \quad i = 1,2,3 \tag{4.42}$$

In summary, one obtains the following result:

$$e_{ijk}\overline{\sigma}_{jk} = \frac{1}{V}\sum_{e\in E}\left(e_{imn}r_m^e f_n^e\right) \quad i=1,2,3$$ (4.43)

When $i=3$, Eq. 43 becomes:

$$\overline{\sigma}_{12} - \overline{\sigma}_{21} = \frac{1}{V}\sum_{e\in E}\left(e_{3jk}r_j^e f_k^e\right)$$ (4.44)

Equation 43 implies that the average stress tensor is not necessarily symmetric due to the moments induced by the contact forces on the external boundary surface. The term $\sigma_{ij} - \sigma_{ji}$, which quantifies the magnitude of non-symmetry, decreases as the volume V gets larger. When $V \to \infty$, the surface terms are negligible, and the symmetry of the average stress tensor is restored. In the case of spherical and cylindrical particles, Eq. 43 becomes:

$$\overline{\sigma}_{ij} - \overline{\sigma}_{ji} = \frac{1}{V}\sum_{a\neq b\in B \text{ and } I_a\cap I_b\in I}\left(R_a f_i^{ab} n_j^{ab} - R_b f_i^{ba} n_j^{ba}\right)$$

$$= \frac{1}{V}\sum_{a\neq b\in B}(R_a - R_b)f_i^{ab} n_j^{ab} \quad i \neq j = 1,2,3$$ (4.45)

Equation 45 implies that the symmetry is restored when the particles have the same radius.

4.1.6 Alternate definition of stress

The average stress in a granular medium was previously derived by invoking the rational principle of virtual work. It can also be derived in a different way by using a purely static approach [52].

The average stress in the volume V is defined as the weighted average of the average stress in each particle:

$$\overline{\sigma}_{ij}^* = \frac{1}{V}\sum_{a\in B}\overline{\sigma}_{ij}^a V_a$$ (4.46)

The average stress tensor with particle a is:

$$\overline{\sigma}_{ij}^a = \frac{1}{V_a}\int_{V_a}\sigma_{ij}dV$$ (4.47)

where V_a is the volume of particle a. In order to exhibit the contact forces, the volume integral of Eq. 47 is transformed into a surface integral. The mathematical derivations are as follows. First, one can notice that:

$$\sigma_{ij} = \delta_{ik}\sigma_{kj} = \frac{\partial x_i}{\partial x_k}\sigma_{kj} = \left(x_i\sigma_{kj}\right)_{,k} - x_i\sigma_{kj,k} = \left(x_i\sigma_{kj}\right)_{,k}$$ (4.48)

Therefore, the average stress within particle a is:

$$\overline{\sigma}_{ij}^a = \frac{1}{V_a}\int_{V_a}\left(\sigma_{ij}x_k\right)_{,k}dV = \frac{1}{V_a}\int_{S_a}x_i\sigma_{kj}n_k dS = \frac{1}{V_a}\int_{S_a}x_i T_j dS$$ (4.49)

After replacing the integral by discrete sum at the contact points, one obtains:

$$\bar{\sigma}_{ij}^a = \frac{1}{V_a} \sum_{c \in C_a} x_i^c f_j^c$$

(4.50)

The equilibrium of moments about the origin implies that:

$$\sum_{c \in C_a} x_i^a f_j^c = x_i^a \sum_{c \in C_a} f_j^c = 0$$

(4.51)

For spherical and cylindrical particles, Eq. 50 becomes:

$$\bar{\sigma}_{ij}^a = \frac{1}{V_a} \sum_{c \in C_a} (x_i^a + R_a n_i^c) f_j^c = \frac{1}{\pi R_a} \sum_{c \in C_a} n_i^c f_j^c$$

(4.52)

where R_a is the radius of particle a, and $\tilde{x}_i = x_i^- + R_a n_i^-$. Therefore the average stress with particle a is:

$$\bar{\sigma}_{ij}^a = \frac{1}{\pi R_a} \sum_{c \in C_a} n_i^c f_j^c$$

(4.53)

and the average stress within volume V is:

$$\bar{\sigma}_{ij}^* = \frac{1}{\pi R^2} \sum_{a \in B} R_a \sum_{c \in C_a} n_i^c f_j^c$$

(4.54)

This result coincides with those obtained in [52]. The symmetry of the average stress is examined by considering the following term:

$$e_{ijk} \bar{\sigma}_{jk}^a = \frac{1}{V_a} \sum_{c \in C_a} e_{ijk} x_j^c f_k^c = \frac{1}{V_a} \left(\sum_{c \in C_a} e_{ijk} r_j^c f_k^c + x_j^a \sum_{c \in C_a} e_{ijk} f_k^c \right) = 0 \quad i = 1, 2, 3$$

(4.55)

but

$$\sum_{c \in C_a} f_i^c = 0, \text{ and } \sum_{c \in C_a} e_{ijk} r_j^c f_k^c = 0 \quad i = 1, 2, 3$$

(4.56)

and:

$$x_j^c = x_j^a + r_j^c \quad j = 1, 2, 3$$

(4.57)

Therefore, the stress tensor is symmetric because it is the sum of symmetric tensors.

4.1.7 Comparison of stress

The average stress defined from statics is slightly different from the one obtained from variational principle:

$$\bar{\sigma}_{ij} = \frac{1}{V} \sum_{a \in B} \sum_{c \in C_a} x_i^c f_j^c = \frac{1}{V} \sum_{c \in I} (x_i^a - x_i^b) f_j^c + \frac{1}{V} \sum_{e \in E} x_i^e f_j^e \quad i, j = 1, 2, 3$$

(4.58)

The relation between the two average stresses is:

$$\bar{\sigma}_{ij}^* = \bar{\sigma}_{ij} + \frac{1}{V} \sum_{e \in E} x_i^e f_j^e \quad i, j = 1, 2, 3$$

(4.59)

4.2 Displacement gradient

The average displacement gradient F_{ij} in the volume V is defined as:

$$F_{ij} = \frac{1}{V} \int_V \frac{\partial u_i}{\partial x_j} dV \tag{4.60}$$

where u_i represents the displacement field. Applying again Gauss' divergence theorem, Eq. 60 becomes:

$$F_{ij} = \frac{1}{V} \int_S u_i n_j dS = \frac{1}{V} \int_S u_i n_j dS = \frac{1}{V} \sum_{k=1}^{M} \int_{S_k} u_i n_j dS \tag{4.61}$$

where M is the number of disks intersected by the perimeter S of averaging area V. As shown in Fig. 34, S_k is made of the chords AB and BC of S. In order to evaluate Eq. 61, the displacement field u_i needs to be defined in the interstices between particles. Since the displacement field needs to be continuous, it is assumed to vary linearly between B and C, leading to:

$$\int_{S_k} u_i n_j dS = \frac{AB}{2} n_j^{AB} \left(u_i^A + u_i^B \right) + \frac{BC}{2} n_j^{BC} \left(u_i^B + u_i^C \right) \tag{4.62}$$

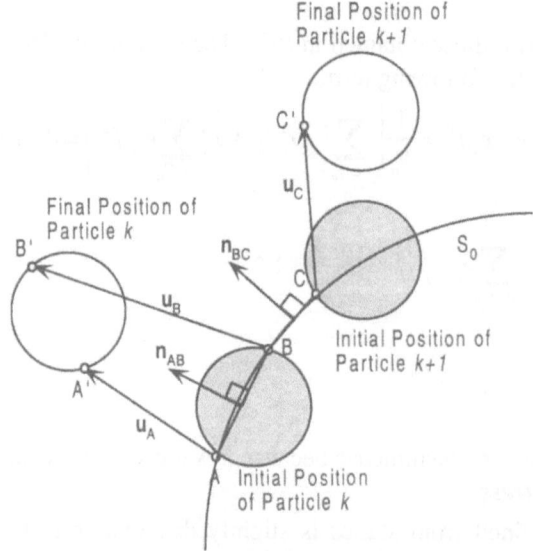

Fig. 34. Calculation of average displacement gradient from the displacements of material points on the boundary.

where u_i^A, u_i^B and u_i^C are the displacement vectors at A, B and C of disks k and k+1, respectively, and n_i^{AB} and n_i^{BC} are the unit vectors normal to segments AB and BC. The gradient of incremental displacement ΔF_{ij} is defined as F_{ij}, but using the incremental displacement Δu_i instead of u_i.

4.3 Alternate definition of displacement gradient

Equation 61 defines the average displacement gradient $\mathbf{F}$ by invoking the divergence theorem. Another technique [8] was proposed to calculate the local displacement gradient $\mathbf{G}$ about a particle. This alternate definition $\mathbf{G}$ was used in describing the rotation and volumetric strain within shear bands.

As shown in Fig. 35, a particle initially at position $\mathbf{X}$ has m neighbors in the reference configuration. Each neighbor is characterized by the relative position of its center $\mathbf{dX_i}$. In the deformed position, the particle center is at position $\mathbf{x}$ while the relative displacements of its neighbors are $\mathbf{dx_i}$. By definition, the local displacement gradient is the linear operator $\mathbf{G}$ that transforms $\mathbf{dX_i}$ into $\mathbf{dx_i}$:

$$\mathbf{dx_i} = \mathbf{G}\ \mathbf{dX_i} \qquad (4.63)$$

Since Eq. 63 is generally not satisfied for all particles when m is larger than 2, $\mathbf{G}$ can be calculated by linear regression. Unlike $\mathbf{F}$, $\mathbf{G}$ ignores the particle rotations since it is defined from the relative displacement of the particle centers.

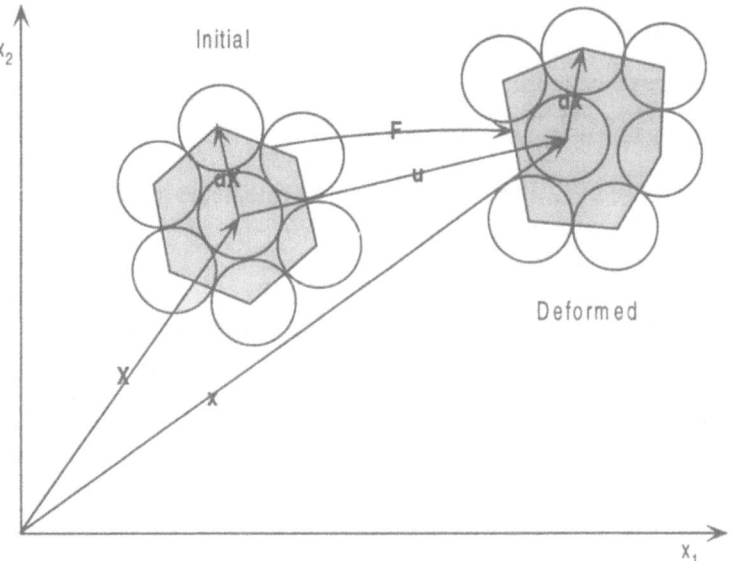

Fig. 35. Calculation of displacement gradient by linear regression.

4.4 Macro-rotations

A macro-rotation characterizes the rigid body rotation of a particle neighborhood. It is calculated by decomposing the deformation gradient:

$$\mathbf{G} = \mathbf{V}\ \mathbf{R} \qquad (4.64)$$

where $\mathbf{R}$ is a rotation matrix (i.e., $\mathbf{R}^T\mathbf{R} = I$) and $\mathbf{V}$ is a symmetric matrix (i.e., $\mathbf{V}^T = \mathbf{V}$). The eigenvalues of $\mathbf{V}$ are found as follows:

$$\begin{vmatrix} V_{11} - \lambda & V_{12} \\ V_{21} & V_{22} - \lambda \end{vmatrix} = \lambda^2 - (V_{11} + V_{22})\lambda + V_{11}V_{22} - V_{12}V_{21} = 0 \tag{4.65}$$

which leads to the characteristic equation:

$$\lambda^2 - trace(\mathbf{V})\lambda + det(\mathbf{V}) = 0 \tag{4.66}$$

By invoking the Cayley-Hamilton theorem, $\mathbf{V}$ satisfies its characteristic equation:

$$\mathbf{V}^2 - trace(\mathbf{V})\mathbf{V} + det(\mathbf{V})\mathbf{1} = 0 \tag{4.67}$$

therefore:

$$\mathbf{V} = \frac{1}{trace(\mathbf{V})}\left(\mathbf{V}^2 + det(\mathbf{V})\mathbf{1}\right) \tag{4.68}$$

After taking the trace of Eq. 67, one obtains:

$$trace(\mathbf{V}) = \sqrt{trace(\mathbf{V}^2) + 2det(\mathbf{V})} \tag{4.69}$$

However, $\mathbf{G}$ and $\mathbf{V}$ are related through:

$$det(\mathbf{G}) = det(\mathbf{VR}) = det(\mathbf{V}) \quad \text{and} \quad \mathbf{V}^2 = \mathbf{GG}^T \tag{4.70}$$

where $\mathbf{G}^T$ is the transpose of $\mathbf{G}$. In summary, $\mathbf{V}$ can be calculated in closed-form from $\mathbf{G}$:

$$\mathbf{V} = \frac{1}{\sqrt{trace(\mathbf{GG}^T) + 2det(\mathbf{G})}}\left(\mathbf{GG}^T + det(\mathbf{G})\mathbf{1}\right) \tag{4.71}$$

The macro-rotation θ of a particle is obtained from the rotation matrix $\mathbf{R}$:

$$\mathbf{R} = \mathbf{V}^{-1}\mathbf{G}\begin{pmatrix} \cos\theta & \sin\theta \\ -\sin\theta & \cos\theta \end{pmatrix} \tag{4.72}$$

In the case of infinitesimal deformation, $\mathbf{R}$ is approximately the antisymmetric part of $\mathbf{G}$ and the infinitesimal macro-rotation θ is:

$$\theta \approx w_{21} = \frac{1}{2}(G_{21} - G_{12}) \tag{4.73}$$

4.5 Strain

In the case of finite deformation, the Lagrangian strain E_{ij} is:

$$E_{ij} = \frac{1}{2}\left(F_{ij} + F_{ji} + F_{ki}F_{kj}\right) \tag{4.74}$$

In the case of small deformation, E_{ij} coincides with the infinitesimal strain ε_{ij}:

$$\varepsilon_{ij} = \frac{1}{2}\left(F_{ij} + F_{ji}\right) \tag{4.75}$$

The incremental strain $d\varepsilon_{ij}$ is :

$$d\varepsilon_{ij} = \frac{1}{2}\left(dF_{ij} + dF_{ji}\right)$$ (4.76)

4.6 Gradient of particle rotation (curvature)

The particle rotations are also characterized by the gradient κ_i, referred to as the curvature tensor in the Cosserat theory [143]:

$$\kappa_i = \theta^c_{,i}$$ (4.77)

In the same way as the local deformation gradient, κ_i is calculated by using a linear regression of the relative rotation of the particles surrounding a given particle.

4.7 Void ratio

The numerical determination of e is inspired from the γ-ray absorption technique that measures the material density in the path of a beam. As shown in Fig. 36, the void ratio e in the circular sample of radius r and center M is:

$$e = \frac{\pi r^2 - \sum_{i=1}^{n} A_i}{\sum_{i=1}^{n} A_i}$$ (4.78)

where n is the total number of particles and A_i is the area of the ith particle covering the sampling area:

$$A_i = \begin{cases} 0 & \text{if } d_i > R_i + r \\ \pi R_i^2 & \text{if } d_i \geq r - R_i \\ (\theta_2 - \sin\theta_2)R_i^2/2 + (\theta_1 - \sin\theta_1)r^2/2 & \text{if } r - R_i < d_i \leq R_i + r \end{cases}$$ (4.79)

where d_i is the distance between M and the center of the ith particle, and where:

$$\theta_2 = 2\cos^{-1}\left\{\frac{d_i^2 + R_i^2 - r^2}{2d_i R_i}\right\} \quad \text{and} \quad \theta_1 = 2\sin^{-1}\left\{\frac{R_i}{r}\sin\frac{\theta_2}{2}\right\}$$ (4.80)

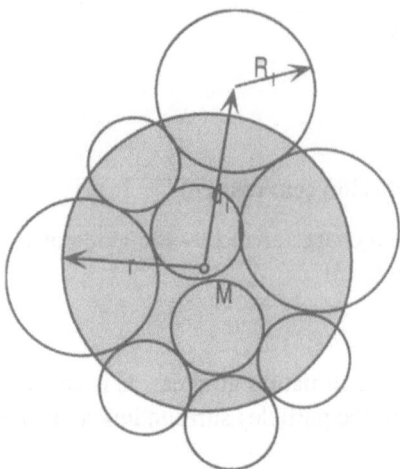

Fig. 36.   Calculation of local void ratio.

4.8 Volumetric Strain

The volumetric strain ε_v is:

$$\varepsilon_v = \frac{\Delta V}{V_0} = \det(\mathbf{F}) - 1 \tag{4.81}$$

where ΔV is the variation in volume of the initial volume V_0. In the case of infinitesimal strain, the volumetric strain is approximated as:

$$\varepsilon_v = \varepsilon_{11} + \varepsilon_{22} \tag{4.82}$$

The volumetric strain can also calculated from the void ratio, similarly to the measurement technique in the triaxial tests of soil mechanics:

$$\varepsilon_v = \frac{\Delta V}{V_0} = \frac{e - e_0}{1 + e_0} \tag{4.83}$$

where e is the present void ratio and e_o is the initial void ratio at $\varepsilon_{22} = 0\%$.

4.9 Interpolated and projected average displacement

In discrete granular media, a continuous displacement field $\mathbf{u}(X,Y)$ is calculated by linearly interpolating the discrete displacement $\mathbf{u}(X_j,Y_j)$ of the particle centers. Besides the linear interpolation, no other approximation is made for defining this continuous field. Any point M initially at position (X,Y) within the sample belongs to a particle or to a void between particles. In any case, it belongs to a triangle generated by three particles. Within the smallest triangle containing M, the x-component of displacement is assumed to be linear:

$$u_x(X,Y) = a X + b Y + c \tag{4.84}$$

where the 3 constants a, b and c are found by solving the following linear system:

$$\begin{cases} u_1 = aX_1 + bY_1 + c \\ u_2 = aX_2 + bY_2 + c \\ u_3 = aX_3 + bY_3 + c \end{cases} \tag{4.85}$$

where (X_1, Y_1), (X_2, Y_2) and (X_3, Y_3) are the initial position of the centers of the 3 particles surrounding M, and u_1, u_2 and u_3 are the x-components of their displacement. The same procedure is repeated in the y-direction.

Once the interpolated displacement is defined, many other quantities can be calculated. For instance, the displacement can be projected perpendicular to the direction **n** as follows:

$$u_\xi(\xi, \eta) = u_x n_y - u_y n_x \tag{4.86}$$

where n_x and n_y represent the unit vector normal **n**.

4.10 Rolling and sliding contacts

In granular media, the particle contacts not only slide but also roll. The amount of sliding and rolling displacement at the contacts between two particles can be determined as follows.

Figure 37 shows two particles that remain in contact between times t and $t + \Delta t$. At time t, the contact point C is at position $\mathbf{X_C}$:

$$\mathbf{X}_c = \frac{1}{R_1 + R_2}(R_1\mathbf{X}_2 + R_2\mathbf{X}_1) \tag{4.87}$$

where $\mathbf{X}_1$ and $\mathbf{X}_2$ are the positions of the particle centers, and R_1 and R_2 are their radii. At time $t + \Delta t$, the contact point C' is at position $\mathbf{X}_c$:

$$\mathbf{X}_c = \frac{1}{R_1 + R_2}(R_1\mathbf{X}_1' + R_2\mathbf{X}_2') \tag{4.88}$$

where $\mathbf{X}_1$ and $\mathbf{X}_2$ are the positions of the particle centers at time $t+\Delta t$. At $t+\Delta t$, the point C is now at C_1 on particle 1, and at C_2 on particle 2; C_1 and C_2 are at positions $\mathbf{X}_{C_1}$ and $\mathbf{X}_{C_2}$:

$$\mathbf{X}_{C_1}' = \mathbf{X}_1' + \mathbf{T}_1(\mathbf{X}_c - \mathbf{X}_1) \quad \text{and} \quad \mathbf{X}_{C_2}' = \mathbf{X}_2' + \mathbf{T}_2(\mathbf{X}_c - \mathbf{X}_2) \tag{4.89}$$

where T_1 and T_2 are the rotation matrices corresponding to the particle rotations $\Delta\theta_1$ and $\Delta\theta_2$ between times t and $t + \Delta t$:

$$\mathbf{T}_1 = \begin{pmatrix} \cos\Delta\theta_1 & -\sin\Delta\theta_1 \\ \sin\Delta\theta_1 & \cos\Delta\theta_1 \end{pmatrix} \text{and } \mathbf{T}_2 = \begin{pmatrix} \cos\Delta\theta_2 & -\sin\Delta\theta_2 \\ \sin\Delta\theta_2 & \cos\Delta\theta_2 \end{pmatrix} \tag{4.90}$$

The unit vector normal to the contact at times t and $t + \Delta t$ are **n** and **n'**, respectively:

$$\mathbf{n} = \frac{1}{R_1 + R_2}(\mathbf{X}_2 - \mathbf{X}_1) \quad \text{and} \quad \mathbf{n}' = \frac{1}{R_1 + R_2}(\mathbf{X}_2' - \mathbf{X}_1') \tag{4.91}$$

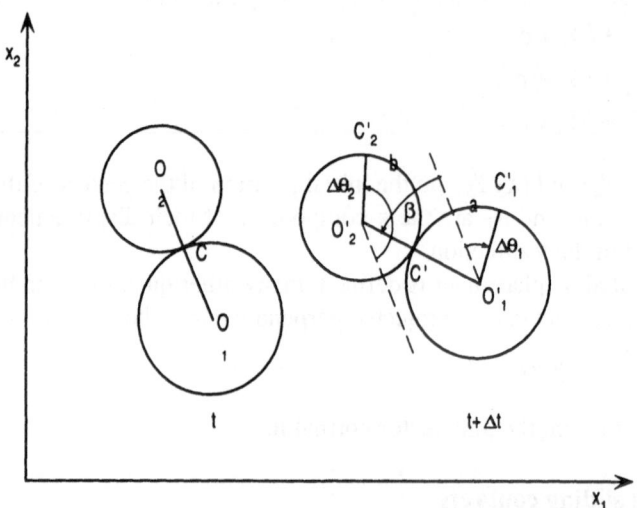

Fig. 37. Definition of a and b for particles in contact at times t and t+Δt.

The arcs CC_1 and CC_2, which represent the displacement of the contact point C on each particle, are denoted a and b, respectively. By using the angles $\Delta\theta_1$, $\Delta\theta_2$, and β shown in Fig. 37:

$$a = R_1|\Delta\theta_1 - \beta| \quad \text{and} \quad b = -R_2|\Delta\theta_2 - \beta|sign\{(\Delta\theta_1 - \beta)(\Delta\theta_2 - \beta)\} \tag{4.92}$$

where a is taken positive by sign convention, and β is the angle between **n** and **n'**. The definitions of Eq. 92 apply to any particle displacements provided that the particles remain in contact between times t and $t + \Delta t$. a and b are uniquely defined from the initial and final particle positions when $|\beta|$ remains smaller than 180°. The general definitions of Eq.92 may be illustrated in two particular cases.

In the case of pure sliding without a particle rotation, i.e., $\Delta\theta_1 = \Delta\theta_2 = 0$,

$$\frac{a}{R_1} = -\frac{b}{R_2} = |\beta| \tag{4.93}$$

As noted in [159], Eq. 93 gives $a=-b$ when $R_1 = R_2$.
In the case of pure rolling without a contact rotation (i.e., $\beta = 0$), the arcs $C'C_1$ and $C'C_2$ have the same length, independently of the particle radii:

$$a = b \tag{4.94}$$

and the particle rotations have opposite signs:

$$R_1 \Delta\theta_1 = -R_2 \Delta\theta_2 \tag{4.95}$$

Purely sliding and purely rolling contacts are particular cases; contacts may be successively and/or simultaneously rolling and sliding. The ratio b/a indicates the relative contact displacement: $b/a=1$ in the case of pure rolling, and $b/a = -R_2/R_1$ in the case of pure sliding.

4.11 Energy dissipation

The relative displacements between particles are responsible for the energy dissipated by frictional contact forces. When the contact action is reduced to a force acting at a point, the energy dissipated by friction forces during a time interval $[t_1, t_2]$ is:

$$\Delta W = \sum_{i=1}^{m} \int_{t_1}^{t_2} F_s^i (da_i - db_i) \qquad (4.96)$$

where m is the number of contacts, F_s^i is the shear forces at the ith contact, and da_i and db_i are the displacements a and b of the ith contact points during a time interval dt. Eq. 96 can be approximated by using a trapezoidal rule:

$$\Delta W = \sum_{i=1}^{m} \frac{1}{2} (F_{s1}^i + F_{s2}^i)(a_i - b_i) \qquad (4.97)$$

where F_{s1}^i and F_{s2}^i are the tangential contact forces at times t_1 and t_2, and $a_i - b_i$ is the total relative displacement of the ith contact point between t_1 and t_2. Similar to fixed contacts ($a = b = 0$), purely rolling contacts ($a = b$) dissipate no energy. In the case of sliding contacts, $a - b$ becomes positive, and compounds with the average shear force to dissipate energy. In general, rolling contacts dissipate much less energy than sliding contacts, except for contrary rolling when particles rotate in the same direction. Equation 96 considers only persistent contacts; it certainly underestimates the dissipated energy by ignoring the deleted and created contacts between t_1 and t_2.

4.12 Fabric tensor

The fabric tensor, sometimes also referred to as anisotropy tensor, is a way to characterize the average orientation of contacts [159, 178, 148]. In the case of cylindrical particles, the fabric tensor is defined as:

$$H_{ij} = \frac{1}{m} \sum_{c=1}^{n} n_i^c n_j^c \qquad (4.98)$$

where n_i^c is the ith component of the cth contact, and m is the total number of contacts. By definition, $H_{11} + H_{22} = 1$ since $(n_1^c)^2 + (n_2^c)^2 = 1$.

5. Acknowledgement

The author acknowledges the financial support of the International Center for Mechanical Sciences (CISM). He thanks B. Cambou for inviting him to lecture at CISM and S. Kaliskzy and his colleagues for their assistance and hospitality in Udine. He also thanks M. Harris for proofreading the manuscript.

6. References

1. **Aizawa, T., S. Tamura, Y. Shibata, J. Okuno, and J. Kihara,** 1993, Powder granular modeling and simulation for agglomeration, *Proceedings of the 6th international symposium on agglomeration,* Nagoya, Japan, 70-75.

2. **Anandarajah, A.,** 1994, Micromechanics of clays evaluated by the discrete element methods, *Proceedings of the 8th International Conference on Computer Methods and Advances in Geomechanics, Morgantown,* VA, H. J. Siriwardane, and M. M. Zaman, Balkema, Rotterdam, The Netherlands, 797-802.

3. **Azarkhin, A.,** 1988, Some history dependent problems for dissimilar cylinders with finite friction, *Journal of Applied Mechanics, ASME,* 55; 81-86.

4. **Bardet, J. P.,** 1994, Numerical simulations of the incremental responses of idealized granular materials, *International Journal of Plasticity,* 10 (8), 879-980.

5. **Bardet, J. P.,** 1994, Observations on the effects of particle rotations on the failure of idealized granular materials, *Mechanics of Materials,* 18, 159-182.

6. **Bardet, J. P.,** 1997, *Experimental Soil Mechanics,* Prentice-Hall, Upper Saddle River, NJ.

7. **Bardet, J. P., and J. Proubet,** 1989, JP2, a program to simulate the behavior of two-dimensional granular materials, *Civil Engineering Department,* University of Southern California, Los Angeles, CA.

8. **Bardet, J. P., and J. Proubet,** 1991, A numerical investigation of the structure of persistent shear bands in granular materials, *Géotechnique,* 41 (4), 599-613.

9. **Bardet, J. P., and J. Proubet,** 1991, Adaptative relaxation technique for the statics of granular materials, *Computers and Structures,* 39 (3/4), 221-229.

10. **Bardet, J. P., and J. Proubet,** 1992, A shear band analysis in idealized granular materials, *Journal of Engineering Mechanics, ASCE,* 118 (2), 397-415.

11. **Bardet, J. P., and Q. Huang,** 1992, Numerical modeling of micropolar effects in idealized granular materials, *in Mechanics of granular materials and powder systems,* M. M. Mehrabadi, ed., *ASME,* 37, 85-92.

12. **Bardet, J. P., and Q. Huang,** 1993, Rotational stiffness of cylindrical particle contacts, *Proceedings of the 2nd International Conference on Micromechanics of Granular Media,* Birmingham, UK, C. Thornton, ed., 39-44.

13. **Bardet, J. P., and R. F. Scott,** 1985, Seismic stability of fractured rock masses with the distinct element method, *Proceedings of the 26th Us Symposium on Rock Mechanics,* Rapid City, SD, E. Ashworth, ed., 139-149.

14. **Bashir, Y. M. and J. D. Goddard,** 1991, A novel simulation method for the quasistatic mechanics of granular assemblages, *J. Rheology,* 35 (5), 849-885.

15. **Bathe, K. J.,** 1996, *Finite element procedures,* Prentice-Hall, Englewood Cliffs, NJ.

16. **Bathurst, R. J., and L. Rothenburg,** 1988, Micromechanical aspects of isotropic granular assemblies with linear contact interactions, *Journal of Applied Mechanics, ASME,* 55, 17-23.

17. **Bathurst, R. J., and L. Rothenburg,** 1988, Note on a random isotropic granular material with negative Poisson's ratio, *International Journal of Engineering Sciences,* 26 (4), 373-383.

18. **Bathurst, R. J., and L. Rothenburg,** 1990, Observations on stress-force-fabric-relationships in idealized granular materials, *Mechanics of Materials,* 9, 65-80.

19. **Behringer , R. P., and J. T. Jenkins,** eds., 1997, *Proceedings of the 3rd International Conference on Powders and Grains,* Durham, NC, Balkema, Rotterdam, the Netherlands.

20. **Bentall, R. H., and K. L. Johnson,** 1967 Slip in the rolling contact of two dissimilar elastic rollers, *Int. J. Mech. Sci.,* 9(55), 389-404.

21. **Biarez, J.,** 1962, Contribution à l' étude des propriétés mécaniques des sols et des materiaux pulvérulents, *Ph.D. Thesis, University of Grenoble, France.*

22. **Biarez, J., and R. Gourves,** eds., 1989, *Proceedings of the 1rst International Conference on Powders and Grains,* Clermont-Ferrand, France, Balkema, Rotterdam, the Netherlands.

23. **Bishop, A. W.,** 1954, Correspondence on shear characteristics of a saturated silt measured in triaxial compression, *Géotechnique,* 4 (1), 43-45.

24. **Bogdanova-Bontcheva, N., and H. Lippmann,** 1975, Rotationssymmetrisches ebenes Fließen eines granularen Modellmaterials, *Acta Mechanica,* 21, in German, 93-113.

25. **Borja, R. I. and J. R. Wren,** 1995a, Micromechanics of granular media, Part I: Generation of overall constitutive equation for assemblies of circular disks, *Comput. Methods Appli. Mech. Engrg.,* 127, 13-36.

26. **Borja, R. I. and J. R. Wren,** 1995b, Micromechanics of continuum models for granular materials, *Proc. 10th Conf. Engineering Mechanics, ASCE,* S. Sture, ed., 11, 497-500.

27. **Brogliato, B.,** 1996, *Nonsmooth Impact Mechanics. Models, Dynamics and Control,* Lecture Notes in Control and Information Sciences, 220, Springer-Verlag,

28. **Bromwell, L. W.,** 1966, The friction of quartz in high vacuum, *M.I.T. Department of Civil Engineering,* Research Report B66-18.

29. **Campbell, C. S., and C. E. Brennen,** 1985, Computer simulation of granular shear flow, *Journal of Fluid Mechanics,* 151, 167-188.

30. **Carter, F.W,** 1926, On the action of a locomotive driving wheel, *Proc. Royal Society,* A112, 151-157.

31. **Chang, C. S.,** 1993, Micromechanical modeling of deformation and failure for granulates with frictional contacts, *Mechanics of Materials,* 16, 13-24.

32. **Chang, C. S., A. Misra, and K. Acheampong,** 1992, Elastoplastic deformation for particulates with frictional contacts, *Journal of Engineering Mechanics, ASCE,* 118 (8), 1692-1707.

33. **Chang, C. S., A. Misra, and S. Sundararam,** 1990, Micromechanical modeling of cemented sands under low amplitude oscillations, *Géotechnique,* 40 (2), 251-263.

34. **Chang, C. S., and A. Misra,** 1989, Computer simulation and modeling of mechanical properties of particulates, *Computers and Geotechnics,* 7 (4), 269-287.
35. **Chang, C. S., and A. Misra,** 1990, Packing structure and mechanical properties of granulates, *Journal of Engineering Mechanics, ASCE,* 116 (5), 1077-1093.
36. **Chang, C. S., and C. S. Liao,** 1990, Constitutive relation for a particulate medium with the effect of particle rotation, *Int. J. Solids and Structures,* 26 (4), 437-455.
37. **Chang, C. S., and J. Gao,** 1995, Second-gradient constitutive theory for granular material with random packing structure, *Int. J. Solids and Structures,* 32 (16), 2279-2293.
38. **Chang, C. S., and L. Ma,** 1991, A micromechanical-based micropolar theory for deformation of granular solids, *Int. J. Solids and Structures,* 28 (1), 67-86.
39. **Chang, C. S., M. G. Kabir, and Y. Chang,** 1992b, Micromechanical modeling for ˙tress-strain behavior of granular soils. II: Evaluation *Journal of Geotechnical Engineering, ASCE,* 118 (12), 1975-1992.
40. **Chang, C. S., S. J. Chao, and Y. Chang,** 1995, Estimates of elastic moduli for granular material with anisotropic random packing structure, *Int. J. Solids and Structures,* 32 (14), 1989-2008.
41. **Chang, C. S., Y. Chang, and M. G. Kabir,** 1992, Micromechanical modeling for stress-strain behavior of granular soils. I: Theory, *Journal of Geotechnical Engineering, ASCE,* 118 (12), 1959-1974.
42. **Chichili, D. R., D. E. Mouton, and M. M. Mehrabadi,** 1993, Simulation of the mechanical behavior of two-dimensional granular assemblies utilizing linear programming, *Mechanics of Materials,* submitted.
43. **Christoffersen, J., M. M. Mehrabadi, and S. Nemat-Nasser,** 1981, A micromechanical description of granular material behavior, *Journal of Applied Mechanics,* ASME, 48, 339-344.
44. **Clément, E., J. Duran and J. Rajchenbach,** 1992, Experimental study of heaping in a two-dimensional sandpile, *Physics Review Letters,* 69, 1189-1192.
45. **Cosserat, E., and F. Cosserat,** 1909, *Théorie des corps déformables,* Hermann, Paris.
46. **Cundall, P. A,** 1982, Adaptative density-scaling for time-explicit calculations, *Proceedings of the 4th International Conference on Numerical methods in geomechanics,* Edmonton, 23-26.
47. **Cundall, P. A.,** 1971, A computer model for simulating progressive large scale movements of blocky rock systems, *Proceedings of the Symposium of the International Society of Rock Mechanics,* Nancy, France, 1, 132-150.
48. **Cundall, P. A.,** 1980, UDEC - A generalized distinct element program for modeling jointed rock, *Final Technical Report to European Research Office,* US Army, Contract No. DAJA 37-79-C-0548, NTIS order No. AD-A087 610/2.
49. **Cundall, P. A.,** 1988, Formulation of a three-dimensional distinct element model - Part I: A scheme to detect and represent contacts in a system composed of many polyhedral blocks, *International Journal of Rock Mechanics,* 25, 107-116.
50. **Cundall, P. A.,** 1989, Numerical experiments on localization in frictional materials, *Ingenieur-Archiv,* 59, 148-159.

51. **Cundall, P. A.,** 1989, Numerical modeling of discontinua, *Proceedings of the 1ʳˢᵗ US Conference on Discrete Element Methods (DEM)*, G. G. W. Mustoe, M. Henriksen,. and H. -P. Huttelmaier, eds., Colorado School of Mines Press, Golden, CO, 1-17.

52. **Cundall, P. A., and O. D. L. Strack,** 1978-1979, The distinct element method as a tool for research in granular media, Parts I and II, *Report to National Science Foundation, Eng. 76-20711,* Department of Civil and Mineral Engineering, University of Minnesota, Minneapolis, MN.

53. **Cundall, P. A., and O. D. L. Strack,** 1979, A discrete numerical model for granular assemblies, *Géotechnique,* 29, 47-65.

54. **Cundall, P.A., A. Drescher, and O.D.L. Strack,** 1982, Numerical experiments on granular assemblies; Measurement and observations, *Proceedings of the IUTAM symposium on deformation and failure of granular materials, Delft,* Balkema Publishers, P.A. Vermeer and H.J. Luger eds., 355-370.

55. **Dantu, P.,** 1957, Contribution à l' étude mécanique et géometrique des milieux pulvérulents, *Proceedings of the 4ᵗʰ International Conference of Soil Mechanics and Foundation Engineering,* London, UK.

56. **Daudon, D., J. Lanier, and M. Jean,** 1997, A micromechanical comparison between experimental results and numerical simulation of a biaxial test on 2D granular materials, *Proceedings of the 3ʳᵈ International Conference on Powders and Grains,* Durham, NC, R. P. Behringer and J. T. Jenkins, eds., Balkema, Rotterdam, the Netherlands, 219-222.

57. **de Josselin de Jong, G.,** 1959, *Statics and kinematics of the failable zone of a granular material,* Vitgeverij Waltman, Delft.

58. **Delassus, E.,** 1917, Mémoire sur la théorie des liaisons finies unilatérales, *Ann. Sci. Ecole Norm. Sup.,* 34, 95-179.

59. **Deresiewicz, H.,** 1958, Mechanics of granular material, *Advd. Appl. Mech.,* 5, 233-306.

60. **Deresiewicz, H.,** 1958, Stress-strain relations for a simple model of a granular medium, *Journal of Applied Mechanics, ASME,* 403-406.

61. **Desrues, J.,** 1984, Localization de la déformation plastique dans les matériaux granulaires, *Thèse d'état, Université de Grenoble..*

62. **Desrues, J., and B. Duthilleul,** 1984, Stereophotogrametric method applied to the determination of plane strain fields, *Journal de Mécanique théorique et appliquée,* 3 (1), 79-103.

63. **Desrues, J., J. Lanier, and P. Stutz,** 1985, Localization of the deformation in tests on sand samples, *Engineering Fracture Mechanics,* 21, 909-921.

64. **Diepolder, W., V. Mannl, and H. Lippman,** 1991, The Cosserat continuum, a model for grain rotations in metals?, *International journal of plasticity,* 7, 313-328.

65. **Dobry, R., and T.-T. Ng,** 1992, Discrete modeling of stress-strain behavior of granular media at small and large strain, *Engineering Computations,* 9, 129-143.

66. **Doménech, A., T. Doménech, and J. Cebrián,** 1987, Introduction to the study of rolling friction, *American Journal of Physics,* 55 (3), 231-235.

67. **Drescher, A.** 1976, An experimental investigation of flow rules for granular materials using optically sensitive glasss particles, *Géotechnique,* 26 (4), 591-601.

68. **Drescher, A. and G. De Josselin de Jong**, 1972, Photoelastic verification of a material model for the flow of a granular material. *J. Mech. Phys. Solids*, 20, 337-351.

69. **Dubujet, P., B. Cambou, F. Dedecker, and F. Emeriault**, 1997, Statistical homogenization for granular media - Application to non linear elastic modeling, *Proceedings of the 3rd International Conference on Powders and Grains*, Durham, NC, R. P. Behringer and J. T. Jenkins, eds., Balkema, Rotterdam, the Netherlands, 199-202.

70. **Duffy, J., and Mindlin, R.D.**, 1957, Stress-strain relations and vibrations of a granular medium, *J. of Applied Mech., ASME*, 79, 585-593.

71. **El-Sohby, A. A. K.**, 1969, Deformation of sand under constant stress ratio, *Proceedings of the 7th international conference in Soil Mechanics and Foundation Engineering*, Mexico, 1, 111-119.

72. **Eringen, A.C.**, 1966, Linear theory of micropolar elasticity, *Journal of Mathematics and Mechanics*, 15 (6, 909-923.

73. **Eringen, A.C.**, 1967, *Mechanics of Continua*, John Wiley, New York, NY.

74. **Fabrikant, V. I.**, 1986, Inclined flat punch of arbitrary shape on an elastic half-space, *Journal of Applied Mechanics, ASME*, 53, 798-806.

75. **Fabrikant, V. I.**, 1988, Elastic field around a circular punch, *Journal of Applied Mechanics, ASME*, 55, 604-605.

76. **Foerster, S., M. Louge, H. Chang and K. Allia**, 1994, Measurements of the collision properties of small spheres, *Phys. Fluids*, 6, 1108-1115.

77. **Frankel, S. P.**, 1950, Convergences rates of iterative treatments of partial differential equations, *Mathl. Tabl. Natn. Res. Coun.*, Washington DC, 4, 65-75.

78. **Germain, P.**, 1973, La méthode des puissances virtuelles en mécanique des milieux continus. Théories due second gradient, *Journal de Mécanique*, 12 (2), 235-274.

79. **Germain, P.**, 1973, The method of virtual power in continuum mechanics. Part 2: Microstructures, *J. of Applied Mathematics, SIAM*, 25 (3), 556-575.

80. **Ghaboussi, J., and R. Barbosa**, 1990, Three-dimensional discrete element method for granular materials, *International Journal for Numerical and Analytical Methods in Geomechanics*, 14, 451-472.

81. **Goddard, J. D.**, 1977, An elastohydrodymanic theory for the rheology of concentrated suspensions of deformable particles, *Journal of Non-Newtonian Fluid mechanics*, 2, 169-189.

82. **Goddard, J. D.**, 1986, Microstructural origins of continuum stress fields - A brief history and some unresolved issues, Chapter 6 in *Recent developments in structured continua*, D. DeKee and P. N. Kaloni, eds., Pitman research notes in mathematics (143, Longman/J. Wiley, p. 179-208.

83. **Goddard, J. D., X. Zhuang, and A. K. Didwania**, 1993, Microcell methods and the adjacency matrix in the simulation of the mechanics of granular media, *Proceedings of the 2nd international conference on discrete element methods*, Massachusetts Institute of Technology, J. R. Williams and G. W. Mustoe, eds., March, 3-14.

84. **Goodman, E. L.**, 1962, Contact stress analysis of normally loaded rough spheres, *J. of Applied Mech., ASME*, 515-522.

85. **Gray, J. E.**, 1960, The relationship between principal stress dilatancy and friction angle of a granular materials, *M. Sc. Thesis*, University of Manchester, UK.

86. **Greening, D. R., G. G. W. Mustoe, and G. L. DePorter,** 1997, Discrete element modeling of fabrication flaw precursors in the compaction of agglomerated ceramic powders, *Proceedings of the 3rd International Conference on Powders and Grains,* Durham, NC, R. P. Behringer and J. T. Jenkins, eds., Balkema, Rotterdam, the Netherlands, 113-116.

87. **Hafiz, M. S.,** 1950, Strength characteristics of sands and gravels in direct shear, *PhD thesis,* University of London, UK.

88. **Hahn, J. K.,** 1988, Realistic animation of rigid bodies, *Computer graphics,* 22. (4), 299-308.

89. **Hart, R. D., P. A. Cundall, and J. Lemos,** 1988, Formulation of a three-dimensional distinct element model - Part II: Mechanical calculations for motion and interaction of a system composed of many polyhedral blocks, *International Journal of Rock Mechanics,* 25, 117-126.

90. **Hertz, H.,** 1882, Über die Berührung fester elastische Körper and über die Harte (On the contact of rigid elastic solids and on hardness), *Verhandlungen des Vereins zur Beförderung des Gewerbefleisses,* Leipzig.

91. **Hill, R.,** 1962, Acceleration waves in solids, *J. Mech. Phys. Solids,* 10, 1-16.

92. **Hill, R.,** 1967, The essential structure of constitutive laws for metal composite and polycrystals, *J. Mech. Physics Solids,* 11, 357-372.

93. **Hocking, G., G. G. W. Mustoe, and J. R. Williams,** 1985, CICE discrete element analysis code - theoretical manual, *Applied Mechanics Inc.,* Lakewood, CO.

94. **Hong, C.-W.,** 1997, Process modeling and design for colloidal powder forming, *Proceedings of the 3rd International Conference on Powders and Grains,* Durham, NC, R. P. Behringer and J. T. Jenkins, eds., Balkema, Rotterdam, the Netherlands, 41-44.

95. **Horn, H. M., and D. V. Deere,** 1962, Frictional characteristics of minerals, *Géotechnique,* 12 (4), 319-335.

96. **Horne, M. R.,** 1965, The behavior of an assembly of rotund, rigid, cohesionless particles (I and II), *Proc. Roy. Soc. London,* A286, 62.

97. **Horne, M. R.,** 1969, The behavior of an assembly of rotund, rigid, cohesionless particles (III), *Proc. Roy. Soc. London,* A310, 21.

98. **Houlsby, G.T.,** 1981, A study of plasticity theory and their applicability to soils, *Ph.D. thesis, University of Cambridge.*

99. **Hughes, T. J. R.,** 1983, Analysis of transient algorithms with particular reference to stability behavior, *Computational methods for transient analysis,* T. Belitschko, and T.J.R. Hughes, eds., North Holland, Amsterdam, Holland, 67-156.

100. **Hughes, T. J. R.,** 1987, *The finite element method,* Prentice-Hall, Englewoods Cliffs, NJ.

101. **Hughes, T. J. R., and T. Belytschko,** 1993, *Nonlinear finite element analysis,* Short course notes, Stanford, CA.

102. **Jagota, A., C. Argento, and S. Mazur,** 1997, Viscoelastic coalescence of spherical particles, *Proceedings of the 3rd International Conference on Powders and Grains,* Durham, NC, R. P. Behringer and J. T. Jenkins, eds., Balkema, Rotterdam, the Netherlands, 31-36.

103. **Jean, M.,** 1995, Frictional contact in collections of rigid or deformable bodies: numerical simulation of geomaterials, in *Mechanics of Geomaterial Interfaces,* eds. A. P. S. Salvadurai and M. J. Boulon, Elsevier Science Publisher, Amsterdam, 463-486.

104. **Jean, M., and J. J. Moreau,** 1996, Numerical treatment of contact and friction: the Contact Dynamics method, *Proc.1996 Engineering Systems Design and Analysis Conference,* eds. A. Lagarde and M. Raous, ASCE, New York, 4, 201-208.

105. **Johnson, D. L., L. M. Schwartz, D. Elata, J. G. Berryman, B. Hornby, and A. N. Norris,** 1997, Linear and nonlinear elasticity of granular media: Stress-induced anisotropy of a random sphere pack, *Proceedings of the 3rd International Conference on Powders and Grains,* Durham, NC, R. P. Behringer and J. T. Jenkins, eds., Balkema, Rotterdam, the Netherlands, 243-246.

106. **Johnson, K. L.,** 1958, The effect of spin upon the rolling motion of an elastic sphere on a plane, *Journal of Applied Mechanics, September,* 332-338.

107. **Johnson, K. L.,** 1958, The effect of tangential contact force upon the rolling motion of an elastic sphere on a plane, *Journal of Applied Mechanics, September,* 339-346.

108. **Johnson, K. L.,** 1985, *Contact Mechanics,* Cambridge University Press, Cambridge, UK.

109. **Kalker, J. J.,** 1970, Transient phenomena in two elastic cylinders rolling over each other with dry friction, *J. of Applied Mech., ASME,* 677-688.

110. **Kalker, J. J.,** 1990, Three-dimensional elastic bodies in rolling contact, *Kluwer Academic Publishers, Dordrecht, The Netherlands,* 59-97.

111. **Kanatani, K.,** 1979, A micro-polar continuum theory for the flow of granular materials, *Int. Journal of Engineering Science,* 17 (4), 419-432.

112. **Kawaguchi, T., Y. Yamamoto, T. Tanaka, and Y. Sudji,** 1995, Numerical simulation of a single rising bubble in a two-dimensional fluidized bed, *Proceedings of the 2nd International Conference on Multiphase Flows,* Kyoto, Japan, FB2-17-22.

113. **Ke, T.-C., and J. Bray,** 1995, Modeling of particulate media using discontinuous deformation analysis, *Journal of Engineering Mechanics,* ASCE, 121 (11), 1234-1243.

114. **Kishino, Y.,** 1988, Disc model analysis of granular media, *in Micromechanics of Granular Materials,* M. Satake and J. T. Jenkins, eds., Elsevier Science Publishers, Amsterdam, The Netherlands, 143-152.

115. **Knight, J. B., H. M. Jaeger and S. R. Nagel,** 1993, Vibration-induced size separation in granular media : the convection connection, *Physics Review Letters,* 70, 3728-3731.

116. **Koenders, M. A.,** 1987, The incremental stiffness of an assembly of particles, *Acta Mechanica,* 70, 31-49.

117. **Krawietz, A.,** 1982, Some features of the gross behavior of granular media from micromechanics, *Proceedings of the IUTAM conference on deformation and failure of granular materials, Delft,* Balkema, Rotterdam, The Netherlands, 29-36.

118. **Laalai, K. Sab, and N. Guérin,** 1995, Micromechanical modelling of ballasted tracks, in *Contact Mechanics,* eds. M. Raous et al., Plenum Press, New York, 363-368.

119. **LaBudde, R. A., and D. Greenspan,** 1976, Energy and momentum conserving methods of arbitrary order for the numerical integration of equation of motion, I. Motion of a single particle, II. Motion of a system of particles, *Numerical Mathematics,* 25, 323-346; 26, 1-16.

120. **Lambe, T. W. and R. V. Whitman,** 1979, *Soil Mechanics,* John Wiley and Sons, New York, N.Y.

121. **Lecornu, L.,** 1905, Sur la loi de Coulomb, *Comptes Rendus Acad. Sci. Paris, 140,* 847-848.

122. **Lee, I. K.,** 1966, Stress dilatancy performance of feldspar, *Journal of the Soil Mechanics and Foundations Division,* ASCE, 92, SM2, 79-103.

123. **Lesburg, L., X. Zhang, L. Vu-Quoc, and O. R. Walton,** 1997, Simplified tangential force-displacement models for a discrete element particle flow code, *Proceedings of the 3rd International Conference on Powders and Grains,* Durham, NC, R. P. Behringer and J. T. Jenkins, eds., Balkema, Rotterdam, the Netherlands, 243-246.

124. **Lian, G., C. Thornton, and M. J. Adams,** 1997, A microscopic simulation of oblique collision of wet agglomerates, *Proceedings of the 3rd International Conference on Powders and Grains,* Durham, NC, R. P. Behringer and J. T. Jenkins, eds., Balkema, Rotterdam, the Netherlands, 159-162.

125. **Lun, C. K. K., and A. A. Bent,** 1993, Computer simulation of simple shear flow of inelastic frictional spheres, *Proceedings of the 2nd International Conference on Micromechanics of Granular Media,* Birmingham, UK, C. Thornton, ed., Balkema, Rotterdam, 301-306.

126. **Manna, S. S., and H. J. Herrmann,** 1991, Precise determination of the fractal dimensions of Appolonian packing and space-filling bearings, *Journal of Physics, A: Math. Gen.,* 24, 481-490.

127. **Matsuoka, H., and S. Yamamoto,** 1994, A microscopic studdy on shear mechanism of granular materials by DEM, *Journal of Geotechnical Engineering, JSCE,* In Japanese, 487, 167-175.

128. **Meakin P., and A. T. Skeltorp,** 1993, Application of experimental and numerical methods to the physics of multiparticle systems, *Advances in Physics,* 42 (1), 1-127.

129. **Meftah, W. P. Evesque, J. Biarez, D. Sornette, and N. E. Abriak,** 1993, Evidence of local 'seisms' of microscopic and macroscopic stress fluctuations during the deformation of packings of grains, in *Powders and Grains,* C. Thornton, ed., Balkema, Rotterdam, 173-178.

130. **Mehrabadi, M. M., B. Loret, and S. Nemat-Nasser,** 1993, Incremental constitutive relations for granular materials based on micromechanics, *Proceedings of the Royal Society,* London, A 441, 443-463.

131. **Mehrabadi, M. M., S. Nemat-Nasser, and M. Oda,** 1982, On statistical description of stress and fabric in granular materials, *Int. Journal for Numerical and Analytical Methods in Geomechanics,* 6, 95-108.

132. **Mindlin, R. D.,** 1949, Compliance of elastic bodies in contact, *Journal of Applied Mechanics,* ASME, 71, 259-268.

133. **Mindlin, R. D., and H. Deresiewicz,** 1953, Elastic spheres in contact under varying oblique forces, *Journal of Applied Mechanics,* ASME, 20, 327-344.

134. **Misra, A.,** 1995, Interfaces in particulate materials, *Mechanics of geomaterial interfaces,* A. P. S. Selvadurai and M. Boulon, eds., Elsevier.

135. **Monteiro Marques, M. D. P.,** 1993, Differential Inclusions, *in Nonsmooth Mechanical Problems: Shocks and Dry Friction,* Berlin, Germany.

136. **Moreau, J. J.,** 1966, Quadratic programming in mechanics: dynamics of one-sided constraints, *SIAM J. Control,* 4, 153-158.

137. **Moreau, J. J.,** 1988, Bounded variation in time, in *Topics in Nonsmooth Mechanics,* J. J. Moreau, P. D. Panagiotopoulos and G. Strang, eds., Berlin, Germany, 1-74.

138. **Moreau, J. J.,** 1988, Unilateral contact and dry friction in finite freedom dynamics, in *Nonsmooth Mechanics and Applications,* J. J. Moreau and P. D. Panagiotopoulos, eds., CISM Courses and Lectures, 302, Springer-Verlag, Wien, New York, 1 - 82.

139. **Moreau, J. J.,** 1993, New computation methods in granular dynamics, *Proceedings of the 2nd International Conference on Micromechanics of Granular Media,* Birmingham, C. Thornton, ed., Balkema, Rotterdam, The Netherlands, 227-232.

140. **Moreau, J. J.,** 1994, Some numerical methods in multibody dynamics: application to granular materials, *Eur. J. Mech. Solids,* 13, 93-114.

141. **Moreau, J. J.,** 1995, Numerical experiments in granular dynamics: vibration-induced size segregation, in *Contact Mechanics,* M. Raous et al., eds., Plenum Press, New York, 347-358.

142. **Moreau, J. J.,** 1996, Numerical investigation, of shear zones in granular materials, *Proceedings of the HLRZ-workshop on Friction, Arching and contact Dynamics,* Jülich, Germany, October.

143. **Mühlhaus, H. B., and I. Vardoulakis,** 1987, The thickness of shear bands in granular materials, *Géotechnique,* 37, 271-283.

144. **Mujinza, A., N. Bicanic, and D. R. J. Owen,** 1993, BSD contact detection algorithm for discrete elements in 2D, *Proceedings of the 2nd international conference on discrete element methods,* Massachusetts Institute of Technology, J. R. Williams and G. W. Mustoe, eds., March, 39-52.

145. **Mustoe, G. G. W., M. Henriksen, and H. -P. Huttelmaier,** eds., 1989, *Proceedings of the 1^{rst} US Conference on Discrete Element Methods (DEM),* Colorado School of Mines Press, Golden, CO.

146. **Nakase, H., T. Annaka, F. Katahira, and T. Kyono,** 1992, An application study of the distinct element method to plane strain compression tests, *Soils and Foundations,* in Japanese, 454, 55-64.

147. **Nemat-Nasser, S., and M. M. Mehrabadi,** 1983, Micromechanically based rate constitutive description for granular materials, *Proceedings of International Conference on Constitutive Laws for Engineering Materials; Theory and Applications,* Tucson, A2.

148. **Nemat-Nasser, S., and M. M. Mehrabadi,** 1984, Micromechanically based rate constitutive descriptions for granular materials, *Mechanics of Engineering Materials,* C.S. Desai and R.H. Gallagher, eds, John Wiley and Sons, New York, 451-463.

149. **Ng, T. -T. and R. Dobry, R.,** 1994, Numerical simulation of monotonic and cyclic loading of granular soil, *Journal of Geotechnical Engineering, ASCE,* 120 (2), 388-403.

150. **Ng, T. -T.,** 1992, A non-linear numerical model for soil mechanics, *International Journal for Numerical and Analytical Methods in Geomechanics*, 16, 247-263.
151. **O'connor, R., J. Gill, and J. R. Williams,** 1993, A linear complexity contact detection algorithm for multi-body analysis, *Proceedings of the 2nd international conference on discrete element methods,* Massachusetts Institute of Technology, J. R. Williams and G. W. Mustoe, eds., 53-64.
152. **Oda, M.,** 1972, Deformation mechanism of sand in triaxial compression tests, *Soils and Foundations*, Japanese Society of Soil Mechanics and Foundation Engineering, 12, 45-63.
153. **Oda, M.,** 1972, Initial fabrics and their relations to mechanical properties of granular material, *Soils and Foundations*, Japanese Society of Soil Mechanics and Foundation Engineering, 12, 18-36.
154. **Oda, M.,** 1972, The mechanism of fabric changes during compressional deformation of sand, *Soils and Foundations*, Japanese Society of Soil Mechanics and Foundation Engineering, 12, 1-17.
155. **Oda, M.,** 1993, Micro-fabric and couple-stress in shear bands of granular materials, *Proceedings of the 2nd International Conference on Micromechanics of Granular Media,* Birmingham, UK, C. Thornton, ed., 161-166.
156. **Oda, M., and J. Konishi,** 1974, Microscopic deformation mechanism of granular material in simple shear, *Soils and Foundations*, Japanese Society of Soil Mechanics and Foundation Engineering, 14, 25-38.
157. **Oda, M., J. Konishi, and S. Nemat-Nasser,** 1980. Some experimentally based fundamental results on the mechanical behavior of granular materials, *Géotechnique,* 30, 479-495.
158. **Oda, M., K. Iwashita, and T. Kakiuchi,** 1997, Importance of particle rotation in the mechanics of granular materials, *Proceedings of the 3rd International Conference on Powders and Grains,* Durham, NC, R. P. Behringer and J. T. Jenkins, eds., Balkema, Rotterdam, the Netherlands, 207-210.
159. **Oda, M., Konishi, J, and S. Nemat-Nasser,** 1982. Experimental micromechanical evaluation of strength of granular materials: effects of particle rolling, *Mechanics of Materials,* 1, 269-283.
160. **Papadrakakis, M.,** 1981, A method for the automatic evaluation of the dynamic relaxation parameters, *Computer methods in applied mechanics and engineering,* 25, 35-48.
161. **Parikh, P. V.,** 1967, The shearing behavior of sand under axisymmetric loading, *PhD thesis,* University of Manchester, UK.
162. **Park, K. C., and P. Underwood,** 1980, A variable-step central difference method for structural dynamic analysis - Part 1. Theoretical aspects, *Computer methods in applied mechanics and engineering,* 22, 241-258.
163. **Penman, A. D. M.,** 1953, Shear characteristics of saturated silts measured in triaxial compression, *Géotechnique,* 3 (4), 312-328.
164. **Pfeiffer, F., and C. Glocker,** 1966, *Multibody Dynamics with Unilateral Contacts,* John Wiley & Sons, New York.

165. **Procter, D. C., and R. R. Barton,** 1974, Measurement of the angle of interparticle friction, *Géotechnique,* 24 (4), 581-604.
166. **Radjai, F., M. Jean, J. J. Moreau and S. Roux,** 1996, Force distributions in dense two-dimensional granular systems, *Physics Review Letters,* 77, 274.
167. **Reynolds, O.,** 1885, On the dilatancy of media composed of rigid particles in contact, with experimental illustration, *Philosophical Magazine,* Series 5, 20, 469-481.
168. **Reynolds, O.,** 1895, On rolling friction, *Phil. Trans. Royal Society,* 166, 155.
169. **Rice, J. R.,** 1976, The localization of plastic deformation, *Theoretical and Applied Mechanics,* Proceedings of the 14th IUTAM Congress, W.T. Koiter, ed., North Holland, New York, 207-220.
170. **Roscoe, K.H. and A.N. Schofield,** 1964. Discussion on P.W. Rowe's paper, Stress-dilatancy, earth pressures and slopes (Proc Paper 3507, May 1963), *J. Soil mechanics and Foundation Engineering, ASCE,* 90(1), 136.
171. **Rosenberg, L., and A. P. S. Selvadurai,** 1981, Micromechanical definition of the cauchy stress tensor for particulate media, In *Mechanics of Structured Media,* A. P. S. Selvadurai, ed., Elsevier, Amsterdam, The Netherlands, 469-486.
172. **Rothenburg, L., and R. J. Bathurst,** 1989, Analytical study of induced anisotropy in idealized granular materials, *Géotechnique,* 39, 601-614.
173. **Rothenburg, L., and R. J. Bathurst,** 1991, Numerical simulation of idealized granular assemblies with plane elliptical particles, *Computers and Geotechnics,* 11, 315-329.
174. **Rothenburg, L., and R. J. Bathurst,** 1992, Micromechanical features of granular assemblies with planar elliptical particles, *Géotechnique,* 42, 79-95.
175. **Rowe, P.W.,** 1962, The stress-dilatancy relation for static equilibrium of an assembly of particles in contact, *Proc. Roy. Soc. London,* A269, 500.
176. **Rudnicki, J.W., and J. R. Rice,** 1975, Conditions for the localization of deformation in pressure-sensitive dilatant materials, *J. Mech. Phys. Solids,* 23, 371-394.
177. **Satake, M.,** 1978, Constitution of mechanics of granular materials through graph representation, in *Theoretical and Applied Mechanics 26,* University of Tokyo Press, 257.
178. **Satake, M.,** 1982, Fabric tensor in granular materials, *Proceedings of the IUTAM symposium on deformation and failure of granular materials, Delft,* Balkema Publishers, P.A. Vermeer & H.J. Luger eds., 63-68.
179. **Savage, S. B., and D. J. Jeffrey,** 1981, The stress tensor in a granular flow at high shear rate, *J. Fluid Mech.,* 110, 255-272.
180. **Schneebli, G.,** 1955, Une analogie mécanique pour les terres sans cohésion, *Proceedings of the Academy of Sciences,* Paris, France, 243, p. 256.
181. **Schofield, A. N., and C. P. Wroth.,** 1968, *Critical State Soil Mechanics.* Mc Graw-Hill, London, UK.
182. **Shi, G. -H,** 1993, Block system modeling by discontinuous deformation analysis, *Topics in engineering Volume 11,* C. A. Brebbia and J. J. Connor, eds., Computational mechanics Publication, Southampton, UK.

183. **Shi, G.-H., and R. E. Goodman,** 1988, Discontinuous deformation analysis - a new method for computing stress, strain and sliding of block systems, in *Key questions in rock mechanics*, Cundall et al., eds., Balkema, Rotterdam, the Netherlands, 381-393.

184. **Shi, G.-H., and R. E. Goodman,** 1989, Generalization of two-dimensional discontinuous deformation analysis for forward modeling, *International Journal for Numerical and Analytical Methods in Geomechanics,* 13(4), 359-380.

185. **Singer I. L., and H. M. Pollock,** 1992, *Fundamentals of friction: Macroscopic and microscopic processes,* Kluwer Academic, Dordrecht, the Netherlands.

186. **Skinner, A. E.,** 1969, A note on the influence of interparticle friction on the shearing strength of a random assembly of spherical particles, *Géotechnique,* 19 (1) 150-157.

187. **Southwell, R. V.** 1940, *Relaxation methods in engineering sciences,* Oxford University Press, Oxford, UK.

188. **Sternberg, E.,** 1968, Couple stress and singular stress concentrations in elastic solid, in Mechanics of generalized continua, *Proceedings of the IUTAM Symposium on the Generalized Cosserat Continuum and the Continuum Theory of Dislocation with Applications,* Edited by E. Kroner, Springer, New York.

189. **Subbash, S. Nemat-Nasser, M. M. Mehrabadi, and H. M. Shodja,** 1991. Experimental investigation of fabric-stress relations in granular materials, *Mechanics of Materials,* 11, 87-106.

190. **Sudji, Y., T. Kawaguchi, and T. Tanaka,** 1993, Discrete particle simulation of two-dimensional fluidized bed, *Powder Technology,* 77, 79-87.

191. **Sudji, Y., T. Tanaka, and T. Ishida,** 1992, Lagrangian numerical simulation of plug flow of cohesionless particles in an horizontal pipe, *Powder Technology,* 71, 239-250.

192. **Sukla, A., and H. P. Rossmanith,** 1982, A photoelastic investigation of dynamic load transfer in granular media, *Acta Mechanica,* 42, 211-225.

193. **Sun, G., and C. Thornton,** 1994, Computer simulation of 3D quasi-static shear deformation of particulate media, *Proceedings of 1rst International Particle technology Forum,* American Institute of Chemical Engineers, Denver, CO, 24-29.

194. **Supel, J. A.,** 1985, Local destruction of granular media caused by crushing a single grain, *Archives of Mechanics,* 37 (4-5), 535-548.

195. **Tabor, D.,** 1955, The mechanism of rolling friction: the elastic range, *Proc. Royal Society,* A251, 198-220.

196. **Tamura, S., and T. Aizawa,** 1993, Mechanical behavior of powder particle on the applied vibration, *International journal of modern physics,* 7 (9-10), 1829-1838.

197. **Tatsuoka, F., S. Nakamura, C., Huang, and K. Tani,** 1990, Strength anisotropy and shear band direction in plane strain tests on sand, *Soils and Foundations,* 30, 35-54.

198. **Taylor, R. L.,** 1977, Computer procedures for finite element analysis. Ch. 24, in: O.C. Zienkiewicz, *The finite element method,* 3rd ed., McGraw-Hill Book Co., London, England.

199. **Thornton, C.,** 1979, The conditions of failure of a face-centered cubic array of uniform rigid spheres, *Géotechnique,* 29, 441-459.

200. **Thornton, C.,** 1994, Micromechanics of elastic sphere assemblies during 3D shear, *Proceedings of Mechanics and Statistical Physics of Particulate Materials*, Institute for Mechanics and Materials, La Jolla, CA, June, 64-67.
201. **Thornton, C., and C. W. Randall,** 1988, Applications of theoretical contact mechanics to solid particle system simulation, *in Micromechanics of Granular Materials*, M. Satake and J. T. Jenkins, eds., Elsevier Science Publishers, Amsterdam, Netherlands, 133-142.
202. **Thornton, C., and G. Sun,** 1994, Numerical simulation of general 3D quasi-static shear deformation of granular media, *Proceedings of Numerical Methods in Geotechnical Engineering*, I. M. Smith, ed., Balkema, Rotterdam, Netherlands, 143-148.
203. **Thornton, C.,** ed., 1993, *Proceedings of the 2nd International Conference on Powders and Grains*, Birmingham, UK, Balkema, Rotterdam, the Netherlands.
204. **Thurston, C. W., and H. Deresiewicz,** 1959, Analysis of a compression test of a model of a granular medium, *Journal of Applied Mechanics, ASME*, 251-258.
205. **Timoshenko, S., and J. N. Goodier,** 1951, *Theory of elasticity*, 3rd edition, McGraw-Hill, New York, NY.
206. **Ting, J. M.,** 1992, A robust algorithm for ellipse-based discrete element modeling of granular materials, *Computers and Geotechnics*.
207. **Ting, J. M., and B. T. Corkum,** 1992, A computational laboratory for discrete element geomechanics, *Journal of Computers in Civil Engineering, ASCE*, 6, 129-146.
208. **Ting, J. M., B. T. Corkum, C. R. Kauffman, and C. Greco,** 1989, Discrete numerical model for soil mechanics, *Journal of Geotechnical Engineering*, ASCE, 115, 379-398.
209. **Ting, J. M., M. Khawaja, L. M. Meachum, and J. D. Rowell,** 1993, An ellipse-based discrete element model for granular materials, *International Journal for Numerical and Analytical Methods in Geomechanics*, 17, 603-623.
210. **Tong, P. Y. L.,** 1970, Plane strain deformation of sands, *PhD thesis*, University of Manchester, UK.
211. **Trent, B. C. and L. G. Margolin,** 1992, A numerical laboratory for granular solids, *Engineering Computations*, 9, 191-197.
212. **Truesdell, C.,** 1985, *The Elements of Continuum Mechanics*, Springer-Verlag, Berlin, Germany.
213. **Tschebotarioff, G. P., and J. D. Welch,** 1948, lateral earth pressures and friction between soil minerals, *Proceedings of the 2nd International Conference on Soil Mechanics and Foundations*, Rotterdam, VII, 135-138.
214. **Tsuji, Y.,** 1997, Discrete particle simulation of dispersed gas-solid flows, *Proceedings of the 3rd International Conference on Powders and Grains*, Durham, NC, R. P. Behringer and J. T. Jenkins, eds., Balkema, Rotterdam, the Netherlands, 25-30.
215. **Underwood, P.,** 1983, Dynamic relaxation, *in Computational methods for transient analysis*, T. Belitschko and T.J.R. Hughes, eds., North Holland, Amsterdam, Holland, 245-265.
216. **Underwood, P., and K. C. Park,** 1980, A variable-step central difference method for structural dynamic analysis, Part 2. Implementation and performance evaluation, *Computer methods in applied mechanics and engineering*, 23, 259-279.

217. **Vardoulakis, I. and B. Graf,** 1985, Calibration of constitutive models for granular materials using data from biaxial experiments, *Géotechnique,* 35, 299-317.

218. **Vardoulakis, I.,** 1988, Theoretical and experimental bounds for shear-band bifurcation strain in biaxial tests on dry sand, *Res Mechanica,* 23, 239-259.

219. **Vermeulen, P. J., and K. L. Johnson,** 1964, Contact of nonspherical elastic bodies transmitting tangential forces, *Journal of Applied Mechanics,* June, 338-340.

220. **Walton, O. R.,** 1980, Particle dynamics modeling of geological materials, *Lawrence Livermore National Laboratory,* Report UCRL-52915.

221. **Walton, O. R.,** 1993, Numerical simulation of inelastic, frictional particle-particle interactions, in *Particulate two-phase flow,* M. C. Roco, ed., Butterworth Heinemann, Boston, 884-910.

222. **Walton, O. R., and R. L. Braun,** 1986, Stress calculations for assemblies of inelastic spheres in uniform shear, *Acta Mechanica,* 63, 73-86.

223. **Walton, O. R., and R. L. Braun,** 1986, Viscosity, granular-temperature, and stress calculations for shearing assemblies of inelastic, frictional disks, *Journal of Rheology,* 30, 949-980.

224. **Walton, O. R., R. L. Braun, R. G. Mallon, and D. M. Cervelli,** 1988, Particle-dynamics calculations of gravity flows of inelastic, frictional sphere, *Micromechanics of granular materials,* M. Satake and J. T. Jenkins, eds., Elsevier Science Publishers, Amsterdam, 153-161.

225. **Weber, J.,** 1996, Recherches concernant les contraintes intergranulaires dans les milieux pulvérulents, *Cahiers du Groupe Français de Rhéologie,* 2, 161-170 (see also *Bulletin de Liaison des Ponts et Chaussées* 20, 3-1 to 3-20).

226. **Wells, J. C.,** 1997, Contact of rough elastic spheres: a simplified analysis, *Proceedings of the 3rd International Conference on Powders and Grains,* Durham, NC, R. P. Behringer and J. T. Jenkins, eds., Balkema, Rotterdam, the Netherlands, 311-314.

227. **Williams, J. R., and G. G. W. Mustoe,** 1987, Modal methods for the analysis of discrete systems, *Computers and Geotechnics,* 4, 1-19.

228. **Williams, J. R., and G. G. W. Mustoe,** eds., 1993, *Proceedings of the 2nd International Conference on Discrete Element Methods (DEM),* The Massachusetts Institute of Technology, Intelligent Engineering System Laboratory Publication, Cambridge, MA.

229. **Witters, J., and D. Duymelinck,** 1986, Rolling and sliding resistive forces on balls moving on a flat surface, *American Journal of Physics,* 54 (1), 80-83.

230. **Zhuang, X. and J. D. Goddard,** 1993, Computer simulation and experiments on the quasi-static mechanics and transport properties of granular materials, *Res. Rep. GR 93-01,* University of California, San Diego, CA, October.

231. **Zhuang, X., K. Didwania, and J. D. Goddard,** 1995, Simulation of the quasi-static mechanics and scalar transport properties of ideal granular assemblages, *Journal of computational physics,* 121, 331-346.

232. **Zienkiewicz, O. C., and R. L. Taylor,** 1991, *The finite element method,* Vol. 2, McGraw-Hill, London, UK.

MICROMECHANICAL APPROACH IN GRANULAR MATERIALS

B. Cambou
Central School of Lyon, Ecully, France

1. Introduction

Granular materials are made up of grains in contact. These kinds of materials are then, highly discontinuous and nonhomogeneous with two or three phases (solids and voids which may be made up of air or air and water). The macroscopic properties of these materials are obviously related to the basic structure and properties of the constituents and their interactions (grains and voids). It therefore seems of great interest to derive the overall behaviour of such materials (discontinuous, inhomogeneous material) from the local properties. This kind of approach has been used extensively and is now usual for nonhomogeneous, continuous materials (fluids or solids) and is known as "the homogenization process". The aim of this kind of approach is, in particular, to derive the constitutive equations of a material to be used at the scale of the boundary problem, from the knowledge of the local behaviour and microstructure. This approach differs greatly from the phenomenological approach in which the constitutive equations are defined considering mathematical formulations whose constants are obtained from experimental tests.

The aim of this chapter is first to present fundamental concepts of the multiple scale approach used in continuous media. Then, some experimental evidence, showing that the evolution of the internal structure in granular media governs the behaviour of such a material, will be analysed as well as the different variables classically used to describe this local internal structure. Finally the application of these concepts for the development of constitutive modelling in discontinuous granular media will be proposed.

2. Fundamental concepts of the multiple scale approach

The multiple scale approach has been intensively used for the constitutive modelling of continuous media whose behaviour is essentially governed by the microstructure and its evolution (composite materials, metallic polycrystals, mixture of fluids, suspensions, ...). Different techniques known as "homogenization processes" have then been developed in recent years and their basics will be presented in this paragraph [21], [32], [33].

2.1 Definition of the two scales

The goal of the multiple scale approach is to use information described at the micro level to build constitutive modelling at the macro level. It is then necessary to define clearly the two scales considered.

2.1.1 Macro scale

The macro scale is the scale which corresponds to the discretization of the boundary problem to be solved, using the global constitutive equations. This corresponds, for example, to the dimension of the smallest element of a mesh in the F.E.M.

2.1.2 Micro scale

The micro scale corresponds to a smaller level at which a microstructure exists and which essentially governs the material global behaviour. The different constituents identified at this level are called the phases of the heterogeneous material. In some cases this microscale is obvious. In the case, for example, of a material with a periodic distribution of inclusions, the local scale corresponds to the periodic local cell (Fig.1). In other cases the choice of the micro scale is not obvious. For example, for metallic polycrystals is the relevant scale the cristallin mesh, the grain size or the grain arrangement ? In these cases the essential difficulty lies in the knowledge of the level at which the evolution of the microstructure governs the main features of the global material behaviour. This level allows the "relevant micro scale" to be defined.

When the micro scale has been defined, it is necessary to consider a "representative volume element (RVE)". The choice of the "RVE" or its modelling determines a first difference between various theories of homogenization. In particular, two classes of homogenization processes can be distinguished :

a) Homogenization methods for periodic media. The basic hypothesis in this case is to consider that the medium may be described by a periodic structure (Fig. 1). The RVE is, in this case, the unit cell. In this approach the treatment is completely determinist. These models are able to account for precise local information such as, for instance, the shape, dimensions and orientation of inclusions.

b) Homogenization methods for media with randomly distributed phases. In this case it is not possible to give a determinist description of the microstructure, and a statistical treatment is then appropriate. The RVE will be, in this case, a statistically homogeneous specimen. In some cases, the modelling of the RVE results from a crude simplification of a complex microstructure. Different hypotheses have been proposed leading to different theories of homogenization (Voigt and Reuss approximations, model of sphere assembly, self consistent scheme, ...). In these cases only a partial information on the distribution of the phases is usually available and it is only, then, possible to derive bounds or estimates for the effective behaviour of such materials.

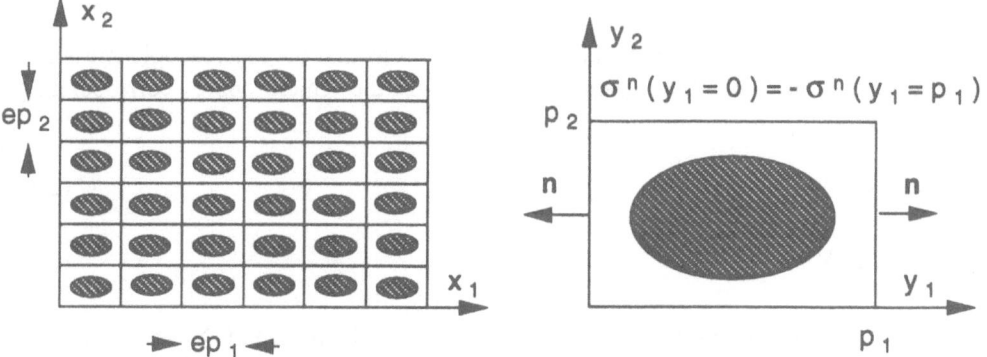

Fig. 1 : Double scale in periodic media

2.2 General scheme of homogenization theories

Generally, the derivation of the overall behaviour from the local properties may be described in a simplified manner by the diagram in Figure 2.

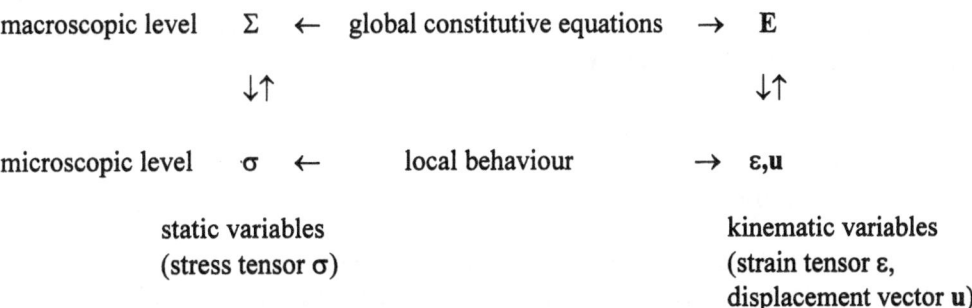

Fig. 2 : General scheme of homogenization theories.

Then the definition of the constitutive equations at the global level can be obtained using information given at the local level. This approach needs the development of two essential elements :
a) a realistic modelling at the micro scale :
   . of the local behaviour,
   . of the microstructure and its evolution,
b) a technique to go between the two levels (micro and macro) : the operation leading from the local level to the global level is called : "averaging process", the inverse operation (global to local level) is called "localization operation".

2.2.1 Averaging operations

It results from classical arguments on oscillating functions that the macroscopic stress and strain tensors must be the spatial averages of the microscopic corresponding quantities [32], [33] :

$$\Sigma_{ij} = \frac{1}{V}\int_V \sigma_{ij}\, dV = <\sigma_{ij}> \tag{2.1}$$

$$E_{ij} = \frac{1}{V}\int_V \varepsilon_{ij}\, dV = <\varepsilon_{ij}> \tag{2.2}$$

Notation $<>$ means volumic average. V is the volume in which E and Σ are defined. Morever, all the mechanical quantities which are usually assumed to be additive functions are averaged in the same way (mass, internal energy, dissipation...). In particular the "Hill macrohomogeneity equality" relates the energy defined at the local level to the energy defined at the global level :

$$E_{ij}\Sigma_{ij} = \frac{1}{V}\int_V \sigma_{ij}\varepsilon_{ij}\, dV = <\sigma_{ij}\varepsilon_{ij}> \tag{2.3}$$

2.2.2 Localization operations

These operations deal with the definition of the microscopic quantities (σ, ε) from macroscopic ones (Σ, E). For this purpose the following system of equations, with data Σ or E is to be solved for σ and u.

$$\left.\begin{array}{l} \text{microscopic constitutive law} \\ \text{div } \sigma = 0 \text{ (micro equilibrium)} \\ <\sigma> = \Sigma \quad \text{or} \quad <\varepsilon(u)> = E \end{array}\right\}$$

This problem turns out to be ill posed, due to the absence of boundary conditions. These boundary conditions must be defined depending on the kind of RVE considered, as will be seen below.

2.3 Basic concepts in homogenization methods for periodic media, application to elastic analysis

The basic hypothesis in this case is to consider that the medium can be described by a periodic structure defined by period Pe, P being a parallelepiped with dimensions p_1, p_2, p_3 and e being a positive small parameter (Fig. 1). Then the geometry and mechanical properties are identical if we consider a displacement which is a multiple of the unit cell ($n\,Pe$). This hypothesis is realistic in particular in manufactured materials (composite materials). Usually the RVE is chosen to be the unit cell and the boundary conditions can be clearly defined. The double scale method can then be used to solve boundary problems directly. But generally the homogenization technique is only used to define the constitutive equations at the global level, as proposed in Figure 2. This will be used in a numerical method (F.E.M.) to solve boundary problems.

In the elastic case it can be shown that, at the local level :

$\sigma\,(x)$, $\varepsilon(x)$ and the elastic constants a_{ijkl} are eP periodic,

$$u(x) = Ex + \tilde{u}(x) \quad \text{with} \quad \tilde{u}(x) \quad eP \quad \text{periodic.} \tag{2.4}$$

The elastic problem can then be defined on the unit cell, adding the following boundary conditions (Fig. 1) :

$\left. \begin{array}{ll} \sigma\,n & : \text{antiperiodic} \\[1ex] \tilde{u} & : \text{periodic} \end{array} \right\}$ boundary conditions on the unit cell .

This problem can be solved and leads to the displacement field :

$$\tilde{u}_i(x) = \chi_{ikl}(x)\, E_{kl} \tag{2.5}$$

Then the local strain field can be deduced :

$$\varepsilon_{ij}(x) = \Gamma_{ijkl}(x)\, E_{kl} \tag{2.6}$$

Tensor $\Gamma_{ijkl}(x)$ which relates the global variable E_{kl} to the local variable $\varepsilon_{ij}(x)$ is called the localisation tensor. Then the global constitutive law obtained by the homogenization technique can be deduced using the local constitutive law, relation (2.6) and relation (2.1) :

$$\sigma_{ij}(x) = a_{ijmn}(x)\, \varepsilon_{mn}(x) = a_{ijmn}(x)\, \Gamma_{mnkl}(x)\, E_{kl}$$

$$\Sigma_{ij} = <\sigma_{ij}(x)> = <a_{ijmn}(x)\, \Gamma_{mnkl}(x)> E_{kl}$$

$$\Sigma_{ij} = A_{ijkl}\, E_{kl} \qquad \text{with} \quad A_{ijkl} = <a_{ijmn}(x)\, \Gamma_{mnkl}(x)> \tag{2.7}$$

Usually the tensor field $\Gamma_{mnkl}(x)$ can only be obtained from a numerical analysis of the unit cell, using for instance the F.E.M. This numerical calculation allows the definition of :

the homogenized law characterized by tensor A,
the localisation operators : $\sigma(x) = c(x)\,\Sigma$ $\qquad$ and $\qquad$ $\varepsilon(x) = \Gamma(x)\,E$.

2.4 Basic concepts in homogenization methods for media with randomly distributed phases [21]

In this case the RVE is a statistically homogeneous specimen. In a first step it is necessary to define the level of accuracy of the statistical description of the phases. Usually the physical information available is rather poor and therefore the statistical description of the phases is not complete. Then it is only possible to derive bounds or estimates for the effective behaviour of such materials. We will present hereafter bounds obtained in the case of elastic modelling.

2.4.1 Voigt approximation (upper bound)

Let us consider the global uniform strain E_{ij} applied on the RVE boundaries.
A local kinematical admissible field can be proposed :

$$\overset{*}{u_i} = E_{ij} x_j \quad \text{then} \quad \overset{*}{\varepsilon_{ij}} = E_{ij} \tag{2.8}$$

This field satisfies the kinematic boundary conditions on ∂V .
The potential energy of this field is :

$$\frac{1}{2} \int_V \overset{*}{\varepsilon} \, a \, \overset{*}{\varepsilon} \, dV$$

a   characterises the local elasticity : $\sigma = a\varepsilon$
The potential energy theorem implies that the solution ε gives a minimum of this potential energy. Then :

$$\frac{1}{2} \int_V \varepsilon \, a \, \varepsilon \, dV \ \leq \ \frac{1}{2} \int_V \overset{*}{\varepsilon} \, a \, \overset{*}{\varepsilon} \, dV \ = \ \frac{1}{2} \int_V E \, a \, E \, dV ,$$

or $< \varepsilon \, a \, \varepsilon > \ \leq \ E < a > E$ (2.9)

The "Hill macrohomogeneity equality" (2.3) implies the equality of virtual work at the two scales, then :

$$< \varepsilon \, a \, \varepsilon > \ = \ EAE \tag{2.10}$$

A   characterizes the macro elasticity : $\Sigma = AE$,

then : $EAE \ \leq \ E < a > E$. (2.11)

Then an upper bound for the global behaviour A is obtained considering $< a >$.

$$A \ \leq \ < a > \tag{2.12}$$

2.4.2 Reuss approximation (lower bound)

The static admissible field $\overset{*}{\sigma} = \Sigma$ is considered.

The complementary energy theorem can be written :

$$\frac{1}{2}\int_V \sigma\, b\, \sigma\; dV \;-\; \frac{1}{2}\int_{\delta V} \sigma\, n\, u\; ds \;\leq\; \frac{1}{2}\int_V \overset{*}{\sigma}\, b\, \overset{*}{\sigma}\; dV \;-\; \frac{1}{2}\int_{\delta V} \overset{*}{\sigma}\, n\, u\; ds \qquad (2.13)$$

b characterizes the local elasticity : $\varepsilon = b\,\sigma$

$$\int_{\delta V} \overset{*}{\sigma}\, n\, u\; ds = V\, \Sigma\, E = \int_{\delta V} \sigma\, n\, u\; ds \;,$$

then : $<\sigma\, b\, \sigma> \;\leq\; \Sigma <b> \Sigma$ $\qquad (2.14)$

From the "Hill macrohomogeneity equality" (2.3) :

$$<\sigma\, b\, \sigma> \;=\; \Sigma\, B\, \Sigma \;\leq\; \Sigma <b> \Sigma \qquad (2.15)$$

B characterizes the macro elasticity : $E = B\Sigma$,

then : $<b> \;\geq\; B$,

as $B = A^{-1}$ and $b = a^{-1}$,
the two approximations of Voigt and Reuss lead to the two bounds of A :

$$<a^{-1}>^{-1} \;\leq\; A \;\leq\; <a> \qquad (2.16)$$

Other methods considering more complex hypotheses have been developed (self consistent method, Haschin-Shtrikman approach...) and have been studied in more specialized papers [21], [31], [32].

3. Experimental evidence at microscale of specific behaviour of granular materials

It is now widely recognized that the mechanical properties of granular materials depend mostly on their microstructure. The goal of this chapter is to present some illustrations of the link between global behaviour and local microstructure.
It can be emphasized first that the distribution of local variables (contact forces, local displacements, local rotations) are highly heterogeneous. In particular, forces are transmitted along discrete paths connecting particles submitted to high levels of contact forces. Between these discrete paths, particles are submitted to low level contact forces [14].

3.1 Global strain and local phenomena

It is usually admitted that global recoverable strain is due to elastic deformations of the contacts between particles and that irrecoverable global strain is linked to sliding and rolling occuring between particles in contact. Breakage of particles can also give rise to large irrecoverable strains but this phenomenon occurs only for large values of stresses which are not considered in this analysis. Considering two rigid neighbouring particles of convex shape, the relative displacement at contact (nearest neighbouring points) and the relative rotation can be defined by (Fig. 3) :

$$\Delta\delta_i^c = \Delta u_i^B - \Delta u_i^A + e_{ijk}\left(\Delta\omega_j^B \, r_k^B - \Delta\omega_j^A \, r_k^A\right)$$

$$\Delta\omega_i^c = \Delta\omega_i^B - \Delta\omega_i^A$$

(3.1)

$\Delta\delta_i^c$ is the displacement of contact point C belonging to particle B minus the displacement of contact point C belonging to particle A ;

e_{ijk}: permutation symbol ;

Δu^A, Δu^B : displacements of centroïds A and B of the two particles ;

$\Delta\omega^A$, $\Delta\omega^B$: rotations of particles A and B ;

$$r^A = \overrightarrow{AC} \quad \text{and} \quad r^B = \overrightarrow{BC}.$$

Considering the local axes defined at contact point C(**n, t**),

if $\Delta\delta_i^c n_i > 0$ a loss of contact occurs (Fig. 4b),

if $\Delta\delta_i^c n_i < 0$ a creation of contact can occur (Fig. 4a).

If during the considered increment of strain $\Delta\delta_i^c n_i = 0$, the relative displacement is only a tangential one which can be related to sliding (Fig. 4d) or rolling (Fig. 4c). These two phenomena may occur simultaneously.

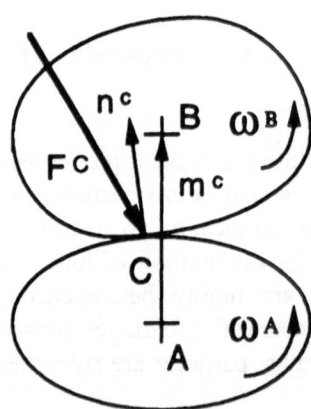

Fig. 3 : Definition of local variables at a contact between two particles.

Many experimental results [4], [11], [25] seem to demonstrate that the prevailing phenomenon to explain irrecoverable global strain is the rolling between particles. The means value of the rotation of particles seems to be of the order of magnitude of the rotation ω_i defined from the tensor gradient of global displacements [4], [11]

$(\Omega_{ij} = \frac{1}{2}(u_{i,j} - u_{j,i})$ and $\omega_i = \frac{1}{2}e_{ijk}\Omega_{jk}$, e_{ijk} being the permutation symbol). It can be noted that, even if this mean value of local rotation ω_i is equal to zero (loading without any change of principal axis orientations), the individual values of rotations are generally significantly different from zero and play an important role on irrecoverable global strains (Fig. 5).

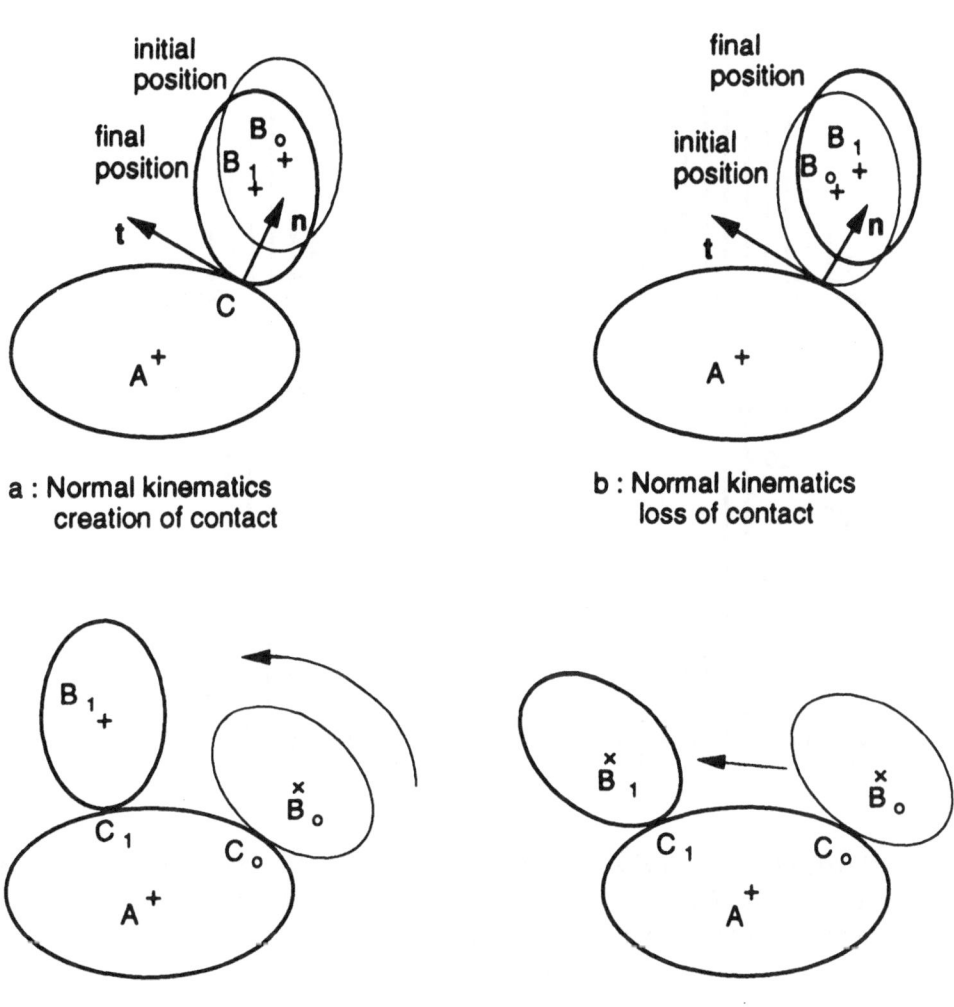

a : Normal kinematics
creation of contact

b : Normal kinematics
loss of contact

c : Tangential kinematics
pure rolling

d : Tangential kinematics
pure slidiing

Fig. 4 : Different kinematics of two neighbouring particles.

All these results have been obtained for two dimensional analogical materials generally with cylindrical shape. The prevalence of rolling has not been clearly demonstrated for natural, angular three dimensional materials.

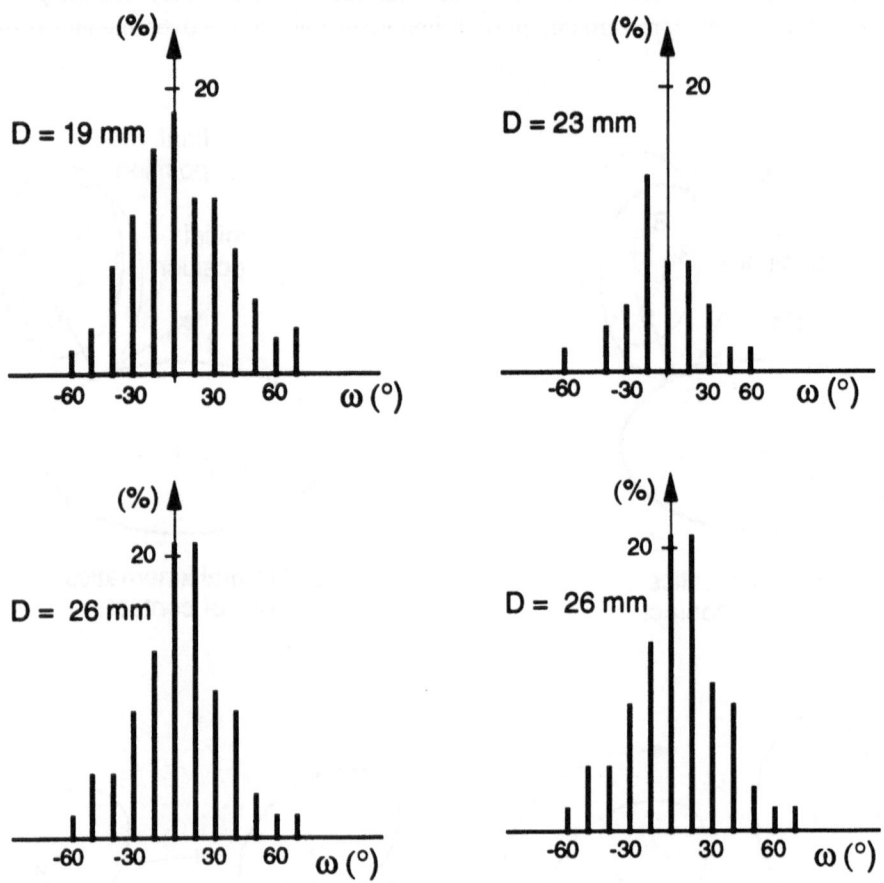

Fig. 5 : Distribution of individual particle rotations ω for the different sizes of particles of a two dimensional analogical material submitted to the test defined in Fig. 6. The rotations are measured between steps 2 and 17 of this compression test.
From Chapuis [11].

3.2 Dilatancy

Dense granular media submitted to deviatoric loadings show an increase of volume. This phenomenon is called dilatancy. Usually this dilatancy vanishes at large strains (Fig. 6a). Tests realized on two dimensional analogical materials allow local information to be collected during the loading. Results obtained by Chapuis [11] very clearly show the local phenomena which lead to this global behaviour. Two points can be clearly emphasized from the results shown in Figure 6 :

. the coordination number (average number of contacts per particle) decreases as the global density of the medium decreases and then is stabilized as the global density remains constant (Fig. 6b),

. when the volume change curve demonstrates the dilatancy phenomenon a significant change in the distribution of the contact orientation occurs : in the major principal stress orientation an increase of the number of contact points can be observed and in the minor principal stress orientation a decrease of this number occurs (Fig. 6c).

It can then be concluded that the global dilatancy is linked at the local level to the decrease of the number of contacts and to their orientation evolutions.

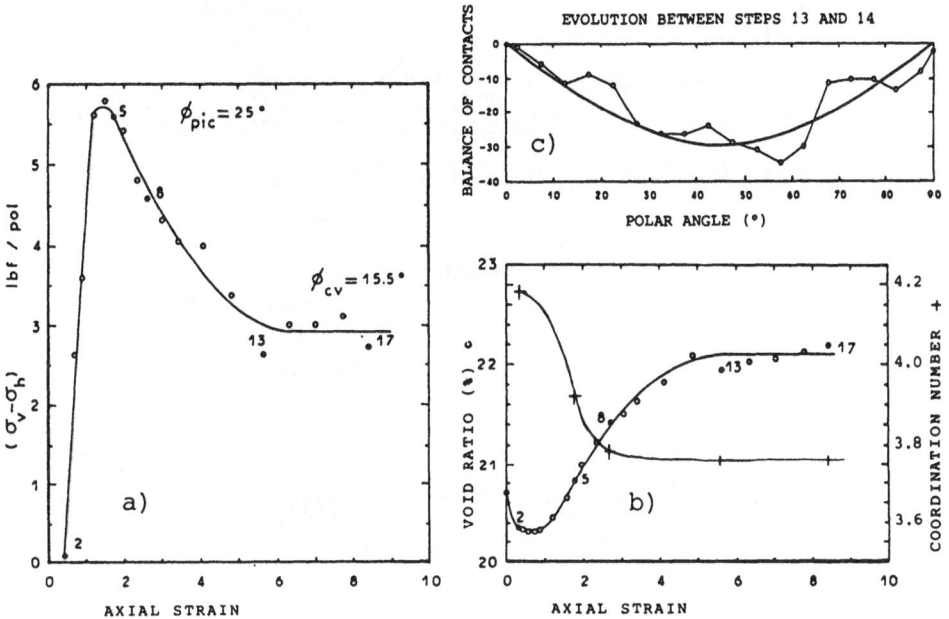

Fig. 6 : Evolution of local variables in a biaxial test realised on two dimensional analogical material. From Chapuis [11]

3.3 Anisotropy

The deposit process of granular materials is usually non isotropic due to the particular direction of the gravity. Furthermore the loading history can induce an evolution of this initial anisotropy. This anisotropy plays an important role in the mechanical behaviour of such a material. Illustrations of this phenomenon have been given by Hicher in the first chapter of this book. This anisotropy can be seen on local measurements such as those performed by Biarez et al [3] (Fig.7). The measurements performed on an analogical two dimensional material show first that the initial distribution of contact orientations is not isotropic (initial anisotropy due to the deposit process) and that the applied strain induces an evolution of the contact orientation. A compressive strain corresponds to an increase in the number of contacts, and an extensive strain corresponds to a decrease in the number of contacts.

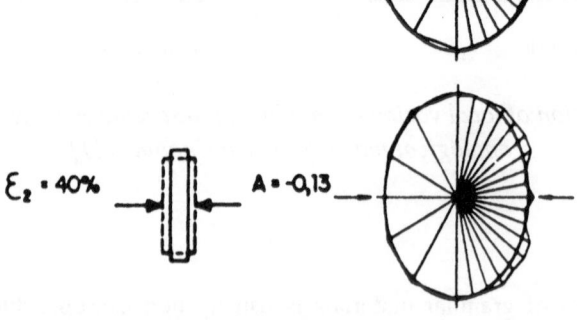

Fig. 7 : Geometrical anisotropy measured on two dimensional analogical material. From Biarez and Wiendieck [3]

These analyses at the local scale show that essential mechanical properties such as irrecoverable strains, dilatancy and induced anisotropy are essentially linked to local phenomena such as relative normal displacements at contacts, rolling, sliding and evolution of the distribution of contacts.

4. Basic concepts in granular media

Granular media are made up of grains in contact. Different local variables can be used to describe this kind of media in particular the local geometry, the local static variables and the local kinematic variables.

Type of packing	Simple cubic	Cubical tetrahedral	Tetragonal sphenoïdal	Pyramidal	Tetraedral
Coordination number N	6	8	10	12	12
Void ratio e	0.909	0.654	0.432	0.350	0.350

Table 1 : Characteristics of regular arrays of identical size spheres.

4.1 Description of the local geometry

Some local variables first allow the particles to be characterized (dimension, shape). These variables are characteristic of the considered medium, and can usually be considered to be constant for any kind of loadings. Their definitions are well known (granulometry curve, shape factors). Other kinds of variables can be used to describe the packing state (number of contacts per particle, orientation of contacts, orientation of particles). These variables will depend on the deposit process and on the history of applied loadings. Their measurements are usually very difficult. The more usual variables of this kind used in micromechanical approaches are proposed below.

4.1.1 Coordination number N

This number is equal to the average number of contacts per particle, it characterizes the density of the packing between particles of a considered granular medium. Different empirical relationships between this number and global density variables (void ratio e, density, porosity n) have been proposed in the literature such as :

$$N = \frac{12}{1+e} \qquad \text{(Field [20]) ,} \qquad (4.1)$$

$$N = 3.183^{(2.469-e)} \qquad \text{(Yanagisawa [31]) ,} \qquad (4.2)$$

$$N = \frac{3}{n} \qquad \text{(Grivas and Harr) ,} \qquad (4.3)$$

N= 13.28 - 8e   (Chang, Misra, Sundaram [8])   (4.4)

Table 1 shows also the values of N and e for regular packings of spheres with identical diameters. This table shows the high correlation between N and e. It can be emphasized that these kinds of very simple formulations, which do not take into account the dimensions and shape of particles, cannot be established from analytical analyses in general cases and have to be considered as only empirically approximated formulations.

4.1.2 Definition of the packing global orientation

In general cases, the global strain in granular media is essentially due to relative displacements and rotations between particles in contact. Therefore the knowledge of the relative locations between particles in contact appears to be of great importance in micromechanical analyses. These relative locations can essentially be described by two vectors (Fig. 3) :

n   : unit vector normal to the tangential contact plane,

m   : branch vector joining the centroïds of two particles in contact ($\overrightarrow{AB}$ in Figure 3).

The distribution of contacts is defined by P(**n**) with :

$\int_\Omega P(\mathbf{n})\, d\Omega = 1$.

$d\Omega$ is a small solid angle with orientation **n** and Ω represents all the possible orientations in the space.

Usually this distribution is represented by radial graphs (Fig. 7). These graphs clearly point out that the deposit process and the applied loadings lead to different states of the internal structure of which P(**n**) gives valuable representations. Global measures of the distributions of **n** and **m** have been proposed in the literature and are known as "fabric tensors" [28] :

contact fabric tensor :

$$H_{ij} = N_o \int_\Omega P(\mathbf{n})\ n_i n_j\ d\Omega = <n_i\, n_j> \ ,$$ (4.5)

branch fabric tensor :

$$H'_{ij} = N_o \int_\Omega P(\mathbf{n})\ m_i(\mathbf{n})\ m_j(\mathbf{n})\ d\Omega = <m_i\, m_j> \ ,$$ (4.6)

combined fabric tensor :

$$H''_{ij} = N_o \int_\Omega P(\mathbf{n})\ n_i\ m_j(\mathbf{n})\ d\Omega = <n_i\, m_j> \ .$$ (4.7)

N_o : number of contacts per unit volume.

H_{ij} and H'_{ij} are symmetric second rank tensors, H''_{ij} is not symmetric.

For isotropic materials and spherical particles (**m** and **n** are then colinear) :

$$P(\mathbf{n}) = \frac{1}{4\pi} \ , \quad H_{ij} = \frac{N_o}{3}\delta_{ij} \ , \quad H'_{ij} = \frac{N_o}{3}l_o{}^2\delta_{ij} \ , \quad H''_{ij} = \frac{N_o}{3}l_o\delta_{ij} \ .$$

l_o = average value of the distance between the centroïds of particles in contact.

δ_{ij} : Kronecker tensor.

These different tensors provide similar information on the anisotropy of the packing. It can be noted that these second rank tensors give a first order description of the considered distributions. Different studies [7] have shown that this description is not sufficient for granular media which have been submitted to complex loading paths. In these cases a second order description using, for example, a fourth rank tensor $< n_i \, n_j \, n_k \, n_l >$ seems necessary (Fig.8).

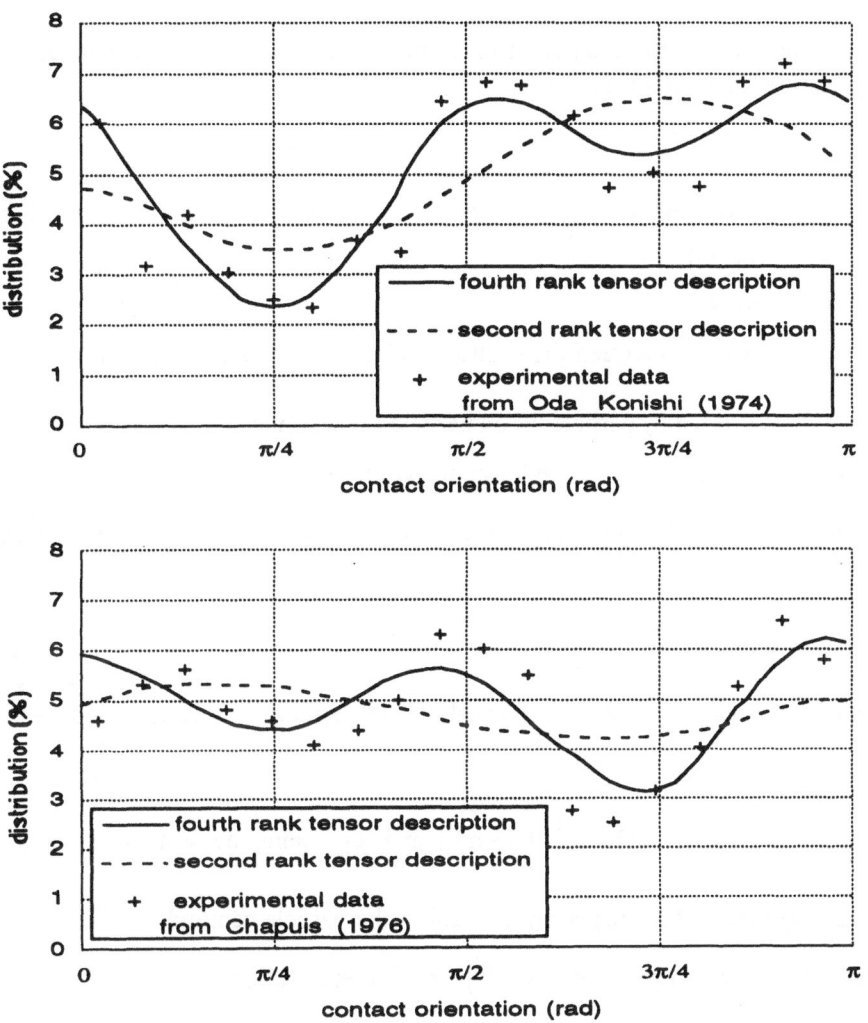

Fig. 8 : Distribution of the contact orientation in a two dimensional analogical material after simple shear test. Description by a second rank tensor and a fourth rank tensor.

4.2 Description of static variables

The usual local static variables considered are the contact forces. Contacts are usually considered to be transmitted at a contact point. The couples which can be transmitted at a contact are usually neglected.

4.2.1 Definition of the stress tensor from local contact forces using the virtual work theorem

The distribution of contact forces in a given volume V can be related to the applied stress tensor by relation [5], [12], [30] :

$$\sigma_{ij} = \frac{1}{V} \sum_c F_i{}^c l_j{}^c \tag{4.8}$$

This relation is written considering the usual convention in solid mechanics $\left(\sigma_{ij}\, n_j > 0 \text{ in extension}\right)$

V : considered volume

$l^c = \overrightarrow{AB}$: branch vector, between the centroïds of two particles in contact at contact C (Fig. 3).
F^c : contact force exerted by particle B on particle A at contact point C (Fig. 3).

This relation (4.8) can be easily derived from the virtual work theorem :

At the global level the following linear virtual increment of displacement field $\Delta^* U(x)$ is considered :

$$\Delta^* U_i(x) = \overset{*}{C_i} + \overset{*}{h_{ij}} x_j \tag{4.9}$$

This field is associated with the virtual global strain : $\quad \Delta^* \varepsilon_{ij} = \frac{1}{2}\left(\overset{*}{h_{ij}} + \overset{*}{h_{ji}}\right)$,

and virtual global rotation :

$$\Delta \overset{*}{\omega}_i = \frac{1}{2} e_{ijk} \Delta \overset{*}{\Omega}_{jk} \quad \text{with} \quad \Delta \overset{*}{\Omega}_{ij} = \frac{1}{2}\left(\overset{*}{h_{ij}} - \overset{*}{h_{ji}}\right) \text{ and } e_{ijk} \text{ being the permutation symbol}.$$

At the local level the virtual displacement field of the particle centroïds $\left(\Delta^* u^A\right)$ and the virtual rotation field of particles $\left(\Delta^* \omega^A\right)$ are chosen as follows :

$$\Delta^* u^A = \Delta^* U\left(x^A\right)$$
$$\Delta^* \omega^A = \Delta^* \omega \tag{4.10}$$

x^A : vector defining the location of the centroïd of particle A.
Then the virtual work theorem can be written as :

$$\sigma_{ij} \overset{*}{\Delta} \varepsilon_{ij} = \frac{1}{V} \sum_c F_i^c \overset{*}{\Delta} \delta_i^c \tag{4.11}$$

F_i^c : Force exerted by particle B on particle A at contact point C .

$\overset{*}{\Delta} \delta_i^c$: increment of virtual local relative displacement at contact C (displacement of C belonging to particle B with respect to the displacement of C belonging to particle A) (Fig. 3).

$\Delta \delta_i^c$ is given by relation (3.1) :

$$\Delta \delta_i^c = \Delta u_i^B - \Delta u_i^A + e_{ijk} \left(\Delta \omega_j^B r_k^B - \Delta \omega_j^A r_k^A \right) .$$

With the particular values of $\overset{*}{\Delta} u^A$, $\overset{*}{\Delta} u^B$, $\overset{*}{\Delta} \omega^A$, $\overset{*}{\Delta} \omega^B$ it follows that :

$$\overset{*}{\Delta} \delta_i^c = \overset{*}{\Delta} U \left(x^B \right) - \overset{*}{\Delta} U \left(x^A \right) - e_{ijk} \overset{*}{\Delta} \omega_j l_k^c , \tag{4.12}$$

with $l^c = \overrightarrow{AB} = r^A - r^B$.

Then $\overset{*}{\Delta} \delta_i^c = \overset{*}{\Delta} \varepsilon_{ij} l_j^c + e_{ijk} \overset{*}{\Delta} \omega l_k^c - e_{ijk} \overset{*}{\Delta} \omega l_k^c = \overset{*}{\Delta} \varepsilon_{ij} l_j^c$, $\tag{4.13}$

thus $\sigma_{ij} \Delta \varepsilon_{ij} = \frac{1}{V} \sum_c F_i^c \overset{*}{\Delta} \varepsilon_{ij} l_j^c$. $\tag{4.14}$

This relation must be satisfied for any value of $\overset{*}{\Delta} \varepsilon_{ij}$ which implies relation (4.8).

4.2.2 Definition of the average stress tensor from local contact forces using Gauss theorem [23]

Similar relationships to relation (4.8) can also be obtained considering the average stress tensor defined in a volume V as (Fig. 9) :

$$\bar{\sigma}_{ij} = \frac{1}{V} \iiint_V \sigma_{ij} dV \tag{4.15}$$

Using Gauss theorem and considering the equilibrium equation : $\sigma_{ij,j} = 0$, it can be deduced that :

$$\bar{\sigma}_{ij} = \frac{1}{V} \iint_S n_k \sigma_{ik} x_j dS \tag{4.16}$$

n_i is the normal unit vector on the boundary S of volume V.

If this boundary does not cut any particle but passes only at contact points, it is then possible to identify $\sigma_{ik} n_k dS$ to all the forces applied at the contacts located on the boundary, called F^β, then relation (4.16) can be written as :

$$\bar{\sigma}_{ij} = \frac{1}{V} \sum_{\beta \in S} F_i^\beta x_j^\beta \tag{4.17}$$

F_i^β : external force acting on the boundary S of volume V, located at x_i^β .

This relation relates external contact forces applied on the boundary of V to the average value of σ defined in V.

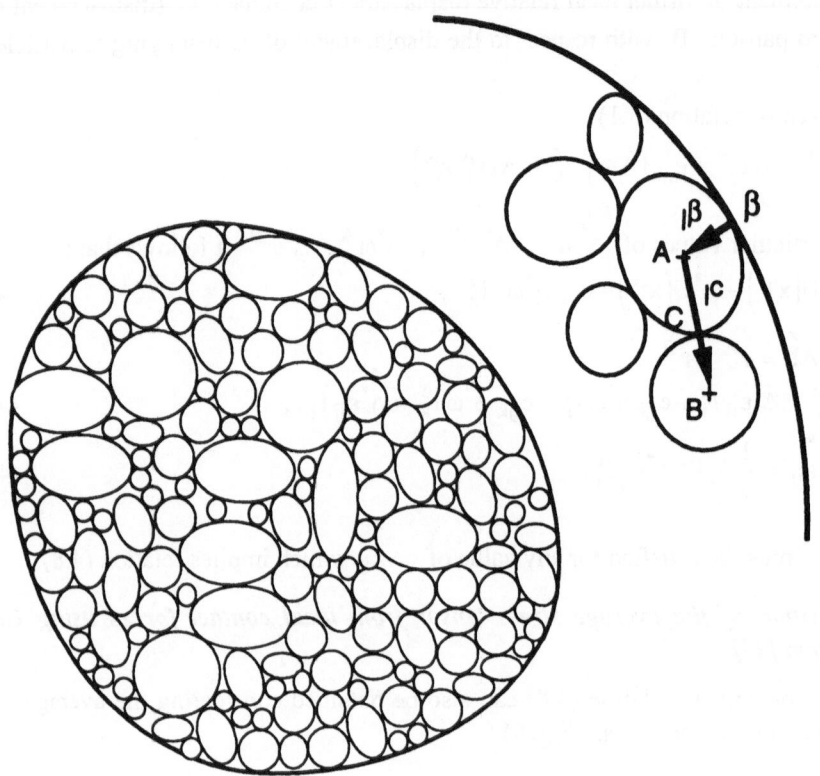

Fig. 9 : Definition of the average stress tensor from local contact forces.

Starting from (4.17) it is also possible to obtain a relation similar to (4.8).
Equilibrium conditions for particle A can be written :

$$\sum_B F_i^{AB} = 0 \tag{4.18}$$

F^{AB} : contact force exerted by B on A $= F^c$
Then :

$$\sum_A \sum_B x_j^A F_i^{AB} = 0 \tag{4.19}$$

x_j^A : center of mass of particle A.

For each boundary contact β with particle A , $x_j^A F_i^{A\beta}$ can be written as :

$$\left(x_j^{A\beta} - 1_j^{A\beta}\right) F_i^{A\beta} \tag{4.20}$$

$1^{A\beta}$: vector connecting the center of mass of A to the boundary contact point located at $x_j^{A\beta}$.

For each contact between two particles A and B two terms will appear in summation (4.19) :

$$x_j^A \, F_i^{AB} + x_j^B \, F_i^{BA} = F_i^c \, 1_j^c \quad , \tag{4.21}$$

with : $F_i^c = F_i^{AB} = -F_i^{BA}$ and $1^c = \overrightarrow{AB}$.

Considering relation (4.21) on internal contacts and relation (4.20) on external contacts, relation (4.19) can be written as :

$$\sum_{\beta \in S} F_i^\beta \, x_j^\beta = \sum_{c \in V} F_i^c \, 1_j^c + \sum_{\beta \in S} F_i^\beta \, 1_j^\beta \quad . \tag{4.22}$$

Then, considering relation (4.17) :

$$\bar{\sigma}_{ij} = \frac{1}{V} \sum_{\beta \in S} F_i^\beta \, x_j^\beta = \left[\sum_{c \in V} F_i^c \, 1_j^c + \sum_{\beta \in S} F_i^\beta \, 1_j^\beta \right] \tag{4.23}$$

Considering that branch vectors connected with boundary contact points $\left(1^\beta\right)$ can be identified to the other branch vectors 1^c , relation (4.23) can be written as :

$$\bar{\sigma}_{ij} = \frac{1}{V} \sum_{c \in V \, \text{and} \, S} F_i^c \, 1_j^c \tag{4.24}$$

It can be noted that the tensor σ_{ij} defined by relations (4.8) is not symmetrical. Different studies [5], [11], [12], [13] have shown that the antisymmetric part of σ_{ij} is linked to the contacts located on the boundaries of the considered volume. If the number of contacts located on the boundaries is small compared to the total number of contacts, the tensor σ_{ij} can be considered to be symmetrical in a first approximation.

4.2.3 Definition of local contact forces from the stress tensor

Relation (4.8) allows the variable σ to be defined from the local variables F^c . This expression can be considered to be well established and is one of the basic relationships in micromechanical studies in granular materials. The inverse relation (F^c from σ) is much more difficult to define. In fact, contact forces are stochastic variables which cannot be defined only from the knowledge of the stress tensor σ. It is then obvious that only average values of contact forces can be evaluated. Considering that the contact orientation **n** is a basic variable for the description of the internal structure, the estimate of the average value of contact forces $\overline{F(n)}$ for a given orientation of contact **n** seems to be of great interest. Different formulations have been proposed in the literature ; we will present two of them.

. Model proposed by Chang et al [10]
Based on a usual formulation in continuous mechanics ($\mathbf{f} = \sigma\mathbf{n}$) Chang et al [10] proposed a similar formulation to define this variable :

$$\overline{F_i(\mathbf{n})} = n_j \, \hat{\sigma}_{ij} \, S \tag{4.25}$$

$\hat{\sigma}_{ij}$: modified stress tensor to be calculated,

S : constant to be evaluated.

Then relation (4.8) can be written in a continuous formulation :

$$\sigma_{ij} = \frac{N}{V} \int_\Omega P(n) \, \overline{F_i(n)} \, l_j(n) d\Omega \tag{4.26}$$

N : number of contacts in volume V,
$P(\mathbf{n})$: Probability of contacts in direction $\mathbf{n}$.

Substituting relation (4.24) in relation (4.25) :

$$\sigma_{im} = N_o \, S \, \hat{\sigma}_{ij} \int_\Omega P(\mathbf{n}) \, n_j \, l_m \, d\Omega = N_o \, S \, \hat{\sigma}_{ij} \, H''_{jm} \;,$$

$N_o = \dfrac{N}{V}$: number of contacts per unit volume,

H''_{jm} is the combined fabric tensor defined by relation (4.6).

Considering an isotropic medium :

$$H''_{jm} = \frac{l_o \, N_o \, \delta_{jm}}{3}$$

l_o : average value of the distance between the centroïds of particles in contact.

$$\text{Then } S = \frac{3}{l_o \, N_o} \qquad \text{and } \sigma_{im} = \hat{\sigma}_{ij} \, \frac{3H''_{jm}}{l_o \, N_o} \quad .$$

$$\hat{\sigma}_{ij} = \sigma_{im} \frac{l_o \, N_o}{3} \, G''_{jm} \;, \tag{4.27}$$

G''_{jm} is the inverse of the combined fabric tensor H''_{im} ,

$$G''_{ik} \, H''_{kj} = \delta_{ij} \quad .$$

The modified stress tensor defined from this relation (4.27) not only takes into account the loading but also the internal fabric of the material.

The average values of contact forces can then be evaluated by :

$$\overline{F_i(n)} = n_j \, \sigma_{im} \, G''_{jm} \quad . \tag{4.28}$$

This formulation clearly shows that the distribution of contact forces depends on the loading and on the internal fabric.

For isotropic materials, $\sigma = \hat{\sigma}$ and $\overline{F_i(n)} = \dfrac{3n_j}{N_o l_o} \sigma_{ij}$

. Model proposed by Cambou et al [6], [18]

A more general formulation can be proposed for the expression of the mean value of contact forces $\overline{F(n)}$ [6].

The local considered variable is :

$$f_i = P(n) \overline{F_i(n)} \dfrac{4\pi N_o l_o}{3} \quad . \tag{4.29}$$

This expression has the dimension of a stress and, in the case of isotropic material, provides a very simple expression of relation (4.25) :

$$f_i = \dfrac{1}{4\pi} \dfrac{3n_j}{N_o l_o} \sigma_{ij} \dfrac{4\pi N_o l_o}{3} = n_j \sigma_{ij} \quad . \tag{4.30}$$

Relation (4.26) can then be written as :

$$\sigma_{ij} = \dfrac{3}{4\pi l_o} \int_\Omega f_i \, l_j \, d\Omega \quad . \tag{4.31}$$

f is considered to be an isotropic function of **n** and σ, linear with respect to σ. The general formulation in this case [6] is :

$$f(n) = A \, \sigma n + B n \, \sigma n \, n + C \, tr(\sigma n) \quad . \tag{4.32}$$

If spherical particles are considered $l_j^c = D^c n_j$ and $l_o = \overline{D}$, then this relation can be substituted in relation (4.32) implying that :

$$A \, \sigma_{ij} + \dfrac{B}{5}\left[tr\sigma_{ij} \, \delta_{ij}\right] + C \, tr\sigma_{ij} \, \delta_{ij} = \delta_{ij} \quad .$$

This relation has to be satisfied for any value of σ_{ij}, then :

$$B = \dfrac{5}{2}(1-A) \quad , \quad C = \dfrac{1}{2}(A-1) \quad .$$

we choose to write $A = \mu$ and then :

$$f(n) = \mu \, \sigma n + \dfrac{1-\mu}{2}(5 \, n \, \sigma \, n - tr\sigma) n \quad . \tag{4.33}$$

This expression has been proposed by Delyon et al [15]. The first term of this expression has the same orientation as the stress vector in a continuum while the second term is normal to the contact plane. The orientation of the distribution of the average contact forces is then characterized by parameter μ :

for $\mu = 0$, the average contact forces are normal,

for $\mu = 1$, the average contact forces are oriented in the σn direction.

This parameter can be considered to be directly linked to the mechanical properties of the contact and is then independent of the loading path. In the two dimensional case, the distribution of contacts is illustrated in Figure 10 which represents vector $\mathbf{f}$ in the $(\mathbf{n}, \mathbf{t})$ plane, the end of $\mathbf{f}$ describes an ellipse whose axes are $\mu(\sigma_1 - \sigma_2)$ in the $\mathbf{t}$ direction, and $(2 - \mu)(\sigma_1 - \sigma_2)$ in the $\mathbf{n}$ direction.

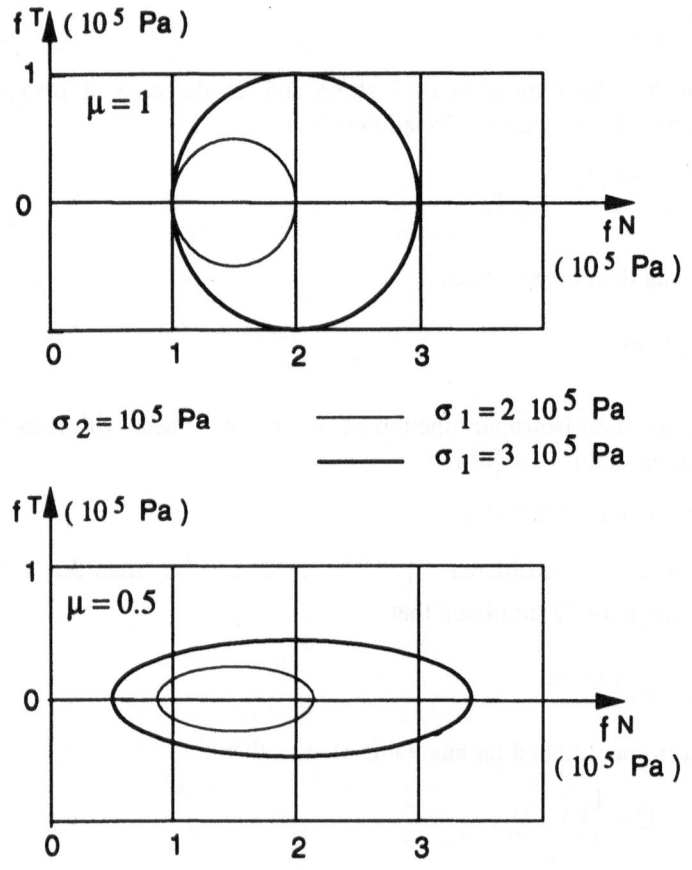

Fig. 10 : Distribution of contact forces for different μ. From Cambou [7].

Of particular interest is the decomposition of the stress tensor σ and its deviatoric part s in two parts σ^N, s^N and σ^T, s^T respectively resulting from the normal and tangential components of $\mathbf{f}$:

$$\sigma = \sigma^N + \sigma^T \qquad\qquad s = s^N + s^T \qquad\qquad\qquad (4.34)$$

It can be easily shown [6], [7] that :

$$s^N = (1 - \mu/2)\, s \qquad\qquad\qquad (4.35)$$

Parameter μ thus also represents the fraction of the deviatoric stress tensor supported by the normal components of contact forces (from 1/2 for $\mu = 1$ to 1 for $\mu = 0$).

The previously described distribution corresponds to a medium whose anisotropy is only described by the distribution $P(n)$. If another form of anisotropy is considered (static anisotropy for example), it is necessary to take into account another internal state variable. For the sake of simplicity, this variable will be chosen as a second rank symmetric tensor e which is the simplest case inducing anisotropy. This internal state variable can be related with one of the "fabric tensors" previously defined (contact fabric tensor, branch fabric tensor, combined fabric tensor). Then the force distribution function f therefore will depend on σ, n and e. The general expansion is rather complex [6] but a first approximation expression has been proposed :

$$f(n) = f_o(n) \;+\; tr\sigma\left[\frac{4}{10}\left(en - e_o(n)n\right)\right] \qquad\qquad\qquad (4.36)$$

$f_o(n)$: expression obtained in (4.33)
$e_o(n) = e_{ij}n_i n_j$

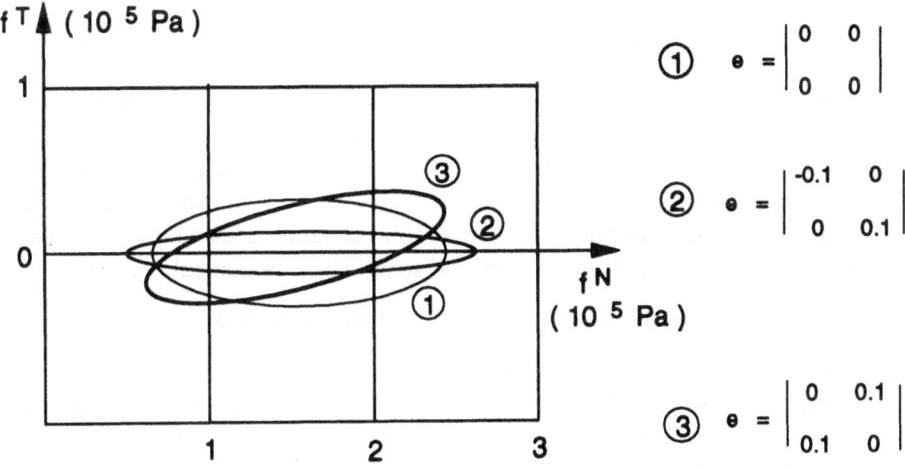

Fig. 11 : *Influence of the anisotropy on the distribution of contact forces in the n, t axes*
($\sigma_1 = \sigma_2 = 2$, $\mu = 0.5$). From Cambou [7].

Figure 11 shows the influence of the internal anisotropy e on the distribution of contact force distribution f in the (n,t) plane. The decomposition defined by relation (4.34) allows the role of the anisotropy to be pointed out. The integration of the tangential components and of the normal components leads to the following relations :

$$s^T = \frac{\mu}{2} s + \frac{tr\,\sigma}{2} e \qquad (4.37)$$

$$s^N = \left(1 - \frac{\mu}{2}\right) s - \frac{tr\,\sigma}{2} e \qquad (4.38)$$

These relations (4.37, 4.38) show that the anisotropy allows tangential components to be transfered to normal components. The anisotropy developed in the distribution of contact forces will allow the bearing of the greater part of the applied deviatoric stress tensor by means of more stable normal contact forces.

4.3 Description of kinematic variables

The local forces in granular media are transmitted at the contacts. The work developed at a contact C is then : $F^c \, \Delta \delta^c$.

$\Delta \delta^c$: relative displacement at contact C. This variable has been defined in relation (3.1) and in Figure 3.

$$\Delta \delta^c_i = (\Delta u^B_i - \Delta u^A_i) + e_{ijk}\left(\Delta\omega^B_j \, r^B_k - \Delta\omega^A_j \, r^A_k\right) \qquad (4.39)$$

It can be easily seen that this relative displacement has two terms. The first is linked to the relative displacement of the two centroïds, and the second is linked to the rotations of the two particles in contact. In particular it is possible to have

$$\Delta \delta^c_i = 0 \quad \text{with} \quad \Delta u^B_i - \Delta u^A_i > 0 \quad \text{and} \quad \left(\Delta u^B_i - \Delta u^A_i\right) = -e_{ijk}\left(\Delta\omega^B_j \, r^B_k - \Delta\omega^A_j \, r^A_k\right)$$

In this case the energy dissipation at contact C will be null even though the local array has changed. Numerical simulations realised on cylindrical particles show that rotations occuring in granular media greatly reduce the dissipation for a given global imposed strain [1], [24]. However, it can be noted that rotations are not so easy in natural granular media made of angular particles. This phenomenon therefore certainly does not play as important a role in natural materials as it does for idealized spherical or cylindrical materials.

4.3.1 Definition of the strain tensor from local kinematic variables

- Definition of the strain tensor from the local relative displacement at contact points.
Corresponding to the stress tensor defined in equation (4.8) the strain tensor can be defined by using the principle of energy conservation. For an increment of strain the equality of external and internal work can be written as [4] :

$$\sigma_{ij} \, \Delta \, \varepsilon_{ij} = \frac{1}{V} \sum_c F^c_i \Delta \, \delta^c_i = \frac{1}{V} \sum_c \frac{F^c_i l^c_j l^c_j \Delta\delta^c_i}{l^c_k l^c_k} \qquad (4.40)$$

$$\sigma_{ij} \Delta \varepsilon_{ij} = \frac{1}{V} \sum_c \left(F_i^c l_j^c \right) \frac{l_j^c \Delta \delta_i^c}{l_k^c l_k^c} \quad . \tag{4.41}$$

Considering the assumption that the two terms appearing in relation (4.41) are uncorrelated, it can be deduced that :

$$\sigma_{ij} \Delta \varepsilon_{ij} = \frac{1}{V} \sum_c \left(F_i^c l_j^c \right) \frac{1}{N} \sum_c \frac{l_j^c \Delta \delta_i^c}{l_k^c l_k^c} \quad , \tag{4.42}$$

then taking into account relation (4.8) and considering that only the symmetric part of the displacement gradient gives rise to work, it follows :

$$\Delta \varepsilon_{ij} = \frac{1}{2N} \sum_c \frac{l_j^c \Delta \delta_i^c + l_i^c \Delta \delta_j^c}{l_k^c l_k^c} \tag{4.43}$$

N : number of contact points in volume V.
In relation (4.43) it can be noted that the global kinematic variable $\Delta \varepsilon_{ij}$ is defined from the local relative displacement at each contact point $\Delta \delta^c$.

- Definition of the strain tensor from the displacement of particle centroïds.
Other formulation can be proposed considering the displacement of particle centroïds. For the sake of simplicity we will consider only a two dimensional analysis. We consider the boundary L of the considered surface S passing throught the particle centroïds (Fig. 12). The average displacement gradient is :

$$\overline{\Delta U_{i,j}} = \frac{1}{S} \iint_S \Delta U_{i,j} dS \tag{4.44}$$

The Gauss theorem leads to :

$$\frac{1}{S} \iint_S \Delta U_{i,j} dS = \frac{1}{L} \int_L \Delta U_i \, n_j dL \tag{4.45}$$

Between two particle centroïds, it is assumed [13] that the displacement field U_i can be considered as a linear function of x_i, then the integral appearing in relation (4.45) can be expressed as a discrete summation :

$$\frac{1}{L} \int_L \Delta U_i \, n_j \, dL = \frac{1}{L} \sum_{AB} \frac{1}{2} \left(\Delta u_i^A + \Delta u_i^B \right) n_j^{AB} |AB| \tag{4.46}$$

Then the average strain tensor can be expressed as the symmetric part of the average displacement gradient (Fig. 12)

$$\overline{\Delta \varepsilon_{ij}} = \frac{1}{L} \sum_{AB} \frac{|AB|}{4} \left[\left(\Delta u_i^A + \Delta u_i^B \right) n_j^{AB} + \left(\Delta u_j^A + \Delta u_j^B \right) n_i^{AB} \right] \tag{4.47}$$

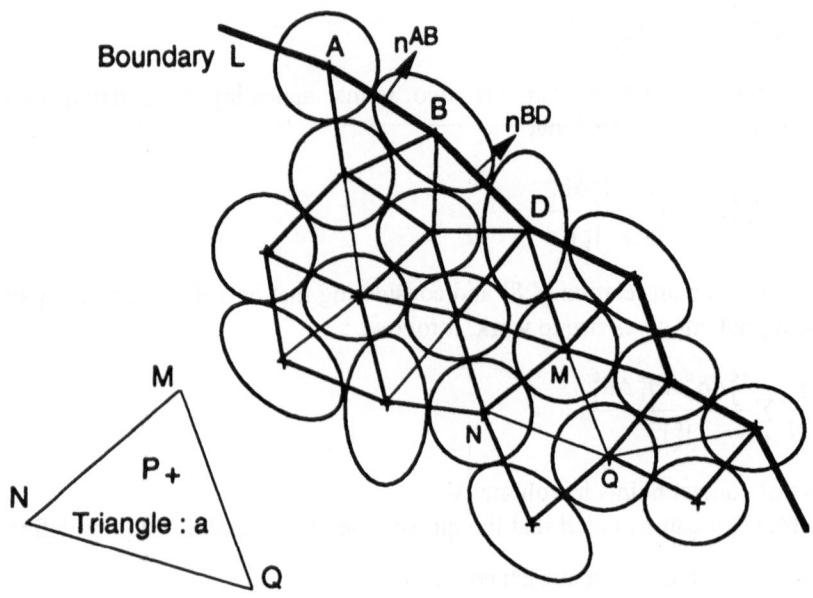

Fig. 12 : Definition of the strain tensor from the displacements of particle centroïds.

Another expression can be obtained, considering that the two dimensional medium can be completely discretized using the Delaunay triangulation (Fig. 12), obtained from the particle centroïds. Then any point P in the medium belongs to a triangle "a" defined by the centroïds of three particles (M, N, Q).

It is assumed that the displacement of point P, ΔU^P is linear with respect to x^P in the triangle MNQ. Then ΔU^P can be expressed from x^P and the displacements of points M, N, Q (Δu^M, Δu^N, Δu^Q). Then it will be easy to define the average value of $\Delta\varepsilon_{ij}$ in each triangular domain "a" :

$$\Delta\varepsilon_{ij}^a = \frac{1}{2}\left(\frac{\partial U_i^P}{\partial x_j} + \frac{\partial U_j^P}{\partial x_i}\right)$$

as ΔU_i^P is linear with respect to x_j, $\Delta\varepsilon_{ij}^a$ is constant in each triangle, depending on Δu^M, Δu^N, Δu^Q and x^M, x^N, x^Q , then :

$$\Delta\bar{\varepsilon}_{ij} = \frac{1}{N}\sum_a \frac{1}{S^a}\varepsilon_{ij}^a\left(\Delta u^M, \Delta u^N, \Delta u^Q, x^M, x^N, x^Q\right) \tag{4.48}$$

N: number of triangles in the considered domain.

It can be pointed out that two kinds of relations have been obtained to express the global variable $\Delta\varepsilon$:

a) relation (4.43) relates $\Delta\varepsilon$ to the local relative displacement at each contact point $\Delta\delta^c$,

b) relations (4.47), (4.48) relate $\Delta\varepsilon$ to the local displacement of particle centroïds Δu^A.

It can be noted that the two considered local variables are related by relation (4.39). This relation shows the difference of these local variables which is directly linked to the rotations of particles ($\Delta\omega^A$, $\Delta\omega^B$). Relations involving the centroïd displacements are certainly more realistic for materials with important local rotations. Relation envolving the local relative displacements does not take into account local rotations and can be only used for materials and loadings compatible with this assumption. Nevertheless this relation is convenient as will be seen further on in the homogenization techniques because, in these approaches, a relation is needed between the local kinematic and static variables. This is usually written between ΔF^c and $\Delta\delta^c$ (the contact law).

- Definition of the strain tensor using particular expressions for local static variables

The previous relations (4.43, 4.47, 4.48) are general. Another way to define $\Delta\varepsilon$ is to consider the principle of energy conservation and particular expressions for local static variables proposed in paragraph 4.2.2.

If relation (4.28) is considered :

$$\overline{F}_i(n) = n_j \sigma_{im} G''_{jm}$$

The principle of energy conservation can be written, considering the means values of contact forces $\overline{F}(n)$ and of the increment of relative displacements $\Delta\overline{\delta}(n)$ for a given orientation of contact plane n.

$$\sigma_{ij} \Delta \varepsilon_{ij} = \frac{N}{V} \int_\Omega P(n)\, \overline{F}_i(n)\, \overline{\Delta\delta_i}(n)\, d\Omega \tag{4.49}$$

Substituting relation (4.28) for relation (4.49) $\Delta\varepsilon_{ij}$ can be evaluated. In fact the strain tensor corresponds only to the symmetric part of the displacement gradient, then :

$$\Delta\varepsilon_{ij} = \frac{G''_{jq}}{2}\left[\int_\Omega P(n)\, n_q\, \overline{\Delta\delta_i}(n)\, d\Omega\right] + \frac{G''_{iq}}{2}\left[\int_\Omega P(n)\, n_q\, \overline{\Delta\delta_j}(n)\, d\Omega\right] \tag{4.50}$$

This formulation includes the fabric tensor $H'' = G''^{-1}$ for a measure of the packing structure.

Another expression of the increment of strain can be obtained considering relation (4.33) in the equality of external and internal work (4.49). The following development is only proposed for spherical particles.

$$\sigma_{ij} \Delta\varepsilon_{ij} = \frac{3}{4\pi} \int_\Omega f_i\, \Delta r_i\, d\Omega \tag{4.51}$$

with $\Delta r_i = \dfrac{\overline{\Delta \delta_i(\mathbf{n})}}{\overline{D}}$ (4.52)

$\overline{D}$ average value of particle diameters $= l_0$.
Then substituting relation (4.33) for (4.51) it is then possible to obtain an expression of $\Delta \varepsilon$ from $\Delta \mathbf{r}$.

$$\Delta \varepsilon_{ij} = \int_\Omega \left[\mu \Delta r_i \ n_j + \frac{1-\mu}{2} \left(5 n_i \ n_j - \delta_{ij} \right) \Delta r_k n_k \right] d\Omega$$ (4.53)

This expression (4.53) will be used further on as averaging operator to define the global strain $\Delta \varepsilon$ from the local variable $\Delta \mathbf{r}$.

4.3.2 Definition of the local kinematic variables from the strain tensor

To define the inverse relation (local variable from the strain tensor), an analysis similar to the static one can be proposed in the simplest case in which variable $\Delta \mathbf{r}(\mathbf{n})$ is assumed to be an isotropic function of the two variables $\mathbf{n}$ and $\Delta \varepsilon$. The representation theorems [6] are used and $\Delta \mathbf{r}$ is written in the following form :
$\Delta \mathbf{r} = A' \Delta \varepsilon \ \mathbf{n} + (B' \mathbf{n} \ \Delta \varepsilon \mathbf{n} + C' \mathrm{tr} \Delta \varepsilon) \ \mathbf{n}$.
This relation is substituted for relation (4.51) which allows two of the three constants (A', B', C') to be defined. We then obtain :

$$\Delta \mathbf{r}(\mathbf{n}) = \left\{ \left[1 + b\left(\frac{3}{5}\mu - 1 \right) \right] \Delta \varepsilon \ \mathbf{n} + b\left(\mathbf{n} \ \Delta \ \varepsilon \ \mathbf{n} - \frac{\mu}{5} \mathrm{tr} \Delta \varepsilon \right) \mathbf{n} \right\}$$ (4.54)

μ : variable defined in the static analysis (4.33),
b : new internal variable appearing in the kinematic analysis.

Physical meanings of variable b can be pointed out considering the normal and tangential components of $\Delta \mathbf{r}$:

$$\Delta r^N = \left[(\mathbf{n} \Delta \varepsilon) \mathbf{n} + \frac{\mu b}{5} [3 \mathbf{n} \Delta \varepsilon \ \mathbf{n} - \mathrm{tr} \Delta \varepsilon] \right]$$ (4.55)

$$\Delta r^T = \left[1 + b\left(\frac{3\mu}{5} - 1 \right) \right] \left[\mathbf{n} \Delta \varepsilon - (\mathbf{n} \Delta \varepsilon \ \mathbf{n}) \mathbf{n} \right]$$ (4.56)

It is easy to draw the extremity of vector $\Delta \mathbf{r}$ in axes Δr^N, Δr^T for the two dimensional case. Figure 13 shows that parameter b will increase the difference between the maximum and the minimum values of the normal relative displacements and will decrease the relative tangential displacement for a global given kinematic $\Delta \varepsilon$. To achieve this kind of global kinematic the special local phenomena will be (Fig. 4) :

for the relative normal displacement, the creation and loss of contacts which are possible in a granular material and are not possible in a continuum,

. for the relative tangential displacement, the possible relative rotation (rolling without sliding) which allows the relative tangential displacement to be decreased in comparison with the displacement in a continuum.

Then the variable b can be considered to be a measure of these special features occuring in granular media : creation and loss of contacts between particles, relative rotation between two particles in contact.

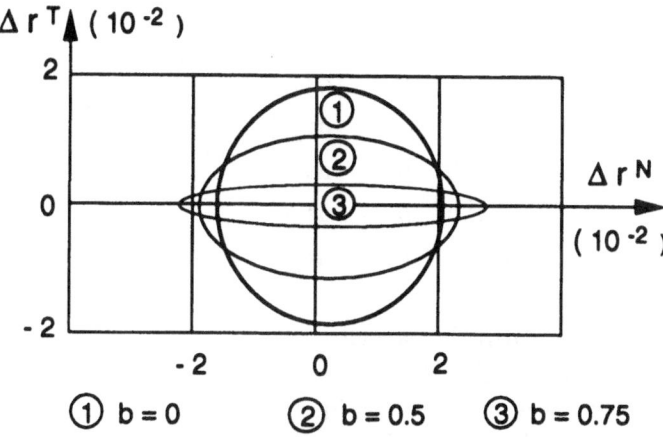

$$\Delta r^T \ (10^{-2})$$

$$\Delta r^N$$

$$(10^{-2})$$

① b = 0   ② b = 0.5   ③ b = 0.75

Fig. 13 : Influence of parameter b on the distribution of Δr in the n, t   axes.
From Cambou [7].

5. Homogenization techniques for granular media

It has been shown in paragraph 1 that essentially two kinds of techniques can be used for the homogenization of materials.

5.1 Homogenization methods for periodic media applied to granular materials.

In general cases particles in granular materials are randomly distributed. Therefore the hypothesis of periodic media is generally non realistic. Nevertheless two cases have been analysed in the literature : regular arrays of spheres and periodic media with a basic cell including a representative number of particles.

5.1.1 Regular arrays of spheres.

Analytical calculations can be achieved on regular arrays of identical spheres. Essentially two kinds of analyses have been performed. The first consists in the definition of elastic characteristics derived from an elastic contact law (essentially the Hertz-Mindlin model) and the second in the analysis of limit states. The procedure proposed by these authors considers a cubical representative cell and defines stress and strain for this cell in a conventional manner. For more complicated packing configurations that do not exhibit cubic symmetry and for general loading conditions, the derivation of the stress-strain relationship becomes very involved and intractable.

. Elastic analysis

Deresiewicks [16] gives results for two regular packings of like spherical granules. For a face centered cubic array (Fig. 12a) submitted to an arbitrary incremental stress applied subsequent to an intial isotropic compressive stress σ_o the stress-strain relationship has been given in an incremental from :

$$
\begin{vmatrix} d\sigma_{11} \\ d\sigma_{22} \\ d\sigma_{33} \\ d\sigma_{23} \\ d\sigma_{13} \\ d\sigma_{12} \end{vmatrix} = \begin{vmatrix} c_1 & c_2 & c_2 & 0 & 0 & 0 \\ c_2 & c_1 & c_2 & 0 & 0 & 0 \\ c_2 & c_2 & c_1 & 0 & 0 & 0 \\ 0 & 0 & 0 & c_2 & 0 & 0 \\ 0 & 0 & 0 & 0 & c_2 & 0 \\ 0 & 0 & 0 & 0 & 0 & c_2 \end{vmatrix} \begin{vmatrix} d\varepsilon_{11} \\ d\varepsilon_{22} \\ d\varepsilon_{33} \\ d\varepsilon_{23} \\ d\varepsilon_{13} \\ d\varepsilon_{12} \end{vmatrix}
$$

with :

$$
c_1 = \frac{4 - 3v_m}{2 - v_m} \left[\frac{3G_m^2 \sigma_o}{2(1 - v_m)} \right]^{1/3}
$$

$$
c_2 = \frac{v_m}{2(4 - 3v_m)} c_1
$$

(5.1)

G_m : shear modulus of the particle material,
v_m : Poisson'ratio of the particle material.

This formulation has been obtained considering that $d\sigma_{ij}$ is small with respect to σ_o. Deresiewicz [16] has also analysed a simple cubic lattice of like spheres where calculations are easier and the strain-stress relationship can be obtained in a more general form :

$$
\varepsilon_{ii} = \frac{3(1 - v_m)R\sigma_o}{2G_m a_o} \left[\left(1 + \frac{\sigma_{ii}}{\sigma_o} \right)^{2/3} - 1 \right]
$$

$$
\varepsilon_{ij} = \frac{3 \tan \psi (2 - v_m) R\sigma_o}{8G_m a_o} \left\{ \begin{array}{l} \frac{\sigma_{ij}}{\tau_{jk}} \left(1 + \frac{\sigma_{ii}}{\sigma_o} \right)^{2/3} \left[1 - \left(1 - \frac{\tau_{jk}}{\tan \psi (\sigma_o + \sigma_{ii})} \right)^{2/3} \right] \\[3mm] + \frac{\sigma_{ji}}{\tau_{ik}} \left(1 + \frac{\sigma_{ij}}{\sigma_o} \right)^{2/3} \left[1 - \left(1 - \frac{\tau_{jk}}{\tan \psi (\sigma_o + \sigma_{ij})} \right)^{2/3} \right] \end{array} \right\}
$$

(5.2)

where τ_{jk} denotes the resultant shear stress on a plane whose normal is parallel to the i direction, i.e. $\tau_{jk} = \left(\sigma_{ij}^2 + \sigma_{ik}^2 \right)^{1/2}$,

σ_o : initial isotropic stress,
σ_{ij} : applied stress leading to the strain ε_{ij},
tan ψ : coefficient of static friction,
R : radius of particles,

a_0 : radius of contact under the initial loading σ_0 : $a_0 = \left[\dfrac{3\left(1 - \nu_m^2\right)}{E_m} \sigma_0 \right]^{1/3} R$

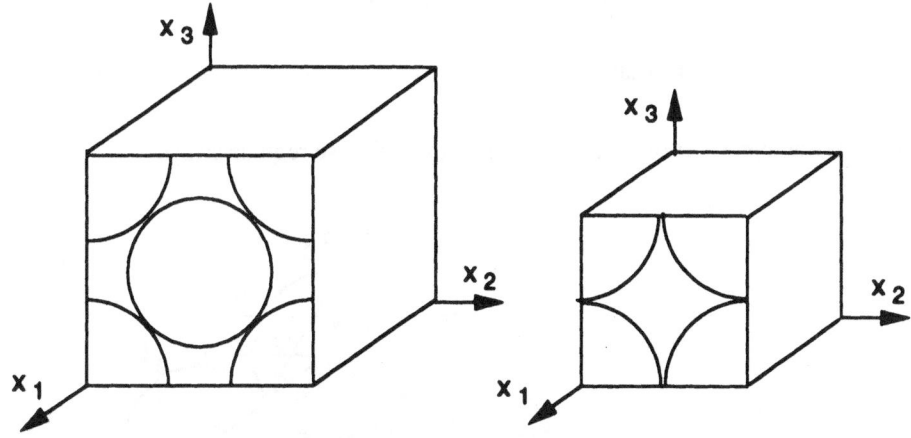

a : face centered cubic array   b : simple cubic array

Fig. 14 : Elementary cube of regular arrays of like spheres.

. Analysis of limit states

The second kind of analysis performed considering regular arrays deals with the definition of limit states. Deresiewicz [16] analyses the failure of simple cubic lattice, under initial isotropic pressure σ_0. When it is compressed by σ in the direction of one of its face diagonals the total stress tensor is :

$$\sigma_{11} = \sigma_{12} = \sigma_{22} = \sigma_0 + \frac{\sigma}{2}$$

$$\sigma_{33} = \sigma_0 \qquad \sigma_{13} = \sigma_{23} = 0$$

The failure occurs when :

$$\frac{\sigma}{\sigma_0} = \frac{2 \tan \psi}{1 - \tan \psi} \tag{5.3}$$

$\tan \psi$: coefficient of friction of the particle material.

For more general loading, σ applied in the direction θ with respect to an axis of the cubic array, the criterion becomes :

$$\frac{\sigma}{\sigma_0} \leq \frac{2 \tan \psi / \sin 2\theta}{1 - \tan \psi \tan \theta} \qquad \text{for} \qquad 0 \leq \theta \leq \pi/4 , \tag{5.4}$$

$$\frac{\sigma}{\sigma_o} \leq \frac{2 \tan \psi / \sin 2\theta}{1 - \tan \psi \cot an\theta} \qquad \text{for} \qquad \pi/4 \leq \theta \leq \pi/2 \ . \tag{5.5}$$

The corresponding smallest stress causing failure is :

$$\frac{\sigma}{\sigma_o} = 2 \tan \psi \left(\tan \psi + \left(1 + \tan^2 \psi\right)^{1/2} \right) \tag{5.6}$$

This criterion clearly shows the high anisotropy of these regular arrays.

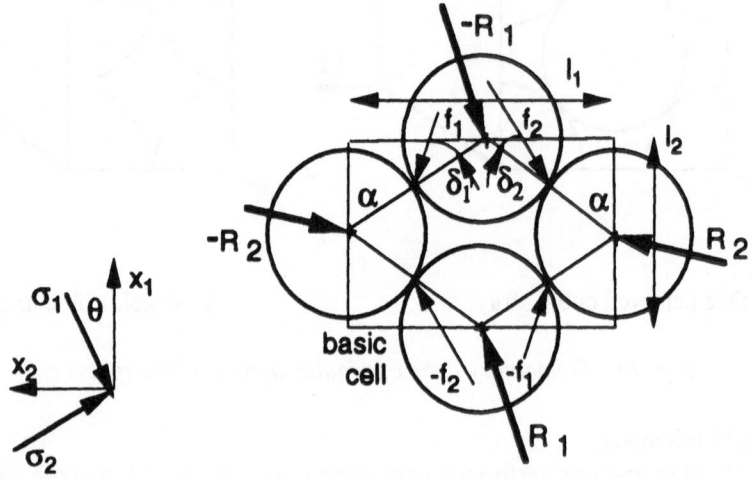

Fig. 15 : Analysis of failure condition for two dimensional regular arrays of like spheres. From Biarez [2].

Biarez [2] has analysed two dimensional regular arrays of like particles whose the structure is characterized by angle α and the direction of principal stresses by angle θ (Fig. 15).
Writing relation (4.8) in the basic cell it is easy to obtain :

$$\frac{\sigma_1}{\sigma_2} = \tan(\theta + \alpha) \tan(\delta_2 + \alpha - \theta) = \tan(\theta - \alpha) \tan(\delta_1 - \alpha - \theta) \ , \tag{5.7}$$

if $\theta = 0$ (loading in the axes of the array)

$$\frac{\sigma_1}{\sigma_2} = \tan(\alpha) \tan(\delta + \alpha) \ ,$$

δ : orientation of contact forces with respect to the contact normal (Fig. 15).
Sliding will occur for $\delta = \Psi$.

In this case $(\theta = 0)$ the sliding will occur for : $\dfrac{\sigma_1}{\sigma_2} = \tan\alpha\,\tan(\psi + \alpha)$. (5.8)

In this case the strain linked to the sliding can be defined as :

$$\frac{\Delta\varepsilon_1}{\Delta\varepsilon_2} = \frac{dl_1\ l_2}{l_1\ dl_2} = \frac{l_2\ dl_1}{l_1\ dl_2} = \tan\alpha\,\tan\alpha = \tan^2\alpha \qquad (5.9)$$

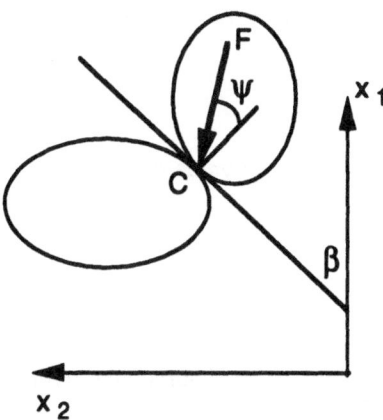

Fig. 16 : Definition of local variables used to obtain the stress-dilatancy relationship [26].

. Rowe stress-dilatancy relationship
Rowe, [26], [27] starting from a very simple analysis of the failure condition at a contact point, obtained a very interesting relationship known as the "Rowe stress dilatancy relationship" valid for a usual triaxial loading $(\sigma_2 = \sigma_3)$.

$$\frac{\sigma_1}{\sigma_2} = \tan^2\left(\frac{\pi}{4} + \frac{\psi}{2}\right)\frac{1}{1 + \dfrac{\Delta\varepsilon_v}{\Delta\varepsilon_1}} \qquad (5.10)$$

$\Delta\varepsilon_1 > 0$ for compressive strain,
$\Delta\varepsilon_2, \Delta\varepsilon_3 > 0$ for extension strain,
$\Delta\varepsilon_v > 0$ for volume increase : $\Delta\varepsilon_v = -\Delta\varepsilon_1 + 2\Delta\varepsilon_3$.

The sliding condition is written at contact C on the contact force F (Fig. 16) :

$$\left|\frac{F_1}{F_2}\right| = \tan(\beta + \psi) \qquad (5.11)$$

$$\left|\frac{dl_1}{dl_2}\right| = \frac{1}{\tan\beta} \;,$$

(5.12)

$$\left|\frac{F_1 dl_1}{F_2 dl_2}\right| = \frac{\tan(\beta + \psi)}{\tan\beta} \;,$$

(5.13)

This energy ratio is considered valid at the macro scale :

$$\frac{\sigma_1 d\varepsilon_1}{2\sigma_3 d\varepsilon_3} = \frac{\tan(\beta + \psi)}{\tan\beta} \;.$$

(5.14)

Rowe considers that sliding occurs for the value of β which minimizes this energy ratio, then :

$$\beta_{min} = \frac{\pi}{4} - \frac{\psi}{2} \;.$$

(5.15)

Then relation (5.10) can be derived. This relation, though it has been obtained considering very crude approximations, is usually in good agreement with experimental results.

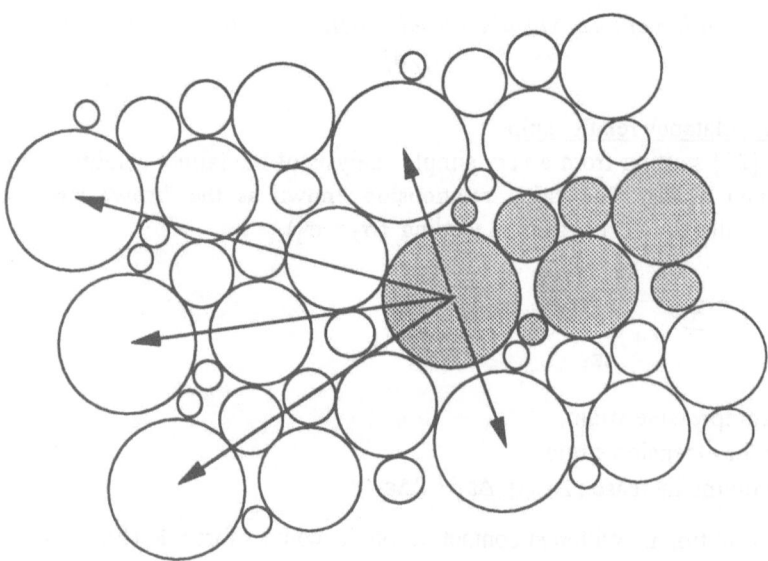

Fig. 17 : Definition of the basic cell of a periodic granular materia. From Caillerie [5].

5.1.2 Granular media constituted of periodic cells including a representative number of particles

Several authors [5], [9] have proposed to consider that the basic cell is composed of a representative number of particles and that the material can be obtained by juxtaposition of like cells (Fig. 17). In this approach the problem lies in the difficulties in solving the mechanical problem at the cell scale. Usually only numerical solutions can be obtained (see chapter on numerical modelling of granular materials by J.P. Bardet).

5.2 Homogenization methods for randomly distributed phases applied to granular materials

It has been shown in paragraph 1 that these approaches do not allow exact solutions to be defined but allow only for estimates or bounds.
A homogenization process includes :
. a localisation operator,
. a local constitutive equation,
. an averaging process.
Usually the principal difficulty lies in the definition of the localisation operator and the solutions proposed in the literature differ on this point. The results proposed in the literature essentially deal with elastic behaviours and failure conditions. We will first present two examples of models dealing with elastic behaviours.

5.2.1 Model proposed by Chang and Liao [10] for the estimate of elastic characteristics

. kinematic hypothesis.
In this model, following the diagram of Figure 2 in chapter 1, the local relative displacement will be defined from the global strain. Then the local contact will allow the contact forces to be defined and finally an averaging process will lead to the global stress from contact forces. The basic assumption lies in the definition of the kinematic localisation operator. The relative displacement at contact C will be estimated from the global kinematics by :

$$\Delta \delta_m^c = \Delta \varepsilon_{mp} \, l_p^c \tag{5.16}$$

I : branch vector between the two particles in contact at contact point C (Fig. 3).
The local contact law can be expressed in an incremental form :

$$\Delta F_i^c = K_{im}^c \Delta \delta_m^c \tag{5.17}$$

K_{im}^c is the contact stiffness tensor.
The averaging process is defined by relation (4.8) :

$$\Delta \sigma_{ij} = \frac{1}{V} \sum_c \Delta F_i^c l_j^c \tag{5.18}$$

Then considering the three relations (5.16), (5.17), (5.18) :

$$\Delta\sigma_{ij} = \frac{1}{V}\sum_c K^c_{im}\,\Delta\varepsilon_{mp}\,l^c_p\,l^c_j$$

$$\Delta\sigma_{ij} = \frac{1}{V}\Delta\varepsilon_{mp}\sum_c K^c_{im}\,l^c_p\,l^c_j$$

The incremental constitutive law is then defined by the fourth rank tensor :

$$C_{ijmp} = \frac{1}{V}\sum_c K^c_{im}\,l^c_p\,l^c_j$$

Chang and Liao [10] give the expression of this tensor for equal size granules and an isotropic medium. The form of the constitutive matrix is found, in this case, to resemble the form of an isotropic elastic medium (5.27) with :

$$E = \frac{4NR^2}{3V}k_n\left(\frac{2+3\alpha}{4+\alpha}\right) \quad, \quad G = \frac{2NR^2}{15V}k_n(2+3\alpha) \quad, \quad v = \frac{1-\alpha}{4-\alpha} \quad, \tag{5.19}$$

with : $\alpha = \dfrac{k_s}{k_n}$,

k_n = contact normal stiffness,
k_s = contact tangential stiffness.

. Static hypothesis

Chang and Liao [10] have proposed another estimate of elastic characteristics of granular media considering the inverse path defined in Figure 2 of Chapter 1. Starting from the global stress, the local forces are defined first, then the local contact displacements are defined and finally the global strain. The basic assumption lies in the definition of local forces from the stress tensor given by relation (4.28) which is considered valid for each contact C :

$$\Delta F^c_i = \Delta\sigma_{im}\,n^c_j\,G''_{jm} \tag{5.20}$$

G''_{jm} is the inverse of the combined fabric tensor H''_{jm} defined from relation (4.7).

This expression of ΔF^c_i has led to the definition of $\Delta\varepsilon_{ij}$ given by relation (4.50) which can also be written as follows :

$$\Delta\varepsilon_{ij} = \frac{1}{2V}\left[G''_{jq}\sum_c n^c_q\,\Delta\delta^c_i\right] + \frac{1}{2V}\left[G''_{iq}\sum_c n^c_q\,\Delta\delta^c_j\right] \tag{5.21}$$

Considering the local contact model :

$$\Delta\delta^c_i = S^c_{ij}\,\Delta F^c_j \quad, \tag{5.22}$$

the three equations (5.20), (5.21), (5.22) lead to :

$$\Delta\varepsilon_{ij} = \frac{1}{2V}\left[G_{jq}^{"}\sum_{c}\left(n_q^c S_{ip}^c n_k^c G_{kl}^{"}\right)\Delta\sigma_{pl}\right] + \frac{1}{2V}\left[G_{iq}^{"}\sum_{k}\left(n_q^c S_{jp}^c n_k^c G_{kl}^{"}\right)\Delta\sigma_{pl}\right]$$ (5.23)

In the case of equal size granules and isotropic medium, the elastic matrix is similar to the matrix of an isotropic medium (5.27) with :

$$E = \frac{20NR^2}{3V}k_n\left(\frac{\alpha}{2+3\alpha}\right) \quad , \quad G = \frac{10NR^2}{3V}k_n\left(\frac{\alpha}{3+2\alpha}\right) \quad , \quad \nu = \frac{1-\alpha}{2+3\alpha} \quad .$$ (5.24)

5.2.2 Model proposed by Cambou et al [6], Emeriault et al [18] for the estimate of elastic characteristics

Variables $f(n)$ et $\Delta r(n)$ defined by relations (4.29) and (4.52) are considered in this analysis. As a first approximation a linear contact model, which is similar to the model used by Chang and Liao is considered :

$$f^n(n) = k_n\,\Delta r^n \qquad \text{and} \qquad f^t(n) = \alpha k_n\,\Delta r^t$$ (5.25)

k_n is the normal local rigidity and αk_n the tangential local rigidity.

. First kinematic hypothesis (Voigt approximation)
The local relative displacement field Δr is approximated from the global strain $\Delta\varepsilon$ by :

$$\Delta r = \frac{3}{4\pi}\varepsilon\,n$$ (5.26)

This approximation is very similar to the Voigt approximation described in paragraph 1 for homogenization processes used for continuous media. Then using relations (5.25), (5.26), (4.8) σ_{ij} can be evaluated from ε_{ij}. If the structure of the granular material is isotropic, distribution $P(n) = \frac{1}{4\pi}$

Then the global behaviour can be evaluated by the classical isotropic elastic model :

$$\sigma_{ij} = \frac{\nu E}{(1-2\nu)(1+\nu)}\varepsilon_{kk}\delta_{ij} + \frac{E}{1+\nu}\varepsilon_{ij}$$ (5.27)

The macroelastic constants E, ν, G, K can then be calculated :

$$E = \frac{3k_n}{4\pi}\frac{2+3\alpha}{4+\alpha} = E_o\frac{4+6\alpha}{4+\alpha} \qquad\qquad \nu = \frac{1-\alpha}{4+\alpha}$$

$$G = \frac{E}{2(1+\nu)} = E_o\frac{4+6\alpha}{10} \qquad\qquad 3K = \frac{E}{(1-2\nu)} = E_o\frac{4+6\alpha}{2+3\alpha}$$ (5.28)

E : Young modulus   G : shear modulus
ν : Poisson's ratio   K : compressibility modulus

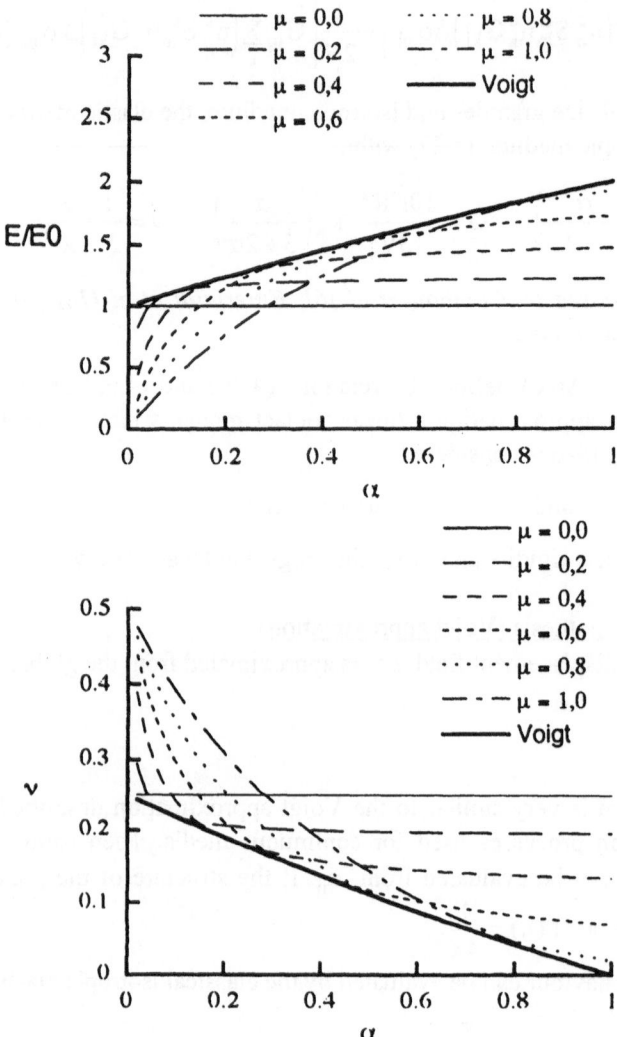

Fig. 18 : Elastic characteristics given by different homogenization processes
(5.28), (5.29). From Cambou et al [6].

<u>Static analysis</u>
The local distribution of contact forces is approximated from the global stress tensor by
relation (4.33) including only one internal variable μ, if the anisotropy of the medium is
only described by the distribution of contact P(n). Then using relations (5.25) and (4.53)
the strain tensor can be expressed from the stress tensor. If the structure of the granular
material is isotropic, distribution $P(n) = \dfrac{1}{4\pi}$. Then the global behaviour can be evaluated

by the classical isotropic elastic model (5.27). The macroelastic constants can then be calculated :

$$E = E_o \frac{\alpha}{\frac{\mu^2}{5} + \alpha\left(1 - \mu + \frac{3\mu^2}{10}\right)} \qquad \text{with} \quad E_o = \frac{3k_n}{8\pi}$$

(5.29)

$$\nu = \frac{2\mu^2 + \alpha\left(5 - 10\mu + 3\mu^2\right)}{4\mu^2 + \alpha\left(20 - 20\mu + 6\mu^2\right)} \qquad G = \frac{10E_o\,\alpha}{6\mu^2 + \alpha\left(25 - 30\mu + 9\mu^2\right)} \qquad 3K = 2E_o$$

These values of ν and E have been represented as function of α in Figure 18.

. <u>Second level of kinematic approximation.</u>

In a previous development, a kinematic localization operator (5.26) was proposed leading to what has been called Voigt's homogenization. This operator is very simple and, unlike the static localization operator does not take into account any internal variable. In this paragraph a more general kinematic localization operator defined by relation (4.54) is considered. This new localization operator, is defined with two internal variables, μ and b. Then using relations (4.54), (5.25), (4.8) the stress tensor σ' can be defined from ε. If the structure of the granular material is isotropic, distribution $P(n) = \frac{1}{4\pi}$ and the global behaviour can be defined by the usual isotropic elastic model (5.27). The macroelastic constants can then be calculated :

$$E = \frac{3k_n}{8\pi} \frac{2 + 6b\frac{\mu}{5} + 3\alpha\left[1 - b\left(1 - 3\frac{\mu}{5}\right)\right]}{2 + b\frac{\mu}{5} + \frac{\alpha}{2}\left[1 - b\left(1 - 3\frac{\mu}{5}\right)\right]} \qquad G = \frac{E_o}{5}\left\{2 + 6\frac{b\mu}{5} + 3\alpha\left[1 - b\left(1 - 3\frac{\mu}{5}\right)\right]\right\}$$

(5.30)

$$\nu = \frac{1 - 2b\frac{\mu}{5} - \alpha\left[1 - b\left(1 - 3\frac{\mu}{5}\right)\right]}{4 + 2b\frac{\mu}{5} + \alpha\left[1 - b\left(1 - 3\frac{\mu}{5}\right)\right]} \qquad 3K = 2E_o$$

It must be noted that the three approaches which have been presented only give estimates of the elastic behaviour which are usually different. More complex formulations have been obtained considering another internal variable **e** considered in the description of the distributions of contact forces and local relative displacements [6], [18].

The previous analysis has been proposed considering a linear contact law characterized by contact rigidity k_n and αk_n. A more realistic contact law, the Hertz-Mindlin model [16] has also been considered, which can be written in an incremental form :

$$\Delta F^n = \left(\frac{\sqrt{3RG_m}}{1-v_m}\right)^{2/3} \left(F^n\right)^{1/3} \Delta \delta^n ,$$

$$\Delta F^t = \left(\frac{\sqrt{3RG_m}}{1-v_m}\right)^{2/3} \frac{2(1-v_m)}{2-v_m} \left(F^n - \frac{F^t}{\tan \psi}\right)^{1/3} \Delta \delta^t ,$$

(5.31)

where G_m = shear modulus of the particle material, v_m = Poisson's ratio of the particle material.

At this point it is possible to deduce the elastic constants of a granular medium from the local contact law (G_m, v_m) and the internal variables of the medium defined in the localisation and averaging operators (μ, or μ and b, e in the anisotropic case).

We will only present the case of isotropic materials.

Considering the elastic law written in an incremental form the Voigt approximation leads to :

$$E = E_o \frac{5-4v_m}{5-3v_m} , \qquad v = \frac{v_m}{10-6v_m} , \qquad G = E_o \frac{10-8v_m}{10-5v_m} , \qquad 3K = E_o \qquad (5.32)$$

$$\text{Where } E_o = \left(\frac{G_m \overline{ND}^3}{\sqrt{6}(1-v_m)}\right)^{2/3} \sigma_o^{1/3} \qquad \text{and } \sigma_o : \text{means stress}$$

The two other approximations (static localisation and second level of kinematic localisation) have also been analysed and lead to more complex expressions [18]. Comparisons with experimental results [17] show that a better agreement is obtained considering the second level of kinematic localisation, in particular for the expression of the Poisson's ratio [18].

5.2.3 Analysis of yielding and failure surfaces

The static localisation operator previously defined can also be used to analyse yielding and failure surfaces from the definition of local criteria of contact instability. Two criteria arise :

i) no loss of contact characterized by $F.n \leq 0$, where F is the force acting through a plane normal to n,

ii) no sliding at the contact characterized by $\left\|F^t\right\| - \left\|F^n\right\| \tan \psi \leq 0$, where F^n and F^t are respectively the normal and tangential components of F and ψ is the intergranular friction angle.

These conditions defined for each local contact point are necessarily valid for the mean values of contact forces and thus for variable f. The considered conditions are obviously necessary but not sufficient for non-reversible displacements. Variable f can be expressed from the global static variable σ using relation (4.33) in the isotropic case and (4.36) for an anisotropic case. From these local criteria, a global yielding surface can be inferred. For a given set of micromechanical parameters μ, e and a given orientation of contact $\mathbf{n}$, two stress domains C_l^n and C_s^n can be defined by :

$$C_s^n = \left\{ \sigma \, , \, \mathbf{f}(\sigma, \mathbf{n}, \mu, e).\mathbf{n} \geq 0 \right\}$$

$$C_s^n = \left\{ \sigma ; \left\| \mathbf{f}^t(\sigma, \mathbf{n}, \mu, e) \right\| - \left\| \mathbf{f}^n(\sigma, \mathbf{n}, \mu, e) \right\| \tan \psi \leq 0 \right\}$$

(5.33)

By linearity of relation (4.33) or (4.36) with σ, C_l^n and C_s^n are two cones with the origin as a vertex. The intersection of the domains C_l^n for all contact orientation $\mathbf{n}$ gives the global 'loss of contact' criterion. C_1. The global 'sliding' criterion C_s is also defined as the intersection of the C_s^n. Both C_1 and C_s depend on the values of the micromechanical parameter μ, e and C_s also depends on the intergranular friction angle ψ. The friction angles in compression ϕ_c and in extension ϕ_e can be derived analytically in the case of an isotropic medium.

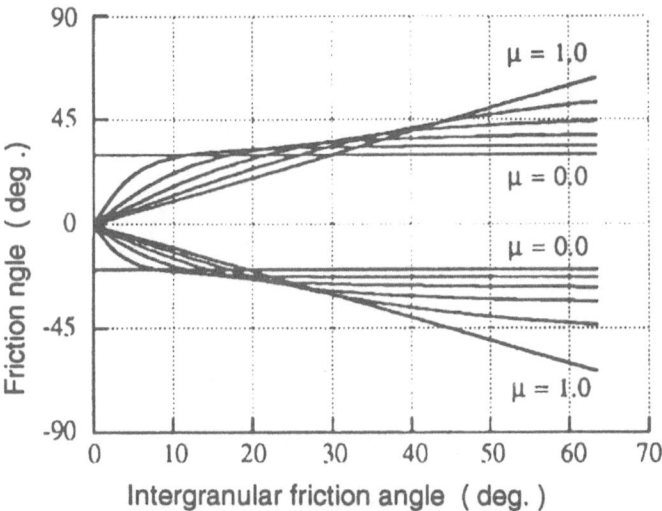

Fig. 19 : Evolution of compression and extension friction angle
with ψ (5.35). From Emeriault et al [19].

For the 'loss of contact' criterion :

$$\sin \phi_c = \frac{1}{2 - \mu} \qquad , \qquad \sin \phi_e = \frac{1}{2\mu - 3}$$

(5.34)

For the 'sliding' criterion :

$$\sin \phi_c = \tan \psi \frac{(3-\mu)\tan \psi + \sqrt{4\mu^2 + \tan^2 \psi (5-3\mu)^2}}{2\mu^2 + \tan^2 \psi \left(4\mu^2 - 13\mu + 11\right) + \tan \psi \sqrt{4\mu^2 + \tan^2 \psi (5-3\mu)^2}}$$

(5.35)

$$\sin \phi_e = \tan \psi \frac{(3-\mu)\tan \psi + \sqrt{4\mu^2 + \tan^2 \psi (5-3\mu)^2}}{2\mu^2 + \tan^2 \psi \left(4\mu^2 - 13\mu + 11\right) - \tan \psi \sqrt{4\mu^2 + \tan^2 \psi (5-3\mu)^2}}$$

For anisotropic media, the friction angle can only be derived numerically.
Figure 19 shows the evolution of both the compression and extension friction angle with
the intergranular friction angle and μ. It can be pointed out that for the values of μ

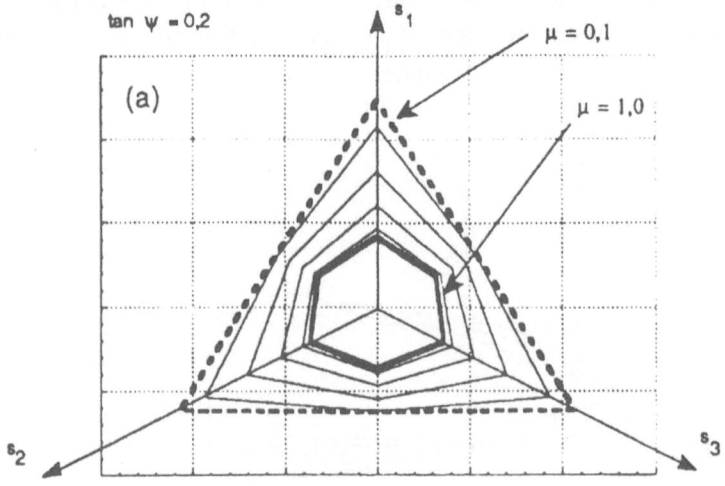

*Fig. 20 : Evolution of the yielding surface in the deviatoric plane for isotropic materials.
From Emeriault et al [19].*

provided by numerical simulations (μ around 0.5), the variation of the friction angle is not
linear with ψ. In particular for values of ψ greater than 20°, the friction angle is slightly
dependent on the local friction angle. This has been verified experimentally and
numerically by the DEM. Moreover, in any case of anisotropy and intergranular friction
angle, the sliding occurs before the loss of contact. Figure 20 is an illustration of the
yielding surfaces in the deviatoric plane for isotropic material. This figure has been
obtained numerically considering different proportional loading paths and noting the point
where one of the two local conditions is not satisfied for each value of n. The influence of
the value of μ is noticeable. If the anisotropy e is considered it affects the shape of the
yielding surfaces. The anisotropy can be considered to be an internal parameter and the

evolution of the yielding surfaces with this parameter is similar to that given by a kinematic hardening.

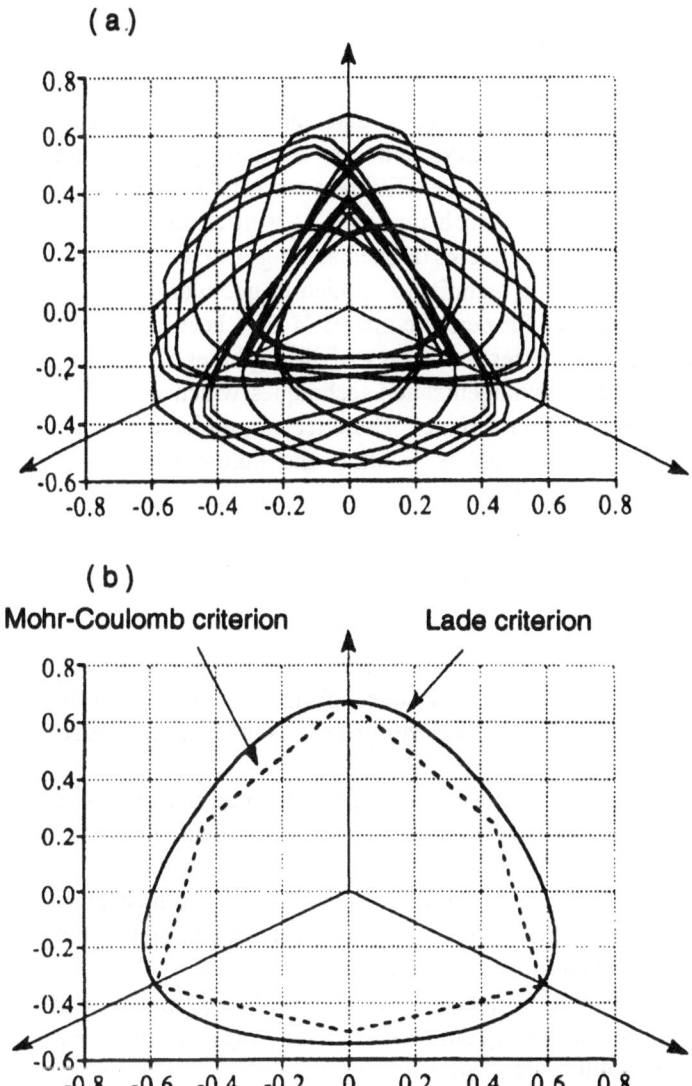

Fig. 21 : Definition of the failure surface and comparison with usual criteria. From Emeriault et al [19].

The final step is the definition of a failure surface as the envelope of the yielding surfaces when it is not possible, for a given value of σ_{ij}, to find a value of anisotropy e allowing the two local conditions to be satisfied for any value of **n**. Figure 21 shows an example of this kind of surface. In order to provide a comparison, the classical Mohr-Coulomb and Lade criteria are also plotted. The Lade criterion fits very well with the predictions defined by the proposed method.

6. CONCLUSIONS

Micromechanical analyses clearly show that complex behaviours of granular material such as non linearity, dilatancy, induced anisotropy are directly linked to the discontinuous nature of the structure. Different tensorial variables have been proposed to describe the internal fabric of such materials as well as relationships relating variables defined at the local level to variables defined at the global level.

A review of homogenization techniques applied to granular material has been proposed. Homogenization processes for periodic materials have been considered for the definition of the elastic behaviour of regular arrays of like spheres. Random arrays can only be analysed from statistical homogenization techniques which only allow the definition of estimates of mechanical characteristics. These techniques have been applied to define non linear elastic behaviour as well as yielding and failure surfaces characteristic of the behaviour of granular materials.

REFERENCES

1   Bardet, J.P. : Observations on the effects of particle rotations on the failure of idealized granular materials, Mechanics of Materials, Vol. 18 (1994), 159-182.

2   Biarez, J. : Contribution à l'étude des propriétés mécaniques des sols et des matériaux pulvérulents, Thesis, Grenoble, (1962).

3   Biarez, J., Wiendieck, K. : La comparaison qualitative entre l'anisotropie mécanique et l'anisotropie de structure, Comptes rendus à l'Académie des Sciences, Paris n° 254, (1963), 2712-2714.

4   Calvetti, F., Combe, G., Lanier, J. : Experimental micromechanical analysis of a 2D granular material : relation between structure evolution and loading path, Mechanics of cohesive frictional materials, vol 2, (1997), 121-163.

5   Caillerie, D. : Evolution quasistatique d'un milieu granulaire, loi incrémentale par homogénéïsation, in : Des géomatériaux aux ouvrages, Hermes, Paris, (1995), 53-80.

6   Cambou, B., Dubujet, Ph., Emeriault, F., Sidoroff, F. : Homogenization for granular materials, European Journal of Mechanics, A/solids, vol. 14 (2), (1995), 225-276.

7   Cambou, B. : From global to local variables in granular materials, Powders and Grains (1993), Thornton ed., Balkema, (1993), 73-86.

8   Chang, C.S., Misra, A., Sundaram, S. : Micromechanical modelling of cemented sands under low amplitude oxcillations, Geotechnique 40, n° 2, (1990), 251-263.

9   Chang, C. S. : Micromechanical modeling of deformation and failure for granulates with frictional contacts, Mechanics of Materials, vol. 16, (1993), 13-24.

10 Chang, C. S., Liao, C. L. : Estimates of elastic modulus for media of randomly packed granules, Applied Mechanics Revue, vol. 47 n° 1, part 2, (1994), 197-206.

11 Chapuis, P. : De la structure géométrique des milieux granulaires en relation avec
    leur comportement mécanique, Thesis, Ecole Polytechnique, Montréal, Canada,
    (1976).

12 Christofferson, J., Mehrabadi, M.M., Nemat-Nasser, S. : A micromechanical
    description of granular material behavior, Journal of Applied Mechanics, ASME, vol.
    48 n° 2, (1981), 339-344.

13 Cundall, P.A., and Strack, O.D.L. : A discrete numerical model for granular
    assemblies, Geotechnique, Vol. 29 (1979), 47-65.

14 Dantu, P. : Contribution à l'étude mécanique et géométrique des milieux
    pulvérulents, Proceedings of the 4th International Conference of Soil Mechanics and
    Foundation Engineering, London, U.K. (1957), 144-148.

15 Delyon, F., Dufresne, D., Levy, Y. : Physique et génie civil, deux illustrations
    simples, Annales des Ponts et Chaussées, numéro spécial : Mécanique des milieux
    granulaires, (1990), 53-60.

16 Deresiewicz, H. : Mechanics of granular matter, in Advances in applied mechanics 5
    Academic Press-New York, (1958), 233-306.

17 El Hosri, M. : Contribution à l'étude des propriétés mécaniques des matériaux,
    Thesis, Université Paris VI, (1984).

18 Emeriault, F., Cambou, B. : Micromechanical modelling of anisotropic non-linear
    elasticity of granular medium, International Journal of Solids and Structures, vol. 33
    n° 18, (1996), 2591-2607.

19 Emeriault, F., Cambou, B. Mahboubi, A. : Homogenization for granular materials,
    non reversible behaviour, Mechanics of Cohesive-Frictional Materials, vol. 1, (1996),
    199-218.

20 Field, W. G. : Towards the statistical definition of a granular mass, proceedings of
    4th Australia-New Zealand conf. on soil mechanics and found. eng., (1963),143-148.

21 François, D., Pineau, A. Zaoui, A. : Comportement mécanique des matériaux, Vol 1,
    Hermes, Paris, 1992.

22 Hill, R. : The essential structure of constitutive laws for metal composites and
    polycrystals, J. Mech. Physics Solids, Vol. 11 (1967), 357-372.

23 Kruyt, N.P., Rothenburg, L. : Micromechanical definition of the strain tensor for
    granular materials. Journal of Applied Mechanics, vol. 118 (1996), 706-711.

24 Oda, M., Iwashita, K., Kakiuchi, T. : Importance of particle rotation in the mechanics
    of granular materials, in Powders and Grain, Behringer and Jenkins eds, Balkema
    publisher, (1997), 207-210.

25 Oda, M. , Konishi, J., Nemat Nasser, S. : Experimental micromechanical evaluation
    of strength of granular materials : effect of particle rolling, Mechanics of materials
    n°1, (1982), 269-283.

26 Rowe, P.W. : The stress-dilatancy, earth pressures, and slopes. ASCE, SM3, (1963),
    37-61.

27 Rowe, P.W. : The stress-dilatancy relation for static equilibrium of an assembly of
    particles in contact, Proc. Roy. Soc. London A269, (1962), 500-527.

28 Satake, M; : Fabric tensor in granular material, Proceedings of the IUTAM symposium on deformation and failure of granular materials, Delft, Balkema Publisher, P.A. Vermeer and H.J. Luger eds.(1982), 63-68.

29 Sidoroff, F., Cambou, B., Mahboubi, A. : Contact force distribution in granular media, Mechanics of materials, vol. 16, (1993), 83-89.

30 Weber, J. : Recherche concernant les contraintes intergranulaires dans les milieux pulvérulents, Bulletin de liaison des Ponts et Chaussées, Paris, n° 20, (1966), 1-20.

31 Yanagisawa, E. : Influence of void ratio and stress condition on the dynamic shear modulus of granular media, in : Advances in the mechanics and the flow of granular materials - M. Shahinpoor Ed. - Gulf Publishing - Houston, (1983), 947-960.

32 Zaoui, A. : Comportement global des polycristaux : passage du monocristal au polycristal, in : Physique et mécanique de la mise en forme des métaux, Presses du CNRS, Paris, (1990), 337-351.

33 Zanchez-Palencia, E., Zaoui, A. : Homogenization techniques for composite media, Springer-Verlag, Berlin, 1987.

COUPLING:
PORE FLUID-GRAINS INTERACTION

D. Kolymbas

University of Innsbruck, Innsbruck, Austria

Dedicated to the memory of Prof. Dr. W. Günther

1 Introduction

The interaction (coupling) between grains and pore fluid is manifested in the static case (no relative motion between grains and pore fluid) as well as in the dynamic case. In the static case (i.e. no relative velocity between grains and pore fluid), the important item is the so-called principle of effective stress. In the dynamic case the interaction is mainly expressed by DARCY's law and the so-called inertia coupling. We shall see that the pertinent relations are of limited validity. Before doing so we shall consider the main equations of the theory of mixtures which are very useful when dealing with composite materials. It should be stressed that we consider here only granular media where the grains form a grain skeleton. Thus, dispersions of grains within a fluid carrier are here not considered.

2 Theory of mixtures

2.1 Definitions

Variables related to the grains are denoted by the index s (for solid); variables related to the pore fluid are denoted by the index f. All variables are mean values of the corresponding quantities gained by averaging over a certain volume. Hereby we distinguish between two different mean values:

If the averaging is carried out over the volume occupied by one phase, we speak of *effec-*

tive values. E.g.

$$\rho^s := \frac{1}{V_s} \int_{V_s} \rho dV \tag{1}$$

is the effective density of the granulate. If the grains are homogeneous ρ^s is identical with the real density of the grain (e.g. for quartz-sand $\rho^s = 2,65$ g/cm^3). The integration domain V_s must be small enough, such that ρ^s can be treated as a local quantity. On the other hand it should be large enough such that the local variations of ρ^s are sufficiently smoothened.

If the averaging is carried out over the total volume, the mean values are denoted by the attribute *partial*. E.g.

$$\rho_s := \frac{1}{V} \int_{V_s} \rho dV \tag{2}$$

is the partial density of the granulate. By defining the volume fractions of a water saturated granulate as:

$$\alpha_s := \frac{V_s}{V}, \quad \alpha_f := \frac{V_f}{V}, \quad \alpha_s + \alpha_f = 1, \tag{3}$$

the density ρ of the mixture can be written as:

$$\rho = \alpha_s \rho^s + \alpha_f \rho^f = \rho_s + \rho_f. \tag{4}$$

In soil mechanics the volume fraction α_f is called *porosity* and is denoted by n, and V_f/V_s is denoted as *void ratio e*. Note that the so-called *filter-* or *superficial velocity* is the partial velocity v_f of the fluid.

The pressure in the pore fluid p, in soil mechanics denoted as pore pressure, is an effective value ($p = p^f$). The total stress σ can in analogy to equation (4) be represented as the sum of the partial stress of the granulate, σ_s, and of the fluid, np:

$$\sigma = \sigma_s + np. \tag{5}$$

Equation (5) reads in tensorial form: $\sigma_{ij} = (\sigma_s)_{ij} + np\delta_{ij}$. Here, however, the indices are skipped for simplicity. In soil mechanics the total stress σ is separated in another manner:

$$\sigma = \sigma' + p. \tag{6}$$

σ' is called the *effective stress* but is a different quantity from the effective stress in the sense of mixture theory. By comparing equation (5) with equation (6) the relation between the effective stress σ' in the sense of classical soil mechanics and the partial stress σ_s can be written as:

$$\sigma_s = \sigma' + (1-n)p \quad \text{or} \quad \sigma' = \sigma_s - (1-n)p. \tag{7}$$

2.2   Conservation equations

The governing equations for the problem are the equations of conservation of mass and momentum for the two phases. In a spatial- or Eulerian description the equations of conservation of mass are:

$$\text{granulate:} \qquad \frac{\partial(\alpha_s \rho^s)}{\partial t} + \text{div}(\alpha_s \rho^s \mathbf{v}^s) \; = \; 0 \tag{8}$$

$$\text{fluid:} \qquad \frac{\partial(\alpha_f \rho^f)}{\partial t} + \text{div}(\alpha_f \rho^f \mathbf{v}^f) \; = \; 0 \tag{9}$$

The equation of conservation of momentum for the granulate can be written either using partial-or effective stresses. Using partial stresses it can be written as :

$$\rho_s \frac{d_s \mathbf{v}^s}{dt} \; = \; \nabla \sigma_s + \mathbf{R} + \rho_s \mathbf{g}, \tag{10}$$

where $\frac{d_s \mathbf{v}^s}{dt} = \frac{\partial \mathbf{v}^s}{\partial t} + \mathbf{v}^s \cdot \nabla \mathbf{v}^s$ is the material derivative of $\mathbf{v}^s$ and $\rho_s \mathbf{g}$ is the volume force (here: gravitation) with $\mathbf{g}$ being the gravitational acceleration.
$\mathbf{R}$ is the interaction force (force per unit volume), which the fluid imposes on the granulate. It will be discussed in the next section. The conservation of momentum for the granulate can be written, using the effective stress, as:

$$\rho_s \frac{d_s \mathbf{v}^s}{dt} \; = \; \nabla \sigma' + (1 - n)\nabla p - p \nabla n + \rho_s \mathbf{g} + \mathbf{R} \quad . \tag{11}$$

Equation (11) was obtained by substituting (7) into (10). The conservation of momentum for the fluid phase is:

$$\rho_f \frac{d_f \mathbf{v}^f}{dt} \; = \; p \nabla n + n \nabla p + \rho_f \mathbf{g} - \mathbf{R} \quad , \tag{12}$$

where $d_f \mathbf{v}/dt$ is equal to $\partial \mathbf{v}^f/\partial t + \mathbf{v}^f \cdot \nabla \mathbf{v}^f$.
The so called REYNOLDS-stress arising from the fluctuations of $\mathbf{v}^f$ is here neglected.

2.2.1   Interaction force R

The interaction force can be expressed as follows:

$$\mathbf{R} = p \nabla n + \kappa(\mathbf{v}^f - \mathbf{v}^s) + \mathbf{A} \tag{13}$$

The first term is also active when the components are at rest ($\mathbf{v}^s = \mathbf{v}^f = 0$). For a detailed derivation see NIGMATULIN [1]. This term can easily be justified by considering equation (12) for the one-dimensional statical case $\mathbf{v}^f \equiv 0$ with the coordinate z pointing upwards. Taking into account that $\rho_f g = n \rho^f g = n \gamma^f$ equation (12) can be written as:

$$p \frac{\partial n}{\partial z} + n \frac{\partial p}{\partial z} - n \gamma^f - R = 0 \tag{14}$$

This equation can be reduced to the known relation for the pore pressure increase with depth $\partial p/\partial z = \gamma^f$ (doing so we assume p negative), if R is set equal to $p\, \partial n/\partial z$. Therefore the term $p\nabla n$ is justified.

The second term $\kappa(\mathbf{v}^f - \mathbf{v}^s)$ is due to the viscosity of the pore fluid and represents the hydrodynamic drag. For the laminar case this force is proportional to the velocity difference $\mathbf{v}^f - \mathbf{v}^s$. κ is related to the coefficient of permeability k through:

$$\kappa := \rho^f g \frac{n^2}{k} \tag{15}$$

The term $\mathbf{A}$ represents a series of additional interaction mechanisms, some of which will be explained in the sequel. It is remarcable to notice, however, that the hydrostatic buyoancy force (according to ARCHIMEDES) needs not to separately appear in $\mathbf{R}$. Muchmore, it results from the conservation of momentum for solids, if we write this equation in terms of effective stress. This can be seen from equations (11) and (13) by considering the static case $\mathbf{v}^s = \mathbf{v}^f = 0$. It then follows (with $\mathbf{A} = 0$):

$$\nabla \sigma' + (1-n)\nabla p + \rho_s \mathbf{g} = 0 \quad .$$

Considering the z-direction and taking into account that $\mathbf{g} = -g\mathbf{e}_z$ we obtain

$$\frac{\partial \sigma'_z}{\partial z} + (1-n)\frac{\partial p}{\partial z} - \rho_s g = 0 \quad .$$

Noting that $\rho_s g$ is equal to the so-called dry unit weight $\gamma_d = \gamma_r - n\gamma_w$ (γ_r is the unit weight of the saturated soil, and γ_w is the familiar notation for the unit weight of water[1]) and that $\partial p/\partial z = \gamma_w$ we obtain

$$\begin{aligned}\frac{\partial \sigma'_z}{\partial z} &= -(1-n)\gamma_w + (\gamma_r - n\gamma_w) \\ &= (\gamma_r - \gamma_w) = \gamma' \quad ,\end{aligned}$$

where γ' is the so-called buoyant unit weight. Note, furthermore, that if we consider the conservation of momentum in terms of partial stress σ_s (see equ. 10), then buoyancy does not play any role, as the equation

$$\frac{\partial(\sigma_s)_z}{\partial z} = \gamma_d$$

holds true (in the static case), no matter whether the soil is saturated or dry.

---

[1] We consider here water as pore fluid, i.e. $\gamma_w \equiv \gamma^f$

2.2.2   Further interaction mechanisms

The term $\mathbf{A}$ represents a series of additional interaction mechanisms such as due to the MAGNUS-effect (that appears if the grains rotate) or due to osmotic effects. Therefore $\mathbf{A}$ can be considered as consisting of several terms, $\mathbf{A} = \mathbf{A}_1 + \mathbf{A}_2 + \dots$. Here we will only consider the interaction which arises due to accelerations of the pore fluid and/or the grains. This interaction is also called inertia coupling. Considering a single solid sphere with radius r_0 and volume $V_0 = \frac{4}{3}\pi r_0^3$ the force exerted by the surrounding non-viscous fluid upon the sphere (see e.g. [2]) reads

$$\rho^f V_0 \frac{\mathrm{d}v_i^f}{\mathrm{d}t} + m_{ik}\frac{\mathrm{d}}{\mathrm{d}t}(v_k^f - v_k^s) \quad . \tag{16}$$

The first term represents a sort of buyoancy which is not due to gravity $\mathbf{g}$ but due to the acceleration $\mathrm{d}_f v^f/\mathrm{d}t$ of the fluid. The second term is due to the fluid mass that is "added" to the solid sphere. The added mass tensor $\mathbf{m}$ (or m_{ik}) relates the momentum of the fluid $\mathbf{P}$ with the velocity of the immersed solid body $\mathbf{v}^s$.

$$\mathbf{P} = \mathbf{m}\mathbf{v}^s \quad (\text{or} \quad P_i = m_{ik}v_k^s) \quad .$$

For a solid sphere with radius r_0 the added mass reads $\frac{2\pi}{3}r_0^3\rho^f\delta_{ik}$.
Considering a two-phase continuum consisting of grains and pore fluid, the first term of expression 16 is taken into account by the equations 12 and 11, which are coupled via ∇p. The added mass, however, should be taken into account by an additional term of the interaction force $\mathbf{R}$, which now read as follows:

$$\mathbf{R} = p\nabla n + \kappa(\mathbf{v}^f - \mathbf{v}^s) + \rho_a\left(\frac{\mathrm{d}_f v^f}{\mathrm{d}t} - \frac{\mathrm{d}_s v^s}{\mathrm{d}t}\right)$$

ρ_a is the density of the added mass. Following GAJO [3] it can be set

$$\rho_a = (\tau - 1)n\rho^f \quad ,$$

where τ is the so-called tortuosity of the grain skeleton. The tortuosity is the length of a pore channel connecting two points divided by the distance of these points. For a statistically homogeneous and isotropic granulate the tortuosity is considered as a scalar quantity. Following NIGMATULIN [1] (equation 1.9.6) ρ_a can be expressed as

$$\rho_a = \frac{1}{2}\chi_m\rho^f n(1 - n) \quad ,$$

where the coefficient $\chi_m(0 < \chi_m < 1)$ takes into account the size distribution and the non-spherical form of the grains.

3 Effective stress

Regarding saturated soils we have to realise that a part of the applied loads can be carried by the pore fluid. This is the case for the so-called undrained conditions. In order
to analyse the pertinent effects we consider a vessel filled with water and a spring (see
Fig. 1).

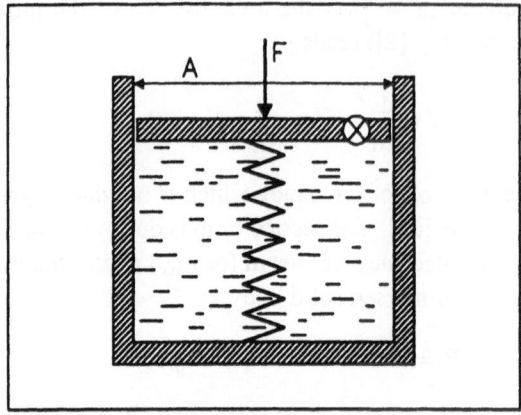

Fig. 1: Principal array for effective stress

The grain skeleton is represented by the spring. Both, water and grains are here assumed
as incompressible, i.e. their volume is not changed by pressure changes. As long as the
valve (drainage) is closed, the spring (grain skeleton) cannot experience any external load.
For, to experience an external load it has to be deformed, which is impossible as long as
the drainage is closed and the water cannot escape. Thus, in this case the pore pressure
must equal the external load. When we open the drainage, the water can gradually escape,
the spring gets deformed and starts carrying part of the load. I.e., the difference between
external (total) pressure and pore pressure has to be carried by the spring (grain skeleton):

$$\sigma' = \sigma - p$$

We also can say that the total stress σ can be decomposed in effective stress σ' and pore
pressure p :

$$\sigma = \sigma' + p \quad . \tag{17}$$

Generalising our consideration of Fig.1 we can infer the principle of effective stress stating that σ' (and not the pore pressure p) is responsible for deformation and strength of
granular media.

For the considered special case of incompressible grains and incompressible pore fluid the principle of effective stress can also be deduced from the internal constraint of incompressibility in the sense of continuum mechanics. Assume that the constitutive relation

$$\dot{\sigma} = f(\sigma, \dot{\epsilon}, e)$$

determines the stiffness and the strength of a drained or dry granular medium. For a water saturated granulate the condition "undrained" imposes the internal constraint of incompressibility. Consequently, the hydrostatic stress part can only be determined up to a constitutively undetermined hydrostatic pressure p (which, of course, equals the pore pressure):

$$\dot{\sigma} = \underbrace{f(\sigma', \dot{\epsilon}, e)}_{\dot{\sigma}'} + \dot{p}$$

Using the index notation we can write

$$\sigma_{ij} = \sigma'_{ij} + p\delta_{ij}.$$

Herewith we assume that the pore pressure p follows the same sign convention as stress[2]. Now we consider the equations of equilibrium for a water saturated soil. The balance of the forces acting upon an elementary volume with the edges dx, dy and dz, together with the assumption that the stresses $\sigma_x, \sigma_y, \sigma_z$ are principal stresses and that the positive z-direction points downwards leads to the equations in the z-direction

$$\frac{\partial \sigma_z}{\partial z} = \gamma_r \quad , \tag{18}$$

and, in horizontal direction

$$\frac{\partial \sigma_x}{\partial x} = 0 \quad . \tag{19}$$

Introducing (17) into (18) and (19) we obtain

$$\frac{\partial \sigma'_z}{\partial z} = \gamma_r - \frac{\partial p}{\partial z} \quad , \tag{20}$$

$$\frac{\partial \sigma'_x}{\partial x} = -\frac{\partial p}{\partial x} \quad . \tag{21}$$

---

[2]Often, compressive stress in the grain skeleton is taken negative but compressive fluid pressure is taken positive. Then the decomposition looks like: $\sigma_{ij} = \sigma'_{ij} - p\delta_{ij}$.

If the pore fluid is at rest

$$\nabla p = \begin{pmatrix} \partial p/\partial x \\ \partial p/\partial y \\ \partial p/\partial z \end{pmatrix} = \begin{pmatrix} 0 \\ 0 \\ \gamma_w \end{pmatrix} . \qquad (22)$$

Thus we have

$$\frac{\partial \sigma'_z}{\partial z} = \gamma_r - \gamma_w$$

Again, $\gamma' = \gamma_r - \gamma_w$ is the buoyant or submerged unit weight of the grain skeleton. For flowing pore fluid we obtain instead of Equ. (22):

$$\nabla p = \begin{pmatrix} 0 \\ 0 \\ \gamma_w \end{pmatrix} - \gamma_w \nabla h$$

Herein, h is the energy head composed of the geodetic head z and the pressure head[3] $-p/\gamma_w$:

$$h = z - p/\gamma_w$$

Thus we obtain from the equilibrium equation (20) that for the case of moving pore fluid the additional volume force (i.e. force per unit volume) $\gamma_w \nabla h$ appears. This force is called seapage force and has the direction of the flow velocity. It has to be taken into account (together with γ') when we consider effective stresses.

3.1 Principle of effective stress

Despite of its primary importance to soil mechanics the principle of effective stress and the conditions of its validity are still vividly discussed, a fact which is manifested by the publication of many related papers [4], [5]. From a logical point of view we can proceed in two ways:

Either we define the effective stress σ' by the equation $\sigma' := \sigma - p$ and subsequently we show by means of experiments that the deformation of the grain skeleton is only influenced by σ',

or we define the effective stress σ' as that stress which is responsible for the deformation (and strength) of soil. Obviously, the effective stress depends on the total stress and on the pore pressure: $\sigma' = F(\sigma, p)$. It is natural to require for $p = 0$ the validity of $\sigma' = \sigma$, therefore it appears reasonable to consider the special case $\sigma' = \sigma - \eta p$. It then remains

---

[3]If the pore pressure p is taken positive for compression, then the energy head reads $h = z + p/\gamma_w$.

to determine the value of η, say, from experiments, or to find out under which conditions $\eta = 1$ holds true. This case constitutes the principle of effective stress, which of course is rather a statement with a constitutive character than a generally valid principle.

Subsequently we shall see that the principle of effective stress (i.e. $\eta = 1$) holds whenever the grains may be considered as incompressible, i.e. when the volume change of the grains due to a change of the pore pressure can be neglected. Herewith, the compressibility of the pore fluid does not play any role. For simplicity we omit indices and denote the stress simply with σ. In a grain skeleton with the porosity n the volume (or area) fraction of grains is $1 - n$ and the volume (or area) fraction of the pores is n. Let σ be the total stress and p the pore pressure[4]. Now we introduce the partial stresses σ_s and p_f. σ_s is the partial pressure of the grain skeleton, i.e. this part of the total stress which is transmitted by the grains (per unit area of the soil), p_f is the partial pressure of the pore fluid, i.e. this part of the total stress which is transmitted by the pore fluid (per unit area of the soil). Obviously

$$p_f = np \quad ,$$

such that from

$$\sigma = \sigma_s + p_f$$

follows

$$\sigma_s = \sigma - np$$

The hydrostatic pressure p which prevails in the pore fluid acts also upon the grains. This means that per unit cross section area the stress p is transmitted. The grains transmit the part $(1 - n)p$. If they are incompressible, this stress part has no influence (i.e. it does not produce any mechanical work). Consequently, only the remaining difference $\sigma_s - (1 - n)p = \sigma - p = \sigma'$ is responsible for the deformation of the grain skeleton. We thus infer that $\eta = 1$ holds true for incompressible grains, no matter whether the pore fluid is incompressible or not. We furthermore infer that clay particles can be considered as incompressible in the above sense, as the principle of effective stress has been shown to be valid for clay [6].

Often, the smallness of the particle contact areas is cited as a condition for the validity of the principle of effective stress. However, we have seen that the area of particle contacts does not play any role as regards the influence of the pore pressure upon the deformation of the grain skeleton. However, a non-vanishing particle contact area means that the force acting between the individual grains (the intergranular force or the intergranular stress) increases when the pore pressure is increased. This increase does not influence the

---

[4]we consider here p and σ positive for compression

deformation of the grain skeleton[5] — but it influences its strength. In fact, SKEMPTON reports [7] that an increase of the cell pressure is observed to induce a small increase of the strength $(\sigma_1 - \sigma_2)_{max}$ of unjacketed soil samples. It can furthermore be derived that the friction angle is still not affected by pore pressure changes, provided that the grain contact forces are normal to the grain surface (this fact has been shown to be valid by methods of micromechanic analysis).

For compressible pore fluid the principle of effective stress is still valid (provided the grains are incompressible). However, for undrained conditions a relation is now needed that determines which part of an external load is carried by the grain skeleton and which part is carried by the pore fluid. The case of compressible pore fluid is encountered in soil mechanics when the pore water contains air or natural gas bubbles. Then we have:

$$\Delta V_p = -\kappa_p V_p \Delta p$$

κ_p is the compressibility of the pore fluid consisting of water and gas bubbles. It depends on the degree of saturation and can be calculated taking into account HENRY's law and the surface tension between air and water [11].

The change of the void ratio e reads (for V_s =const, V_s =Volume of grains):

$$\Delta e = \frac{\Delta V_p}{V_s} = -\frac{\kappa_p V_p \Delta p}{V_s} = -\kappa_p e \Delta p \quad . \tag{23}$$

On the other hand we have:

$$\Delta e = -C_c \frac{\Delta \sigma'}{\sigma'} \quad . \tag{24}$$

Setting (23) equal to (24) yields:

$$\Delta \sigma' = \frac{\sigma' \kappa_p e}{C_c} \Delta p \quad .$$

From $\Delta \sigma = \Delta \sigma' + \Delta p$ follows then

$$\Delta \sigma = \left(1 + \frac{\sigma' \kappa_p e}{C_c}\right) \Delta p$$

and finally

$$\Delta p = B \Delta \sigma \quad \text{with} \quad B = \frac{1}{1 + \sigma' \kappa_p e / C_c} \tag{25}$$

---

[5]We only need to require that the pore pressure p also acts entirely within the grains

The B-factor has been introduced by SKEMPTON. For an incompressible pore fluid ($\kappa_p = 0$) $B = 1$ holds true.

With the aid of Equ. (25) we can calculate the pore pressure generation due to cyclic loading induced by applications of the total stress $\Delta\sigma$ and the subsequent unloadings with $-\Delta\sigma$. Doing so we may approximately consider σ' and e as constants. However, for unloading we have to replace C_c with $C_s (< C_c)$. We then obtain a pore pressure increase for every cycle

$$\Delta p = \frac{C_c - C_s}{C_c C_s / (\sigma' \kappa_p e) + C_c + C_s + \sigma' \kappa_p e} > 0 \quad . \tag{26}$$

This pore pressure generation can lead to liquefaction of a cyclically loaded soil the pores of which are filled with a compressible fluid [8].

It is clear from the above argumentation that the principle of effective stress is no more valid if we consider compressible grains. In this case we have to expect that not only changes of $\sigma' = \sigma - p$ but also changes of p will cause a deformation ε of the grain skeleton. The relation

$$\varepsilon = f(\sigma', p)$$

however, has not been investigated so far. Some researchers use a "revised" principle of effective stresses by considering relations of the type

$$\varepsilon = f(\sigma'')$$

with a sort of a revised effective stress defined as

$$\sigma'' := \sigma - \eta p \quad , \quad \eta < 1 \quad .$$

Of course, it is reasonable to relate η to the compressibility κ_s of the grains. It is likewise reasonable to non-dimensionalize κ_s by dividing through κ, with κ being the compressibility of the granular matrix[6]. It should however be kept in mind that relations such as the one introduced by BIOT [9]

$$\sigma'' = \sigma - \eta p \quad \text{with} \quad \eta = 1 - \frac{\kappa_s}{\kappa}$$

are, for soils, nothing but constitutive assumptions.

---

[6]Note that, for granular media, κ is not a material constant but depends pronouncedly on the actual stress state

3.1.1   Effective stress in partly saturated soils

We consider a partially saturated granulate (soil) with the volume fractions α_g (granulate), α_w (water) and α_a (air). The degree of saturation S is defined as

$$S = \frac{\alpha_w}{\alpha_w + \alpha_a} = \frac{\text{water volume}}{\text{pore volume}}$$

and ranges between 0 and 1. Consequently, α_w ranges between 0 and n: $\alpha_w = nS$. In the range $0 < S < S_1$ the water is concentrated in isolated menisci, whereas for $S_1 < S < 1$ the water forms a connected phase. Analogously, for $0 < S < S_2$ the air forms a connected phase, and for $S_2 < S < 1$ air is contained within isolated bubbles[7].

Before considering the phenomena within a granulate we shall look at an isolated air bubble within water. An adiabatic change of energy E is obtained from

$$dE = -\Delta p dV + \gamma dA \quad ,$$

where Δp is the overpressure within the bubble with respect to the pressure of the surrounding water, γ is the surface energy of the water (with respect to the adjacent air), A is the surface of the bubble and V is the volume of the bubble. If there is no external energy supply, we have $dE = 0$, and it follows

$$\Delta p = \gamma \frac{dA}{dV} = \gamma \frac{dA/dr}{dV/dr}$$

With $A = 4\pi r^2$ and $V = \frac{4}{3}\pi r^3$ it follows the equation of LAPLACE

$$\Delta p = 2\gamma/r$$

Subsequently we consider (as a representant of the grain skeleton) a unique pipe of radius r_0 filled with water. This pipe contains air bubbles with various radii r_b (see Fig. 2). Obviously, the relation $\Delta p = 2\gamma/r_b$ holds true only for $r_b \leq r_0$. If we plot Δp over the degree of saturation of the pipe then we obtain the curve shown in Fig. 3.

The pores of a soil of a given density are a collection of capillary tubes with various diameters and many interconnections. In the sense of a macroscopic-phenomenologic consideration it is useless to consider isolated bubbles. Instead, we consider all three phases as homogeneously spread over the whole soil body. The so-called *capillary pressure curve* describes the mean pressure difference between the air and liquid phase in dependence of the degree of saturation.

We can average the water pressure and the air pressure over the entire pore space. Let p^a and p^w denote the real values of these pressures, then the pore-averages read:

$$p_a := (1 - S)p^a \quad \text{and} \quad p_w := Sp^w$$

---

[7]It is yet unclear whether $S_1 = S_2$ holds true

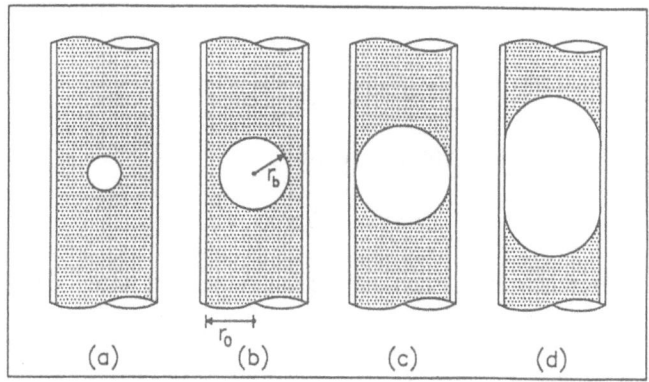

Fig. 2: Air bubbles with various diameters within a pipe

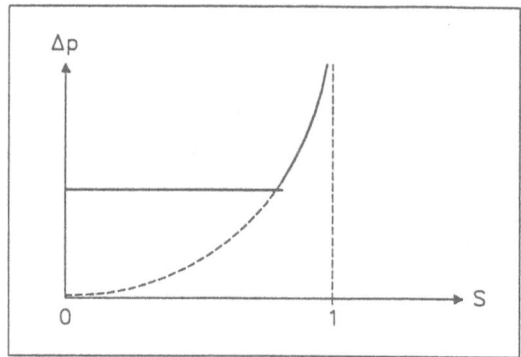

Fig. 3: Relation between the bubble overpressure Δp and the degree of saturation S of a capillary

The mean pore pressure is then obtained as

$$p := p_a + p_w$$

For $S_a \leq S \leq 1$ (case 1: Air in isolated bubbles) the water pressure is distributed hydrostatically over the depth z (if the water does not move):

$$p^w = \gamma_w z.$$

Then, the air pressure is obtained from

$$p^a = p^w + \Delta p(S).$$

For $0 \leq S \leq S_w$ (case 2: Water in isolated menisci) the air pressure is equal to the atmospheric pressure

$$p^a = p_{atm},$$

and the pore water pressure is obtained as suction

$$p^w = p_{atm} - \Delta p(S).$$

If we take p_{atm} as reference value, then the suction in the pore water is given by

$$p^w = -\Delta p(S).$$

Δp is often called *matrix suction*.

In both cases we can evaluate the mean pore pressure p. From the total stress[8] σ we can then obtain the effective stress as

$$
\begin{aligned}
\sigma' \quad &= \sigma - p \\
&= \sigma - (1 - S)p^a - Sp^w \\
&= \sigma - p^a + S(p^a - p^w)
\end{aligned}
\tag{27}
$$

Equation 27 is identical to the equation of BISHOP,

$$\sigma' = \sigma - p^a + \chi(p^a - p^w),$$

if we put $\chi = S$.

Let us consider the case $0 \leq S \leq S_w$ with $\sigma \approx 0$. Then

$$\sigma' = -p = S\Delta p(S),$$

i.e. a considerable effective stress acts upon the grain skeleton and increases thus its stiffness and strength. If we apply an appropriate external load we can registrate a considerable tensile strength of the soil. The tensile and shear strength due to capillarity (so-called *apparent cohesion*) vanishes for $S = 0$ and $S = 1$ and obtains in between a maximum value [10].

For the case $S_a \leq S \leq 1$ we can consider the pore fluid as water which is compressible due to the enclosed air bubbles. The compressibility is determined not only from the compressibility of air (according to the law $pV^\gamma = $ const for adiabatic and $pV = $ const for isothermal compression), but also from the fact that increased mass of air gets dissolved

---

[8]Compressive stresses and pressures in pore fluid and in soil skeleton are taken as positive

if the pressure is increased, and also from changes of the surface energy of the air-water interface [11].

Therefore, the compressibility of the pore fluid (i.e. the mixture of air and water) depends not only on the degree of saturation but also on the pore size distribution.

The usual experimental methods to determine the capillary pressure curve of a grain skeleton is to withdraw either the air or the water from the pores by the application of a pressure on the boundary of the considered granulare body. However, in the usual watering and dewatering tests the degree of saturation is not homogeneously distributed over the sample. Therefore the obtained *watering* and *dewatering curves* are virtually inappropriate as capillary pressure curves. This fact is very often overlooked.

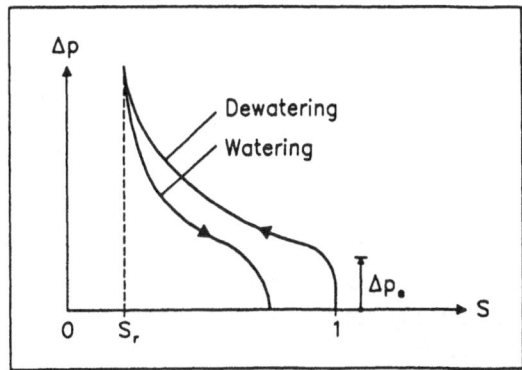

Fig. 4: Watering and dewatering curves. $S_r =$ is the residual degree of saturation

The initial steep increase Δp_e of the dewatering curve is due to the fact that air can only enter into a pore having the diameter $d = 2r$ if the air overpressure has obtained the value $\Delta p_e = 2\gamma/r$ (so-called *air entry value* or *bubbling pressure*).

4 Darcy's law

DARCY's law has been conceived for water-saturated soils.

According to DARCY's law from 1856 the superficial velocity v is proportional to the energy head Δh which is consumed to cover the length Δl:

$$v = k\frac{\Delta h}{\Delta l}$$ (28)

The meaning of Δh and Δl is explained in Fig. 5. The dimensionless quantity $\Delta h/\Delta l$ is denoted as the hydraulic gradient i. Thus DARCY's law obtains the form

$$v = ki$$ (29)

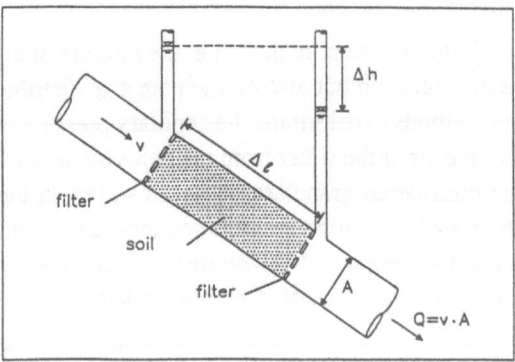

Fig. 5: Principal array for Equ. (28)

k is called the coefficient of permeability or hydraulic conductivity. DARCY's law is valid under the following limitations:

1. The grain skeleton (i.e. the soil) is isotropic, i.e. the permeability is equal in all spatial directions. This is, of course, not the case if the soil is composed of equally oriented platelets, like e.g. pre-loaded clays. In such cases, DARCY's law should be written in tensorial form, i.e.

$$v_i = -k_{ij} \frac{\partial h}{\partial x_j} \quad .$$

2. DARCY's law is not valid for turbulent flow. Then, according to FORCHHEIMER, a quadratic law holds true :

$$i = Av + Bv^2 \quad .$$

The transition to turbulent flow is controlled by the REYNOLDS number defined as

$$Re = vd/\mu \quad ,$$

where d is an appropriate grain diameter (indicative for a characteristic pore diameter) and μ is the viscosity of the pore fluid. Following PAVLOVSKI the transition to turbulent flow occurs at

$$Re = \frac{1}{0,75n + 0,23} \frac{vd_{10}}{\mu} \approx 7 \dots 9 \quad .$$

Other transition criteria are mentioned by KÉZDI [12] and others [13].

3. In many clayey soils seapage sets on only if the hydraulic gradient exceeds a characteristic value i_0 called the *stagnation gradient*. Then, DARCY's law is to be modified as follows:

$$v = k(i - i_0) \quad .$$

4. DARCY's law in the form $v = ki$ or $nv^f = ki$ presupposes that the grains are at rest. However it is important to realise that, virtually, it is the *relative* velocity between grains and pore fluid which has to appear in an interaction equation. The pertinent modification is known as the law of DARCY-GERSEVANOV and reads

$$n(v^f - v^s) = ki$$

or, with $nv^f = v$:

$$v - nv^s = ki \quad . \tag{30}$$

Obviously, Equ. 30 coincides with DARCY's law for $v^s = 0$.

4.1 Permeability

The coefficient of permeability k has the dimension of velocity. It only can be roughly determined (up to an order of magnitude).

The value of k depends on several factors, one of which is the viscosity (and, thus the temperature[9]) of the pore fluid. The void ratio e is also of importance. However, it is interesting to note that the two soils of Fig. 6 have identical void ratios but quite different

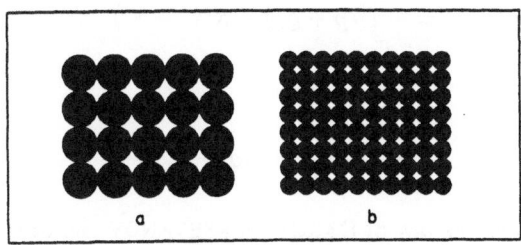

Fig. 6: Soil a has the same void ratio but a larger permeability than soil b

permeabilities. For an estimation of the influencial factors we consider the known law of

---

[9]Note that the viscosity μ depends on the temperature. For water $\mu = 1,31 \cdot 10^{-3}$ Ns/m² at $T = 10°$C and $\mu = 1,00 \cdot 10^{-3}$ Ns/m² at $T = 20°$C.

HAGEN-POISEUILLE which is valid for laminar flow within a pipe of radius r and length Δs. The driving pressure difference is Δp. The flux v is then determined by the equation

$$v = \frac{r^2}{4\mu} \overbrace{\frac{\Delta p}{\Delta s}}^{k} = \frac{r^2 \gamma_w}{4\mu} \overbrace{\frac{\Delta(p/\gamma_w)}{\Delta s}}^{i} \tag{31}$$

From Equ. 31 follows that the permeability depends on the ratio μ/γ_w and thus on the properties of the pore fluid. The dependence of k on the diameter d is quadratic. This is confirmed by the empirical fomula of HAZEN which holds true for uniform loose sands and reads:

$$k\,[\mathrm{cm/s}] \approx 100 \cdot (d_{10}[\mathrm{cm}])^2$$

The dependence of k on the porosity n can be expressed [14] with the equation:

$$k = \mathrm{const} \frac{n^3}{(1-n)^2}$$

5 Applicability of DARCY's law for velocity fields with dv/dt ≠ 0

Rigorously speaking, DARCY's law is only applicable to acceleration-free velocity fields, i.e. to steady ($\partial v^f/\partial t = 0$) and homogeneous ($\mathrm{grad} v^f = 0$) velocity fields. For the general case of $d_f v^f/dt \neq 0$ we have to take account of the full equation expressing the balance of linear momentum of the pore fluid:

$$\varrho_f \frac{d_f v^f}{dt} = -n\nabla p + \varrho_f \mathbf{g} - \kappa(\mathbf{v}^f - \mathbf{v}^s) \quad , \tag{32}$$

with

$$\kappa = \varrho^f g \frac{n^2}{k} = \varrho_f g \frac{n}{k} \quad ,$$

as a special case of which the law of DARCY-GERSEVANOV can be obtained for

$$d_f v^f/dt = 0 \quad .$$

If the z-coordinate points upwards, then we have:

$$\frac{1}{n}\varrho_f \mathbf{g} = \varrho^f \mathbf{g} = -\nabla(\gamma_w z)$$

Herewith and with $h = \dfrac{-p}{\gamma_w} + z$ and $\mathbf{v}^s \approx 0$ it follows from (32):

$$\frac{k}{g}\frac{d_f \mathbf{v}^f}{dt} = k\nabla h - \mathbf{v}_f \quad ,$$

or

$$\mathbf{v}_f = k\nabla h - \frac{k}{g}\frac{d_f \mathbf{v}^f}{dt} \quad , \tag{33}$$

wherefrom the difference to DARCY's equation[10] ($\mathbf{v}_f = k\nabla h$) becomes obvious.
The condition for the applicability of DARCY's law can be specified as follows [15]: We
set $\mathbf{v}_f = \mathbf{v}_D + \mathbf{v}_1$ with $\mathbf{v}_D := k\nabla h$ being DARCY's velocity, into equ. (33) and obtain the
following expression for $\mathbf{v}_1$ ($\mathbf{v}_1$ is the deviation from the velocity according to DARCY's
law):

$$\mathbf{v}_1 = -\frac{k}{ng}\frac{d_f}{dt}(k\nabla h + \mathbf{v}_1) \tag{34}$$

If $\frac{d_f}{dt}(k\nabla h) \approx 0$ holds true, then we obtain from equ. (34) the result that $\mathbf{v}_1$ decreases
exponentially, i.e. with $e^{-ngt/k}$.

Acknowledgement

The author is indepted to Alessandro Gajo, Trento, for many valuable remarks.

---

[10]If p is set positive for compression, then DARCY's law reads $\mathbf{v}_f = -k\nabla h$.

References

[1] Nigmatulin, R.: Dynamics of Multiphase Media, Hemisphere Publishing Corporation, 1991.

[2] L.D. Landau, E.M. Lifschitz: Lehrbuch der theoretischen Physik, VI Hydrodynamik, Akademie Verlag 1991.

[3] A. Gajo: The effects of inertia coupling in the interpretation of dynamic soil tests. Géotechnique 46, No. 2, 245-257 (1996).

[4] P.V. Lade, R. de Boer: The concept of effective stress for soil, concrete and rock, Géotechnique 47, 1 (1997), 61-78.

[5] F. Oka: Validity and limits of the effective stress concept in geomechanics, Mechanics of Cohesive-Frictional Materials, 1, 1996, 219-234.

[6] L. Rendulic: Ein Grundgesetz der Tonmechanik und sein experimenteller Beweis. Der Bauingenieur, 18. Jahrgang, 31/32, 459-467, 1937.

[7] A.W. Skempton: Effective stress in soils, concrete and rock. Proceed. Conf. on Pore Pressure and Suction in Soils, Butterworth, 1960, 4-16.

[8] I. Vardoulakis: Compression induced liquefaction of water-saturated granular media. Constitutive Laws for Engineering Materials: Theory and Applications. C.S. Desai et al., Editors, Elsevier Amsterdam, 1987, 647-656.

[9] M.A. Biot: Theory of propagation of elastic waves in a fluid-saturated porous solid. II. Higher frequency range. J. Acoust. Soc. Am. 28, No. 2, 179-191 (1956).

[10] W.A. Mikulitsch, G. Gudehus: Uniaxial tension, biaxial loading and wetting tests on loess. 1st Int. Conf. on Unsaturated Soils, Paris, 1995.

[11] Schuurman, Ir.E.: The compressibility of an air/water mixture and a theoretical relation between the air and water pressures. Géotechnique 16 (1966), 269-281.

[12] A. Kézdi: Handbuch der Bodenmechanik, Band 1, S. 132 ff, VEB Verlag für Bauwesen, Berlin, 1969.

[13] W. Herth and E. Arndts: Theorie und Praxis der Grundwasserabsenkung, Ernst & Sohn, Berlin 1985.

[14] G. Mattheß und K. Ubell: Allgemeine Hydrogeologie, Grundwasserhaushalt, Lehrbuch der Hydrogeologie Band 1, Gebrüder Borntraeger, 1983.

[15] P.Ya. Polubarinova-Kochina: Theory of Ground Water Movement. Princeton University Press, 1962, S. 23.

HYPOPLASTICITY:
A FRAMEWORK TO MODEL GRANULAR MATERIALS

D. Kolymbas and I. Herle
University of Innsbruck, Innsbruck, Austria

1 Introduction

1.1 What is the meaning of a constitutive model?

To solve a series of problems, engineers are interested in the mechanical behaviour of materials (and geomaterials are, probably, the most fascinating materials). What are our approaches to detect the mechanical behaviour of materials? We approach by observation, experiments and theories. Constitutive models are theories. A common opinion is that experiments are of primary importance. As soon as the experimental results are available, the formulation of an appropriate theory is straightforward. Another common opinion is that constitutive theories (models) are only valuable to feed sophisticated FEM-codes. Both opinions are not completely true and are, thus, misleading. It is very important to realize that a theory (besides serving as a tool for better FEM-simulations) helps to understand nature. C. H. DARWIN said that *'All observation must be for or against some view, if it is to be of any service'*. Besides of this, we have to recognize that experiments are burdened with a series of errors.

It should be added that a well-accepted theory creates a series of preoccupations. Thus, a new theory can help to recognize and — possibly — overcome such preoccupations. E.g., the general validity of the MOHR-COULOMB failure criterion has never been fully corroborated by experiments. We are still not sure whether the friction angle is the same for triaxial compression and extension and for shearing. Another preoccupation is the common (and erroneous) belief in the existence of an elastic regime for soils, or in softening as being completely a material property.

1.2 Notation

To refer to a tensor, we either use the index notation σ_{ij} which represents the whole matrix

$$\begin{pmatrix} \sigma_{11} & \sigma_{12} & \sigma_{13} \\ \sigma_{21} & \sigma_{22} & \sigma_{23} \\ \sigma_{31} & \sigma_{32} & \sigma_{33} \end{pmatrix}$$

or we use the symbolic notation. Following the NLFT [38], we denote the CAUCHY stress tensor by $\mathbf{T}$. Note that, depending on the chosen bases, there are several other stress tensors conceivable. The great number of definitions and choices is particularly striking with respect to the strain tensor. In hypoplasticity we circumvent this source of confusion by using only the stretching $\mathbf{D}$, which is the symmetric part of the velocity gradient $\mathbf{L} := \operatorname{grad} \mathbf{v} = \partial v_i / \partial x_j$. For special conditions [56] $\mathbf{D}$ may be set equal to the time rate of logarithmic strain $\dot{\epsilon}$.

1.3 Rate equations

A constitutive equation is expected to represent stress as it results from a strain (or deformation) history starting from some specified reference state. If we represent stress as a *function* of strain, this automatically means that the stress does not depend on the deformation history. This special case is called (by definition) elastic behaviour. Soil is not elastic, so we have to find another type of relation. How can we represent strain history? Some people tried to introduce integral transformations using appropriate kernels. This approach is not useful for soils in general. A general way to introduce history (or path) dependence in physics is to use non-integrable differential forms (or PFAFFean forms), i.e. to represent y by the differential equation

$$dy = a_1 dx_1 + a_2 dx_2 + \ldots a_n dx_n.$$

This equation connects increments $dx_1, dx_2, \ldots$ with dy (or $dy_1, dy_2, \ldots$) in such a way that there is no closed form representation of $y(x)$. I.e. the relation (which is called 'incremental' as it relates increments) $dy = f(dx_i)$ is not integrable. This is the way we proceed in soil mechanics when we represent the stress increment as a non-integrable function of the strain increment:

$$d\sigma = f(d\epsilon)$$

This approach is common to the theories of plasticity and hypoplasticity.
Now we can divide all increments by dt and obtain time rates:

$$\dot{\sigma} = d\sigma/dt \quad , \quad \dot{\epsilon} = d\epsilon/dt \quad , \qquad \text{etc.}$$

Thus, an equation between increments is also representable as an equation between rates, as long as we refer to so-called rate-independent materials. An equation of the form $\dot{\sigma} = f(\dot{\epsilon})$ is called a rate-equation. It does not imply the existence of an equation $\sigma = f_1(\epsilon)$.

2   Requirements on rate equations

2.1   Incremental nonlinearity

$d\sigma/d\epsilon = \dot{\sigma}/\dot{\epsilon}$ represents the incremental stiffness of the material considered (see Fig. 1). Since with anelastic (plastic) materials $|d\sigma|$ is much larger at unloading than at loading

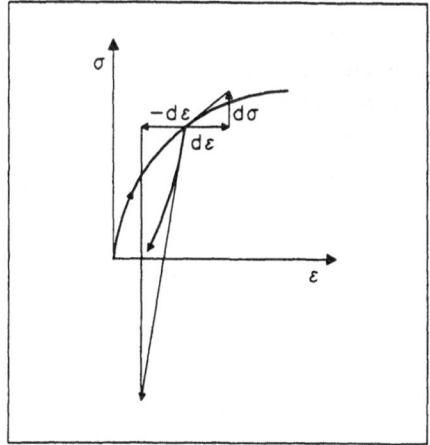

Fig. 1: Different stiffness at loading and unloading

(i.e. the stiffness is much larger at unloading than at loading), we infer that for such a material the function $d\sigma = f(d\epsilon)$ or $\dot{\sigma} = f(\dot{\epsilon})$ must be nonlinear in $\dot{\epsilon}$ (or $d\epsilon$). This non-linearity remains, no matter how small $d\epsilon$ is. Therefore it is called 'non-linearity in the small' or 'incremental non-linearity'. Note that incremental non-linearity has nothing to do with the curved form of the stress-strain curve for loading. This curve can be, of course, linearized for small $|d\epsilon|$, a fact which led many people to believe that in physics every relation can be linearized 'in the small'. Thus, all elastoplastic and hypoplastic relations are incrementally non-linear.

2.2   Objective stress rate

The time derivative of the stress tensor, $\dot{T}$, is a quantity which depends on rotations of either the considered body or the observer. This means that it does not necessarily express stress changes due to deformation. Since a constitutive equation aims to express only stress changes due to deformation, it must use so-called objective stress rates, i.e. stress rates registered by a co-rotated observer. This can be achieved by the so-called co-rotated (or JAUMANN) stress rate

$$\overset{\circ}{\mathbf{T}} = \dot{\mathbf{T}} - \mathbf{WT} + \mathbf{TW},$$

where the spin-tensor $\mathbf{W}$ is the antimetric part of the velocity gradient. It should be noted that there are many other objective stress rates, and a large part of the literature is devoted to the question of which of the objective stress rates is the proper one to be used in constitutive modelling.

2.3 Rate independence

Rate-independent materials are defined as materials without an internal time scale. I.e., the rate of deformation is immaterial for the final stress. In other words, rate-independent materials are invariant with respect to changes of time scale. If we deform a rate-independent material twice as fast, then the stress rate will also be doubled. Considering constitutive equations of the rate-type, this means that the stress rate $\overset{\circ}{\mathbf{T}}$ is positively homogeneous of the first degree with respect to $\mathbf{D}$:

$$\mathbf{h}(\mathbf{T}, \lambda \mathbf{D}) = \lambda \mathbf{h}(\mathbf{T}, \mathbf{D}) \quad \text{for} \quad \lambda > 0$$

Note that this homogeneity does by no means imply linearity (cf. the relation $y = |x|$, which is homogeneous in the above sense, but not linear). Soils are not exactly rate-independent. Clays are more pronouncedly rate dependent than sands, but also sands exhibit rate dependence [8]. However, for a first approximation we can consider soils as rate-independent materials.

2.4 Homogeneity in stress

Assume that the relation $\overset{\circ}{\mathbf{T}} = \mathbf{h}(\mathbf{T}, \mathbf{D})$ is homogeneous in $\mathbf{T}$, i.e.

$$\mathbf{h}(\lambda \mathbf{T}, \mathbf{D}) = \lambda^{n} \mathbf{h}(\mathbf{T}, \mathbf{D})$$

Let us investigate the consequences of this assumption. Consider a stress state $\mathbf{T}_1$. We now determine the stretching in such a way that $\overset{\circ}{\mathbf{T}} = \mathbf{h}(\mathbf{T}_1, \mathbf{D}_1) = \lambda \mathbf{T}_1$. If we then continuously apply $\mathbf{D}_1$, then we shall obtain a stress path which is a straight line passing through the origin of stress space. This follows from our assumption, because

$$\overset{\circ}{\mathbf{T}}(t + dt) = \mathbf{h}(\mathbf{T}_1 + \lambda \mathbf{T}_1 dt, \mathbf{D}_1) = (1 + \lambda dt)^{n} \mathbf{h}(\mathbf{T}_1, \mathbf{D}_1) = (1 + \lambda dt)^{n} \overset{\circ}{\mathbf{T}}(t)$$

In other words, our assumption implies that proportional strain paths (i.e. paths with $\mathbf{D} = \text{const}$) are connected with proportional stress paths (i.e. straight stress paths passing through the origin of the stress space).

Note that proportional stress paths must be limited within a fan, because we know that there are inaccessible (unfeasible) stress states.

Let us now consider the degree of homogeneity. Knowing that $d\sigma/d\epsilon = \dot\sigma/\dot\epsilon$ or $\dot{\mathbf{T}}/\mathbf{D}$ is the stiffness, we infer that $\dot{\mathbf{T}}/\mathbf{D}|_{\lambda\mathbf{T}} = \lambda^n \dot{\mathbf{T}}/\mathbf{D}|_{\mathbf{T}}$. In other words, if we increase the stress by a factor λ, the stiffness is increased by the factor λ^n. Experimentalists in soil mechanics often remark that *normalized* stress-strain curves coincide (this is in particular the case with normally consolidated clays). The consequence is $n = 1$. Setting $n = 1$ would imply that the friction angle is invariant with respect to the stress level. This is an acceptable approximation to start with. Later we shall see that if the changes of stress level at a given void ratio are considerable, then the corresponding variation of friction (and dilatancy) angles may not be neglected.

3 Types of rate equations

3.1 Hypoelasticity

TRUESDELL [39] has introduced constitutive relations of the form $\overset{\circ}{\mathbf{T}} = \mathbf{h}(\mathbf{T}, \mathbf{D})$. He required that the function $\mathbf{h}()$ be linear in $\mathbf{T}$ and in $\mathbf{D}$. He introduced the name hypoelasticity for such relations. Hypoelastic constitutive equations may produce curved stress-strain curves, and in some cases these stress-strain curves reach a horizontal plateau and can thus model yielding. However, the imposed incremental linearity implies equal stiffness for loading and unloading and thus renders hypoelastic relations inappropriate to describe anelastic (plastic) materials. Despite this, some hypoelastic relations have been launched in soil mechanics (e.g. by DAVIS and MULLENGER [7]). In order to overcome the equal stiffness at loading and unloading, they are (in most cases tacitly) endowed with additional stress-strain relations holding for unloading. Strictly speaking, these relations (regarded as a whole) are not linear any more, i.e. they are not hypoelastic.

3.2 Elastoplasticity

We have seen that the problem of describing different stiffness at loading and unloading can be treated by introducing at least two different linear relations between $\dot\sigma$ and $\dot\epsilon$, of which one holds for loading and one for unloading. This is the approach of the theory of elastoplasticity. It requires a series of precautions. First, what should be considered as loading and what as unloading has to be defined. This is accomplished by the introduction of the so-called yield surface, a surface in the stress space. Only such stress increments which start from this surface and point outwards are considered as loading, the remaining being unloading stress increments. Another precaution refers to the expectation that in the transition between loading and unloading the response must be continuous. This is accomplished by the so-called consistency condition. The common theories of elastoplasticity require that the behaviour is elastic inside the yield surface, an assumption which is not realistic for soils. Another point of concern in elastoplasticity is how the yield surface changes its shape and position with loading. It is typical for the theory of elastoplasticity to consider a decomposition of strain into elastic and plastic strain which cannot be in fact distinguished in experiments. The stress-strain relation for loading is determined by the so-called flow rule, which states that the increment (or rate) of the plastic strain is always normal to a so-called plastic potential surface. A special case arises if the plastic potential surface is set equal to the yield surface. This special case is called the normality condition. A set of very useful theorems has been formulated for materials obeying the normality condition. However, normality is not realistic for frictional soils as it would imply that the dilatancy angle is equal to the friction angle, which is not the case. After all, we can summarize: Elastoplastic constitutive laws consist of two or more linear relations between $d\epsilon$ and $d\sigma$. As a whole, they are incrementally non-linear.

3.3 Hypoplasticity

Elastoplastic and hypoplastic equations are both of the general form

$$\mathbf{T} = \mathbf{h}(\mathbf{T}, \mathbf{D})$$

Starting from the fact that every function $\mathbf{h}(\mathbf{T}, \mathbf{D})$ can be represented according to the general representation theorem,

$$\mathbf{h}(\mathbf{T}, \mathbf{D}) = \psi_1 \mathbf{1} + \psi_2 \mathbf{T} + \psi_3 \mathbf{D} + \psi_4 \mathbf{T}^2 + \psi_5 \mathbf{D}^2 + \psi_6 (\mathbf{TD} + \mathbf{DT})$$
$$+ \psi_7 (\mathbf{TD}^2 + \mathbf{D}^2 \mathbf{T}) + \psi_8 (\mathbf{T}^2 \mathbf{D} + \mathbf{DT}^2) + \psi_9 (\mathbf{T}^2 \mathbf{D}^2 + \mathbf{D}^2 \mathbf{T}^2)$$

(ψ_i are scalar functions of invariants and joint invariants of $\mathbf{T}$ and $\mathbf{D}$), the experiment was undertaken to find such a unique function which appropriately describes the mechanical properties of soils [22]. In order to avoid the shortcomings of hypoelasticity, this function has to be non-linear in $\mathbf{D}$. On the other hand, it should be homogeneous of the first degree in $\mathbf{D}$ in order to describe rate independent materials, and homogeneous in $\mathbf{T}$ in order to describe proportional stress paths in case of proportional strain paths. Therefore, the design of such a function had to proceed along the above stated representation theorem and some general mathematical restrictions:
- non-linearity in $\mathbf{D}$
- homogeneity in $\mathbf{D}$ and $\mathbf{T}$
with avoidance of any recourse to notions from the theory of elastoplasticity such as yield functions, decomposition of strain etc.
This experiment (every theory is, virtually, an experiment) was more or less success-ful, as by trial and error a function was found which was able to describe many aspects of soil behaviour. Thus, a new approach to constitutive modelling was created. The name 'hypoplastic equation' fits very well, as the relation between hypoplasticity and

(elasto)plasticity is the same as the one between hypoelasticity and elasticity: The theories with "hypo" do not have potential. It should be mentioned that DAFALIAS [5] coined the term hypoplasticity earlier for something else, which can be considered as a general case of what we call hypoplasticity.

Let us now have a look at some hypoplastic equations. Most of them consist of 4 tensorial terms combined together with 4 material parameters, e.g. [50]:

$$\overset{\circ}{\mathbf{T}} = C_1(\operatorname{tr}\mathbf{T})\mathbf{D} + C_2\frac{\operatorname{tr}(\mathbf{TD})}{\operatorname{tr}\mathbf{T}}\mathbf{T} + C_3\frac{\mathbf{T}^2}{\operatorname{tr}\mathbf{T}}\sqrt{\operatorname{tr}\mathbf{D}^2} + C_4\frac{\mathbf{T}^{*2}}{\operatorname{tr}\mathbf{T}}\sqrt{\operatorname{tr}\mathbf{D}^2} \tag{1}$$

with the deviatoric stress $\mathbf{T}^*$ defined as

$$\mathbf{T}^* = \mathbf{T} - \frac{1}{3}(\operatorname{tr}\mathbf{T})\mathbf{1}$$

An alternative representation of hypoplastic constitutive equations is to summarize the linear terms by $\mathbf{L}[\mathbf{D}]$ and the non-linear terms by $\mathbf{N}||\mathbf{D}||$ with $||\mathbf{D}|| := \sqrt{\operatorname{tr}\mathbf{D}^2}$. Then, a hypoplastic equation assumes the general form

$$\overset{\circ}{\mathbf{T}} = \mathbf{L}[\mathbf{D}] + \mathbf{N}||\mathbf{D}||. \tag{2}$$

As already stated, the experiment to describe soil behaviour with a sort of hypoelastic equation, extended to comprise terms which are non-linear though homogeneous of the first degree in $\mathbf{D}$, was successful. Several equations could be found with only 4 material parameters C_1, C_2, C_3, C_4 [19, 45, 47, 50] which were capable to describe

- the triaxial test as characterized by a stiffness decreasing down to zero at the limit state and a corresponding volumetric strain curve exhibiting first contractancy and then dilatancy

- incrementally non-linear behaviour, i.e. unloading stiffness much larger than at loading

- realistic asymptotic properties (referring to proportional paths).

However, void ratio was not taken into account, and, therefore, such simple hypoplastic constitutive models were not capable of describing the difference of friction angle and stiffness between dense and loose samples, or the decrease of the peak friction angle to the residual one with increasing volumetric strain (softening). But this was also not expected from such simple constitutive models. To achieve this, in more recent versions elaborated in Karlsruhe [2, 11, 24, 46, 48] several tensorial terms are multiplied with scalar factors which aim to model the influence of density and stress level as well as the transition to the so-called critical state. Of course, such factors increase the intricacy of the model.

Hypoplastic constitutive relations are directly presented without any reference to any sort of surfaces in stress space. However, various surfaces can be derived from a hypoplastic equation, as will be explained in Section 4.5.

4   Testing of rate equations

4.1   Response envelopes

The response envelopes introduced by GUDEHUS [10] to model axisymmetric response
are very useful in discussing properties of constitutive models. GUDEHUS showed that re-
sponse envelopes are ellipses for linear relations between $\dot{\sigma}$ and $\dot{\epsilon}$ (e.g. for hypoelasticity).
Setting

$$
\mathbf{D} = \begin{pmatrix} -\sin\alpha & 0 & 0 \\ 0 & -\cos\alpha/\sqrt{2} & 0 \\ 0 & 0 & -\cos\alpha/\sqrt{2} \end{pmatrix}
$$

with $0° \le \alpha \le 360°$ (i.e. $\|\mathbf{D}\| = 1$), a linear constitutive equation $\mathring{\mathbf{T}} = f(\mathbf{D})$ assumes the
form

$$
\dot{\sigma}_1 = a_{11}\sin\alpha + a_{12}\cos\alpha
$$
$$
\dot{\sigma}_2 = a_{21}\sin\alpha + a_{22}\cos\alpha
$$

which is the parametric representation of an ellipse in the $\dot{\sigma}_1$-$\dot{\sigma}_2$-space. For hypoplasticity
the response envelopes are also ellipses, but the reference state is no more the centre of
the ellipse. Thus, limit states can be modelled.
It is interesting to see that there are several ways to model limit states (see Fig. 2, 3, 4).

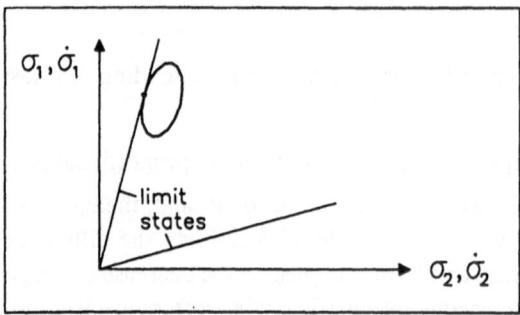

Fig. 2: Response envelope at limit state (hypoplasticity)

Each of them should provide at least one vanishing stress rate. With respect to response
envelopes this means that the reference stress state must coincide with at least one point of
the periphery of the response envelope. With hypoplastic and elastoplastic relations this is
the case for only one stretching $\mathbf{D}$, whereas the Grenoble type (CloE) envelopes shrink to
a single point. This has the disadvantage that the limit state cannot be abandoned through

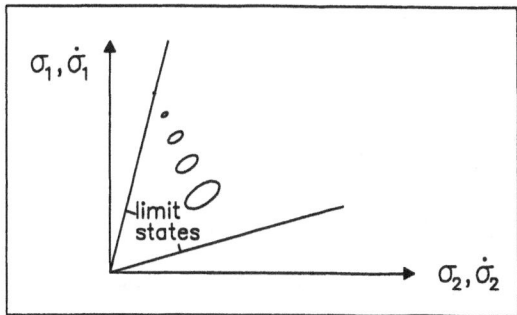

Fig. 3: Response envelopes approaching the limit state for CLoE model

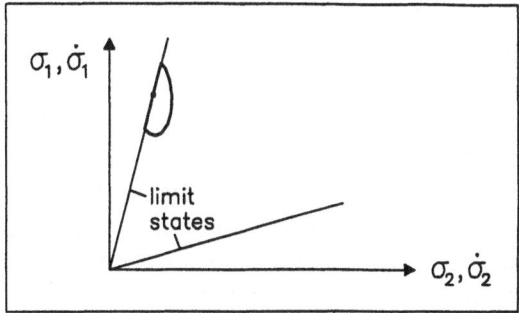

Fig. 4: Response envelope at limit state (elastoplasticity)

unloading — unless another response is defined for unloading. However, this leads to a sort of discontinuous response which is also typical for some 'hypoelastic' constitutive equations, as they have been suggested by DAVIS & MULLENGER [7] and others.
It is worth noting that in elastoplasticity the response envelopes are composed of several ellipses such that the envelope is continuous but not smooth (see Fig. 5).

4.2   Element tests

Whereas constitutive equations relate stresses with strains, in the laboratory we only can measure forces and displacements. Thus, in order to check a constitutive relation, we need tests with homogeneous (i.e. constant) distribution of stress and strain within the sample. If homogeneity is given, then we can easily obtain stresses and strains from boundary forces and displacements. For inhomogeneously deformed samples the knowledge of the distribution of stresses and strains within the sample is necessary. This is however impossible (except for some highly sophisticated tests with x-rays or other sort of transparent samples), and we are only able to numerically calculate the stress and strain fields by

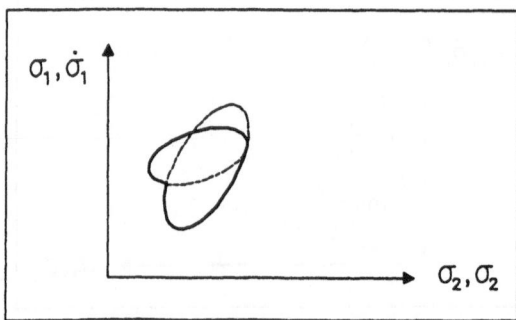

Fig. 5: Response envelope of an elastoplastic relation

means of the constitutive equation we want to check. Thus, constitutive equations have to be based on element tests, i.e. tests with homogeneous sample deformation. In the Russian literature element tests are called "zero-dimensional" tests. Such tests are connected with extraordinarily high experimental difficulties.

Despite all efforts, experimentalists have to admit that in the course of a test a sample starts getting inhomogeneously deformed, no matter what measures have been undertaken to prevent this. In other words, inhomogeneous deformation is inevitable. This fact adds to soil mechanics the same headache as the transition from laminar to turbulent flow in hydromechanics does. Thus, we see that with respect to laboratory tests aiming to support constitutive modelling, (i) we have to spend provisions to achieve homogeneous deformation, and (ii) even so the homogeneous deformation can only be realized for part of the test duration. In other words, nature allows us to look upon homogeneously deformed samples only within a time-window. At that, we do not know exactly the boundaries of this window. Unfortunately, it is a common practice in soil mechanics to ignore inhomogeneous deformation and to evaluate tests as if they were homogeneous. The results are, of course, useless.

How can we obtain simulations of laboratory element tests by using an equation of the rate type? First, we have to start from a state with known stress state. If the test to be simulated has kinematical boundary conditions, then the stretching $\mathbf{D}$ is known (e.g. in case of the oedometer test all but one components of $\mathbf{D}$ are equal zero). With knowledge of $\mathbf{T}$ and $\mathbf{D}$ the constitutive equation $\overset{\circ}{\mathbf{T}} = \mathbf{h}(\mathbf{T}, \mathbf{D})$ makes possible to evaluate $\overset{\circ}{\mathbf{T}}$. Multiplying $\overset{\circ}{\mathbf{T}}$ with a sufficiently small time step Δt gives $\Delta \mathbf{T} \approx \overset{\circ}{\mathbf{T}} \Delta t$. The new stress state is then obtained to $\mathbf{T} + \Delta \mathbf{T}$. This process can be continued and corresponds to a numerical integration of the evolution equation $\overset{\circ}{\mathbf{T}} = \mathbf{h}(\mathbf{T}, \mathbf{D})$ (so-called EULER-forwards integration). The procedure is a little more difficult if not all of the boundary conditions are of the kinematic type. In case of a static boundary condition (e.g. $\sigma_2 = \sigma_3 = $ const for triaxial test), the component

d_2 of $\mathbf{D}$ must be determined by solving the algebraic equation $\dot{\sigma}_2(d_2) = 0$.

4.3   Calibration

A constitutive relation is of no use if the parameters involved cannot be adapted to a particular material. The values of these parameters constitute the identity card of this material with respect to a particular constitutive model. Moreover, a particular parameter is useless unless it is embedded within a constitutive model. The process of the determination of the values of the parameters of a constitutive model is called "calibration" or "parameter identification". In the overwhelming number of publications on constitutive models the calibration is simply omitted as being not worth mentioning. In fact it is — with the majority of models — a task which can take up to several months of work! Considering hypoplastic constitutive equations, the calibration is straightforward by fitting the equation to the outcomes of one or several (say triaxial) tests [21]. Knowing the strain increment and the stress increment at a particular stress state from experiments, the only unknowns in the equation are the material constants. To solve a system of four linear equations is no problem nowadays.

More recently the calibration has been considerably facilitated, as parameters appearing in hypoplastic equations of the new generation can be directly related to simple properties (mainly granulometric ones) of the soil considered. This will be shown in Section 5.

4.4   Initial stress

Equation (2) is of the rate type, i.e. it is an evolution equation which makes possible to calculate the stress changes due to a given increment of deformation. The initial stress has to be known or assumed. Thus, the problem of determining the stress can only be back-stepped but not entirely solved by means of equations of the rate type. This fact is, of course, not very pleasant, but there is no means how to circumvent it. We only can hope that the influence of the initial state fades with increasing length of the history. Besides this fact there are some cases (e.g. one-dimensional consolidation following sedimentation) where we can determine the initial stress by reasoning.

The problem of determination of the initial stress is traditionally hidden by elastoplasticity where it is always tacitly assumed that the initial stress results from the theory of elasticity. The latter has to be applied for a deformation starting from a stress free state: The gravity is 'switched on' and the transition to the deformed state is considered to be elastic. This simplification is reflected in almost all existing finite element codes. At least for soils, it is not realistic.

4.5   Limit states

A very important property of granular materials is their ability to flow (or yield), i.e. to undergo large deformations without stress change, as soon as the stresses and the void ratio obtain their critical values. This sort of flow should be attributed as 'plastic' flow and distinguished from the flow of fluids. The latter has, a pronounced viscous (rate-dependent) character.

Plastic flow occurs as soon as the stress state $\mathbf{T}$ and the strain rate $\mathbf{D}$ fulfil the condition $\mathbf{h}(\mathbf{T}, \mathbf{D}) = 0$. In the theory of elastoplasticity the first condition is called the yield (limit) surface, and the second condition is called the flow rule. 'Flow' means that stiffness vanishes for particular $\mathbf{T}$ and $\mathbf{D}$ values.

In elastoplasticity the yield function is the starting point and the mathematical relation $\mathbf{h}(\mathbf{T}, \mathbf{D})$ is based upon this yield function. In contrast, it can be shown that a yield function is contained in a hypoplastic formulation $\overset{\circ}{\mathbf{T}} = \mathbf{h}(\mathbf{T}, \mathbf{D})$, i.e. the yield function $f(\mathbf{T})$ can be derived from the constitutive relation. To this purpose we rewrite (following a proposition of DESRUES and CHAMBON [57]) the equation (2) in the form

$$\overset{\circ}{\mathbf{T}} = \mathbf{L}(\mathbf{T})[\mathbf{D}] + \mathbf{N}(\mathbf{T})||\mathbf{D}|| = \mathbf{L}(\mathbf{T}) \left[\mathbf{D} + \mathbf{B}||\mathbf{D}|| \right] \quad ,$$

with $\mathbf{L}(\mathbf{T})$ being a matrix operator applied to its tensorial argument. It is obvious that $\mathbf{h}(\mathbf{T}, \mathbf{D}) = 0$ occurs for

$$\mathbf{D}^0 = \mathbf{D}/||\mathbf{D}|| = -\mathbf{B} \quad .$$

Consequently, the function $f(\mathbf{T})$ reads

$$f(\mathbf{T}) = \text{tr}\mathbf{B}^2 - 1 \quad ,$$

with $\mathbf{B}$ being a function of $\mathbf{T}$. In other words, the limit surface reads:

$$f(\mathbf{T}) = \text{tr}\mathbf{B}^2 - 1 = 0 \quad .$$

For the constitutive equation 1 $\mathbf{B}$ reads as follows:

$$\mathbf{B} = \mathbf{L}^{-1}\mathbf{N} = \frac{C_3 \mathbf{T}^2}{C_1 (\text{tr}\mathbf{T})^2} + \frac{C_4 \mathbf{T}^{*2}}{C_1 (\text{tr}\mathbf{T})^2} - \frac{C_2 C_3}{C_1} \frac{\text{tr}(\mathbf{T}^3)}{(\text{tr}\mathbf{T})^2} \cdot \frac{\mathbf{T}}{C_1 (\text{tr}\mathbf{T})^2 + C_2 \text{tr}(\mathbf{T}^2)}$$
$$- \frac{C_2 C_4}{C_1} \frac{\text{tr}(\mathbf{T}\mathbf{T}^{*2})}{(\text{tr}\mathbf{T})^2} \cdot \frac{\mathbf{T}}{C_1 (\text{tr}\mathbf{T})^2 + C_2 \text{tr}(\mathbf{T}^2)}$$

Due to the homogeneity of $\mathbf{h}(\mathbf{T}, \mathbf{D})$ (and consequently also of $\mathbf{B}$) in $\mathbf{T}$, the surface $f(\mathbf{T}) = 0$ is a cone with apex at the origin $\mathbf{T} = \mathbf{0}$. The cross section of this cone with the deviatoric plane reveals the influence of the intermediate principal stress, i.e. the yield surface differs from the MOHR-COULOMB criterion.

In classical soil mechanics the yield surface is considered as a limit, i.e. it is expected that no stress path can transcend this surface. According to our understanding, this is an unnecessary and also unrealistic restriction. There is in fact no reason (either conceptual or experimental) why this surface should be conceived as a limit which encloses all feasible stress states. Experiments show that the yield surface can be transcended [48, 55]. This fact can easily be modelled by constitutive equations of the type (2). Moreover, such equations make possible to determine a surface in the stress space which encloses all feasible stress states. This surface is called bound surface. Of course, the bound surface lies outside the yield surface.

4.6   Invertibility and controllability

In kinematically controlled tests (such as oedometric test or undrained triaxial test) the stretching $\mathbf{D}$ is prescribed and the stress rate $\overset{\circ}{\mathbf{T}}$ can be uniquely determined by means of the hypoplastic constitutive equation. What about the unique determination of $\mathbf{D}$ when $\overset{\circ}{\mathbf{T}}$ is prescribed? To answer this question of unique invertibility we multiply the equation $\overset{\circ}{\mathbf{T}}= \mathbf{L}\mathbf{D} + \mathbf{N}\|\mathbf{D}\|$ with the inverse operator $\mathbf{L}^{-1}$ and obtain

$$\mathbf{A} := \mathbf{L}^{-1}\,\overset{\circ}{\mathbf{T}}= \mathbf{D}+\mathbf{L}^{-1}\mathbf{N}\|\mathbf{D}\|$$

or

$$\mathbf{D} = \mathbf{A} - \mathbf{B}\|\mathbf{D}\| \tag{3}$$

with $\mathbf{B} := \mathbf{L}^{-1}\mathbf{N}$. With the notation $\mathbf{X} \cdot \mathbf{Y} := \mathrm{tr}(\mathbf{X}\mathbf{Y})$ we obtain from (3):

$$\mathbf{D} \cdot \mathbf{D} = (\mathbf{A} - \mathbf{B}\|\mathbf{D}\|) \cdot (\mathbf{A} - \mathbf{B}\|\mathbf{D}\|) = \mathbf{A} \cdot \mathbf{A} - 2\mathbf{A} \cdot \mathbf{B}\|\mathbf{D}\| + \mathbf{B} \cdot \mathbf{B}\|\mathbf{D}\|^2 \quad . \tag{4}$$

Noting that $\mathbf{D} \cdot \mathbf{D} \equiv \|\mathbf{D}\|^2$ we observe that (4) is a quadratic equation for $x := \|\mathbf{D}\|$. Its solution reads

$$x_{1/2} = \frac{2\mathbf{A} \cdot \mathbf{B} \pm \sqrt{4(\mathbf{A} \cdot \mathbf{B})^2 - 4\mathbf{A} \cdot \mathbf{A}(\mathbf{B} \cdot \mathbf{B} - 1)}}{2(\mathbf{B} \cdot \mathbf{B} - 1)}$$

Since x is a modulus, only a solution $x > 0$ is meaningful. Moreover, in order to obtain a unique solution we have to require that only one solution is positive, i.e. $x_1 \cdot x_2 < 0$:

$$x_1 \cdot x_2 = \frac{4(\mathbf{A} \cdot \mathbf{B})^2 - 4(\mathbf{A} \cdot \mathbf{B})^2 + 4\mathbf{A} \cdot \mathbf{A}(\mathbf{B} \cdot \mathbf{B} - 1)}{4(\mathbf{B} \cdot \mathbf{B} - 1)^2} = \frac{\mathbf{A} \cdot \mathbf{A}}{\mathbf{B} \cdot \mathbf{B} - 1} < 0$$

Since $\mathbf{A} \cdot \mathbf{A} > 0$ we infer [30] that invertibility is given for $\mathbf{B} \cdot \mathbf{B} - 1 < 0$, i.e. for all stress states inside the limit surface $\mathbf{B} \cdot \mathbf{B} - 1 = 0$, as already pointed by CHAMBON [5]. A more subtle question on unique solutions of the constitutive equation arises if some (say k) components of $\mathbf{D}$ and $6 - k$ components of $\overset{\circ}{\mathbf{T}}$ are prescribed, and the remaining components have to be determined[1]. The unique solution of this problems (which corresponds e.g. to the conventional triaxial test with the mixed conditions $D_{11} = -1$ in axial direction and $\overset{\circ}{T}_{22} = \overset{\circ}{T}_{33} = 0$ in lateral directions) is called controllability [30]. For simplicity we consider $\mathbf{D}$ and $\overset{\circ}{\mathbf{T}}$ as column or row vectors $\mathbf{x}$ and $\mathbf{y}$, i.e. we take $x_1 := D_{11}$, $x_2 := D_{12}, \ldots$ and similarly $y_1 := \overset{\circ}{T}_{11}, y_2 := \overset{\circ}{T}_{12}, \ldots$.

The selection of the independent and dependent components can be accomplished by the partition matrices $\mathbf{P}$ and $\mathbf{Q}$, the components of which vanish for $i \neq j$. Their diagonal components are either 1 or 0. $\mathbf{P}$ and $\mathbf{Q}$ are related by $\mathbf{P} + \mathbf{Q} = \mathbf{1}$, with $\mathbf{1}$ being the unit matrix. E.g.

$$
\mathbf{P} = \begin{pmatrix} 1 & 0 & 0 & 0 & 0 & 0 \\ 0 & 0 & 0 & 0 & 0 & 0 \\ 0 & 0 & 0 & 0 & 0 & 0 \\ 0 & 0 & 0 & 1 & 0 & 0 \\ 0 & 0 & 0 & 0 & 0 & 0 \\ 0 & 0 & 0 & 0 & 0 & 1 \end{pmatrix}, \quad
\mathbf{Q} = \begin{pmatrix} 0 & 0 & 0 & 0 & 0 & 0 \\ 0 & 1 & 0 & 0 & 0 & 0 \\ 0 & 0 & 1 & 0 & 0 & 0 \\ 0 & 0 & 0 & 0 & 0 & 0 \\ 0 & 0 & 0 & 0 & 1 & 0 \\ 0 & 0 & 0 & 0 & 0 & 0 \end{pmatrix}
$$

We can now obtain the independent (or controlling) variable $\mathbf{X}$ of a problem with mixed conditions as

$$\mathbf{X} = \mathbf{Qx} + \mathbf{Py} \tag{5}$$

and, likewise, the dependent variable $\mathbf{Y}$ as:

$$\mathbf{Y} = \mathbf{Px} + \mathbf{Qy} \tag{6}$$

E.g.

$$
\begin{aligned}
\mathbf{X} &= (X_1, X_2, \ldots)^T &&= (D_{11}, \overset{\circ}{T}_{12}, \overset{\circ}{T}_{13}, D_{22}, \ldots)^T \\
\mathbf{Y} &= (Y_1, Y_2, \ldots)^T &&= (\overset{\circ}{T}_{11}, D_{12}, D_{13}, \overset{\circ}{T}_{22}, \ldots)^T
\end{aligned}
$$

Inverting the system of equations (5) and (6) and using $(\mathbf{1} - 2\mathbf{P})^{-1} = (\mathbf{1} - 2\mathbf{P})$ we obtain

$$\mathbf{x} = \mathbf{QX} + \mathbf{PY} \tag{7}$$
$$\mathbf{y} = \mathbf{PX} + \mathbf{QY} \tag{8}$$

---

[1] As NOVA points out, the most general case of test control (e.g. a test with $T_1 + T_2 + T_3 = \text{const}$, $D_2 = D_3, D_1 = 1$) is obtained if we replace $\dot{\mathbf{T}}$ by $\dot{\mathbf{T}}' := \mathbf{S}\dot{\mathbf{T}}$ and $\mathbf{D}$ by $\mathbf{D}' := \mathbf{S}^{-1}\mathbf{D}$ with some appropriately chosen non-singular matrix $\mathbf{S}$. Obviously, $\mathbf{T}'$ and $\mathbf{D}$ are energy conjugated in the sense that $\text{tr}(\mathbf{TD}) = \text{tr}(\mathbf{T}'\mathbf{D}')$ or $\text{tr}(\dot{\mathbf{T}}\mathbf{D}) = \text{tr}(\dot{\mathbf{T}}'\mathbf{D}')$.

Inserting (7) into the constitutive equation $\mathbf{y} = \mathbf{h}(\mathbf{x})$ or $\mathbf{y} - \mathbf{h}(\mathbf{x}) = 0$ we obtain an implicit relation between $\mathbf{X}$ and $\mathbf{Y}$:

$$F(\mathbf{X}, \mathbf{Y}) := \mathbf{PX} + \mathbf{QY} - \mathbf{h}(\mathbf{QX} + \mathbf{PY}) = 0 \qquad (9)$$

A unique determination of $\mathbf{Y}$ from (9) (i.e. controllability) is possible if $\det(\partial \mathbf{F}/\partial \mathbf{Y}) = \det(\partial F_i/\partial Y_j) \neq 0$. This means that

$$\det\left(\mathbf{Q} - \mathbf{P}\frac{\partial \mathbf{h}}{\partial \mathbf{x}}\right) \neq 0 \quad .$$

$\dfrac{\partial \mathbf{h}}{\partial \mathbf{x}}$ is the stiffness matrix. For a hypoplastic constitutive equation $\mathbf{h}(\mathbf{x}) := \mathbf{Lx} + \mathbf{N}|\mathbf{x}|$ the stiffness matrix reads

$$\mathbf{H}(\mathbf{x}) := \frac{\partial \mathbf{h}}{\partial \mathbf{x}} = \mathbf{L} + \mathbf{N} \otimes \frac{\mathbf{x}}{|\mathbf{x}|} = \mathbf{L} + \mathbf{N} \otimes \mathbf{x}^0 \quad , \qquad (10)$$

i.e. the stiffness matrix depends on the direction of $\mathbf{x}$. This is to be contrasted with elasto-plastic formulations where

$$\mathbf{H} = \begin{cases} \mathbf{H}_{elastic} & \text{for unloading or inside the yield surface} \\ \mathbf{H}_{plastic} & \text{for loading} \end{cases}$$

The application of the operator $\mathbf{P}$ to $\mathbf{H}(\mathbf{x})$ selects from $\mathbf{H}(\mathbf{x})$ only those rows which have a non-vanishing $\mathbf{P}$-component. E.g. for[2]

$$\mathbf{P} = \begin{pmatrix} 1 & 0 & 0 \\ 0 & 1 & 0 \\ 0 & 0 & 0 \end{pmatrix} \quad , \quad \mathbf{Q} = 1 - \mathbf{P} = \begin{pmatrix} 0 & 0 & 0 \\ 0 & 0 & 0 \\ 0 & 0 & 1 \end{pmatrix}$$

we obtain

$$\mathbf{PH}(\mathbf{x}) = \begin{pmatrix} H_{11} & H_{12} & H_{13} \\ H_{21} & H_{22} & H_{23} \\ 0 & 0 & 0 \end{pmatrix}$$

Subtracting this from $\mathbf{Q}$ we obtain

$$\mathbf{Q} - \mathbf{PH}(\mathbf{x}) = - \begin{pmatrix} H_{11} & H_{12} & H_{13} \\ H_{21} & H_{22} & H_{23} \\ 0 & 0 & -1 \end{pmatrix}$$

---

[2]for simplicity only 3×3 matrices are considered here

such that

$$\det(\mathbf{Q} - \mathbf{PH}(\mathbf{x})) = \det \begin{pmatrix} H_{11} & H_{12} \\ H_{21} & H_{22} \end{pmatrix}$$

We thus see that controllability is given if the determinants of all conceivable symmetric minors of $\mathbf{H}(\mathbf{x})$ are positive. This is the case if $\mathbf{H}(\mathbf{x})$ fulfils the condition[3].

$$\mathbf{x} \cdot \mathbf{H}(\mathbf{x})\mathbf{x} > 0 \quad \text{for} \quad \forall \mathbf{x} \quad . \tag{11}$$

This can be easily seen by transforming $\mathbf{H}(\mathbf{x})$ into its principal directions. Then, each eigenvalue is a symmetric minor of $\mathbf{H}(\mathbf{x})$. Note that $\mathbf{x} \cdot \mathbf{H}(\mathbf{x})\mathbf{x}$ represents the so-called second order work $\mathrm{tr}(\overset{\circ}{\mathbf{T}} \mathbf{D})$. In other words, positive second order work implies controllability.

It is interesting to note that, with hypoplastic constitutive equations, the condition $\mathrm{tr}(\overset{\circ}{\mathbf{T}} \mathbf{D}) = 0$ is in fact encountered *before* the peak. More specifically, the condition $\mathrm{tr}(\overset{\circ}{\mathbf{T}} \mathbf{D}) = 0$ (i.e. vanishing second order work) constitutes a surface in the stress space. Since this surface is connected with possible loss of uniqueness, we call it "bifurcation surface". It can be easily determined if we insert the hypoplastic equation into the equation $\mathrm{tr}(\overset{\circ}{\mathbf{T}} \mathbf{D}) = 0$. For simplicity we consider only rectilinear extensions, i.e. we restrict the dimension of column vectors $\mathbf{x}$ and $\mathbf{y}$ to 3. We then obtain:

$$F(\mathbf{x}) = \mathrm{tr}(\overset{\circ}{\mathbf{T}} \mathbf{D}) = L_{ij}x_i x_j + N_i x_j \cdot g(\mathbf{x}) = 0 \tag{12}$$

with

$$G(\mathbf{x}) = \sqrt{x_1^2 + x_2^2 + x_3^2} = 1 \quad . \tag{13}$$

The bifurcation surface is defined as a surface in the stress space. It consists of stress states for which the equation $\mathrm{tr}(\overset{\circ}{\mathbf{T}} \mathbf{D}) = 0$ possesses only one solution. This means that the surfaces $F(\mathbf{x}) = 0$ and $G(\mathbf{x}) - 1 = 0$ have a point with a common tangent plane, i.e. the following equations must be fulfilled:

$$\begin{aligned} \nabla F + \lambda \nabla G &= 0 \\ G - 1 &= 0 \end{aligned} \tag{14}$$

The stress states for which the system (14) possesses one solution constitute the bifurcation surface.

---

[3]The more general case $\mathbf{y} \cdot \mathbf{H}(\mathbf{x})\mathbf{y} > 0$ for $\forall \mathbf{x}, \mathbf{y}$ is not considered here.

4.7 Dilatancy and pore pressure

The tendency of soils to contract or dilate at shearing is·known as a peculiar feature of granular materials. Some people believe that dilatancy can be modelled even with a linear-elastic material provided POISSON's ratio is properly chosen. However, this is not true, as in linear elastic materials the hydrostatic and the deviatoric stresses and strains are completely uncoupled. It is in this coupling where dilatancy resides, since it means that the volumetric strain is affected by deviatoric stress and vice versa. In the course of conventional triaxial tests with dense sand, an initial contractancy (or negative dilatancy) is followed by dilatancy. The initial contractancy is often attributed to compressibility since the volumetric decrease is accompanied with an increase of the hydrostatic stress. However, the initial volume decrease occurs also in tests with constant hydrostatic stress, so that "contractancy" appears to be a suitable name.

A reference is often made to *the* angle of dilatancy. However, this angle is by no means a material constant. Much more it depends on the deformation mode and stage, on the density and on the stress level. We should get rid of the preoccupation that the dilatancy angle (as well as the friction angle) are material constants.

Let us now consider suppressed dilatancy (or contractancy) in case of undrained deformations of water saturated soil. An often cited expression is "the soil wants to decrease its volume but it cannot; consequently the hydrostatic effective stress is decreased". Of course, such an explanation is not very satisfactory. A much better approach is to consider pore water as imposing the internal constraint of incompressibility. Then, the porewater pressure p is constitutively indeterminate and can only be determined from static boundary conditions. The constitutive equation is now changed to

$$\overset{\circ}{\mathbf{T}} = \mathbf{h}(\mathbf{T}, \mathbf{D}) - \dot{p}\mathbf{1} \quad .$$

Here, $\mathbf{T}$ should be understood as the effective stress. Thus, the pore pressure built up in the course of undrained tests can be easily and realistically modelled by means of hypoplasticity.

4.8 Barotropy and pyknotropy

These words of Greek origin are introduced in order to express the fact that the soil behaviour depends on stress level (i.e. on $\operatorname{tr}\mathbf{T}$) and on density [23]. Contrary to an initial assumption valid mainly in the mechanics of normal consolidated clays, normalized stress strain curves referring to tests on sands do *not* coincide. The experimental observation, that the friction and dilatancy angles are gradually decreased by increased stress levels, points to the fact that the function $\mathbf{h}(\mathbf{T}, \mathbf{D})$ is not exactly homogeneous with respect to $\mathbf{T}$. Thus, sand is not a self-similar material, i.e. it cannot model itself at geometrically reduced scale model tests.

Pyknotropy stems from the obvious fact that the behaviour of granular materials depends on density (void ratio). It is a very astonishing fact that granulates (contrary to other atomistic materials like gases) do not have a unique relation between density and stress level. Even more, at one and the same stress level the void ratio may range between two limits. As GUDEHUS [11] pointed out, these limits must be stress-level dependent.

4.9 Softening

It is one of the open questions in soil mechanics, whether softening (see Fig. 6) exists or not. Traditionally, softening was considered as a principal part of soil behaviour. Later on it became "fashionable" to deny softening as being only an apparent effect due to the inhomogeneous deformation of the samples. The view in hypoplasticity is that a large amount of the registered softening is due to the inhomogeneous sample deformation. However, the 'material' softening, i.e. the softening which would be exhibited by a fictitious sample undergoing homogeneous deformation, is also there. We have to admit that the onset of inhomogeneous deformation makes the experimental approach unfeasible. We can, however, proceed by reasoning. It is a matter of fact that dense samples have a higher strength (i.e. peak stress deviator) than loose ones. In the course of deformation, dilatancy transforms a sample from dense to loose. Consequently, its strength must decrease and this is material softening.

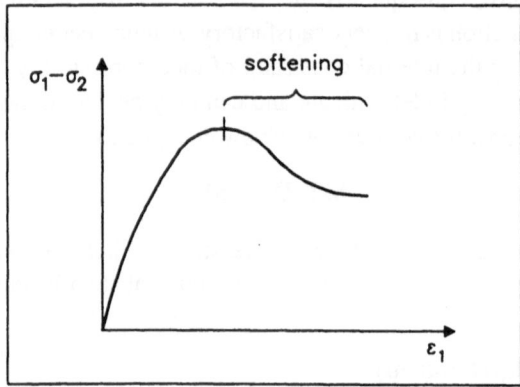

Fig. 6: Softening in triaxial test deformation

4.10 Shear Banding

A typical pattern of inhomogeneous deformation is the localization of deformation within a narrow zone called shear band (see Fig. 7). Such shear bands constitute one of the most fascinating phenomena in geomechanics. Due to the work of DESRUES based on tomography, we know now that also apparently non-localized inhomogeneous deformation modes are actually localized. The transition to localized deformation may occur on either

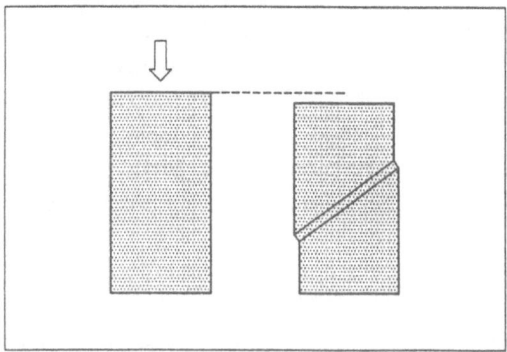

Fig. 7: Soil sample before and after shear-banding

gradually or suddenly. In the latter case it consists in a drastic change of the deformation direction, as experiments by VARDOULAKIS show. If a constitutive model is capable of realistically describing also this new direction, then it will be possible to predict when and under which inclination a shear band can occur.

This ability is not self-evident since many constitutive models (e.g. the elastoplastic ones) are suggested or tested only for some particular fans of deformation directions. It is therefore a good check of constitutive relations to try to predict the formation of shear bands. This test has been passed by several versions of the hypoplastic relations [17, 20, 25, 34, 36, 37, 53, 54].

4.11 Rate dependence of soils

Referring to rate dependence, we should distinguish related notions such as viscosity, relaxation, time dependence. All these notions have in common that they imply one or more material parameters bearing the dimension of time. An explicit appearance of time t in the constitutive equation implies that there is no invariance with respect to the time coordinate. In other words, the material exhibits ageing. This can only be defined within the framework of particular constitutive relations.

In soil mechanics, rate dependence is more pronounced for clays and is mainly manifest as creep (secondary consolidation) or increase of shear stresses by increasing the deformation rate, respectively. Such rate dependence resides in the grain skeleton and not in the free pore fluid (cf. consolidation according to TERZAGHI's theory).

Another phenomenon is relaxation, i.e. the decrease of stress with time when the deformation remains constant. The measurement of relaxation is very difficult as the measurement of forces by means of load cells requires a minute deformation of the load cell and, consequently, also of the sample such that the condition $\epsilon = $ const is not perfectly fulfilled.

The often assumed correspondence between relaxation and creep is directly applicable to viscoelastic materials only. For the general case, it is a very difficult task to incorporate relaxation in a constitutive relation. Perhaps this task is also of minor importance since the stress drop due to relaxation is quickly recovered as soon as the deformation is resumed. There have been several attempts to model rate dependence in hypoplasticity using a constitutive relation $\mathbf{h}(\mathbf{T}, \mathbf{D})$ which is no more homogeneous with respect to $\mathbf{D}$ [16, 49, 11] A review of these relations together with a promising proposal was given by NIEMUNIS [28].

4.12   Cyclic loading, ratcheting, shake-down

Cyclic loading is recognized as one of the most difficult fields in soil mechanics. From a mathematical viewpoint, it is very difficult to model the transition from the plastic to the quasi-elastic behaviour. Elastoplasticity and hypoplasticity bear some inherent deficiencies which become more important in the case of cyclic loading. In the realm of classical elastoplasticity all unloading-reloading cycles are completely elastic, a feature which is not realistic. On the other hand, in (the initial versions of) hypoplasticity the first and subsequent unloading-reloading cycles were identical to the virgin loading-unloading [27]. This shortcoming is called "ratcheting effect" and is due to the fact that in simple hypoplasticity the stress is the only memory parameter.

In reality either a gradual transition from plastic to elastic behaviour (so-called "shake-down") takes place or deformation increases unbounded with the number of cycles (so-called "incremental collapse"). Regarding shake-down, soil dynamics proves that the behaviour of soil never becomes completely elastic, as every cycle is connected with dissipation of energy, a fact which is modelled by a fictitious viscous damping.

It turns out that the quality of the modelling of cyclic behaviour depends on whether the stress amplitudes are small or large. If the unloading extends to the extension side (stress deviator changes sign), then the hypoplastic models work satisfactorily. Furthermore, the proper incorporation of barotropy and pyknotropy by the advanced hypoplastic models enables that cyclic loading with constant load produces gradually a high density (i.e. small void ratio) which cannot be exceeded by additional cycles.

A more general representation of the cyclic behaviour in hypoplasticity requires an additional state variable such as a structure tensor [20] which involves the recent deformation history. A 'memory function' [3] or an 'intergranular strain' [29] were proposed for this purpose.

5 Development of hypoplastic equations

The first hypoplastic equation which satisfied the requirements given in Section 2 was published in 1985 [18]:

$$\overset{\circ}{\mathbf{T}} = C_1 \frac{1}{2}(\mathbf{T} \cdot \mathbf{D} + \mathbf{D} \cdot \mathbf{T}) + C_2 \mathrm{tr}(\mathbf{T} \cdot \mathbf{D})\mathbf{1} + \left[C_3 \mathbf{T} + C_4 \frac{\mathbf{T} \cdot \mathbf{T}}{\mathrm{tr}\,\mathbf{T}} \right] \|\mathbf{D}\| \,.$$

In order to remove some deficiencies, WU WEI [45] changed the tensorial terms and obtained the aforementioned relation

$$\overset{\circ}{\mathbf{T}} = C_1(\mathrm{tr}\,\mathbf{T})\mathbf{D} + C_2 \frac{\mathrm{tr}\,(\mathbf{TD})}{\mathrm{tr}\,\mathbf{T}}\mathbf{T} + C_3 \frac{\mathbf{T}^2}{\mathrm{tr}\,\mathbf{T}}\sqrt{\mathrm{tr}\,\mathbf{D}^2} + C_4 \frac{\mathbf{T}^{*2}}{\mathrm{tr}\,\mathbf{T}}\sqrt{\mathrm{tr}\,\mathbf{D}^2} \,.$$

Next, he multiplied the non-linear terms (i.e. the last two terms) by the factor I_e which depends on the void ratio e and becomes equal to 1 when $e = e_{\mathrm{crit}}$:

$$I_e = (1 - a)\frac{e - e_{\min}}{e_{\mathrm{crit}} - e_{\min}} + a$$

The material constants C_1, C_2, C_3, C_4 can be determined from the critical state which can be reached in the course of monotonous shearing. He thus succeeded to model pyknotropy. Moreover, by taking into account that the critical void ratio e_{crit} depends on the stress level he also managed to model barotropy [46].

The subsequent research focused on two goals:

- to improve the performance (i.e. the predictive capacity) of the equation for deviatoric directions different from the one corresponding to conventional triaxial tests

- to make the calibration easier. GUDEHUS required that the determination of the material parameters should be as far as possible performed on the basis of granulometric properties of the soil.

Another recent achievement is the choice of the 4 tensorial terms and of the constants which relate them with each other following some general requirements, e.g. that the obtained critical limit state coincides with some prescribed curves given in the literature [1, 2, 40].

As for the deviatoric performance, BAUER [1, 2] and VON WOLFFERSDORFF [40] established relations between the material constants C_1, C_2, C_3, C_4 by adapting the deviatoric yield curve to some known yield curves which are well-established in literature (e.g. the one by MATSUOKA-NAKAI). With the constants C_1, C_2 and $C_3 = C_4$ depending on e and $\mathbf{T}$ and introducing the abbreviation

$$\hat{\mathbf{T}} := \frac{\mathbf{T}}{\mathrm{tr}\,\mathbf{T}}$$

the recent hypoplastic equation assumes the form:

$$\mathring{\mathbf{T}} = f_b f_e \frac{1}{\operatorname{tr} \hat{\mathbf{T}}^2} \left[F^2 \mathbf{D} + a^2 \hat{\mathbf{T}} \operatorname{tr}(\hat{\mathbf{T}} \mathbf{D}) + f_d a F(\hat{\mathbf{T}} + \hat{\mathbf{T}}^*) \sqrt{\operatorname{tr} \mathbf{D}^2} \right]$$

(15)

with

$$a := \frac{\sqrt{3}(3 - \sin \varphi_c)}{2\sqrt{2} \sin \varphi_c} \quad ,$$

$$F := \sqrt{\frac{1}{8} \tan^2 \psi + \frac{2 - \tan^2 \psi}{2 + \sqrt{2} \tan \psi \cos 3\vartheta} - \frac{1}{2\sqrt{2}} \tan \psi} \quad ,$$

$$\tan \psi = \sqrt{3 \operatorname{tr} \hat{\mathbf{T}}^{*2}} \quad , \qquad \cos 3\vartheta = -\sqrt{6} \frac{\operatorname{tr} \hat{\mathbf{T}}^{*3}}{\left[\operatorname{tr} \hat{\mathbf{T}}^{*2} \right]^{3/2}} \quad ,$$

$$f_d := \left(\frac{e - e_d}{e_c - e_d} \right)^\alpha \quad ,$$

$$f_e := \left(\frac{e_c}{e} \right)^\beta \quad ,$$

$$f_b := \frac{h_s}{n} \left(\frac{e_{i0}}{e_{c0}} \right)^\beta \frac{1 + e_i}{e_i} \left(\frac{3 p_s}{h_s} \right)^{1-n} \left[3 + a^2 - a\sqrt{3} \left(\frac{e_{i0} - e_{d0}}{e_{c0} - e_{d0}} \right)^\alpha \right]^{-1}$$

The pressure-dependent void ratios e_i and e_d bound the admissible states in plane e vs. $p_s = \operatorname{tr} \mathbf{T}/3$ and depend (together with the critical void ratio e_c) on p_s according to relation

$$\frac{e_i}{e_{i0}} = \frac{e_c}{e_{c0}} = \frac{e_d}{e_{d0}} = \exp \left[-\left(\frac{3 p_s}{h_s} \right)^n \right] \quad .$$

There are 8 constants in the hypoplastic equation (15). They can be easily determined from simple index and/or element tests [13]. The constants are state-independent, at least in a particular pressure range, thus enabling the application of the hypoplastic equation in boundary value problems with pressure and density variation. The critical friction angle φ_c can be estimated from the angle of repose; $e_{d0} \approx e_{min}$, $e_{c0} \approx e_{max}$ and $e_{i0} \approx 1.2 e_{max}$. The exponent n can be calculated from the results of an oedometer test with an initially loose specimen in the pressure range between p_{s1} and p_{s2} using

$$n = \ln \left(\frac{e_1 \lambda_2}{e_2 \lambda_1} \right) \Big/ \ln \left(\frac{p_{s2}}{p_{s1}} \right) \quad ,$$

see Fig. 8 (left), with $\lambda = \Delta e / \Delta \ln(p_s/p_{s0})$. The so-called granulate hardness h_s can then be obtained with n and any λ corresponding to $p_{s1} \leq p_s \leq p_{s2}$:

$$h_s = 3\,p_s \left(\frac{n\,e}{\lambda}\right)^{1/n}$$

The validity range of h_s and n is thus limited by the experimental pressure range. One should not reach pressures that cause an extensive grain breakage (e.g. $p_s > 1$ MPa for quartz sand), otherwise a modification of the constitutive equation is necessary. Evaluating many oedometer tests on various sands, a distinct relation between n, the nonuniformity coefficient $U = d_{60}/d_{10}$ and the mean grain diameter $d = d_{50}$ could be found.

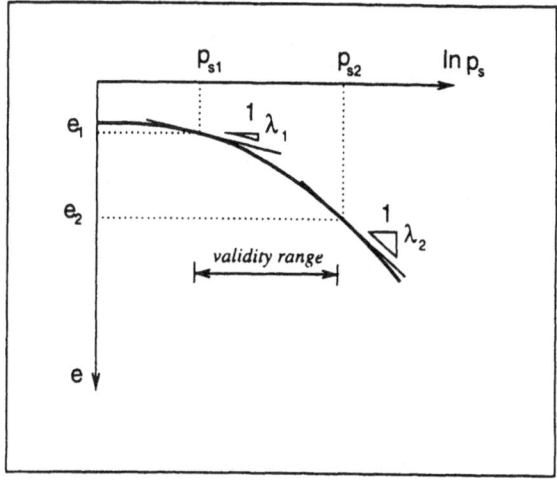

Fig. 8: *Determination of n from an oedometer test with an initially loose specimen*

The exponent α can be determined from the peak friction angle φ_p in a standard drained triaxial test with an initially dense specimen. For the known $r_e = (e - e_d)/(e_c - e_d)$ at the peak one can calculate α from the constitutive equation. φ_p can be also estimated from numerous correlations.

The exponent β can be calculated from a compression coefficient of an initially dense specimen. Thus an additional oedometer test should be performed. An analysis of laboratory tests with various sands and a numerical study show that $\beta \approx 1$.

Fig. 9 shows a comparison between experiments and numerical simulations of standard drained triaxial tests with Lausitz sand at various pressures andensities. The sand parameters are summarized in Tab. 1.

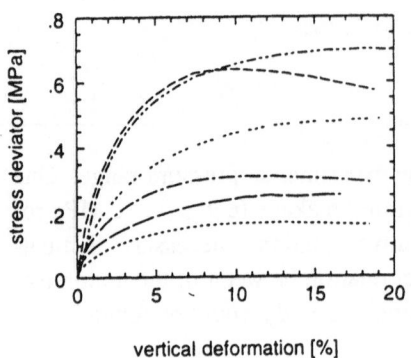
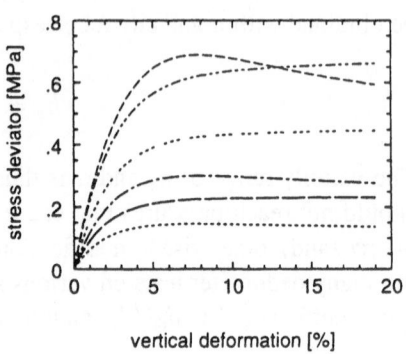

vertical deformation [%]   vertical deformation [%]

Fig. 9: *Measured (left) and calculated (right) stress-strain curves of triaxial tests with Lausitz sand at various pressures and densities (50 kPa $\leq \sigma_2 \leq$ 300 kPa and 0.56 $\leq e \leq$ 0.76).*

φ_c [°]	h_s [MPa]	n	e_{d0}	e_{c0}	e_{i0}	α	β
33	1600	0.19	0.44	0.85	1.0	0.25	1.0

Table 1: Hypoplastic parameters of Lausitz sand.

6 Final remarks

6.1 Relation to other theories

Hypoplasticity is not the only attempt to overcome the formalism of "classical" elasto-plasticity. There are other theories with similar aims and with pronounced similarities:
The arc-length theory of ILYUSHIN, RIVLIN and PIPKIN, the endochronic theory initiated by VALANIS [42] and the several models created in Grenoble by DARVE and CHAMBON and others. Besides this, hypoplasticity has to be compared with all other elastoplastic theories. Three international competitions in Montreal, Grenoble and Cleveland were devoted to the effort to find out which theory was superior to the others. A series of experimental data were communicated to the participants in order to let them calibrate their models. Based on this, the participants were invited to predict some other tests, the outcomes of which were at that time kept secret.

Such competitions appear reasonable, however none of them was conclusive. The reason is that there is no objective measure for the quality of the predictions. And, much more, there is no measure to rate the calibration effort versus the quality of results. After all, philosophers found long ago that the value of a theory cannot be measured by theoretical means.

6.2 FEM-Implementations

One of the main fields of application (but by no means the only one) of constitutive models is the implementation into FEM-codes. This was also the case with a series of dissertation thesis', mainly in Karlsruhe [55, 43, 33, 31, 32, 36, 26, 6, 9, 15].

Despite the fact that hypoplasticity is rather new and cannot ressort to the instruments already developed for elastoplasticity (e.g. radial return algorithms), the performance of hypoplastic laws in FEM-calculations and the comparison with measured results is satisfactory [15, 41, 14, 44].

Acknowledgement

The authors are indebted to Dr. P. Wagner, Innsbruck, for valuable remarks.

References

[1] E. Bauer. Constitutive Modelling of Critical States in Hypoplasticity. *Proceedings of the Fifth International Symposium on Numerical Models in Geomechanics, NUMOG V*, Davos, Switzerland, Balkema, 15-20, 1995.

[2] E. Bauer. Calibration of a comprehensive hypoplastic model for granular materials. *Soils and Foundations*, 36(1):13–26, 1996.

[3] E. Bauer and W. Wu. A hypoplastic model for granular soils under cyclic loading. In D. Kolymbas, editor, *Modern Approaches to Plasticity*, pages 247–258. Elsevier, 1993.

[4] E. Bauer and W. Wu. A hypoplastic constitutive model for cohesive powders. *Powder Technology*, 85:1–9, 1995.

[5] R. Chambon. Une classe de lois de comportement incrementalement non lineaire pour les sols non visqueux, resolution de quelques problemes de coherence. C. R. Acad. Sci. Paris t. 3087 serie II, p. 1571–1576 (1989).

[6] Y.F. Dafalias. Bounding surface plasticity. I: Mathematical foundation and hypoplasticity. *J. Eng. Mech. ASCE*, Vol. 112, 966-987, 1986.

[7] F. Dahlhaus. Stochastische Untersuchungen von Silobeanspruchungen. Schriftenreihe des Institutes füür Massivbau und Baustofftechnologie der Universität Fridericiana Karlsruhe, Heft 25, 1995.

[8] R.O. Davis, G. Mullenger. A rate-type constitutive model for soil with a critical state. *International Journal of Numerical and Analytical Methods in Geomechanics*, Vol. 2, 255–282, 1978.

[9] J. Desrues, R. Chambon. A new rate type Constitutive Model for Geomaterials: CloE. In: D. Kolymbas (ed) Modern Approaches to Plasticity, Elsevier, 1993.

[10] C. di Prisco, S. Imposimato. Time dependent mechanical behaviour of loose sand. *Mech. of Cohesive-Frictional Materials and Structures*, Vol. 1, 45-73, 1996.

[11] H.J. Feise. Modellierung des mechanischen Verhaltens von Schüttgütern. Veröffentlichung 23, Dissertation, Mechanik-Zentrum der Technischen Universität Carolo-Wilhelmina in Braunschweig, 1996.

[12] G. Gudehus. A comparison of some constitutive laws for soils under radially symmetric loading and unloading. *Proc. 3th Int. Conf. Num. Meth. Geom.*, Aachen, ed. Balkema, 1979.

[13] G. Gudehus. A comprehensive constitutive equation for granular materials. *Soils and Foundations*, 36(1):1–12, 1996.

[14] G. Gudehus, D. Kolymbas. Numerical testing of constitutive relations for soils. *Proc. 5th Int. Conf. Num. Meth. Geomech.*, Nagoya, 1985.

[15] M.E. Gurtin, K. Spear. On the relationship between the logarithmic strain rate and the stretching tensor. Int. J. Solids Structures, Vol. 19, No. 5, pp. 437-444 (1983).

[16] I. Herle. Hypoplastizität und Granulometrie von Korngerüsten. *Publ. Series of Institut für Bodenmechanik und Felsmechanik der Universität Fridericiana in Karlsruhe*, No. 142, 1997.

[17] I. Herle, J. Tejchman. Effect of grain size and pressure level on bearing capacity of footings on sand. *Int. Symp. on Deformation and Progressive Failure in Geomechanics*, Nagoya, 1997.

[18] H.M. Hügel. Prognose von Bodenverformungen. *Publ. Series of Institut für Bodenmechanik und Felsmechanik der Universität Fridericiana in Karlsruhe*, No. 136, 1995.

[19] D. Kolymbas. A rate-dependent constitutive equation for soils. *Mech. Res. Comm.*, 4:367–372, 1977.

[20] D. Kolymbas. Bifurcation analysis for sand samples with a non-linear constitutive equation. *Ingenieur-Archiv*, **50**, *131–140*, 1981.

[21] D. Kolymbas. A generalized hypoelastic constitutive law. In *Proc. XI Int. Conf. Soil Mechanics and Foundation Engineering*, volume 5, page 2626, San Francisco, 1985. Balkema.

[22] D. Kolymbas. A novel constitutive law for soils. *Second Int. Conf. on Constitutive Laws For Engineering Materials: Theory and Applications, Tucson, Arizona, January 1987*, Elsevier, 1987.

[23] D. Kolymbas. Eine konstitutive Theorie für Böden und andere körnige Stoffe. *Publ. Series of Institut für Bodenmechanik und Felsmechanik der Universität Fridericiana in Karlsruhe*, Vol. 109, 1988.

[24] D. Kolymbas. Generalized hypoelastic constitutive equation. In Saada and Bianchini, editors, *Constitutive Equations for Granular Non-Cohesive Soils*, pages 349–366. Balkema, 1988.

[25] D. Kolymbas. Computer-aided design of constitutive laws. *International Journal for Numerical and Analytical Methods in Geomechanics*, **15**, 593-604 (1991).

[26] D. Kolymbas. An outline of hypoplasticity. *Archive of Applied Mechanics*, 61:143–151, 1991.

[27] D. Kolymbas, I. Herle, P.-A. v. Wolffersdorff. Hypoplastic constitutive equation with back stress. *International Journal of Numerical and Analytical Methods in Geomechanics*, 19(6):415–446, 1995.

[28] D. Kolymbas, G. Rombach. Shear band formation in generalized hypoelasticity. *Ingenieur-Archiv*, **59**, *177–186*, 1989.

[29] C. Lyle. Spannungsfelder in Silos mit starren, koaxialen Einbauten. Diss., Fakultät für Maschinenbau und Elektrotechnik der TU Carolo-Wilhelmina zu Braunschweig, 1993.

[30] A. Niemunis. Hypoplasticity vs. elastoplasticity, selected topics. In D. Kolymbas, editor, *Modern Approaches to Plasticity*, pages 278–307. Elsevier, 1993.

[31] A. Niemunis. A visco-plastic model for clay and its FE-implementation. In *XI Colloque Franco-Polonais en Mécanique des Sols et des Roches Appliquée*, E. Dembicki, W. Cichy, L. Balachowski (Eds.), University of Gdańsk, 1996.

[32] A. Niemunis, I. Herle. Hypoplastic model for cohesionless soils with elastic strain range. *Mechanics of Cohesive-Frictional Materials*, Vol. 2, 279–299 (1997).

[33] R. Nova. Controllability of the incremental response of soil specimens subjected to arbitrary loading programmes, In: Journal of the Mechanical Behaviour of Materials, Vol. 5, n2, 193-201.

[34] G.A. Rombach. Schüttguteinwirkungen auf Silozellen, Exzentrische Entleerung. Veröffentlichungen Heft 14, Dissertation, Institut für Massivbau und Baustofftechnologie der Universität Fridericiana in Karlsruhe, 1991.

[35] C. Ruckenbrod. Statische und dynamische Phänomene bei der Entleerung von Silozellen. Schriftenreihe des Institutes für Massivbau und Baustofftechnologie der Universit"t Fridericiana in Karlsruhe, Heft 26, 1995.

[36] Z. Sikora. Hypoplastic flow of granular materials. A numerical approach. *Publ. Series of Institut für Bodenmechanik und Felsmechanik der Universität Fridericiana in Karlsruhe*, Vol. 123, 1992.

[37] Z. Sikora, W. Wu. Shear band formation in biaxial tests. *Proc. Int. Conf. on Constitutive Laws for Engineering Materials*, Tucson, Arisona, USA, 1991.

[38] J. Tejchman. Shear banding and autogeneous dynamics in granular bodies. *Publ. Series of Institut für Bodenmechanik und Felsmechanik der Universität Fridericiana in Karlsruhe*, No. 140, 1995.

[39] J. Tejchman, E. Bauer. Numerical simulation of shear band formation with a polar hypoplastic constitutive model. *Computers and Geotechnics*, Vol. 19, No. 3, 221-244, 1996.

[40] J. Tejchman, W. Wu. Numerical simulation of shear band formation with a hypoplastic constitutive model. *Computers and Geotechnics*, 18(1):71–84, 1996.

[41] C. Truesdell, W. Noll. The non-linear field theories of mechanics. Handbuch der Physik III/c. Springer-Verlag, 1965.

[42] C. Truesdell. Hypo-elasticity. *J. Rational Mech. Anal.*, Vol. 4, 83–133, 1955. Springer-Verlag, 1965.

[43] P.-A. von Wolffersdorff. A hypoplastic relation for granular materials with a predefined limit state surface. *Mechanics of Cohesive-Frictional Materials*, 1:251–271, 1996.

[44] P.-A. von Wolffersdorff. Verformungsprognosen für Stützkonstruktionen. *Publ. Series of Institut für Bodenmechanik und Felsmechanik der Universität Fridericiana in Karlsruhe*, Vol. 141, 1997.

[45] K.C. Valanis. An endochronic geomechanical model for soils. *IUTAM Conference on Deformation and Failure of Granular Materials*. Balkema, 159–165, 1982.

[46] J. Weidner. Vergleich von Stoffgesetzen granularer Schüttgüter zur Silodruckermittlung. Veröffentlichungen Heft 10, Dissertation, Institut für Massivbau und Baustofftechnologie der Universität Fridericiana in Karlsruhe, 1990.

[47] W. Wehr, J. Tejchman, I. Herle, G. Gudehus. Sand anchors – a shear zone problem. *Int. Symp. on Deformation and Progressive Failure in Geomechanics*, Nagoya, 1997.

[48] W. Wu. Hypoplastizität als mathematisches Modell zum mechanischen Verhalten granularer Stoffe. *Publ. Series of Institut für Bodenmechanik und Felsmechanik der Universität Fridericiana in Karlsruhe*, Vol. 129, 1992.

[49] W. Wu, E. Bauer. A hypoplastic model for barotropy and pyknotropy of granular soils. In D. Kolymbas, editor, *Modern Approaches to Plasticity*, 225-245. Elsevier, 1993.

[50] W. Wu, E. Bauer. A simple hypoplastic constitutive model for sand. *International Journal of Numerical and Analytical Methods in Geomechanics*, 18:833–862, 1994.

[51] W. Wu, E. Bauer, D. Kolymbas. Hypoplastic constitutive model with critical state for granuar materials. *Mechanics of Materials*, 23:45–69, 1996.

[52] W. Wu, E. Bauer, A. Niemunis, I. Herle. Visco-hypoplastic models for cohesive soils. In D. Kolymbas, editor, *Modern Approaches to Plasticity*, 365-383. Elsevier, 1993.

[53] W. Wu, D. Kolymbas. Numerical testing of the stability criterion for hypoplastic constitutive equations. *Mechanics of Materials*, 9:245–253, 1990.

[54] W. Wu, A. Niemunis. Failure criterion, flow rule and dissipation function derived from hypoplasticity. *Mechanics of Cohesive-Frictional Materials*, 1:145–163, 1996.

[55] W. Wu, A. Niemunis. Beyond Failure in Granular Materials. *Int. J. for Numerical and Analytical Methods in Geomechanics*, Vol. 21, No. 2, 153–174, 1997.

[56] W. Wu, Z. Sikora. Localized bifurcation in hypoplasticity. *International Journal of Engineering Science*, 29(2):195–201,1991.

[57] W. Wu, Z. Sikora. Localized Bifurcation of Pressure Sensitive Dilatant Granular Materials. *Mechanics Research Communications*, Vol. 29, 289-299, 1992.

[58] M. Ziegler. Berechnung des verschiebungsabhängigen Erddrucks in Sand. Veröffentlichungen Heft 101, Institut für Bodenmechanik undFelsmechanik der Universität Fridericiana in Karlsruhe, 1986.

ELASTOPLASTIC AND VISCOPLASTIC CONSTITUTIVE MODELS FOR GRANULAR MATERIALS

Z. Mróz

Institute of Fundamental Technological Research, Warsaw, Poland

1. Introduction

The present work is aimed at discussing constitutive relations for soils and the hardening rules which follow from extensions of the classical plasticity theory. When viscous effects are neglected this theory results in constitutive equations between stress and strain increments (or rates) with instanteneous stiffness or compliance matrices being dependent on actual stress or strain states and the previous deformation history. It is also important that this theory allows for distinction between loading and unloading or reverse loading events for which different incremental relations are valid.

For soils under monotonic loading, the simple elastoplastic or non-linear elastic material models can be used to simulate the deformational response of a material with sufficient accuracy. For a perfectly plastic model both hardening and softening phenomena are neglected and the flow law is associated with the yield condition or the plastic potential by the gradient rule. For an isotropic hardening model, the irreversible void ratio or density is assumed to be a state parameter. Using this model, one is able to predict hardening, softening and critical state response. However, for more complex loading programs involving one-way loading followed by unloading, or for repetitive action of loads, more complex hardening rules should be examined in order to describe the effects occurring for these loading conditions more realistically.

When using the concepts of the theory of plasticity we have to formulate
1) the yield condition defining elastic and inelastic deformation domains;
2) the flow rule relating the increments or rates of stress and irreversible strain;
3) the hardening rule specifying the evolution of yield surface in the course of plastic deformation and the evolution of hardening parameters defining the state of the material.

The formulation of proper evolution rules is a difficult part of constitutive modeling, since memory of particular loading events and progressive cyclic hardening or softening phenomena should be incorporated in the model.

The need for a more accurate description if inelastic behaviour follows for several reasons, namely
1) a more complex model of a wide range of applicability, including monotonic and variable loading, allows for more rational experimental work in which identification and verification programs can be selected and further improvements of the model can be introduced in subsequent steps,

2) qualitative interpretation of material behaviour for various loading histories provided by a more complex model is useful in assessing and formulating design rules,
3) a simpler version of the model for numerical implementation could be obtained by neglecting some interactions; the degree of simplification should be related to the type of problems investigated.

It is therefore important that the model should be formulated in a „modular" way, that is, allowing for neglect or consideration of some effects without affecting other assumptions, or identified material parameters.

2. Non-linear elasticity and hypoelasticity models

The application of *linear* elasticity in soil mechanics provides only the first approximation valid for stress states lying far below the limit state. In fact, the linear elastic analysis does not describe the considerable stress redistribution occurring in the non-linear range prior to failure and underestimates deformations of soil or settlement of structures. A more realistic description is therefore obtained by assuming *non-linear* response under compression or shear and a non-linear relation between volumetric strain and applied pressure. For monotonically increasing stress components the material can be assumed as non-linearly elastic, so there is no need to decompose strains into elastic and plastic portions. Under load reversal, the elastic stress-strain response, is the same and there is no residual deformation after unloading, Figure 1 (a).

The total stress state σ' is decomposed, as usually, into the effective stress σ and the pore pressure p_w thus

$$\sigma' = \sigma + p_w \delta \tag{2.1}$$

where δ denotes the unit tensor (or Kronecker delta). The compressibility moduls of fluid is denoted by K_f and that of soil skeleton by K_s. For soils, it is usually assumed that $K_f \to \infty$ and $K_s \to \infty$ and the compressibility of soil is only due to void closure of growth. The constitutive equations of soils are then expressed in terms of effective stresses. The effective stress and strain deviators $s_{ij} = \sigma_{ij} - \frac{1}{3}\sigma_{kk}\delta_{ij}$, $e_{ij} = \varepsilon_{ij} - \frac{1}{3}\varepsilon_{kk}\delta_{ij}$ provide the invariants

$$\sigma_e = \left(\tfrac{3}{2}s \cdot s\right)^{\frac{1}{2}} = 3J_2' \qquad \varepsilon_e = \left(\tfrac{3}{2}e \cdot e\right)^{\frac{1}{2}} = \left(3I_2'\right)^{\frac{1}{2}} \tag{2.2}$$

where J_2' and I_2' are the second invariants of s_{ij} and e_{ij} and the dot between two vectors or tensors of the same order denotes their scalar product. Similarly, denote

$$p = -\tfrac{1}{3}J_1 \qquad \varepsilon_v = -I_1 \tag{2.3}$$

where $J_1 = \sigma_{kk}$ and $I_1 = \varepsilon_{kk}$ are the first invariants of the stress and strain tensors.

Assume that the non-linear relations between stress and strain have the same form as linear relations for an isotropic material, that is

$$s_{ij} = 2G_s e_{ij} , \qquad p = K_s \varepsilon_v \tag{2.4}$$

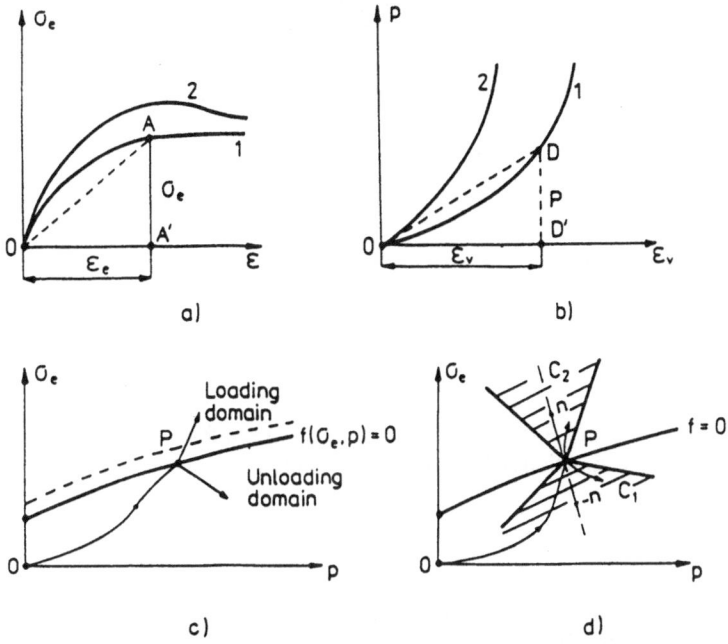

*Figure .1   Non-linear response of a normally consolidated (curves 1)
and over-consolidated (curves 2) soil; (a) shear (or deviatoric) stress-strain response,
(b) isotropic compression curves, c) loading-unloading domains
in the σ_e, p - plane, d) cones of admissible loading paths satisfying convexity condition.*

where

$$G_s = \frac{\sigma_e}{3\varepsilon_e} = G_s(\varepsilon_e, \varepsilon_v), \qquad K_s = \frac{p}{\varepsilon_v} = K_s(\varepsilon_e, \varepsilon_v) \qquad (2.5)$$

are the *secant moduli* for deviatoric and spherical components of the stress and strain tensors. These moduli can easily be identified from the non-linear response curves between σ_e, ε_e and p, ε_v, Figures 1 (a), (b).

Requiring the relations (2.4) to present a non-linear elastic material it should be demonstrated that they follow from the elastic strain or stress potentials $U(\varepsilon_{ij})$ and $W(\sigma_{ij})$ namely

$$\sigma_{ij} = \frac{\partial U}{\partial \varepsilon_{ij}}, \qquad \varepsilon_{ij} = \frac{\partial W}{\partial \sigma_{ij}} \qquad (2.6)$$

For an isotropic material there is $U = U(\varepsilon_e, \varepsilon_v)$ and

$$s_{ij} = \frac{\partial U}{\partial \varepsilon_{\bullet}} \frac{\partial \varepsilon_{\bullet}}{\partial e_{ij}} = \frac{\partial U}{\partial \varepsilon_{\bullet}} \frac{2}{3} \frac{e_{ij}}{\varepsilon_{\bullet}}, \qquad p = \frac{\partial U}{\partial \varepsilon_{v}} \tag{2.7}$$

The secant moduli can now be related to the specific strain energy

$$G_{s} = \frac{1}{3} \frac{\partial U}{\partial \varepsilon_{\bullet}} \frac{1}{\varepsilon_{\bullet}}, \qquad K_{s} = \frac{\partial U}{\partial \varepsilon_{v}} \frac{1}{\varepsilon_{v}} \tag{2.8}$$

and from (2.8) it follows that

$$\frac{1}{\varepsilon_{v}} \frac{\partial G_{s}}{\partial \varepsilon_{v}} = \frac{1}{3\varepsilon_{\bullet}} \frac{\partial K_{s}}{\partial \varepsilon_{\bullet}} \tag{2.9}$$

The potentiality relations (2.9) impose the restrictions on the forms (2.5), so the shear moduli cannot be the arbitrary functions of $\varepsilon_{\bullet}$ and ε_{v}. In particular when

$$G_{s} = G_{s}(\varepsilon_{\bullet}), \qquad K_{s} = K_{s}(\varepsilon_{v}) \tag{2.10}$$

the conditions (2.9) are obviously satisfied. The elastic strain energy is then a sum of deviatoric and volumetric strain energies.

In order to determine the tangent stiffness matrix, let us differentiate the finite relations (2.4) and obtain

$$\begin{bmatrix} \dot{s}_{ij} \\ \dot{p} \end{bmatrix} = \begin{bmatrix} A_{dd} & A_{ds} \\ A_{sd} & A_{ss} \end{bmatrix} \begin{bmatrix} \dot{e}_{kl} \\ \dot{\varepsilon}_{v} \end{bmatrix} \tag{2.11}$$

where

$$A_{dd} = 2G_{s}\delta_{ij} + \frac{4}{3} \frac{\partial G_{s}}{\partial \varepsilon_{\bullet}} e_{ij} \frac{e_{kl}}{\varepsilon_{\bullet}}, \qquad A_{ss} = K_{s} + \frac{\partial K_{s}}{\partial \varepsilon_{v}} \varepsilon_{v}$$

$$A_{ds} = 2\frac{\partial G_{s}}{\partial \varepsilon_{v}} e_{ij}, \qquad A_{sd} = \frac{2}{3} \frac{\partial K_{s}}{\partial \varepsilon_{\bullet}} \varepsilon_{v} \frac{e_{kl}}{\varepsilon_{\bullet}} \tag{2.12}$$

The off-diagonal terms A_{sd} and A_{ds} representing the coupling between deviatoric and volumetric stress and strain changes are not equal in general, that is $A_{sd} \neq A_{ds}$. The equality $A_{sd} = A_{ds}$ occurs only when the potential relations (2.9) are satisfied In particular, when (2.10) occurs, the off-diagonal terms vanish and we have

$$\begin{bmatrix} \dot{s}_{ij} \\ \dot{p} \end{bmatrix} = \begin{bmatrix} A_{dd} & 0 \\ 0 & A_{ss} \end{bmatrix} \begin{bmatrix} \dot{e}_{kl} \\ \dot{\varepsilon}_{v} \end{bmatrix} \tag{2.13}$$

or simply

$$\dot{s}_{ij} = 2G_{t}\dot{e}_{ij}, \qquad \dot{p} = K_{t}\dot{\varepsilon}_{v} \tag{2.14}$$

where $G_t = \frac{1}{2} A_{dd}$, $K_t = A_{ss}$ are the *tangent* shear and bulk moduli. The relations (2.14) represent the linear isotropic response in rates (or increments) of stress and strain whereas the relations (2.11) represent an anisotropic behaviour with anisotropy induced by the strain tensor. As the relations (2.12) are linear and homogeneous of degree one in stress and strain rates, not necessarily following form the elastic potential, they can be regarded as *hypoelastic* constitutive relations. In fact, we could start from the rate form (2.11) without any recourse to the finite relations (2.4) and identify the constitutive matrix coefficients $A_{dd}, A_{ss}, A_{ds}, A_{sd}$ from experimental data. In this case, these coefficients can be assumed as either strain or stress dependent.

A particular form of (2.14) was developed by Kondner and Zelasko [22, 23] and later by Duncan and Chang [17] who assumed a hyperbolic relation for a form of the compression curve in a triaxial test. For a general stress state, this form can be expressed in terms of σ_e and ε_e, namely

$$\sigma_e = \frac{\varepsilon_e}{A + B\varepsilon_e} \qquad (2.15)$$

where A and B are the material parameters. The tangent elastic bulk modulus can easily be identified from the isotropic compression test and it is usually assumed to depend on the mean hydrostatic stress,

$$K_t = k_o\, p_a \left(\frac{p}{p_a}\right)^m \qquad (2.16)$$

where k_o and m. are the material parameters and p_a denotes the atmospheric pressure. However, in order to account for the dilatancy effect, it can be assumed that there is an additional volumetric strain rate $\dot{\varepsilon}_v^d$ related directly to the deviatoric strain rate, $\dot{\varepsilon}_v^d = D_d \dot{\varepsilon}_e$ where D_d denotes the dilatancy parameter, cf Byrne and Eldridge [5].

Writing

$$\dot{p} = K_t \left(\dot{\varepsilon}_v - D_d\, \dot{\varepsilon}_v^d\right) = K_t\, \dot{\varepsilon}_v - K_t\, D_d \frac{2}{3} \frac{e_{kl}\, \dot{e}_{kl}}{\varepsilon_e} \qquad (2.17)$$

the form (2.11) of rate equations is obtained for which

$$A_{ds} = 0, \qquad A_{sd} = -K_t\, D_d \frac{2}{3} \frac{e_{kl}}{\varepsilon_e} \qquad (2.18)$$

So far, the rate equations (2.11) or (2.14) are assumed to apply for monotonic loading though the criterion of unloading has not been formulated. For instance, it can be assumed that these equations apply only when $\dot{f} = \dot{\sigma}_e - \alpha\, \dot{p} > 0$ whereas for $\dot{f} < 0$, different rate equations are to be formulated, Figure 1 c). This kind of approach was used by Baron and Nelson [2] and recently by Dickin and King [9]. For instance, in [2] it was assumed that

$$\begin{aligned} s_{ij} &= 2G_t\, \dot{e}_{ij} \quad \text{for} \quad \dot{\sigma}_e > 0, \quad \dot{p} = K_t\, \dot{\varepsilon}_v \quad \text{for} \quad \dot{p} > 0 \\ \dot{s}_{ij} &= 2G_u\, \dot{e}_{ij} \quad \text{for} \quad \dot{\sigma}_e < 0, \quad \dot{p} = K_u\, \dot{\varepsilon}_v \quad \text{for} \quad \dot{p} < 0 \end{aligned} \qquad (2.19)$$

where G_u and K_u are the new tangent moduli for the unloading paths. In [9], only one loading condition was applied following the Coulomb yield condition. Such formulations of loading-unloading conditions associated with linear relations (2.11) or (2.14) possess unfortunately the deficiency, namely they violate the continuity condition for neutral stress paths satisfying the equality $\dot{f} = 0$. Thus, for instance, for stress paths satisfying $\dot{\sigma}_e = 0$ the relations (2.18) provide

$$\dot{e}_1 = \frac{1}{2G_t}\dot{s}, \qquad \dot{e}_2 = \frac{1}{2G_u}\dot{s}$$

that is different strain rates for the same deviatoric stress rate. This local non-uniqueness introduces discontinuity in the material response and may result in the global non-uniqueness of solution of boundary value problems.

It can be argued that the proposed rate equations, different in loading and unloading domains and violating the continuity conditions on the separating boundary, are applicable only in some conical sub-domains C_1 and C_2 in the stress space, such that the admissible stress path cannot lie on surface $f = 0$. Such admissibility conditions were discussed by Mróz [28] by starting from the condition of local *convexity of transformation* between rates of stress and strain , namely

$$\left(\dot{\sigma}_2 - \dot{\sigma}_1\right)\cdot\left(\dot{\varepsilon}_2 - \dot{\varepsilon}_1\right) > 0 \qquad\qquad (2.20)$$

for any two pairs of rates $\dot{\sigma}_2, \dot{\varepsilon}_1$ and $\dot{\sigma}_2, \dot{\varepsilon}_2$ interconnected by the constitutive relation. Inequality (2.20) provides the sufficient condition for *uniqueness* of the boundary-value problems. It was shown in [28] that (2.20) implies the continuity of material response on separating surface $f = 0$. However, when this continuity condition is not satisfied, then (2.20) provides the admissible domains C_1 and C_2 in the stress space, such as those shown in Figure 1 (d). The applicability of the rate relations is then limited to such loading paths that belong to the admissible domains. In the next section, we shall discuss the loading-unloading conditions in terms of the plasticity theory.

3. Rate (or incremental) constitutive relations

3.1 Uniaxial stress-strain response

The rate relations discussed in the previous section were referred to total strain rates and there was no distinction between elastic and irreversible portions. However, such distinction is necessary when both the loading and unloading process are to be incorporated into one formulation.

Referring to Figure 2, presenting the uniaxial compression curve, it is seen that when the stress is removed after loading to A, the recoverable strain is ε_r and irreversible strain equals ε_i. When elastic behaviour is linear and the unloading curve AB does not depart from linearity, then unloading is elastic and the irreversible strain equals plastic strain, $\varepsilon_i = \varepsilon^p$. However, when the unloading curve is not linear and during reloading there is a hysteresis loop (ABC in Figure 2a), we conclude that reverse plastic strains also develop on the unloading portion AB. To determine the elastic unloading curve, small loading-unloading cycles should be executed at consecutive points on AB. Assuming that the initial stiffness modulus on the unloading or reloading branches represents elastic stiffness, the elastic unloading curve such as AB' can be constructed, Figure 2c). Now the total strain is composed of three portions, namely

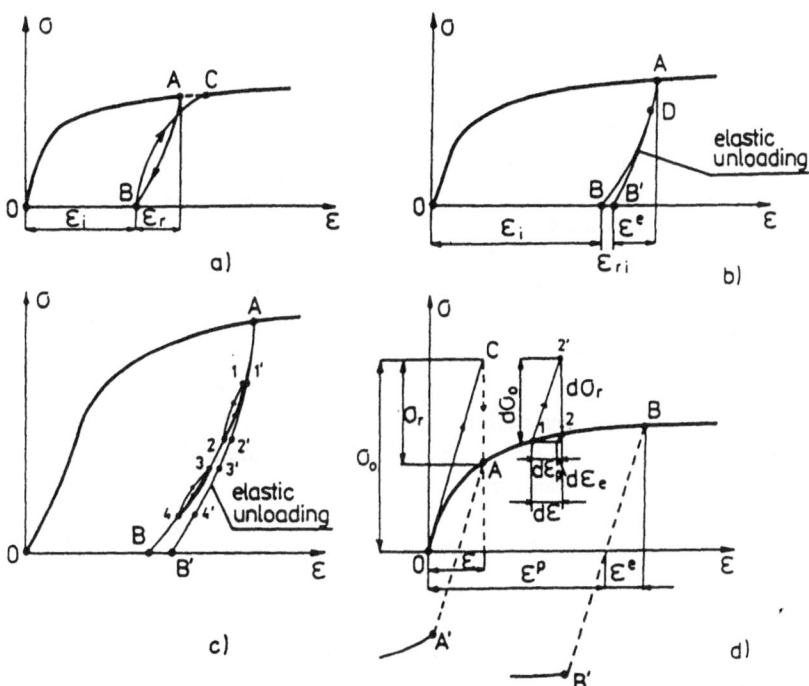

Figure 2 a) Typical triaxial compression or shear curve for loading, unloading, and reloading, b) determination of the yield point D on the unloading curve (departure from the elastic rebound AB'), c) determination of the elastic unloading curve by executing small cycles 1-2 or 3-4 at various stress levels, d) idealized model: irreversible strains beneath the prestress level are neglected and the linear response is elastic

$$\varepsilon = \varepsilon_i + \varepsilon_n + \varepsilon^e \qquad (3.1)$$

where ε_i is the irreversible strain after unloading ε_n is the reverse plastic strain and ε^e denotes the elastic strain. When the unloading branch ADB departs from the elastic curve AB' at D, this point can be assumed as the yield limit during unloading. In actuality, the determination of D depends on the conventional offset limit and for some idealizations the plastic strains developed during unloading and reloading are neglected. Such is the case presented in Figure 2d) where the line AA' is assumed as elastic and during reverse loading the plastic strain increments are assumed to develop at A'.

It is thus seen that the size of the elastic domain after initial loading much depends on the definition of the yield point, that is on the value of plastic strain defining the onset of yielding. For small offset values, the point D will be very close to A or coincide with it, whereas for large offset values the elastic domain will be large such as that in Figure 2d). For cyclic loading problems, when the description of hysteresis loops is important, more accurate constitutive models should be used and the elastic domain is then usually assumed as small or vanishing. The reverse plastic strain then occurs on the whole unloading path AB.

Instead of decomposing the total strain into elastic and plastic portions, an alternative decomposition can be carried out for stress corresponding to the prescribed strain, namely

$$\sigma = \sigma° - \sigma'$$ (3.2)

when $\sigma°$ denotes the elastic stress corresponding to the prescribed strain and σ' is the relaxation stress, Figure 2 d). In the case of linear elastic behaviour, the elastic stress $\sigma°$ is calculated from Hooke's law, whereas the relaxation stress σ' represents the departure from non-linearity and reduces the elastic stress to the actual stress. Similar decomposition can also be carried out for the unloading curve. The yield point can therefore be also defined by the magnitude of the relaxation stress instead of permanent strain. For increments or rates of stress and strain the similar decomposition occurs, namely

$$d\varepsilon = d\varepsilon + d\varepsilon^p, \qquad\qquad d\sigma = d\sigma° - d\sigma'$$ (3.3)

Let us first describe the material response presented in Figure 2 for the uniaxial case. Considering the loading and unloading at point A on the *stable* (hardening) portion of stress-strain curve, Figure 2a), we can write

$$d\varepsilon = L_1 \, d\sigma \quad \text{for} \quad d\sigma > 0$$
$$d\varepsilon = L_u \, d\sigma \quad \text{for} \quad d\sigma < 0$$ (3.4)

where L_1 and L_u denote the loading and unloading tangent compliance moduli. Note that the relations (3.4) are linear with respect to increments both for loading and unloading. An alternative way is to replace (3.4) by a single *non-linear* relation

$$d\varepsilon = B \, d\sigma + B_1 |d\sigma|$$ (3.5)

in which

$$B = \tfrac{1}{2}(L_1 + L_u), \qquad B_1 = \tfrac{1}{2}(L_1 - L_u)$$ (3.6)

In fact, the relation (3.5) has the forms

$$d\varepsilon = (B + B_1) d\sigma \quad \text{for} \ d\sigma > 0$$
$$d\varepsilon = (B - B_1) d\sigma \quad \text{for} \ d\sigma < 0$$ (3.7)

and in view of (3.6), the relations (3.4) are obtained. Alternatively, it can be written

$$d\sigma = E \, d\varepsilon + E_1 |d\varepsilon|$$ (3.8)

and again, we have

$$d\sigma = (E + E_1) d\varepsilon \quad \text{for} \ d\varepsilon > 0,$$
$$d\sigma = (E - E_1) d\varepsilon \quad \text{for} \ d\varepsilon > 0$$ (3.9)

Note that the relations (3.8) or (3.9) apply for both *hardening and softening* portions of the stress-strain curve. Requiring

$$E = \frac{1}{2}\left(\frac{1}{L_l} + \frac{1}{L_u}\right) = \tfrac{1}{2}(K_l + K_u),$$

$$E_1 = \frac{1}{2}\left(\frac{1}{L_l} - \frac{1}{L_u}\right) = \tfrac{1}{2}(K_l - K_u)$$

(3.10)

where $K_l = L_l^{-1}$ and $K_u = L_u^{-1}$ denote the tangent stiffness and unloading moduli, we obtain the relations

$$d\sigma = K_l\, d\varepsilon \quad \text{for } d\varepsilon > 0$$
$$d\sigma = K_u\, d\varepsilon \quad \text{for } d\varepsilon < 0$$

(3.11)

which is the inverse form (3.4). Note that $E_1 < 0$, since $K_l < K_u$, Figure 3. It is thus seen that the local loading-unloading behaviour can be described either by two linear relations accompanied by inequality conditions or by single non-linear relations in stress or strain increments. These relations are prototypes for more general plasticity and endochronic models.
    A more familiar form can be ascribed to relations (3.5) or (3.8) by assuming that their first terms represent the elastic behaviour . Thus, (3.5) takes the form $d\varepsilon = d\varepsilon^e + d\varepsilon^p$ where $d\varepsilon^e = Bd\sigma, d\varepsilon^p = B_1|d\sigma|$ and B denotes the elastic compliance modulus. Similarly, relation (3.8) takes the form $d\sigma = d\sigma^° - d\sigma'$ where $d\sigma^°$ is the elastic stress increment and $d\sigma'$ denotes the relaxation stress increment and E is the elastic stiffness modulus. Thus, we have

$$d\varepsilon^e = Bd\sigma, \qquad d\varepsilon^p = B_1|d\sigma|, \qquad d\varepsilon = d\varepsilon^e + d\varepsilon^p$$
$$d\sigma^° = Ed\varepsilon, \qquad d\sigma' = -E_1|d\varepsilon|, \qquad d\sigma = d\sigma^° - d\sigma'$$

(3.12)

and then it follows immediately that the formulations (3.5) and (3.8) imply the larger unloading stiffness modulus than the elastic modulus. In fact, since $E_1 < 0$, the unloading stiffness modulus equals $E - E_1$ and is greater than E , Figure 3 b).
    The predictions of the presented relations for small stress and strain cycles are illustrated in Figure 3. In Figure 3 a) small stress or strain cycles are applied from the state A and the prediction of (3.4) or (3.11) is presented. Since the unloading stiffness modulus is greater than the loading modulus, that is $K_u > K_1$ or $L_u < L_1$, after each cycle the permanent strain $A - 2, 2 - 4$, etc. accumulates. Similarly, for the prescribed strain cycle, the progressive stress relaxation $A - 2', 2' - 4'$, etc. develops. Identifying the unloading stiffness modulus with the elastic modulus, it is seen that for each cycle the incremental work is negative, that is

$$\oint_\sigma (\sigma - \sigma_A)d\varepsilon = \tfrac{1}{2}d\sigma \cdot d\varepsilon^p = \tfrac{1}{2}(d\sigma)^2(L_u - L_1)$$

$$= \tfrac{1}{2}(d\sigma)^2\left(\frac{1}{K_u} - \frac{1}{K_1}\right) < 0$$

(3.13)

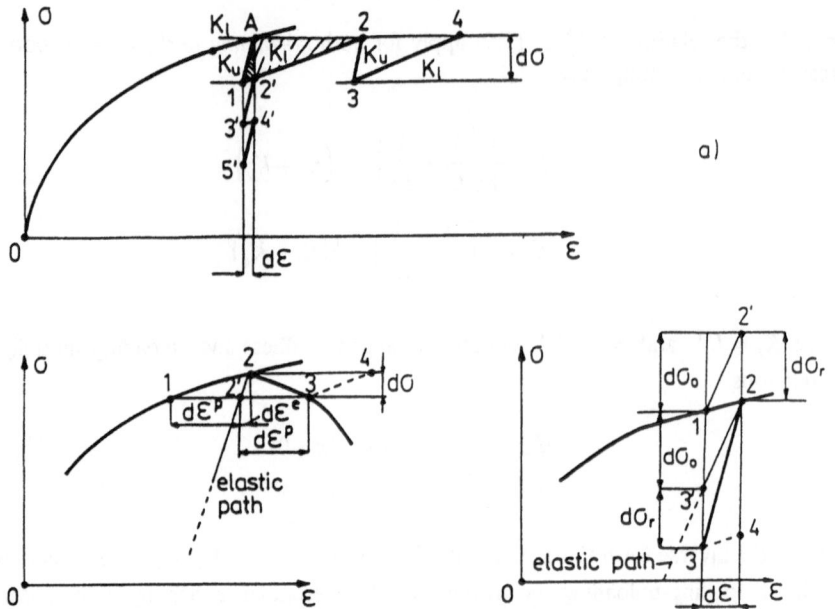

Figure 3   Loading-unloading conditions corresponding to linear and non-linear incremental relations: a) (3.4), b) (3.5), and c) (3.8). Progressive ratchetting and relaxation phenomena for infinitesimal cycles

and similarly, each strain cycle is associated with the negative incremental work

$$\int_\varepsilon (\sigma - \sigma_A) d\varepsilon = \tfrac{1}{2} d\varepsilon \cdot d\sigma' = -\tfrac{1}{2} (d\varepsilon)^2 (K_u - K_j) \tag{3.14}$$

These inequalities imply *cyclic instability* of the behaviour in the sense of Drucker or Ilyushin postulates. In fact, (3.13) corresponds to Drucker and (3.14) to Ilyushin postulates requiring positive incremental work on closed stress or strain cycles. Violation of these postulates would lead to finite strain accumulation or finite stress relaxation for stress or strain cycles infinitesimally departing from the initial state.

Figure 3 b) presents the material response predicted by the non-linear relations (3.5) for application and removal of the stress-cycle. Identifying the second term of (3.5) with the plastic strain increment, it is seen that when $B_1 > B$, that is $\left|d\varepsilon_p\right| > \left|d\varepsilon_e\right|$, the unloading stiffness modulus is negative and $d\sigma \cdot d\varepsilon < 0$ for the unloading path 2-3. Further, the accumulation effect presented in Figure 3 a) obviously occurs. Thus, the relation (3.5) violates the stability conditions both for cyclic and radial paths issuing from A. Similarly, Figure 3 d) presents the strain cycle 1-2-3, predicted by the non-linear relation (3.8). Since now the relaxation stress increment is positive both for loading and unloading, the unloading stiffness modulus is greater than the elastic modulus. However, the cyclic stability condition of Ilyushin is violated since (3.14) is negative.

It is seen that the presented three propositions for loading-unloading conditions are *defective* as they do not correspond to actual response and introduce instability for monotonic or cyclic loading paths. However, it is easy to improve the relations (3.4) or (3.11) by introducing the *memory* of particular loading events. Consider the process of loading to the prescribed stress A, unloading B, and subsequent reloading to A. It can be required that

$$
\begin{aligned}
d\varepsilon &= L_l\, d\sigma & \text{for } d\sigma > 0 - \text{loading} \quad \text{to} \quad A \\
d\varepsilon &= L_u\, d\sigma & \text{for } d\sigma < 0,\ \sigma < \sigma_A - \text{unloading} \ \text{from} \ A \\
d\varepsilon &= L_r\, d\sigma & \text{for } d\sigma > 0,\ \sigma > \sigma_B,\ \sigma < \sigma_A - \text{reloading} \ \text{from} \ B \\
d\varepsilon &= L_l\, d\sigma & \text{for } d\sigma < 0,\ \sigma > \sigma_A - \text{loading} \ \text{from} \ A
\end{aligned}
\tag{3.15}
$$

where compliance modulus L_r may be equal to the unloading modulus K_u or different. In this way, not only the sign of $d\sigma$ defines the loading-unloading events but also the memory of points where stress increments change their sign. We shall call σ_A *the maximum prestress* and σ_B the *stress reversal* in our loading history.

   The non-linear relations (3.5) and (3.8) do not require this definition of particular loading events and are much simpler. On the other hand, they possess disadvantages discussed in this section. In particular (3.8) is a uniaxial prototype of endochronic theories [3,48] where the „intrinsic time" measure is expressed as the absolute value of the total strain, and of non-linear models in which the stress rate is a non-linear function of the strain rate [6]. In what follows, we shall extend our discussion to the multiaxial case and discuss loading-unloading conditions expressed in terms of stress or strain increments.

3.2 Dual incremental elasto-plastic relations

In the classical theory of plasticity it is usually assumed that a material exhibits elastic behaviour for stress states corresponding to the interior of the yield surface, whereas plastic deformation occurs for stress increments directed into the exterior of this surface. Thus, each deformation process can be divided into an active or loading process involving variation of plastic strain and an unloading or elastic process corresponding to elastic behaviour. However, geological materials such as clay, sand or rock do not exhibit purely elastic behaviour during unloading and the yield surface, when defined by a small offset value, usually encloses an elastic domain lying in the vicinity of the loading point. Indeed, in some cases the yield surface may not exist at all. Therefore, a more extended formulation should allow for nullifying the elastic domain in the limiting case and for defining the subsequent reverse loading event.

   Consider first the rate-independent behaviour of the material for which rates (or increments) of stress and strain are interrelated by the equations

$$
\dot{\sigma} = \mathbf{C}\dot{\varepsilon} \quad \text{or} \quad \dot{\varepsilon} = \mathbf{D}\dot{\sigma}
\tag{3.16}
$$

where the matrices $\mathbf{C}$ and $\mathbf{D}$ depend on stress, strain and hardening parameters but not on the stress or strain rates. Thus the relations (3.16) are *linear* and *homogeneous* in rates of stress and strain and the material behaviour does not depend on the natural time scale.
Our analysis will be limited to small strain theory, but all conclusions remain valid when account is taken of translation and rotation of the element in defining the strain rate.
Let there exist at each stage of the deformation process a yield (or loading) surface separating domains of active loading and elastic unloading (or reverse loading). In stress space, this surface is represented by the equation

$$f(\sigma,\alpha) = 0 \tag{3.17}$$

For strain-controlled processes, it is convenient to consider the loading surface in strain space, thus

$$\phi\left(\varepsilon, \varepsilon^p, \alpha\right) = 0 \tag{3.18}$$

where σ and ε are stress and strain tensors and the usual decomposition

$$\varepsilon = \varepsilon^e + \varepsilon^p \tag{3.19}$$

into elastic and plastic portions occurs in small strain theory. The symbol α collectively denotes the hardening parameters. Since different rate relations occur for active loading and unloading trajectories, we can write

$$\dot{\sigma} = C_2 \dot{\varepsilon} \quad \text{for} \quad \Phi = 0, \quad \dot{\Phi}_\varepsilon \equiv \left(\frac{\partial \phi}{\partial \varepsilon}\right)^T \cdot \dot{\varepsilon} > 0, \quad \dot{\varepsilon} \in V_{\varepsilon 2},$$

$$\dot{\sigma} = C_1 \dot{\varepsilon} \quad \text{for} \quad \Phi < 0 \quad \text{or} \quad \dot{\phi}_\varepsilon < 0, \quad \Phi = 0, \quad \dot{\varepsilon} \in V_{\varepsilon 1} \tag{3.20}$$

and similar relations in stress rate space

$$\dot{\varepsilon} = D_2 \dot{\sigma} \quad \text{for} \quad f = 0, \quad \dot{f}_\sigma \equiv \left(\frac{\partial f}{\partial \sigma}\right)^T \cdot \dot{\sigma} > 0, \quad \dot{\sigma} \in V_{\sigma 2},$$

$$\dot{\varepsilon} = D_1 \dot{\sigma} \quad \text{for} \quad f < 0 \quad \text{or} \quad \dot{f}_\sigma < 0, \quad f = 0, \quad \dot{\sigma} \in V_{\sigma 1} \tag{3.21}$$

In above we use dot to denote scalar product, or a trace operator of a matrix product.
Thus, there are two different matrices for stress or strain rates directed into two semispaces separated by the hyperplanes Π_Φ and Π_f tangential to the yield surfaces, Figure 4 (a,b).
Assume now that the continuity condition is satisfied; that is, for stress or strain rates directed tangentially to yield surface

$$C_2 \dot{\varepsilon} = C_1 \dot{\varepsilon} \quad \text{for} \quad \phi = \dot{\phi}_\varepsilon = 0$$

$$D_2 \dot{\sigma} = D_1 \dot{\sigma} \quad \text{for} \quad f = \dot{f}_\sigma = 0 \tag{3.22}$$

The continuity condition (3.22) imposes an essential constraint on the constitutive matrices which, must be interrelated: viz. [23,27]

$$\dot{\sigma}_2 = C_2 \dot{\varepsilon} = C_1 \dot{\varepsilon} + h\left(n_\Phi^T \cdot \dot{\varepsilon}\right) = \dot{\sigma}_1 - \dot{\sigma}^r$$

$$\dot{\varepsilon}_2 = D_2 \dot{\sigma} = D_1 \dot{\sigma} + g\left(n_f^T \cdot \dot{\sigma}\right) = \dot{\varepsilon}_1 + \dot{\varepsilon}^p \tag{3.23}$$

where h and g are arbitrary and n_Φ, n_f are the normalized gradients, i.e.

$$\mathbf{n}_\Phi = \frac{\partial \phi / \partial \varepsilon}{\left[(\partial \phi / \partial \varepsilon)^T \cdot (\partial \phi / \partial \varepsilon)\right]^{\frac{1}{2}}}, \qquad \mathbf{n}_f = \frac{\partial f / \partial \sigma}{\left[(\partial f / \partial \sigma)^T \cdot (\partial f / \partial \sigma)\right]^{\frac{1}{2}}} \tag{3.24}$$

The relations can be given an interpretation which is familiar in plasticity theory: identifying the matrices $\mathbf{C}_1$ and $\mathbf{D}_1$ with elastic stiffness and compliance matrices, we have $\dot{\varepsilon}^e = \dot{\varepsilon}_1 = \mathbf{D}_1 \dot{\sigma}$; and $\dot{\sigma}^e = \dot{\sigma}_1 = \mathbf{C}_1 \dot{\varepsilon}$. The terms $\dot{\varepsilon}^p$ and $\dot{\sigma}^r$ now represent the plastic strain rate and the relaxation stress rate superimposed upon the elastic stress rate. Denoting

$$h = -M \mathbf{n}_h, \qquad g = \frac{1}{K} \mathbf{n}_g \tag{3.25}$$

where $\mathbf{n}_h, \mathbf{n}_g$ are tensors normalized according to (3.24) (or unit vectors in vector space), and M, K are scalar functions, we can rewrite (3.23) as follows:

$$\dot{\varepsilon} = \dot{\varepsilon}^e + \dot{\varepsilon}^p = \mathbf{D}_1 \dot{\sigma} + \frac{1}{K} \mathbf{n}_g (\mathbf{n}_f^T \cdot \dot{\sigma}), \qquad f = 0, \quad \dot{f}_\sigma > 0, \tag{3.26}$$

$$\dot{\sigma} = \dot{\sigma}^e - \dot{\sigma}^r = \mathbf{C}_1 \dot{\varepsilon} - M \mathbf{n}_h (\mathbf{n}_\Phi^T \cdot \dot{\varepsilon}), \qquad \Phi = 0, \quad \dot{\phi}_\varepsilon > 0, \tag{3.27}$$

The scalar functions K and M will be called respectively the *hardening* and *relaxation moduli*. Similarly, the relations

$$\dot{\varepsilon}^p = \frac{1}{K} \mathbf{n}_g (\mathbf{n}_f^T \cdot \dot{\sigma}), \qquad \sigma^r = M \mathbf{n}_h (\mathbf{n}_\Phi^T \cdot \dot{\varepsilon}) \tag{3.28}$$

are respectively the non-associated *flow rule* and the non-associated *relaxation rule*. From (3.28) it follows that

$$K = \frac{(\mathbf{n}_f^T \cdot \dot{\sigma})}{(\dot{\varepsilon}^p \cdot \dot{\varepsilon}^p)^{\frac{1}{2}}}, \qquad M = \frac{(\dot{\sigma}^p \cdot \dot{\sigma}^p)^{\frac{1}{2}}}{(\mathbf{n}_\Phi^T \cdot \dot{\varepsilon})} \tag{3.29}$$

Since, in general, $\mathbf{n}_g \ne \mathbf{n}_f$ and $\mathbf{n}_h \ne \mathbf{n}_\Phi$ in (3.28), the plastic strain rate vector $\dot{\varepsilon}^p$ departs from the direction of the exterior normal vector $\mathbf{n}_f$ to the yield surface, and similarly the direction of $\dot{\sigma}^r$ does not coincide with that of $\mathbf{n}_\Phi$ in the strain space. The case of *associated flow* and *relaxation rules* is obtained by postulating that $\mathbf{n}_g = \mathbf{n}_f$ and $\mathbf{n}_h = \mathbf{n}_\Phi$, which gives

$$\dot{\varepsilon}^p = \frac{1}{K} \mathbf{n}_f (\mathbf{n}_f^T \cdot \dot{\sigma}), \qquad \dot{\sigma}^r = M \mathbf{n}_\Phi (\mathbf{n}_\Phi^T \cdot \dot{\varepsilon}) \tag{3.30}$$

The relations (3.30) are dual representations of the plastic deformation rule: whereas the flow rule (3.30) specifies the plastic strain rate corresponding to given stress rate, the relaxation rule (3.30) specifies that portion of the stress rate which should be subtracted form the elastic stress rate associated with a given rate of strain. The hardening and relaxation moduli can be interrelated by inverting the rate equation (3.26) and identifying it with (3.27). We have

$$\dot{\sigma} = C_1\left(\dot{\varepsilon} - \dot{\varepsilon}^p\right) = \dot{\sigma}^e - \dot{\sigma}^r = \left[C_1 - \frac{C_1\,n_f n_f^T \cdot C_1}{K + n_f^T \cdot C_1\,n_f}\right]\dot{\varepsilon} \tag{3.31}$$

limiting our discussion to the associated flow rule.

Since the yield condition can be expressed in terms of total strain as

$$\Phi\left(\varepsilon,\,\varepsilon^p,\alpha\right) = f\left[C_1\left(\varepsilon - \varepsilon^p\right),\alpha\right] = 0 \tag{3.32}$$

we have

$$n_\Phi = \frac{C_1\,n_f}{\left[\left(C_1\,n_f\right)^T \cdot \left(C_1\,n_f\right)\right]^{\frac{1}{2}}} \tag{3.33}$$

and the relation (3.31) can be rewritten as follows:

$$\dot{\sigma} = \dot{\sigma}^e - \dot{\sigma}^r, \qquad \dot{\sigma}^e = C_1\,\dot{\varepsilon}, \qquad \dot{\sigma}^r = M\,n_\Phi\left(n_\Phi^T \cdot \dot{\varepsilon}\right) \tag{3.34}$$

where

$$M = \frac{1}{K' + n_\Phi^T \cdot D_1 n_\Phi}, \qquad C_1 = D_1^{-1} \qquad K' = \frac{K}{\left(C_1\,n_f\right)^T \cdot \left(C_1\,n_f\right)} \tag{3.35}$$

Equations (3.35) provide the relationship between the hardening and relaxation moduli. Figure 4 presents the yield surfaces in stress and strain spaces. The hardening modulus K is obtained by projecting the stress rate onto the normal vector n_f and dividing by the modulus of $\dot{\varepsilon}^p$.

Similarly, the relaxation modulus M. is defined as the ratio of the modulus of $\dot{\sigma}^r$ and the projection of $\dot{\varepsilon}$ onto the normal vector n_Φ. Figure 5 shows the uniaxial stress-strain curve and the variation of the tangent, plastic and relaxation moduli E_t, E_p and E_r. Since

$$E_t = \frac{d\sigma}{d\varepsilon}, \qquad E_p = \frac{d\sigma}{d\varepsilon^p}, \qquad E_r = \frac{d\sigma^r}{d\varepsilon}$$

we have $\tag{3.36}$

$$E = E_r + E_t, \qquad \frac{E_r}{E} = \frac{1}{1 + E_p / E}$$

and the hardening modulus K corresponds to E_p in the uniaxial case.

　　　Whereas the representation of the yield surface in stress space is common, the use of total strains rather than stresses does have certain advantages. The relaxation rule (3.30) is then characterized by a normality property in strain space, and the relaxation modulus specifies the rate of „relaxation" of stress with respect to an elastic solution. This stress relaxation process is simulated numerically by means of the iterative „initial stress" or „initial strain" techniques applied to a boundary value problem.

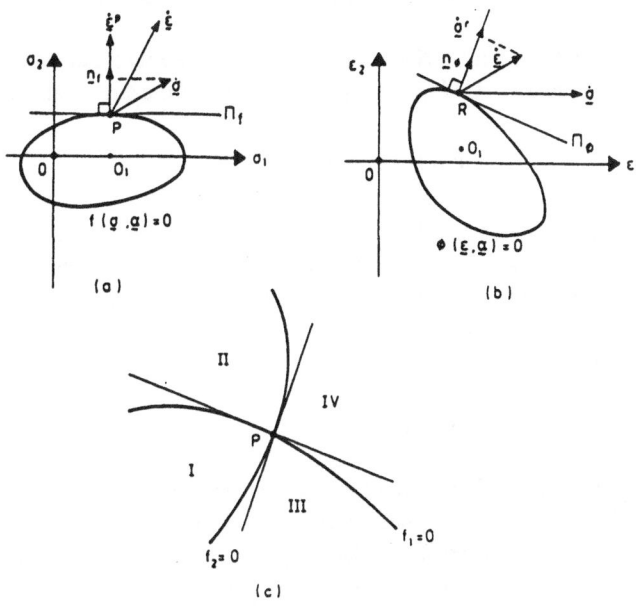

Figure 4 (a, b) Yield surfaces in a) stress and b) strain space;
c) intersection of two analytical surface

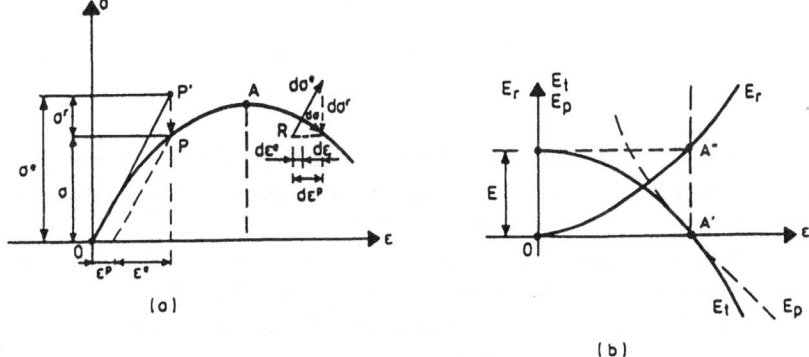

Fig. 5 a) Stress-strain curve; b) variation of tangent, plastic and relaxation moduli

The rate equations (3.26) and (3.27) are the most general linear relations occuring for a regular loading surface when there is a unique vector at the loading point. A singular regime occurs when two or more analytical surface intersect at the loading point. Figure 4 c) presents the case when there are four regions I, II, III, and IV, where different constitutive matrices may occur in stress space. Following (3.23), we may write

$$
\begin{aligned}
\dot{\varepsilon}_1 &= C_1 \dot{\sigma}, & \dot{\sigma} \in V_I \\
\dot{\varepsilon}_2 &= C_2 \dot{\sigma} = C_1 \dot{\sigma} + g_1\, n_{f_1}^T \cdot \dot{\sigma}, & \dot{\sigma} \in V_{II} \\
\dot{\varepsilon}_3 &= C_3 \dot{\sigma} = C_1 \dot{\sigma} + g_2\, n_{f_2}^T \cdot \dot{\sigma}, & \dot{\sigma} \in V_{III} \\
\dot{\varepsilon}_4 &= C_4 \dot{\sigma} = C_1 \dot{\sigma} + g_1\, n_{f_1}^T \cdot \dot{\sigma} + g_2\, n_{f_2}^T \cdot \dot{\sigma}, & \dot{\sigma} \in V_{IV}
\end{aligned}
$$

(3.37)

and the continuity conditions between particular subdomains are satisfied.

As the number of intersecting surfaces tends to infinity, the rate equations become *non-linear* and the constitutive matrix **C** depends on the direction of the stress rate vector. Generally,

$$\dot{\varepsilon} = \mathbf{D}\left(\sigma, \alpha, \dot{\sigma}\right)\dot{\sigma} \tag{3.38}$$

where **D** is a homogeneous function of the stress rate of order zero. Such non-linearity in rate equations will occur also when the yield surface shrinks to a point and an infinitesimal stress reversal produces an inelastic strain. Such non-linear equations arising in the limiting case will be discussed in the following section.

3.3 Loading, unloading and reloading conditions.

There are numerous possibilities to generalize the uniaxial conditions to general multiaxial case. Figure 6 illustrates some of these possibilities. Consider the stress path OP in the stress space and associate with it a *loading* surface $f_l = 0$ such that the stress point remains on this surface. Thus for a hardening material any stress path directed in the exterior of this surface corresponds to loading process whereas the stress path directed into the interior of $f_l = 0$ corresponds to *unloading process* (not necessarily elastic). Different rate relations apply for loading and unloading, so we have

$$\dot{\varepsilon} = \mathbf{D}_1\,\dot{\sigma} \quad \text{for} \quad \dot{\sigma}\cdot\mathbf{n}_1 > 0$$
$$\dot{\varepsilon} = \mathbf{D}_2\,\dot{\sigma} \quad \text{for} \quad \dot{\sigma}\cdot\mathbf{n}_1 < 0 \tag{3.39}$$

where $\mathbf{D}_1$ and $\mathbf{D}_2$ are the tangent compliance matrices depending on the stress state and on a set of state parameters, but not on the stress rate. The relation (3.39) are therefore *linear* in respective semi-spaces $\dot{\sigma}\cdot\mathbf{n}_1 > 0$ and $\dot{\sigma}\cdot\mathbf{n}_1 < 0$.

To define the unloading or reloading processes occurring within the loading surface $f_l = 0$, let us introduce the *unloading surface* $f_u = 0$ lying within the domain enclosed by $f_l = 0$. The unloading process will continue when the stress path is directed the interior of the unloading surface $f_u = 0$ and reloading occurs for the stress trajectory directed into exterior of $f_u = 0$, thus

$$\text{unloading:} \quad \dot{\varepsilon} = \mathbf{D}_2\,\dot{\sigma} \quad \text{for} \quad \dot{\sigma}\cdot\mathbf{n}_u < 0,\ f_l < 0$$
$$\text{reloading:} \quad \dot{\varepsilon} = \mathbf{D}_3\,\dot{\sigma} \quad \text{for} \quad \dot{\sigma}\cdot\mathbf{n}_u > 0,\ f_l < 0 \tag{3.40}$$

where $\mathbf{n}_u$ is the unit normal vector directed into the exterior of $f_u = 0$. Figures 6 a) and 6 b) present two cases of unloading surfaces: whereas in Figure 6 a) the unloading surface shirinks and expands with respect to a fixed point C which can be its symmetry centre, the unloading surface in Figure 6 b) possesses its transformation centre at A so it may translate, shrink, and expand during the loading process. More generally, it can be assumed that unloading surface may translate and shrink in a more complex way, depending on the plastic strain.

As seen in Figures 6 a,b) for stress path, such as PS emanating from P, at some point S the path passes into the exterior of $f_u = 0$ and then the reloading event commences for which the surface $f_u = 0$ expands with the stress point Thus

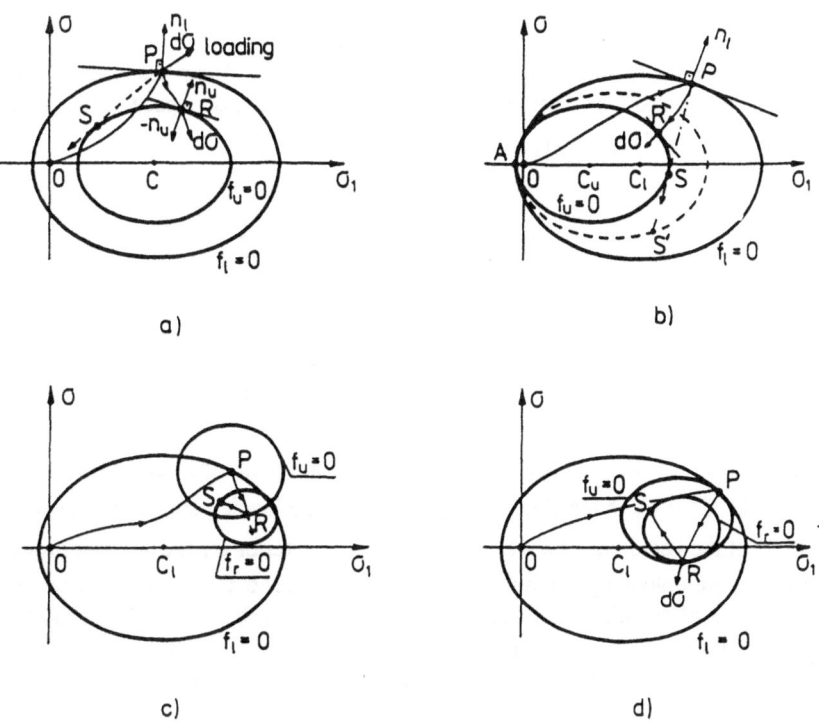

a)

b)

c)

d)

Figure 6.   Loading, unloading, or reloading surfaces in the stress-space. (a) surface shrinking toward fixed centre C, (b) unloading surface shrinking Unloading toward fixed point A, (c) unloading surface expanding from the stress reversal point P, (d) unloading surface tangential at P to the maximal loading surface

the same surface is used to define both unloading and reloading events. This definition has the disadvantage as there is no memory of path loading events and only three kinds of loading processes exist: loading when the stress point is on the surface $f_l = 0$, unloading or reloading when the stress point is on the surface $f_u = 0$. On the other hand, the memory of maximal prestress is incorporated into the model by specifying the maximal loading surface $f_l = 0$.
Figures 6 c) and 6 d) present two other possibilities of defining the unloading process. The new unloading surface $f_u = 0$ grows in size and the unloading process terminates when the stress point reaches the surface $f_l = 0$. We have therefore

$$\text{unloading:} \quad \dot{\varepsilon} = \mathbf{D}_2 \dot{\sigma} \quad \text{for} \quad \dot{\sigma} \cdot \mathbf{n}_u > 0 \quad f_l < 0$$
$$\text{reloading:} \quad \dot{\varepsilon} = \mathbf{D}_3 \dot{\sigma} \quad \text{for} \quad \dot{\sigma} \cdot \mathbf{n}_u > 0 \quad f_l < 0 \tag{3.41}$$

where $\mathbf{n}_u$ is the unit normal vector directed into the exterior of $f_u = 0$. Figure 6 c) the unloading surface is centred at P and intersects the surface $f_l = 0$, whereas in Figure 6 d) the unloading surface is tangential at P to the initial loading surface. When the reloading occurs, a new stress reversal surface develops from the stress reversal point R, . Let us note that not only the particular loading events are defined, but the model possesses the memory of stress reversal points such as P and R. This memory is erased by subsequent loading events of greater intensity, that is when the stress point reaches the previous loading or stress reversal surface.

Let us note that the loading-unloading conditions (3.20) expressed in terms of total strain rates can also be retransformed into the stress space. In fact, since $f(\sigma, \alpha) = f[C^*(\varepsilon - \varepsilon^p), \alpha] = \phi(\varepsilon, \alpha) = 0$ there is

$$C^* \frac{\partial f}{\partial \sigma} = \frac{\partial \phi}{\partial \varepsilon}, \qquad \frac{\partial \phi}{\partial \varepsilon} \cdot \dot{\varepsilon} = \frac{\partial \phi}{\partial \sigma} \cdot \dot{\sigma}^o \qquad (3.42)$$

and the loading-unloading conditions in the stress space are

$$f_1 = 0, \qquad \mathbf{n}_f \cdot \dot{\sigma}^o > 0, \quad \text{and} \quad f_1 = 0, \qquad \mathbf{n}_f \cdot \dot{\sigma}^o < 0 \qquad (3.43)$$

where $\dot{\sigma}^o = C^* \dot{\varepsilon}$ is the elastic stress rate occurring in the decomposition (3.23): $\dot{\sigma} = \dot{\sigma}^o - \dot{\sigma}'$ and $C^* = C_1$ is the elasticity matrix. Thus, the loading surface in the stress space may shrink, but the vector $\dot{\sigma}^o$ is always directed into its exterior for the loading process.

The loading-unloading conditions illustrated in Figures 6 a) - d) may now be referred to various proposals of constitutive models. Thus, Figures 6 a), b) may be referred to loading-unloading conditions used in hypoelastic models by Romano[44] or Davis and Mullenger [16] who assumed the sign of work rate as specifying loading and unloading, namely

$$\begin{aligned} \text{unloading:} \quad & \dot{\varepsilon} = C_1 \dot{\sigma} \quad \text{for} \quad \sigma \cdot \dot{\varepsilon} > 0 \\ \text{reloading:} \quad & \dot{\varepsilon} = C_2 \dot{\sigma} \quad \text{for} \quad \sigma \cdot \dot{\varepsilon} < 0 \end{aligned} \qquad (3.44)$$

It was shown in [29] that these loading-unloading conditions can be geometrically interpreted by introducing the unloading surface $f_u = 0$ and the unloading process terminates when the stress path passes in the exterior of this surface. The typical response in uniaxial tension or compression is that presented in Figures 3 a), b), namely, for stress cycles there is a continuing ratchetting effect and for small strain cycles the continuing relaxation effect occurs. Similarly as for non-linear relations (3.5) or (3.8) the present loading-unloading conditions induce cyclic instability and violate both Drucker and Ilyushin postulates and also the convexity condition (2.20)

The unloading surface of type shown in Figure 6 b) was recently applied by Hashiguchi [20], and Dafalias and Herrmann [13]. The unloading surface shown in Figure 6 c) was applied by Hueckel and Nova [21] and by Bazant and Kim [4] in modelling hysteric behaviour of rock and concrete. The disadvantage of these formulations lies in violation of the continuity condition for neutral paths $f_1 = 0$ or $f_u = 0$ and hence violation of the convexity (or uniqueness) condition (2.20)

The unloading surface which is tangential to the initial loading surface $f_1 = 0$ at the prestress point P, Figure 6 d), was applied by Mróz et al. [31, 35, 37] in developing their multisurface model and its simplified versions. This model will be discussed further in subsequent sections.

Finally, we should mention of loading-unloading conditions specified in particular subdomains of the stress space. For instance, Darve, Boulon, and Chambon [6] used piecewise linear rate relations applicable in subdomains corresponding to positive or negative signs of the principal stress rates. More generally, instead of a single loading-unloading condition, several subdomains in the stress space can be introduced corresponding to various signs of linear stress functions $f_1, f_2,...$, that is (cf. Figure 7)

$$
\begin{aligned}
& \text{1.} \quad \dot{\varepsilon} = \mathbf{C}_1 \dot{\sigma} \quad \text{for} \quad f_1 > 0, \ f_2 > 0 \\
& \text{2.} \quad \dot{\varepsilon} = \mathbf{C}_2 \dot{\sigma} \quad \text{for} \quad f_1 > 0, \ f_2 < 0 \\
& \text{3.} \quad \dot{\varepsilon} = \mathbf{C}_3 \dot{\sigma} \quad \text{for} \quad f_1 < 0, \ f_2 < 0 \\
& \text{4.} \quad \dot{\varepsilon} = \mathbf{C}_4 \dot{\sigma} \quad \text{for} \quad f_1 < 0, \ f_2 > 0
\end{aligned}
\tag{3.45}
$$

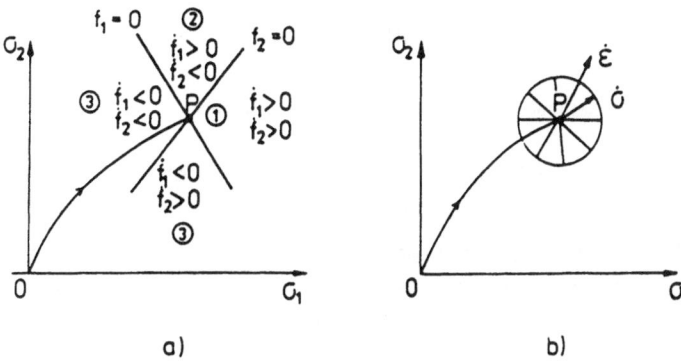

Figure 7 a) Singular loading surfaces and b) non-linear rate relations as a limiting case of singular loading conditions.

with proper continuity conditions between the subdomains 1, 2, 3, 4. When the number of subdomains tends to infinity, the piecewise linear relations (3.53) become *non-linear* relations, thus

$$
\dot{\sigma} = \mathbf{C}(\dot{\sigma})\dot{\sigma}, \quad \dot{\varepsilon} = \mathbf{D}(\dot{\varepsilon})\dot{\varepsilon},
\tag{3.46}
$$

where $\mathbf{C}(\dot{\sigma})$ and $\mathbf{D}(\dot{\varepsilon})$ are homogeneous functions of stress and strain rates, respectively of order zero. Such non-linear rate relations were recently considered by Kolymbas [45] and Benedetto and Darve [49]. For instance, a simple generalization of (3.8) can be presented in a form

$$
\dot{\sigma} = \mathbf{C}'\dot{\varepsilon} - \mathbf{d}(\dot{\varepsilon}\cdot\dot{\varepsilon})^{\frac{1}{2}} = \dot{\sigma}^\circ - \dot{\sigma}^r,
\tag{3.47}
$$

where $\mathbf{D}'$ is a constant elastic stiffness matrix and $\mathbf{d}$ is a second order tensor, not depending on the strain rate. The relation (3.47) can be identified with a familiar decomposition of the stress rate into elastic and relaxation portions. Similar structures to (3.47) can be ascribed to endochronic theories of plasticity [3, 48]. The disadvantage of such formulations lies in violating the uniqueness (or convexity) and cyclic stability conditions, discussed in the uniaxial case, (cf. also Sandler [46] and Rivlin [47] articles concerned with endochronic theories of plasticity).

case, (cf. also Sandler [46] and Rivlin [47] articles concerned with endochronic theories of plasticity).

3.4 Control of deformation processes.

The incremental relations discussed in the previous section can be applied to study the material response for specified loading or deformation programs. When stress components are imposed on the boundary of a specimen, the controllability is lost when the limit point is reached in the stress space and the loading parameter cannot increase beyond the limit value. Similar situation occurs for strain controlled processes. In most experimental programs a mixed control is applied, that is some components of stress and the complementary components of strain are controlled.

Let us briefly discuss the controllability of a deformation process for different selection of control variables cf. Mróz and Rodzik [56]. Consider an elasto-plastic material for which the regular yield surface has the form

$$F\left(\sigma, \ \varepsilon^p\right) \leq 0 \tag{3.48}$$

and the elastic and plastic strain rates are specified as follows

$$\dot{\varepsilon}^e = \mathbf{D}\,\dot{\sigma}, \quad \dot{\varepsilon}^p = \dot{\lambda}\,\mathbf{p}\left(\sigma, \ \varepsilon^p\right), \quad \dot{\varepsilon} = \dot{\varepsilon}^e + \dot{\varepsilon}^p, \tag{3.49}$$

where $\dot{\lambda}$ is the positive plastic multiplier whose form depends on the process control, $\mathbf{C} = \mathbf{D}^{-1}$ denotes the elasticity matrix and dot over the symbol denotes the derivative with respect to time-like evolution parameter. The usual small strain theory assumptions are used. Expand the stress and plastic strain into Taylor series with respect to the evolution parameter t, thus

$$\sigma(t) = \sigma\left(t_o\right) + \dot{\sigma}\left(t_o\right)\Delta t + \tfrac{1}{2}\ddot{\sigma}\left(t_o\right)\Delta t^2 + \dots\dots$$
$$\varepsilon^p(t) = \varepsilon^p\left(t_o\right) + \dot{\varepsilon}^p\left(t_o\right)\Delta t + \tfrac{1}{2}\ddot{\varepsilon}^p\left(t_o\right)\Delta t^2 + \dots\dots \tag{3.50}$$
$$\lambda(t) = \lambda\left(t_o\right) + \dot{\lambda}\left(t_o\right)\Delta t + \tfrac{1}{2}\ddot{\lambda}\left(t_o\right)\Delta t^2 + \dots\dots$$

where $\sigma\left(t_o\right) = \sigma_o$, $\varepsilon^p\left(t_o\right) = \varepsilon_o^p$ is the initial state. Similar expansion applies to the total strain For a regular process, the derivatives occurring in (3.49) do not exhibit discontinuities. Denote

$$\frac{\partial F}{\partial \sigma} = \mathbf{n}\left(\sigma, \ \varepsilon^p\right), \quad H = -\frac{\partial F}{\partial \varepsilon^p}\cdot\mathbf{p}, \quad \frac{\partial \mathbf{p}}{\partial \sigma} = \tilde{\mathbf{A}}\left(\sigma, \ \varepsilon^p\right), \quad \frac{\partial \mathbf{n}}{\partial \sigma} = \mathbf{A}\left(\sigma, \ \varepsilon^p\right) \tag{3.51}$$

where dot between two symbols denotes the scalar product.

3.4.1 Stress control

Consider first the stress control. The first order constitutive equation and consistency condition are

or in the matrix form, we have

$$\begin{bmatrix} 1 & -p \\ 0 & H \end{bmatrix} \begin{bmatrix} \dot{\varepsilon} \\ \dot{\lambda} \end{bmatrix} = \begin{bmatrix} D\,\dot{\sigma} \\ n\cdot\dot{\sigma} \end{bmatrix}$$ (3.53)

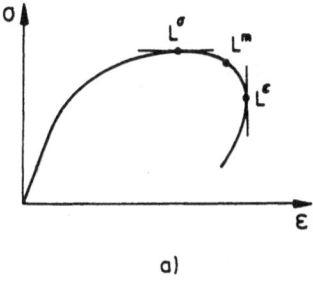
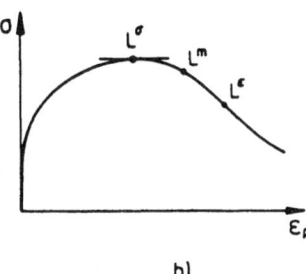

a)   b)

Figure 8. Controllability of the deformation process: a) limit points
L^{σ}, L^{ε} and L^{m} under stress, strain and mixed control; b) disappearance
of limit points under plastic strain control.

The set of equations (3.53) provides $\dot{\varepsilon}$ and $\dot{\lambda}$ in terms of the stress rate. Referring to Figure
8 a) it is seen that the process controllability is lost at the limit point L^{σ} when the determinant
of the state dependent matrix S^{σ} vanishes, thus

$$\det S^{\sigma} = H = 0$$ (3.54)

and the controllability occurs for $H>0$, implying $n\cdot\dot{\sigma} > 0$ for $\dot{\lambda} > 0$.

3.4.2. Strain control

Consider now a total strain controlled process. We have the elasticity relation

$$\sigma = C(\varepsilon - \varepsilon^{p})$$ (3.55)

The yield condition now is

$$F(\sigma, \varepsilon^{p}) = F[C(\varepsilon - \varepsilon^{p}), \varepsilon^{p}] = \Phi(\varepsilon, \varepsilon^{p}) \leq 0$$ (3.56)

and the consistency condition requires that

$$\frac{\partial \Phi}{\partial \varepsilon}\cdot\dot{\varepsilon} + \frac{\partial \Phi}{\partial \varepsilon^{p}}\cdot\dot{\varepsilon}^{p} = 0$$ (3.57)

and the consistency condition requires that

$$\frac{\partial \Phi}{\partial \varepsilon} \cdot \dot{\varepsilon} + \frac{\partial \Phi}{\partial \varepsilon^p} \cdot \dot{\varepsilon}^p = 0 \qquad (3.57)$$

Since in view of (3.42) there is

$$\frac{\partial \Phi}{\partial \varepsilon} = C \frac{\partial F}{\partial \sigma} = Cn, \quad \frac{\partial \Phi}{\partial \varepsilon^p} = -Cn + \frac{\partial F}{\partial \varepsilon^p} \qquad (3.58)$$

then using (3.57), (3.58), we obtain the set of equations in the matrix form

$$\begin{bmatrix} 1 & Cp \\ 0 & p \cdot Cn + H \end{bmatrix} \begin{bmatrix} \dot{\sigma} \\ \dot{\lambda} \end{bmatrix} = \begin{bmatrix} C\dot{\varepsilon} \\ Cn \cdot \dot{\varepsilon} \end{bmatrix} \qquad (3.59)$$

This set specifies $\dot{\sigma}$ and $\dot{\lambda}$ in terms of $\dot{\varepsilon}$. The controllability condition now is

$$\det S^\varepsilon = p \cdot Cn + H > 0 \qquad (3.60)$$

and the limit state corresponds to the condition $p \cdot Cn + H = 0$. The respective strain limit point L^ε is shown in Figure 8 a).

3.4.3 Mixed control

Following the previous study by Klisiński et al. [55] let us decompose the stress and strain tensor components into two mutually orthogonal sets: $\sigma = [\sigma_1, \sigma_2]^T$, $\varepsilon = [\varepsilon_1, \varepsilon_2]^T$ such that

$$\sigma \cdot \varepsilon = \sigma_1 \cdot \varepsilon_1 + \sigma_2 \cdot \varepsilon_2$$
$$\sigma_1 \cdot \varepsilon_2 = \sigma_2 \cdot \varepsilon_1 = 0 \qquad (3.61)$$

Assume ε_1, σ_2 to be the control variables and σ_1, ε_2 the response variables. Writing the constitutive equations in the form

$$\begin{bmatrix} \dot{\sigma}_1 \\ \dot{\sigma}_2 \end{bmatrix} = \begin{bmatrix} C_{11} & C_{12} \\ C_{21} & C_{22} \end{bmatrix} \begin{bmatrix} \dot{\varepsilon}_1 \\ \dot{\varepsilon}_2 \end{bmatrix} - \dot{\lambda} \begin{bmatrix} C_{11} & C_{12} \\ C_{21} & C_{22} \end{bmatrix} \begin{bmatrix} p_1 \\ p_2 \end{bmatrix} \qquad (3.62)$$

where C_{ij} are respective submatrices of C, Eq.(3.62) may be rewritten in a more specific form

$$\begin{bmatrix} 1 & -C_{12} \\ 0 & C_{22} \end{bmatrix} \begin{bmatrix} \dot{\sigma}_1 \\ \dot{\varepsilon}_2 \end{bmatrix} + \dot{\lambda} \begin{bmatrix} C_{11} & C_{12} \\ -C_{21} & -C_{22} \end{bmatrix} \begin{bmatrix} p_1 \\ p_2 \end{bmatrix} = \begin{bmatrix} C_{11} \dot{\varepsilon}_1 \\ -C_{21} \dot{\varepsilon}_1 + \dot{\sigma}_2 \end{bmatrix} \qquad (3.63)$$

In order to specify $\dot{\lambda}$, let us present the yield condition in the form

$$F(\sigma, \varepsilon^p) = F(\sigma_1, \sigma_2, \varepsilon^p) = \Phi(\varepsilon_1, \sigma_2, \varepsilon_1^p, \varepsilon_2^p) \qquad (3.64)$$

where σ_1, with assumption $\det C_{22} \neq 0$, has been eliminated from the elasticity relations

$$\sigma_1 = \left(C_{11} - C_{12}C_{22}^{-1}C_{21}\right)\left(\varepsilon_1 - \varepsilon_1^p\right) + C_{12}C_{22}^{-1}\sigma_2 \tag{3.65}$$

The consistency condition:

$$\frac{\partial \Phi}{\partial \varepsilon_1} \cdot \dot{\varepsilon}_1 + \frac{\partial \Phi}{\partial \sigma_2} \cdot \dot{\sigma}_2 + \frac{\partial \Phi}{\partial \varepsilon_1^p} \cdot \dot{\varepsilon}_1^p + \frac{\partial \Phi}{\partial \varepsilon_2^p} \cdot \dot{\varepsilon}_2^p = 0 \tag{3.66}$$

now provides the relation

$$\mathbf{n}_1\left(C_{11} - C_{12}C_{22}^{-1}C_{21}\right) \cdot \dot{\varepsilon}_1 + \left(n_1\, C_{12}C_{22}^{-1} + n_2\right) \cdot \dot{\sigma}_2 -$$
$$\left[n_1\left(C_{11} - C_{12}C_{22}^{-1}C_{21}\right) \cdot \mathbf{p}_1 + H\right]\dot{\lambda} = 0 \tag{3.67}$$

where $\mathbf{n}_i = \dfrac{\partial F}{\partial \sigma_i}$.

Equation (3.67) implies the limit point condition. In fact, as the two first terms of (3.67) are positive, the controllability occurs when

$$\mathbf{n}_1\left(C_{11} - C_{12}C_{22}^{-1}C_{21}\right) \cdot \mathbf{p}_1 + H > 0 \tag{3.68}$$

with a limit state specified by the condition $\mathbf{n}_1\left(C_{11} - C_{12}C_{22}^{-1}C_{21}\right) \cdot \mathbf{p}_1 + H = 0$, cf. (limit point L^m in Figure 8 a). It was shown by Klisiński et al. [55] thet the stress and strain controls provide lower and upper bounds on the range of plastic strain for which limit points occur for mixed control.

3.4.4 Plastic strain control

To derive the equations associated with plastic control, we use the dissipation function. Consider an isotropic hardening material model for which the yield surface is specified in a form

$$F\left(\sigma, \sigma_p\right) = 0 \tag{3.69}$$

or in a more specific form

$$F = f\left(\sigma\right) - \sigma_p\left(\lambda\right) = 0 \tag{3.70}$$

where $\sigma_p(\lambda)$ is the yield limit. The associated flow rule is obtained from (3.49) by setting $\mathbf{n} = \mathbf{p}$, thus

$$\dot{\varepsilon}^p = \lambda \frac{\partial F}{\partial \sigma} = \lambda \frac{\partial f}{\partial \sigma} = \lambda\, \mathbf{n} \tag{3.71}$$

In the following, it will be assumed that f is a homogeneous function of degree one. The inverse relations (3.71) can be derived by introducing the dissipation function

$$D = D\left(\sigma_p, \dot{\lambda}\right) = \sigma \cdot \dot{\varepsilon}^p = \dot{\lambda}\sigma \cdot \frac{\partial f}{\partial \sigma} = \sigma_p(\lambda)\dot{\lambda} \tag{3.72}$$

where the Euler theorem for homogeneous function was used. The function D represents the specific dissipation rate when the hidden elastic energy in a homogeneously strained element is neglected. The inverse relations to (3.72) are

$$\sigma = \frac{\partial D}{\partial \dot{\varepsilon}^p} = \sigma_p \frac{\partial \dot{\lambda}}{\partial \dot{\varepsilon}^p} \tag{3.73}$$

where $\dot{\lambda}$ is a homogeneous function of plastic strain rate of degree one. Let us now consider the rates of (3.71) and (3.73). We have

$$\ddot{\varepsilon}^p = \dot{\lambda}\frac{\partial^2 f}{\partial\sigma\partial\sigma}\dot{\sigma} + \ddot{\lambda}\frac{\partial f}{\partial\sigma} = \dot{\lambda}\mathbf{A}\,\dot{\sigma} + \ddot{\lambda}\frac{\partial f}{\partial\sigma} = \ddot{\varepsilon}^{p'} + \ddot{\varepsilon}^{p''} \tag{3.74}$$

and

$$\dot{\sigma} = \sigma_p \frac{\partial^2\dot{\lambda}}{\partial\dot{\varepsilon}^p\,\partial\dot{\varepsilon}^p}\ddot{\varepsilon}^p + \sigma_p'(\lambda)\frac{\partial\dot{\lambda}}{\partial\dot{\varepsilon}^p}\dot{\lambda} = \sigma_p\mathbf{B}\ddot{\varepsilon}^p + \frac{\sigma_p'}{\sigma_p}\dot{\lambda}\sigma = \dot{\sigma}' + \dot{\sigma}'' \tag{3.75}$$

The matrices

$$\mathbf{A}\left(\sigma\right) = \frac{\partial^2 f}{\partial\sigma\partial\sigma} = \frac{\partial n}{\partial\sigma} \quad \text{or} \quad A_{ijkl} = \frac{\partial^2 f}{\partial\sigma_{ij}\,\partial\sigma_{kl}}$$

$$\mathbf{B}\left(\dot{\varepsilon}^p\right) = \frac{\partial^2\dot{\lambda}}{\partial\dot{\varepsilon}^p\,\partial\dot{\varepsilon}^p} \quad \text{or} \quad B_{ijkl} = \frac{\partial^2\dot{\lambda}}{\partial\dot{\varepsilon}_{ij}^p\,\partial\dot{\varepsilon}_{kl}^p} \tag{3.76}$$

are homogeneous of degree minus one respectively of stress and plastic strain rate. Using the Euler theorem for homogeneous function of degree n

$$\mathbf{x}\frac{\partial f}{\partial\mathbf{x}} = x_i\frac{\partial f}{\partial x_i} = nf \quad \text{when} \quad f(t\,\mathbf{x}) = t^n f(\mathbf{x}),\ t > 0 \tag{3.77}$$

we obtain

$$\left(\mathbf{A}\sigma\right)_{ij} = A_{ijkl}\,\sigma_{kl} = \frac{\partial^2 f}{\partial\sigma_{ij}\partial\sigma_{kl}}\sigma_{kl} = \frac{\partial}{\partial\sigma_{kl}}\left(\frac{\partial f}{\partial\sigma_{ij}}\right)\sigma_{kl} = 0$$

$$\left(\mathbf{B}\dot{\varepsilon}^p\right)_{ij} = B_{ijkl}\,\dot{\varepsilon}_{kl}^p = \frac{\partial^2\dot{\lambda}}{\partial\dot{\varepsilon}_{ij}^p\,\partial\dot{\varepsilon}_{kl}^p}\dot{\varepsilon}_{kl}^p = \frac{\partial}{\partial\dot{\varepsilon}_{kl}^p}\left(\frac{\partial\dot{\lambda}}{\partial\dot{\varepsilon}_{ij}^p}\right)\dot{\varepsilon}_{kl}^p = 0 \tag{3.78}$$

The following property of the matrices $\mathbf{A}$ and $\mathbf{B}$ can now be stated

Property 1:
i) The matrix $\mathbf{A}$ possesses the eigenvector σ at zero eigenvalue.

ii) Similarly, the matrix **B** possesses the eigenvector $\dot{\varepsilon}^P$ at zero eigenvalue.
In fact, the eigenvalue problem $\mathbf{A}\mathbf{x} = \Lambda\mathbf{x}$ for the eigenvalue $\Lambda = 0$ becomes $\mathbf{A}\mathbf{x} = 0$ and corresponds to (3.78). The matrices **A** and **B** are therefore singular in view of their homogeneity of degree minus one with respect to stress and plastic strain rate.
The following properties of the matrices A and B follow from Property 1, namely

<u>Property 2</u>: The transformation $\mathbf{A}\dot{\sigma}$ represents the vector lying the plane Π_D normal to σ.
In fact,

$$\sigma \cdot \mathbf{A}\dot{\sigma} = \dot{\sigma} \cdot \mathbf{A}\sigma = 0$$
$$\sigma_{ij} A_{ijkl} \dot{\sigma}_{kl} = \dot{\sigma}_{kl} A_{ijkl} \sigma_{ij} = 0 \tag{3.79}$$

where property $\mathbf{A} = \mathbf{A}^T$ was used.

<u>Property 3</u>: The transformation $\mathbf{B}\ddot{\varepsilon}^P$ represents the vector lying in the plane Π_F normal to $\dot{\varepsilon}^P$.
In fact, in view of the symmetry $\mathbf{B} = \mathbf{B}^T$, there is

$$\dot{\varepsilon}^P \cdot \mathbf{B}\ddot{\varepsilon}^P = \ddot{\varepsilon}^P \cdot \mathbf{B}\dot{\varepsilon}^P = 0$$
$$\dot{\varepsilon}^P_{ij} B_{ijkl} \ddot{\varepsilon}^P_{kl} = \ddot{\varepsilon}^P_{kl} B_{klij} \dot{\varepsilon}^P_{ij} = 0 \tag{3.80}$$

The following equalities therefore occur in view of (3.74), (3.75) and (3.79), (3.80).

$$\mathbf{A}\sigma = \mathbf{A}\frac{\partial\dot{\lambda}}{\partial\dot{\varepsilon}^P} = 0, \qquad \mathbf{A}\dot{\sigma} = \sigma_p \mathbf{A}\mathbf{B}\ddot{\varepsilon}^P$$
$$\mathbf{B}\dot{\varepsilon}^P = \mathbf{B}\frac{\partial f}{\partial\sigma} = 0, \qquad \mathbf{B}\ddot{\varepsilon}^P = \dot{\lambda}\,\mathbf{B}\mathbf{A}\dot{\sigma} \tag{3.81}$$

We can now discuss more precisely the constitutive relations (3.74) and (3.75). In view of (3.81), there is

$$\mathbf{A}\dot{\sigma} = \mathbf{A}\dot{\sigma}' + \mathbf{A}\dot{\sigma}'' = \mathbf{A}\dot{\sigma}'$$
$$\mathbf{B}\ddot{\varepsilon}^P = \mathbf{B}\ddot{\varepsilon}^{P'} + \mathbf{B}\ddot{\varepsilon}^{P''} = \mathbf{B}\ddot{\varepsilon}^{P'} \tag{3.82}$$

The operators **A** and **B** therefore act on respective components of $\dot{\sigma}$ in the $\Pi_D \Pi_F$ plane and of $\ddot{\varepsilon}^P$ in the $\Pi_D \Pi_F$ plane.
Figure 9 a) presents the decomposition of $\ddot{\varepsilon}^P$ in the stress space where the plane Π_F is tangential to the yield surface and Figure 9 b) presents decomposition of the stress rate in the plastic strain rate space where Π_D is tangential to the dissipation surface $D = const.$

Let us note that the matrix **B** could be used in both hardening and softening regimes as it does not depend on the hardening rule. Hence, there are *no limit points on the deformation path*, cf. Figure 8 b), provided the plastic strain control is executed.

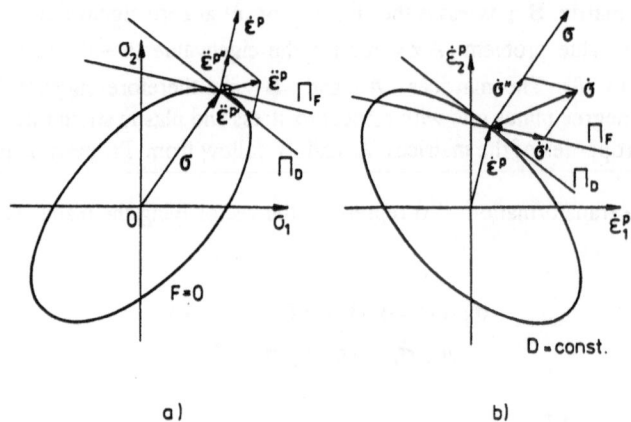

Figure 9 a) decomposition of second-order plastic strain rate $\ddot{\varepsilon}^{P}$ in the
stress space, b) decomposition of the stress rate $\ddot{\sigma}$ in the plastic strain rate space.

4. Isotropic and anisotropic hardening rules

The present discussion of hardening rules will be carried out in the context of soil mechanics.
Let the tensor of effective stress be σ_{ij} and the pore pressure be denoted by p_w so that

$$\sigma'_{ij} = \sigma_{ij} + p_w \delta_{ij} \tag{4.1}$$

where δ_{ij} denotes the Kronecker delta. The compressive stresses and contractive strains will be
assumed positive. The yield condition and constitutive relations can now be expressed in terms
of the effective stress and hardening parameters or in terms of the total stress; thus

$$f(\sigma, \alpha) = 0 \tag{4.2}$$

or

$$f(\sigma', p_w, \alpha) = 0 \tag{4.3}$$

where p_w now becomes a new hardening function. In fact, when the pore pressure varies, the
yield surface in total stress space translates along the hydrostatic axis and the material hardens
or softens.

    The total compressibility of soil depends on both the bulk modulus of the solid material
and of the fluid, but a simple description is obtained when both the solid material and the fluid
are regarded as incompressible and the macroscopic volume variation is due to closing or
opening of voids and flow of fluid. Denoting the volumes of voids, material and the total
volume of a representative element by V_v, V_m and V_t, the void ratio and the relative density
are expressed as follows:

$$e = \frac{V_v}{V_m}, \qquad \eta = \frac{\rho}{\rho_m} = \frac{V_m}{V_t} = \frac{1}{1+e} \tag{4.4}$$

where ρ is the bulk density and ρ_m denotes the material density. Since $\dot{V}_m = 0$, $\dot{V}_v = \dot{V}_t$, the rate of variation of e and η is

$$\dot{e}^p = -(1+e)\operatorname{tr}\dot{\varepsilon}^p, \quad \dot{e}^e = -(1+e)\operatorname{tr}\dot{\varepsilon}^e, \quad \dot{e} = \dot{e}^p + \dot{e}^e,$$
$$\dot{\eta}^p = \eta\operatorname{tr}\dot{\varepsilon}^p, \quad\quad \dot{\eta}^e = \eta\operatorname{tr}\dot{\varepsilon}^e,$$

Thus the macroscopic volume variation is related to the variation of void ratio or relative density and the usual decomposition into elastic and plastic components applies.

4.1 Isotropic hardening rules

a) Critical state model (density-hardening)

A simple hardening rule occurs when the yield condition (4.2) depends on one or several hardening variables whose evolution is directly related to the variation of plastic strain and no distinction is made between particular loading or reverse loading events. Let us discuss the simple assumption applicable to clays that the maximal consolidation pressure p_c or irreversible void ratio variation is the only hardening parameter; thus

$$f(\sigma, e^p) = 0 \quad \text{or} \quad f(\sigma, p_c) = 0 \tag{4.5}$$

where $p_c = \frac{1}{3}\sigma_{kk}$. The two forms (4.5) are equivalent, since for an isotropic consolidation process and subsequent unloading the irreversible void ratio change can be related to the maximal consolidation pressure.

Assuming the associated flow rule

$$\dot{\varepsilon}^P = \frac{1}{K}\mathbf{n}(\dot{\sigma}^T \cdot \mathbf{n}), \quad f = 0, \quad \dot{\sigma}^T \cdot \mathbf{n} > 0 \tag{4.6}$$

and using the consistency condition

$$\left(\frac{\partial f}{\partial \sigma}\right)^T \cdot \dot{\sigma} + \frac{\partial f}{\partial e^p}\dot{e}^p = 0 \tag{4.7}$$

we obtain the expression for the hardening modulus in the form

$$K = \frac{\partial f}{\partial e^p}(1+e)\operatorname{tr}\mathbf{n}\left[\left(\frac{\partial f}{\partial \sigma}\right)^T \cdot \left(\frac{\partial f}{\partial \sigma}\right)\right]^{-\frac{1}{2}} \tag{4.8}$$

where $\mathbf{n} = \mathbf{n}_f$ is the normalized gradient tensor $\dfrac{\partial f}{\partial \sigma}$ From (4.8) it follows that K varies along the yield surface, depending on the value of $\operatorname{tr}\mathbf{n} = \mathbf{n}_{kk}$.

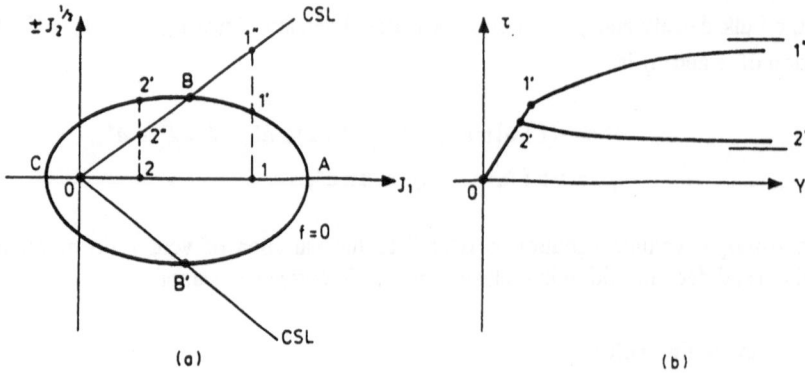

Figure 10. Isotropic hardening model : (a) yield surface; (b) material response under shear

Assuming that

$$\frac{\partial f}{\partial e^p} > 0 \tag{4.9}$$

it is seen that K has the same sign as tr $\mathbf{n}$. Figure 10 (a) presents a typical yield condition in the plane $\left(J_1, \pm J_2^{\frac{1}{2}}\right)$ for a particular value of e^p. Here J_1 and J_2 are respectively the first invariant of the effective stress tensor and the second invariant of the stress deviator. For a closed yield surface there exists a consolidation region for which tr $\mathbf{n} > 0$ (domain OBAB' in Figure 10. (a)) , a dilatancy or softening region for which tr $\mathbf{n} < 0$ (domain OBCB' in Figure 10. (a)) and the critical state line for which tr $\mathbf{n} = 0$ (lines OB and OB'). The hardening modulus K is positive in the consolidation domain OBAB', negative in the softening domain OBCB' and zero for the critical state lines OB or OB'. Figure 10. (b) shows the stress-strain response for paths 1-1'-1'' and 2-2'-2'' when a shear stress is imposed upon an initial hydrostatic stress $p = \frac{1}{3}J_1$.

The rate equations (4.6) should be supplemented by adding constitutive relations between the elastic strain rate $\dot{\boldsymbol{\varepsilon}}^e$ and the effective stress rate $\dot{\boldsymbol{\sigma}}$. For instance, as is usually assumed for clays, the rate equations for deviatoric and volumetric components take the form

$$\dot{e}_{ij}^e = \frac{\dot{s}_{ij}}{2G}, \qquad \dot{\varepsilon}_v^e = k\frac{\dot{p}}{p}\frac{1}{1+e} \tag{4.10}$$

which is valid for $p > 0$. Here $\mathbf{e}$ and $\mathbf{s}$ denote deviatoric strain and stress components. The volumetric rate relation follows the isotropic, elastic compression curve $e^e = e_o - k \ln(p / p_o)$ where k is a constant parameter and e_o, p_o are the reference void ratio and pressure. In writing (4.10) it is assumed that the shear modulus G is constant, hence the rate equations (4.10) can be regarded as being derived from the elastic potential,

$$W = \frac{s_{ij} s_{ij}}{4G} + \frac{1}{1+e} p\left(\ln\frac{p}{p_o} - 1\right) \tag{4.11}$$

However, it turns out that a more accurate description of elastic response is obtained by postulating that the shear modulus depends on effective pressure, $G = G(p)$. Substituting this function into (4.11) and applying the potential rule, we obtain

$$e_{ij}^{e} = \frac{S_{ij}}{2G(p)}, \qquad \varepsilon_{v}^{e} = \frac{-S_{ij}S_{ij}}{4G^{2}}\frac{dG}{dp} + \frac{k}{1+e}\ln\frac{p}{p_{o}} \qquad (4.12)$$

Thus, an additional dilatancy term occurs due to shear modulus variation as usually $dG/dp' > 0$. Differentiating (4.12) we obtain the elastic strain rates expressed in a more complex way, namely

$$\begin{bmatrix} \dot{e}_{ij}^{e} \\ \dot{\varepsilon}_{v}^{e} \end{bmatrix} = \begin{bmatrix} \alpha_{ss} & \alpha_{sp} \\ \alpha_{sp} & \alpha_{ss} \end{bmatrix} \begin{bmatrix} \dot{S}_{ij} \\ \dot{p} \end{bmatrix} \qquad (4.13)$$

where the terms α_{ss}, α_{sp}, α_{pp} are obtained by differentiating (4.12). The use of more general relations of this type for a granular material has been discussed recently in Ref [55], where the existence of a coupling effect due to α_{sp} is demonstrated. On the other hand, when using the simpler relations (4.10) with $G = G(p)$, it must be remembered that they describe recoverable but in general non-elastic (or hypo-elastic) strain rates and that energy dissipation would result for closed circuit stress paths.

The class of hardening models described by (4.5), (4.6) and (4.10) is applicable to clays under monotonic loading and can generally be described as *critical state (or density-hardening) models*. The particular formulations, such as the Cam-clay model [50] or cap model [56] differ only in the form of the yield condition, which is most important in the quantitative description of a stress-strain response. The disadvantage of this model lies in its inability to simulate properly the softening response and dilatancy as well as pore pressure variation in undrained test on overconsolidated clays.

b)   Combined (deviatoric and density) hardening model

As overconsolidated clays and dense sands exhibit stable behaviour despite dilatancy until the maximum stress is reached, one may expect that a better description would result from assuming that additional hardening occurs due to shear action. Let us introduce the combined hardening parameter κ, whose rate is expressed as follows:

$$\dot{\kappa} = \beta\left(\dot{e}^{pl}\cdot\dot{e}^{p}\right)^{1/2} + \dot{\eta}^{p} = \beta\dot{\lambda} + \dot{\eta}^{p} \qquad (4.14)$$

where the first term represents the deviatoric strain and the second corresponds to the irreversible density variation. Using the consistency condition (4.7), we obtain for the yield condition $f(\sigma, \kappa) = 0$

$$K = \left(-\frac{\partial f}{\partial \kappa}\right)\left[\beta\left(dev^{T}\,\mathbf{n}\cdot dev\,\mathbf{n}\right)^{1/2} + \eta\,tr\,\mathbf{n}\right]\left[\left(\frac{\partial f}{\partial \sigma}\right)^{T}\cdot\left(\frac{\partial f}{\partial \sigma}\right)\right]^{-\frac{1}{2}} \qquad (4.15)$$

where dev denotes the deviatoric portion of **n** , and β is a constant parameter. As $\partial f / \partial \kappa < 0$, the hardening modulus is positive when the second bracketed term is positive.

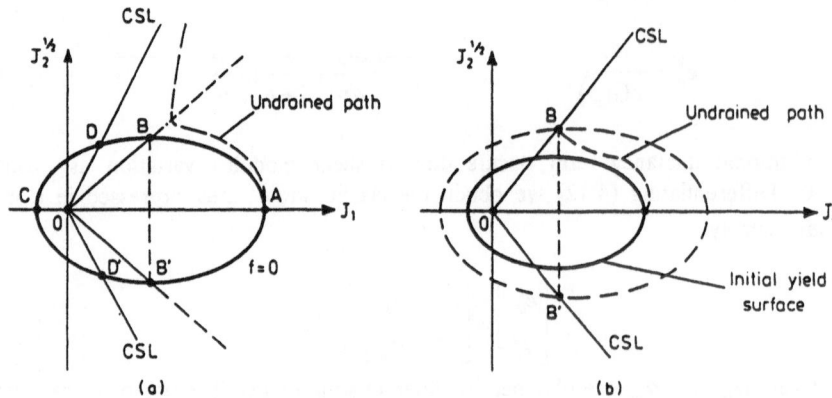

Figure 11. Isotropic hardening model with deviatoric and volumetric hardening; critical state line: (a) differs from zero dilatancy line; (b) coincides with zero dilatancy line

Thus, in the consolidation domain, $K \geq 0$ for

$$\operatorname{tr}\mathbf{n} \geq -\frac{\beta}{\eta}\left(\operatorname{dev}^{T}\mathbf{n} \cdot \operatorname{dev}\mathbf{n}\right)^{1/2} \tag{4.16}$$

and the critical state line, defined as a locus of points of vanishing hardening modulus, satisfies (4.16) as equality. In Figure 11. (a) the consolidation domain ODD'A is bounded by the critical state lines OD and OD' which do not coincide with the zero-dilatancy lines. The shear curves in consolidation and softening domains would be the same as in Figure 11. (b). The undrained stress path would then tend asymptotically to the critical state line from below, which is a generally observed fact for dense sands. A combined hardening parameter of this type was applied by Nova and Wood [36] and Wilde [51] to describe the inelastic response of sands.

If, instead of the parameter κ, we used the plastic work as a hardening parameter, that is

$$\dot{W} = \mathbf{s}^{T} \cdot \dot{\mathbf{e}}^{p} + p\dot{\varepsilon}_{v}^{p} \tag{4.17}$$

the hardening modulus would be expressed as follows:

$$K = \left(-\frac{\partial f}{\partial w_{p}}\right)\left[\mathbf{s}^{T} \cdot \operatorname{dev}\mathbf{n} + p \operatorname{tr}\mathbf{n}\right]\left[\left(\frac{\partial f}{\partial \sigma}\right)^{T} \cdot \left(\frac{\partial f}{\partial \sigma}\right)\right]^{-\frac{1}{2}} \tag{4.18}$$

and the consolidation domain would be described by the inequality

$$K \geq 0 \quad \text{for} \quad \operatorname{tr}\mathbf{n} \geq \frac{-\mathbf{s}^{T} \cdot \operatorname{dev}\mathbf{n}}{p} \tag{4.19}$$

Thus, the critical state line would be positioned in the same way as in Figure 11 (a), that is, departing from the zero-dilatancy line. The two hardening parameters κ and w_p would, therefore, yield similar results when applied in simulating an inelastic sand response. The plastic work w_p was used extensively as a hardening parameter by Lade [27] and Lade and Duncan [26] in modelling deformation of sands. The use of κ however, is more convenient as there is more flexibility associated with the value of β; for $\beta = 0$ we obtain a critical state model, and the deviation between the critical state line and the zero-dilatancy line can be directly associated with the value of β.

Assume finally that λ and η^p act independently in the yield condition

$$f(\sigma, \lambda, \eta^p) = 0 \qquad\qquad (4.20)$$

and the stress-strain curve tends asymptotically to a steady value for any hydrostatic pressure. The initial yield locus and the critical state line for this type of hardening model are shown schematically in Figure 11. (b). A two-parameter hardening rule was considered by Prevost and Hoeg [42].

As follows from our brief review, the combined parameter or two-parameter description may improve the accuracy for monotonic loading paths in sands which cannot be fitted into the density-hardening model. However, the real improvement is achieved both for monotonoc and cyclic loading histories when an anisotropic hardening model is used and the memory of particular loading events is properly incorporated.

4.2 Anisotropic hardening rules.

For monotonic loading processes, the isotropic hardening model can be successfully applied in solving boundary value problems. However, for time-varying loads, and, in particular, for cyclic loading processes, when hysteretic phenomena are of essential importance, we must look for a more accurate description of material behaviour. In fact, the isotropic hardening surface is usually defined by a large offset value and its interior domain is regarded as elastic. However, when after initial pre-consolidation the stress is slightly decreased, reverse plastic flow is usually observed and the unloading stress-strain curve departs significantly from the elastic curve. Therefore, the elastic domain enclosed by the yield surface $f_o = 0$, defined by a small offset value (say $\varepsilon = 10^{-4}$), is much smaller than that corresponding to the isotropic hardening surface, or may not exist at all. Moreover, this yield surface exhibits the material anisotropy (analogous to the Bauschinger effect in metals) which is induced by initial consolidation due to a stress state other than pure hydrostatic pressure.

Another important aspect of a more general material model is its memory of particular loading events, with the possibility of remembering „important" events and erasing „unimportant" events. In the case of isotropic hardening these are two types only, that is, active loading and elastic unloading. However, in a more general case, we have to make a distinction between events of larger „intensity" and those of lesser „intensity". The following fundamental property of material memory will be assumed: *loading events of given intensity can only be erased from material memory by events of larger intensity.*

Figure 12. illustrates this memory property in the uniaxial case. Let the absolute value of a function $P = P(t)$ be assumed as its intensity measure. The first loading event   0-1 terminates at 1 with its maximal intensity P_1; the second loading event commences at 1 with decreasing intensity and terminates at 2 with $P_2 > P_1$, thus erasing the memory of the event 0-1. Similarly, the event 2-3 with $P_3 > P_2$ erases the memory of events 0-1 and 1-2. However, the subsequent events of decreasing or constant intensities $P_4, P_5, P_6, \ldots\ldots$ remain incorporated in the material memory until the event 8-9, with $P_9 > P_i$ $(i = 1, 2, \ldots 8)$, erases the effect of previous load changes. This memory property is an important component of the constitutive model and it is based on the physical assumption that large prestress determines a material structure (i.e.topology of distribution of contact forces) that can only be partially modified by stress variations of small amplitude. We shall use the term *discrete memory*, as only a discrete set of load reversal points needs to be remembered due to continuing erasure of less important events from the material memory.

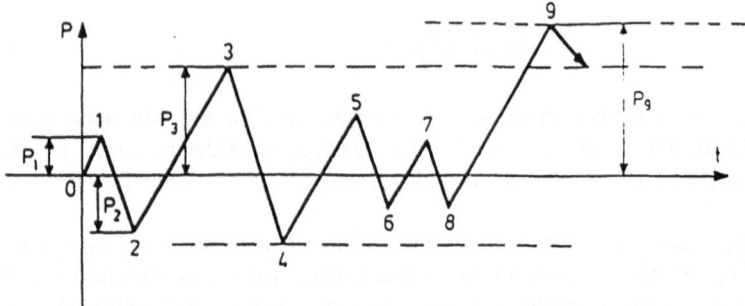

Figure 12. Memory rule for soil: large intensity loading event $\left(P_3\right)$ cannot be

erased by subsequent smaller loading events $\left(P_4, P_5, P_6, P_7, P_8\right)$

In Figure 12. the load level P_3 is kept fixed for all subsequent events of small intensity, thus becoming a memory level that can only be changed by a load of larger intensity. However, it may be assumed that this level is gradually modified by load cycles of smaller amplitude:

$$\overline{P}_3\left(\tau_f\right) = f(\mu), \quad \text{where} \quad \mu = \pi \int_o^{\tau_f} |\dot{P}| d\tau \quad \text{for} \quad |P| \le P_3 \tag{4.21}$$

where $\overline{P}_3(0) = P_3$ and $\tau = t - t_3$; that is, integration applies to all loading events of smaller intensity following the maximal load $P = P_3$ In this way, all cycles of smaller intensity, though they are later erased by larger loading events, can affect the memory level by means of (4.21).

Let us now discuss three versions of the anisotropic hardening model that possesses this multi-level memory rule.

a) A multi-surface hardening model

Assume that the yield surface, if it exists (as defined by a small offset value), may translate and expand or contract in stress space; thus

$$f_o\left(\sigma - \alpha^{(0)}, e^p\right) = 0 \tag{4.22}$$

and the associated flow rule (4.6) applies. In order to construct the complete set of model equations the translation rule of the yield surface and its dependence on the irreversible void ratio should be specified. It turns out, however, that this translation rule is a multi-valued function of plastic strain, and in order to incorporate sufficient memory into the model a set of nesting surfaces in stress space should be introduced to specify particular loading events and memory levels.

Assume that besides the yield surface $f_o = 0$ enclosing the elastic domain there exists a *consolidation* or *bounding* surface defined by the degree of material consolidation. For clays the consolidation surface is developed during the initial consolidation process, whereas for sands it depends on material densification in the formation of the speciment. The domain enclosed by the consolidation surface is not elastic and for stress trajectories within this surface plastic flow occurs once the yield condition $f_o = 0$ is satisfied. Assuming that the consolidation surface expands or contracts isotropically, we can identify it with the isotropic hardening surface discussed in the previous section; thus

$$F_c = f_o(\sigma, e^p) = 0 \tag{4.23}$$

It is further assumed that the hardening modulus on the consolidation surface varies according to (4.8), whereas the hardening modulus on the yield surface depends on the relative configuration of these two surfaces. In order to provide a more precise description of the variation of this modulus within the domain contained between the two surfaces $F = 0$ and $f_o = 0$, let us introduce a set of nesting surfaces that can translate and expand or contract due to density variation: that is

$$f_1(\sigma - \alpha^{(1)}, e^p) = 0, \quad f_2(\sigma - \alpha^{(2)}, e^p) = 0, \ldots, f_k(\sigma - \alpha^{(k)}, e^p) = 0 \tag{4.24}$$

Assuming that all surfaces are similar, we can write

$$f_i(\sigma - \alpha^{(i)}) - [a^{(i)}(e^p)]^2 = 0, \quad i = 0, 1, 2, \ldots, n \tag{4.25}$$

where f_i is a homogeneous function of degree two. For any point P_o on $f_o = 0$, there exists, therefore, a set of conjugate points $P_1, P_2, \ldots P_n = R$ on $f_1 = f_2 = f_n = 0$, for which the normal vector has the same direction. It is assumed that the particular surfaces do not intersect but may contact each other at the conjugate points.

The deformation process from any elastic state can thus be imagined as follows. The stress point first reaches the yield surface at P_o and this surface translates toward a conjugate point P_1 on the nesting surface $f_1 = 0$. Before their contact, the hardening modulus K_o applies in the flow rule (4.6); however, when $f_o = 0$ engages $f_1 = 0$, the first nesting surface becomes the active loading surface and the hardening modulus K_1 occurs in the flow rule. For subsequent contacts of constitutive nesting surfaces, now corresponding values of hardening moduli apply until the boundary surface is reached. Since, for any point P_o on $f_o = 0$, there exists a set of conjugate points $P_1, P_2, \ldots P_n = R$ on nesting surfaces and the moduli on $F_c = 0$ and $f_o = 0$ are known, an interpolation rule is postulated that will specify the variation of K at conjugate points. For instance, it can be assumed that

$$K_K = K_R + \left(\frac{n-k}{n}\right)^\alpha (K_o - K_R), \quad k = 0,1,2,\ldots\ldots n, \qquad (4.26)$$

where K_K is the hardening modulus on the surface $f_k = 0$, K_R is the modulus at R on the consolidation surface and α denotes a material constant. Since K_R varies on the consolidation surface according to (4.8), the values K_K also depend on the position of P_K on the nesting surface. The configuration of nesting surfaces and the interpolation rule (4.26) define the *field of hardening moduli*, and the material response for any loading history may be studied by following the evolution of this field.

To formulate the translation rule for the yield surface or the active loading surface, assume that the surfaces $f_o = 0$ to $f_l = 0$ are in contact at the point P_l and the surface $f_l = 0$ translated towards the conjugate point P_{l+1} (Figure 13). The position of the conjugate point is defined from the

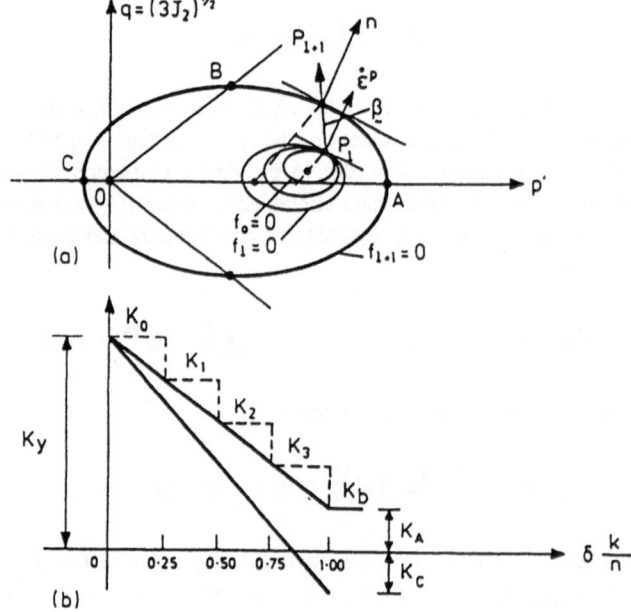

Figure 13. Anisotropic multi-surface hardening model: (a) yield surfaces and nesting surfaces; (b) variation of hardening moduli at conjugate points on nesting surfaces

proportionality condition

$$\sigma_P^{(l)} - \alpha^{(l+1)} = \frac{a^{(l+1)}}{a^{(i)}} \left(\sigma_P^{(l)} - \alpha^{(l)}\right) \qquad (4.27)$$

and the vector connecting the points P_{l+1} and P_l is expressed as follows:

$$\beta = \frac{1}{a^{(l)}}\left[\left(a^{(l+1)} - a^{(l)}\right)\sigma_P^{(l)} - \left(a^{(l+1)}\alpha^{(l)} - a^{(l)}\alpha^{(l+1)}\right)\right] \qquad (4.28)$$

The relative motion of P_l with respect to P_{l+1} is assumed to occur along $P_l - P_{l+1}$; that is

$$\sigma_P^{(l)} - \sigma_P^{(l+1)} = \beta\dot\mu$$

where $\dot\mu$ is a scalar factor. Since

$$\dot\sigma_P^{(l)} - \dot\alpha^{(l)} + \left(\sigma_P^{(l)} - \alpha^{(l)}\right)\frac{\dot a^{(l)}}{a^{(l)}}$$

$$\dot\sigma_P^{(l+1)} = \dot\alpha^{(l+1)} + \left(\sigma_P^{(l+1)} - \alpha^{(l)+1}\right)\frac{\dot a^{(l+1)}}{a^{(l+1)}} \qquad (4.29)$$

we obtain

$$\dot\alpha^{(l)} = \dot\alpha^{(l+1)} + \beta\,\dot\mu + \frac{\dot a^{(l+1)} - \dot a^{(l)}}{a^{(l)}}\left(\sigma_P^{(l)} - \alpha^{(l)}\right) \qquad (4.30)$$

and the scalar $\dot\mu$ can be determined from the consistency condition (4.7).

If all the surfaces $f_o = 0$, $f_1 = 0,.......f_{l-1} = 0$ are in contact with the surface $f_l = 0$ and move with the stress point P_l their translation is governed by the motion of P_l and we have

$$\frac{\sigma_P^{(l)} - \alpha^{(k)}}{\sigma_P^{(l)} - \alpha^{(l)}} = \frac{a^{(k)}}{a^{(l)}}, \qquad k = 0,1,2,.....,l \qquad (4.31)$$

The values of all $a^{(k)}$ can be determined from equation (4.25).

A more detailed discussion of this model can be found in Ref. [31]. It constitutes an extension of the concept of a field of hardening moduli first developed for metals by Mróz[30] The case of undrained deformation of sills by using a similar model was treated by Prevost[41] Petersson and Popov [57] applied a modified version of a multi-surface model to study the cyclic response of structural elements.

b) A two-surface hardening model

The model presented possesses a multi-level memory structure, since for cyclically varying stress only a certain number of surfaces undergo translation; the other surfaces may change only due to density variation. For practical purposes it is possible to simplify this model, and we shall discuss two such simplified versions. The first possibility is to consider only the yield surface $f_o = 0$ and the consolidation surface $F_c = 0$ to which the translation rule (4.30) is applied. Instead of the interpolation rule (4.26) specifying the hardening modulus on consecutive nesting surfaces, the field of hardening moduli is now described by prescribing the variation of K with the distance $\delta = PR$ between the stress point P on $f_o = 0$ and the conjugate point R on $F_c = 0$. Such a modified description of the field of hardening moduli for metals was developed independently by Krieg [25] and Dafalias and Popov [11] and here we shall extend this idea to soils. Further development of this model can be found in Dafalias and Herrmann [13] and Bardet [67]. A more complex description with rotating and translating loading surface was proposed by di Prisco et al. [68].

Consider the initial consolidation process OA, after which the consolidation and yield surfaces are tangential to each other at A. The semi-diameters of these surface are respectively a_c and a_o, so that the maximal distance between the surface is $\delta_o = P_oR = 2(a_c - a_o)$, Figure 14(a). If now the stress point moves along APD, the yield surface translates toward the conjugate point R on $F_c = 0$ and the hardening modulus is a function of δ and δ_o, so

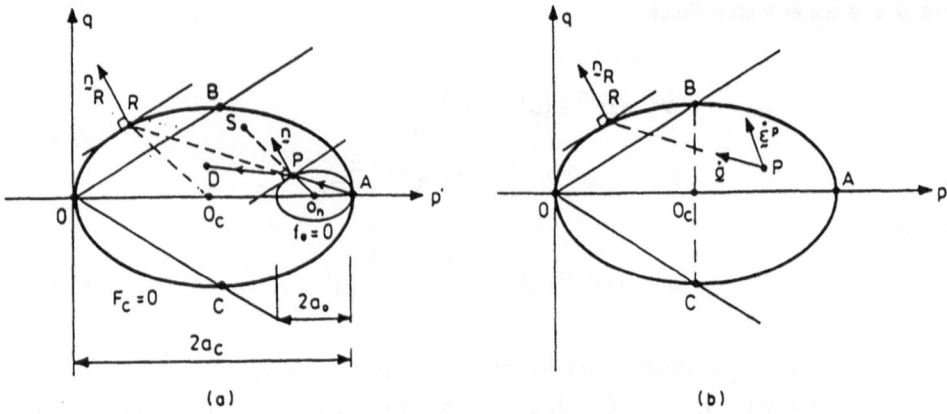

Figure 14. Two surface model: (a) yield and consolidation surfaces; (b) yield surface reduced to point P.

that

$$K = K(\delta, \delta_o),$$
$$K = K_R \quad for \quad \delta = 0,$$
$$K = K_y \quad for \quad \delta = \delta_o$$

(4.32)

where $\delta = f(\sigma_R - \sigma_P)^{1/2}$ is a scaled distance between P and R. For instance, it can be assumed that

$$K = K_R + (K_y - K_R)\left(\frac{\delta}{\delta_o}\right)^{\gamma}$$

(4.33)

so that the conditions (4.32) are satisfied; here γ is a constant parameter. During the plastic deformation process the maximal distance δ_o changes only slightly due to density changes, whereas δ depends on the instantaneous position of yield and consolidation surfaces. When the stress point reaches the consolidation surface, the flow and hardening rules corresponding to this surface are used.

The existence of the yield surface $f_o = 0$ is not essential in our analysis, and in the limiting case it can be assumed that $a_o = 0$. The conjugate point R then lies on the intersection of the direction of the stress rate vector with the consolidation surface, and the flow and translation rules become

$$\dot{\varepsilon}^P = \frac{1}{K}\mathbf{n}_R(\dot{\sigma}^T \cdot \mathbf{n}_R), \qquad \dot{\alpha} = \dot{\sigma}$$

(4.34)

where n_R denotes the normal vector at R (Figure 14. (b)). It is seen that changing the direction of $\dot{\sigma}$ involves variation of the direction of n_R and $\dot{\varepsilon}^P$. In order words, the components of n_R depend on the direction of $\dot{\sigma}$ and the flow rule (4.34) is *non-linear* in rate of stress. For a yield surface which is small with respect to the stress increment we would obtain a strong sensitivity of the plastic strain increment direction to the variation of the stress increment direction. Though we have not formulated corner flow rules we obtain similar effects which will be called called *pseudo-corner effects*. Only for stress states on the consolidation surface does the linear flow rule occur.

c)   *A hardening rule with an infinite number of loading surfaces*

An alternative simplified description of the evolution of hardening is provided by assuming that the field of hardening moduli inside the consolidation surface is represented by an infinite number of nesting surfaces, and that the hardening modulus depends on the ratio of the diameters of the instantaneous loading surface and the consolidation surface. Consider the case shown in Figure 15. (a), where, after initial consolidation along OP_c, the stress path is next directed into the interior of the consolidation surface $F_c(\sigma, e^P) = 0$. We assume that the elastic domain is reduced to a point and that reverse plastic flow occurs immediately for any stress increment directed from P_c into the interior of the domain enclosed by the surface $F_c = 0$. A set of nesting surfaces created after reaching P_c is transformed by the reverse loading P_cP_1. At P the stress point touches the surface $F_{11} = 0$, which now becomes the *active* reverse loading surface and all surface within $F_{11} = 0$ are translated with the stress point. The plastic response at P_1 is described by the flow rule (4.6) applied to $f_{11} = 0$, with the plastic modulus governed by a relation similar to (4.33), namely

$$K = K_R + \left(K_y - K_R\right)R_1^\gamma \qquad (4.35)$$

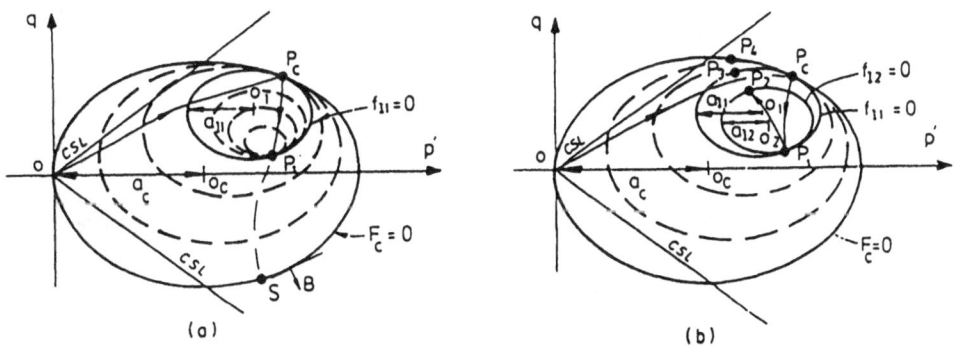

Figure 15. Model with an infinite number of surfaces: (a) first reverse loading; (b) second reverse loading

where

$$R_1 = \frac{a_c - a_{11}}{a_c} = 1 - \frac{a_{11}}{a_c} \qquad (4.36)$$

and K_y is the initial modulus at P_c that is, for $a_{11} = 0$. The semidiameters of the active loading and consolidation surfaces are a_{11} and a_c, respectively, and γ is a material parameter which will be assumed to vary during cyclic loading.

The *first reverse* loading (or unloading) program occurs provided a_{11} continues to increase. Thus when P reaches the consolidation surface at S, the surfaces $f_{11} = 0$ and $F_c = 0$ coincide and $K = K_s$. If, on the other hand, at P_1 the stress path reverses and is directed into the interior of the reverse loading surface $f_{11} = 0$, then the *second reverse* loading (or reloading) program commences. The active loading surface $f_{12} = 0$ for this program is a similar ellipse tangential at P_1 to the surface $f_{11} = 0$, which now becomes the stress-reversal surface, passing through the stress point P_2, Figure 15 (b). If the semi-diameter of this ellipse in denoted by a_{12}, the flow rule (4.6) and the interpolation rule (4.35) with $R_2 = 1 - a_{12}/a_c$ can be applied. If the second reverse loading process continues, then a_{12} increases and at P_3 the surface $f_{12} = 0$ and $f_{11} = 0$ coincide. Thus, for stress paths in the exterior of $f_{11} = 0$, the second reverse loading process $P_1 - P_2 - P_3$ is erased from the material memory and the first reverse loading process is continued until the stress point reaches the consolidation surface at P_4. In this way the memory of any reverse loading event is erased by a subsequent loading event of sufficiently large amplitude.

As is seen from the foregoing discussion, the stress point always remains on the active loading surface $f_{11} = 0$ which characterizes a particular loading event. The memory of past loading events is incorporated into a stack of stress reversal surfaces. Thus for a process $O - P_c - P_1 - P_2$, the stress reversal surfaces are $F_c = 0$ and $f_{11} = 0$, whereas $f_{12} = 0$ is the active loading surface provided the stress trajectory is directed into the exterior of this surface. Thus the i -th reverse loading event is defined by the conditions

$$f_{1i} = 0, \qquad \left(\frac{\partial f_{1i}}{\partial \sigma}\right)^T \cdot \dot{\sigma} \geq 0 \qquad\qquad (4.37)$$

and the condition that the surfaces $f_{1i} = 0$ and $f_{1(i-1)} = 0$ be tangential at the stress reversal point P_{i-1}; that is

$$\alpha_i - \sigma_P = \left(\alpha_{i-1} - \sigma_P\right)\frac{a_{1i}}{a_{1(i-1)}} \qquad\qquad (4.38)$$

where σ_P is the position of the stress reversal point P_{i-1} and α_i, α_{i-1} are the centres of the surfaces $f_{1i} = 0$ and $f_{1(i-1)} = 0$. When the consolidation surface $F_c = 0$ is expanding or contracting, the positions of the stress reversal points $P_c, P_1, P_2, \ldots$ are translated appropriately.

5. A non-associated isotropic hardening model for sand: static liquefaction study

In this section, we shall discuss a refined isotropic hardening model for sands using the non-associated flow rule. The need for use of such rule follows from the undrained sand response for which the *static liquefaction effect* is observed. The undrained path reaches the maximum shear stress with subsequent decrease of stress and progressive failure at large strains, cf. [63], [64], [65].

The deposition of a granular material generates an initial grain configuration associated with the bulk density ρ or void ratio e. The subsequent application of the consolidating pressure p_c induces variation of density affected by the initial deposition density.

The model formulation is based on the following assumptions;cf. Drescher and Mróz [66]:

i) there exists a critical state for which the effective pressure is uniquely related to the void ratio

ii) the material exhibits *volumetric hardening* associated with actual void ratio and *configuration hardening* associated with evolution toward critical density value corresponding to actual effective pressure.

Let us briefly discuss these model assumptions. Consider the initial state of material after initial deposition associated with the void ratio e_o. Consider also the material at the critical void ratio e_o^*, Fig. 16, and subject two materials to isotropic consolidation. The initial void ratio difference $\psi = e_o^* - e_o$ can be regarded as an irreversible volumetric variation. Hence, after applying pressure along AC, Fig. 16. a), and unloading along CB, the volume variation ψ is generated. We can calculate the fictitious pressure p_c^* associated with the void ratio difference ψ. Writing

$$e - e_o^* = -\lambda \ln \frac{p_c^*}{p_c}, \qquad e - e_o = -k \ln \frac{p_c^*}{p_1} \qquad (5.1)$$

we obtain

$$\frac{p_c^*}{p_1} = \exp\left(\frac{\psi}{\lambda - k}\right) \qquad (5.2)$$

Note that the relation (5.2) occurs for any values of pressures p_c^* and p_1. In particular, $p_1 = p_c$, that is p_c is the actual consolidating pressure, and p_c^* is the fictitious pressure associated with the porosity difference ψ. Similar relations occur for states along the critical state line but differing in void ratio values. Denoting by $\bar{p}$ and $\bar{p}^*$ the respective states on the critical state line, we have, cf. Fig. 16. b)

$$\frac{\bar{p}^*}{\bar{p}} = \exp\left(\frac{\psi}{\lambda - k}\right) \qquad (5.3)$$

The loading surface intersects the critical state line for the pressure value $\bar{p}$ and the irreversible void ratio e^P. As the critical void ratio is $\bar{e}^P = e^P + \psi$, we may assume that

$$p_c = p_c\left(e^p, \psi\right) \tag{5.4}$$

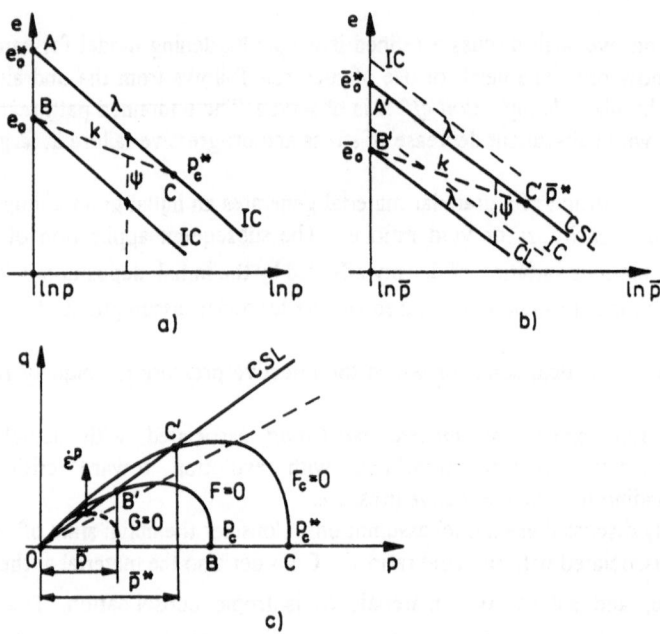

Figure 16. a) Isotropic consolidation lines for sand of initial void ratio e_o
and the critical void ratio e_o^*, b) variation of void ratios on the
critical state line, c) representation in the p, q -plane.

An alternative way to specify the relations (5.4) is possible by introducing the respective pressures. Consider a surface similar to the loading surface but passing through the point $\bar{p}^*, \bar{q}^*$ on the critical state line and specifying the pressure p_c^* on the pressure axis, Fig. 16. c). As $\bar{p}^*$ and $\bar{p}$ are uniquely related to $\bar{e}^*$ and $\bar{e}$, we can write

$$p_c = p_c\left(e^p, \frac{\bar{p}^x}{\bar{p}}\right), \quad \text{or} \quad p_c = p_c\left(e^p, \frac{p_c^*}{p_c}\right) \tag{5.5}$$

where p_c^* / p or $\bar{p}_c^* / \bar{p}$ are expressed in terms of ψ by the relations (5.2) and (5.3).

The initial *configuration surface*

$$F_c = \left(p, q, p_c^*\right) = 0 \tag{5.6}$$

was introduced by Mróz and Pietruszczak [35], Jarzębowski and Mróz [61] and used in model formulation for sands of varying initial densities. The parameter ψ was utilized by Wood et al. [62] in constructing a simplified model for drained sand response.

Denoting the specific volume by $v = 1 + e$, void ratio by e, we have the relations for rates

$$\dot{e} = -(1+e)\dot{\varepsilon}_v, \qquad \dot{e}^P = -(1+e)\dot{\varepsilon}_v^P, \qquad \dot{e}^e = -(1+e)\dot{\varepsilon}_v^e$$
$$\dot{v} = -v\dot{\varepsilon}_v, \qquad \dot{v}^P = -v\dot{\varepsilon}_v^P, \qquad \dot{v}^e = v\dot{\varepsilon}_v^e$$

(5.7)

where the usual notation used for triaxial testing is applied, thus

$$p = \frac{1}{3}(\sigma_1 + 2\sigma_2), \quad q = \sigma_1 - \sigma_2, \quad \varepsilon_v = \varepsilon_1 + 2\varepsilon_2, \quad \varepsilon_q = \frac{2}{3}(\varepsilon_1 - \varepsilon_2)$$

(5.8)

The total strain rates are

$$\dot{\varepsilon}_v = \dot{\varepsilon}_v^e + \dot{\varepsilon}_v^P, \qquad \dot{\varepsilon}_q = \dot{\varepsilon}_q^e + \dot{\varepsilon}_q^e$$

(5.9)

Assume the yield condition and plastic potential in the form identical to that assumed by Drescher et al. [60], namely

$$F(p, q, p_c) = |q| - Lp\sqrt{3}\left(1 - \frac{p}{p_c}\right)^{\frac{1}{2}} \leq 0$$

$$G(p, q, p_l) = |q| - Mp\sqrt{3}\left(1 - \frac{p}{p_l}\right)^{\frac{1}{2}} \leq 0$$

(5.10)

where the parameter p_l is determined from the condition that $G = 0$ and $F = 0$ correspond to the actual stress point, thus

$$1 - \frac{p}{p_l} = \frac{L^2}{M^2}\left(1 - \frac{p}{p_c}\right)$$

(5.11)

The flow rule now provides

$$\dot{\varepsilon}_q^P = \dot{\lambda}\frac{\partial G}{\partial q} = \dot{\lambda}\,\text{sgn}(q)$$

$$\dot{\varepsilon}_v^P = \dot{\lambda}\frac{\partial G}{\partial p} = \frac{3}{2}\dot{\lambda}\frac{M^2 - \eta^2}{\eta} = 2\dot{\lambda}\frac{M^2}{L\sqrt{3}}\frac{1 - 3\frac{L^2}{M^2}\left(1 - \frac{p}{p_c}\right)}{\left(1 - \frac{p}{p_c}\right)^{\frac{1}{2}}}$$

(5.12)

where $\dot{\lambda} > 0$ and $\eta = q/p$. The dilatancy rule follows from (5.12), thus

$$\frac{\dot{\varepsilon}_v^P}{\dot{\varepsilon}_q^P} = \frac{3}{2}\frac{M^2 - \eta^2}{\eta}$$

(5.13)

The zero dilatancy line is specified by the relations

$$\frac{\bar{p}}{p_c} = 1 - \frac{M^2}{3L^2}, \qquad \bar{q} = M\bar{p} \tag{5.14}$$

where $\bar{p}, \bar{q}$ are points on zero dilatancy line which is assumed to coincide with the critical state line.

Let us now discuss the hardening rule. In [60] it was assumed, similarly as for clays that $p_c = p_c(v^p)$, that is the effect of initial density was neglected. Now, however, we assume the following hardening rule

$$\dot{p}_c = p_o' \dot{v}^p + \left| \dot{\varepsilon}^p \right| p_c \left\langle \frac{p_c^*}{p_c} - 1 \right\rangle^n \tag{5.15}$$

or

$$\dot{p}_c = p_c' \dot{v}^p + \left| \dot{\varepsilon}^p \right| p_c \left[\exp \frac{\langle \psi \rangle}{\lambda - k} - 1 \right]^n \tag{5.16}$$

where

$$\dot{\varepsilon}^p = \left[\alpha^2 \left(\dot{\varepsilon}_q^p \right)^2 + \left(\dot{\varepsilon}_v^p \right)^2 \right]^{\frac{1}{2}}, \qquad \alpha > 0 \tag{5.17}$$

and α is a weighting factor. The symbols p_o', p_c' denote derivatives with respect to v^p.

The first term of (5.16) or (5.17) corresponds to the usual volumetric hardening the second term represents the configuration hardening associated with evolution toward the critical state. Using the flow rule (5.12), we have

$$\left| \dot{\varepsilon}^p \right| = \dot{\lambda} K(\eta), \quad K(\eta) = \left[\alpha^2 + \frac{9}{4} \frac{\left(M^2 - \eta^2 \right)^2}{\eta^2} \right]^{\frac{1}{2}} \tag{5.18}$$

The consistency condition

$$\frac{\partial F}{\partial p} \dot{p} + \frac{\partial F}{\partial q} \dot{q} + \frac{\partial F}{\partial p_c} \dot{p}_c = 0 \tag{5.19}$$

combined with (5.16) and (5.17) now provides

$$\dot{\lambda} = \frac{\frac{\partial F}{\partial p} \dot{p} + \frac{\partial F}{\partial q} \dot{q}}{H}, \qquad H = \frac{\partial F}{\partial p_c} \left[p_c' v \frac{\partial G}{\partial p} - K(\eta) p_c f(\psi) \right] \tag{5.20}$$

where

$$f(\psi) = \left[\exp\frac{\langle\psi\rangle}{\lambda - k} - 1 \right]^n \tag{5.21}$$

The elastic strain rates are

$$\dot{\varepsilon}_v^e = \frac{k}{v}\frac{\dot{p}}{p}, \qquad \dot{\varepsilon}_q^e = \frac{1}{3G}\dot{q} \tag{5.22}$$

so the incremental constitutive relations are

$$\dot{\varepsilon} = \left(\mathbf{D}^e + \mathbf{D}^p\right)\dot{\sigma} = \mathbf{D}\dot{\sigma} \tag{5.23}$$

or

$$\begin{bmatrix} \dot{\varepsilon}_q \\ \dot{\varepsilon}_v \end{bmatrix} = \begin{bmatrix} \dfrac{1}{3G} + \dfrac{1}{H}\dfrac{\partial G}{\partial q}\dfrac{\partial F}{\partial q} & \dfrac{1}{H}\dfrac{\partial G}{\partial q}\dfrac{\partial F}{\partial p} \\[2ex] \dfrac{1}{H}\dfrac{\partial G}{\partial p}\dfrac{\partial F}{\partial q} & \dfrac{k}{vp} + \dfrac{1}{H}\dfrac{\partial G}{\partial p}\dfrac{\partial F}{\partial p} \end{bmatrix} \begin{bmatrix} \dot{q} \\ \dot{p} \end{bmatrix} \tag{5.24}$$

Since

$$p_c' = -\frac{p_c}{\lambda - k}, \qquad \frac{\partial F}{\partial p_c} = -\frac{1}{GL^2}\frac{\left(3L^2 - \eta^2\right)^2}{\eta} \tag{5.25}$$

the hardening modulus H equals

$$H = \frac{1}{GL^2}\frac{\left(3L^2 - \eta^2\right)^2}{\eta}\left[\frac{p_c v}{\lambda - k}\frac{3}{2}\frac{M^2 - \eta^2}{\eta} + K(\eta)p_c f(\psi) \right] \tag{5.26}$$

The asymptotic state $H = 0$ occurs when

$$\frac{\bar{p}}{p_c} = 1 - \frac{M^2}{3L^2}, \quad \eta = M, \quad \bar{q} = M\bar{p}, \quad \psi = 0, \quad f(\psi) = 0 \tag{5.27}$$

so the critical state line coincides with the zero dilatancy line.

5.1 Analysis of undrained tests

Consider the undrained test of fully saturated sand. We assume that

$$\dot{\varepsilon}_v = 0 \tag{5.28}$$

The constitutive equations (5.24) now provide

$$\frac{dq}{dp} = \frac{\dot q}{\dot p} = -\frac{H\dfrac{k}{vp} + \dfrac{\partial G}{\partial p}\dfrac{\partial F}{\partial p}}{\dfrac{\partial G}{\partial p}\dfrac{\partial F}{\partial q}}$$

$$\frac{dq}{d\varepsilon_q} = \frac{\dot q}{\dot \varepsilon_q} = -\frac{H\dfrac{k}{vp} + \dfrac{\partial G}{\partial p}\dfrac{\partial F}{\partial p}}{\dfrac{kH}{3Gvp} + \dfrac{1}{3G}\dfrac{\partial G}{\partial p}\dfrac{\partial F}{\partial p} + \dfrac{k}{vp}\dfrac{\partial G}{\partial q}\dfrac{\partial F}{\partial q}}$$

(5.29)

The maximal point on the q-p path and the limit point on the $q - \varepsilon_q$ path are specified by the same condition

$$H\frac{k}{vp} + \frac{\partial G}{\partial p}\frac{\partial F}{\partial p} = 0$$

(5.30)

Neglecting configuration hardening, $\psi = 0$, in (5.26) we obtain

$$\frac{p_m}{p_c} = \frac{2(\lambda - k)}{3\lambda - 2k}, \qquad \eta_m = \frac{q_m}{p_m} = L\sqrt{\frac{3\lambda}{3\lambda - 2k}}$$

(5.31)

where p_m, q_m are the limit state values on the undrained path

$$q = Lp\sqrt{3}\left[\left(1 - \left(\frac{p}{p_{ci}}\right)^{\frac{\lambda}{\lambda - k}}\right)\right]^{\frac{1}{2}}$$

(5.32)

where p_{ci} is the initial consolidating pressure.

For the combined hardening mode, the first Eq. (5.28) takes the form

$$\frac{dq}{dp} = -L\sqrt{3}\frac{\dfrac{3\lambda - 2k}{\lambda - k}\dfrac{p}{p_c} - 2}{\left(1 - \dfrac{p}{p_c}\right)^{\frac{1}{2}}} - \frac{1}{3}\frac{k}{v}\frac{1}{M^2 - \eta^2}K(n)f(\psi)$$

(5.33)

Figure 17 a) presents the predicted response curves for isotropic consolidation starting from different initial pressures and different initial void ratios. The undrained response is illustrated in Figs. 17 (b,c). The undrained curves $q - \varepsilon_q$ exhibit maxima with subsequent softening and consecutive hardening near the critical state line.

The following material parameters were selected: $\lambda = 0.1$, $k = 0.05$, $m = 100$, $n = 2$, $L = 0.7$, $M = 1.1$

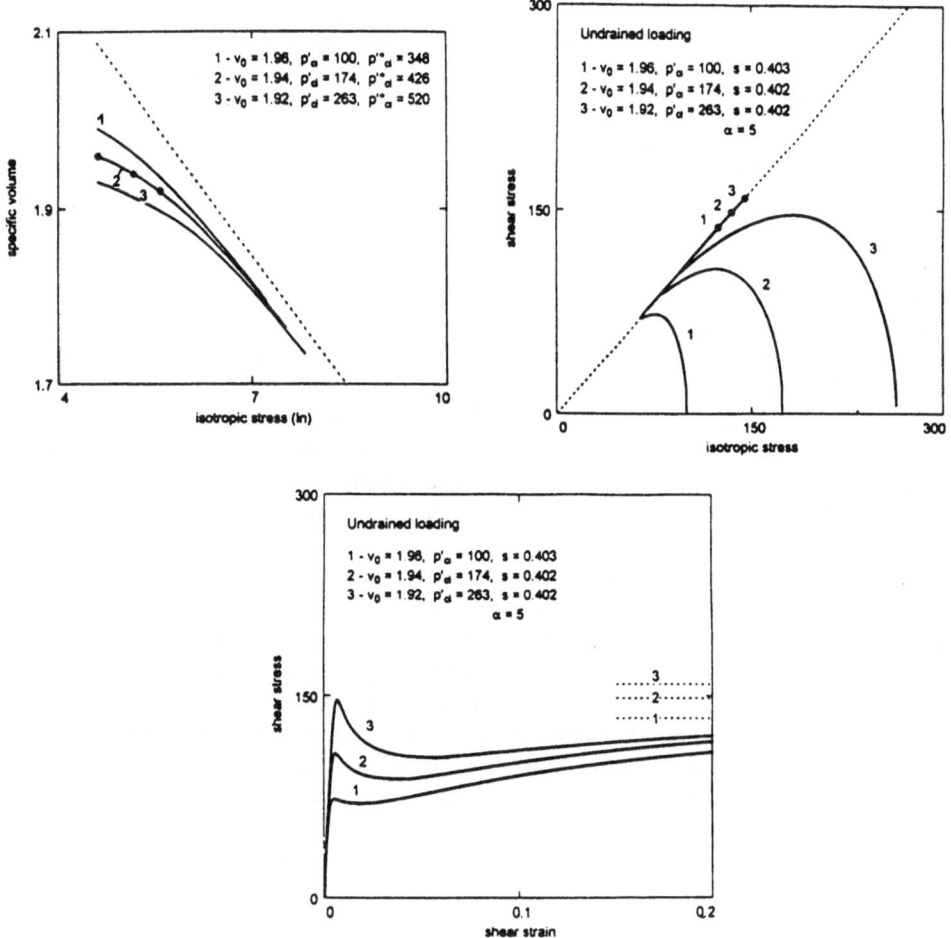

Figure 17. Predicted response curves: a) isotropic consolidation, b) undrained p-q paths, c) deformation response for undrained compression

It is seen that all stress paths exhibit a turn above the $\eta = M$ line, beyond which the paths gradually approach this line from above. The point at which the stress path reaches eventually the $\eta = M$ line depends on the initial specific volume. The lower is this volume the more distant is the terminal point. This matches qualitatively the experimentally observed response of sands with increasing deposition densities. Instead of (5.15), a modified evolution rule was applied, namely

$$\dot{p}_c = \frac{p_c \, v}{\lambda - k} \dot{\varepsilon}_v^p + \left| \dot{\varepsilon}^p \right| p_c \langle f(\beta) \rangle,$$

$$f(\beta) = m \left(\frac{p_c^*}{p_c} - \eta \, s - 1 \right)^n$$

(5.34)

where m, n, s are constant parameters. The parameter s is used to assure continuation of the deformation process near the critical state line and avoid singularities.

6. Viscoplastic model for rapid flow of granular materials

6.1 Introduction

Granular material flow in quasistatic conditions is usually described by the elasto-plastic or rigid-plastic material models. The hardening or softening response is associated with density variation and the critical state is specified by a unique relation between the effective pressure and density or void ratio. During the progresive flow, localised deformation zones develop with the associated dilatancy and softening phenomena. The treatment of boundary-value problems assuming existence of localised shear zones was discussed by Michałowski [79] and Mróz and Maciejewski [82] who treated these zones as material interfaces with strain softening. The phenomenon of flow mode switching in problems of punch or wall penetration into a soil was exhibited and confirmed experimentally. The constitutive models for interface layer was proposed by Mróz and Jarzębowski [81],assuming contact state evolution toward a critical steady state and following the previous formulations of anisotropic hardening rules for sands,cf. Mróz and Norris [34], Mróz and Zienkiewicz.[80] and Jarzębowski and Mróz [61].

Besides the plasticity based models, an alternative approach was developed in the description of the rapid granular material flow. The kinetic theory of gases was modified and adapted in deriving the governing equations. Hard sphere models previously developed for dense fluids were used and the dissipative effect due to contact friction was accounted for, Jenkins and Savage [75], Lun et al. [77], Goldsthein and Shapiro [74], et al. The general representation of constitutive models for granular materials was presented by Goodman and Cowin [72, 73] and modified and improved by numerous authors, for example, Yalamanchili et al. [78]. The usual assumption is that Cauchy stress s depends on the volume fraction of particles n gradient ∇v, and the strain rate tensor $\dot{\varepsilon}$, thus

$$\sigma = \sigma(v, \nabla v, \dot{\varepsilon})$$

The representation indicates strain rate dependence and results in specific forms containing numerous material parameters.

In this paper, an alternative approach will be presented. Neglecting density gradient dependence, we assume that the equilibrium states correspond to the critical state model equations, but the viscoplastic flow develops for stress states exceeding the instantaneous yield surface. The static and dynamic dilatancy effects will be predicted and identification of material parameters will be obtained from tests of rapid shear of granular materials in annular shear apparatus. The model can be applied to study flows of granular materials or suspensions and also in the analysis of post-liquefaction response of soils when rapid shear deformation occurs in localized zones.

6.2. Critical state model of the interface layer.

Consider an interface surface element S specified by the unit normal vector **n** . The traction vector **t** = s**n** can be decomposed into normal and tangential components

$$\mathbf{t} = \mathbf{t}_N + \mathbf{t}_T \tag{6.1}$$

where

$$t_N = (n \otimes n)t = (t \cdot n)n = -\sigma_n n$$

(6.2)

$$t_T = (1 - n \otimes n)t = t - t_N n = \tau_n m$$

where m is the unit vector in the tangential direction. Here σ_n denotes the compressive normal stress to the interface and τ_n is the shear stress. Denote by $\bar{v}$
the relative velocity at the interface

$$\bar{v} = v^+ - v^-$$

(6.3)

where v^+ and v^- denote the velocities on both sides of the interface. Decompose $\bar{v}$ into normal and tangential components, namely

$$\bar{v}_N = (n \otimes n)\bar{v} = (\bar{v} \cdot n)n = \bar{v}_N n$$

(6.4)

$$\bar{v}_T = (1 - n \otimes n)\bar{v} = \bar{v}_N n = \bar{v}_T m$$

Let us denote the compactive normal velocity component by $v_N = -\bar{v}_N$. We use the notation $v_N = -\bar{v}_N$, $v_T = \bar{v}_T$ and decompose the relative velocities into elastic and inelastic components, thus

$$v_N = v_N^e + v_N^p, \quad v_T = v_T^e + v_T^p$$

(6.5)

The interface surface element can now be associated with the interface layer of thickness h several times greater than the average grain size, but small compared to a typical dimension of the problem.
Denote by $\sigma_n, \tau_n, \sigma_t$ the stress components within the layer and by $\dot{\varepsilon}_n, \dot{\gamma}_n, \dot{\varepsilon}_t$ the respective strain rate components. The external components σ_n, τ_n are generated by the traction vector t, the internal component σ_t acts within the interface layer on the plane normal to the interface. Similarly, the external strain rate components $\dot{\varepsilon}_n, \dot{\gamma}_n$ are specified in terms of relative velocity vector, namely

$$\dot{\gamma}_n = \frac{v_T}{h} , \quad \dot{\varepsilon}_n = \frac{v_N}{h}$$

(6.6)

and the internal component $\dot{\varepsilon}_t$ specifies the stretching of the interface layer. Let us note that the external strain rate components do not induce any interface stretching. The specific rate of dissipation per unit area of the interface equals

$$D = (\sigma_n \dot{\varepsilon}_n + \tau_n \dot{\gamma}_n)h = \sigma_n v_N + \tau_n v_T$$

(6.7)

In formulating the constitutive equations of the interface layer, two approaches can be distinguished
i)   the response of the layer is governed only by the external stress components associated with tractions σ_n, τ_n and the conjugate strain rates $\dot{\varepsilon}_n, \dot{\gamma}_n$, or relative velocities v_N, v_T.
ii) the response of the layer is also affected by the internal stress and strain components σ_t, ε_t, playing the role of internal state variables.
In this paper, we shall be concerned with interfaces of the first kind for which the surface tractions and relative velocities specify the interface layer response.

Let us now discuss the density hardening model implying existence of the critical state. In the plane σ_n, τ_n consider a set of elliptical yield loci depending on varying material density and generated by similarity mapping with respect to the origin O, Fig.18. The yield locus has the form

$$F(\sigma_n, \tau_n, c, a) = \left[(\sigma_n - c)^2 + \frac{\tau_n^2}{m^2} \right]^{\frac{1}{2}} - a = 0 \qquad (6.8)$$

where m, a, and c are the geometrical parameters, namely a is the larger semiaxis, c is the position of ellipse centre and m is the ratio of semiaxes. The similarity coefficient α is specified as follows

$$\alpha = \frac{a-c}{a+c} \quad \text{or} \quad c = \frac{1-\alpha}{1+\alpha} a = ka, \qquad 0 \le k \le 1 \qquad (6.9)$$

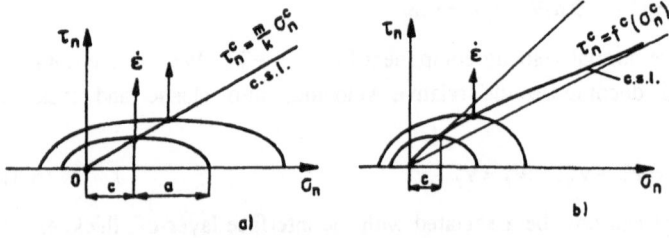

Figure 18. Yield surface and critical state lines for varying material density :
a) c=ka, k=const. b) c = k(ρ)a. The notation c.s.l. means critical state line.

Consider first the rigid-plastic model and assume that m=const. and α=const. Then $c = ka$ and only one material function $a = a(\rho)$ specifies the yield condition . Here ρ denotes the bulk density of a granular material. Note that for $\alpha = 0$ there is $k = 1$ and $c = 0$ so the yield ellipses pass through the origin O. On the other hand, when $\alpha = 1$, then $k = 0$, $c = a$ and ellipse centres coincide with the origin O.

The associated flow rule now provides

$$\dot{\gamma}_n = \frac{v_T}{h} = \frac{1}{h} \dot{\lambda} \frac{\partial F}{\partial \tau_n} = \frac{\dot{\lambda}}{h} \frac{\tau_n}{am^2}, \qquad \dot{\lambda} > 0$$

$$ \tag{6.10}$$

$$\dot{\varepsilon}_n = \frac{v_N}{h} = \frac{1}{h} \dot{\lambda} \frac{\partial F}{\partial \sigma_n} = \frac{\dot{\lambda}}{h} \frac{(\sigma_n - ka)}{a}$$

where

$$\dot{\lambda} = \left[v_N^2 + m^2 v_T^2 \right]^{\frac{1}{2}} = h \left[\dot{\varepsilon}_n^2 + m^2 \gamma_n^2 \right]^{\frac{1}{2}} \qquad (6.11)$$

The dissipation function has the form

$$D = a(\rho)\left\{ kv_N + \left[v_N^2 + m^2 v_T^2\right]^{\frac{1}{2}} \right\}$$ (6.12)

where $a(\rho)$ is a function of density that can be assumed

$$a = a_m \left(\frac{\rho - \rho_{min}}{\rho_{max} - \rho_{min}}\right)^p$$ (6.13)

where ρ_{min} and ρ_{max} denote minimum and maximum density values for a granular material and a_m corresponds to the maximal density ρ_{max}. The rate of variation of a is expressed as follows

$$\dot{a} = a'(\rho)\rho\dot{\varepsilon}_n = a'(\rho)\rho\lambda \frac{\partial F}{\partial \sigma_n}$$ (6.14)

where $a' = da/d\rho > 0$. Using (14), the consistency condition imposed on (6.8) provides the expression of $\dot{\lambda}$ in terms of the stress rate, thus

$$\dot{\lambda} = \frac{\dfrac{\partial F}{\partial \tau_n}\dot{\tau}_n + \dfrac{\partial F}{\partial \sigma_n}\dot{\sigma}_n}{\left(\dfrac{\partial F}{\partial \sigma_n} k + 1\right) a'(\rho)\dfrac{\partial F}{\partial \sigma_n}} = \frac{\dfrac{\tau_n}{m^2}\dot{\tau}_n + (\sigma_n - ka)\dot{\sigma}_n}{H}$$ (6.15)

where

$$H = \frac{\left[k\sigma_n + a(1 - k^2)\right]\left[\sigma_n - ka\right]}{a}\rho a'(\rho)$$ (6.16)

is the hardening modulus. Let us note that $H > 0$ when $\sigma_n > ka$, $H < 0$ when $\sigma_n < ka$ and the critical state line $H=0$ corresponds to $\sigma_n^c = c = ka$, $\tau_n^c = ma = \dfrac{m}{k}\sigma_n^c$. In Fig.18 it is represented by a straight line.

A more general model can be proposed by assuming that k depends on the varying material density, thus

$$c = k(\rho)a(\rho), \quad \sigma_n^c = c(\rho), \quad \tau_n^c = ma(\rho)$$ (6.17)

Assume that $k = \rho/\rho_{max}$. In view of (6.13), the critical state line is then described by the equation

$$m\sigma_n = \tau_n \left[\frac{\rho_{min}}{\rho_{max}} + \left(1 - \frac{\rho_{min}}{\rho_{max}}\right)\left(\frac{\tau_n}{ma_o}\right)^{\frac{1}{p}}\right]$$ (6.18)

and is represented by a curve in the (τ_n, σ_n)-plane, Fig.18.b.

6.3. Dynamic dilatancy description

Consider now the constitutive relations for a granular material applicable in the range of fairly large rates of flow where the dynamic collision forces between grains play equal or even major role with respect to contact forces. The granular material then exhibits viscous effects with stress dependent on rate of straining and dilatancy increasing with the strain rate. The experiments carried out by Savage and Mc Keown [84] in the rotating shear apparatus clearly exhibited this effect. The constant volume test demonstrated that the normal stress increases with the rate of straining. On the other hand, the tests carried at constant normal stress indicated growth of volume with the rate of shearing. To describe this dynamic dilatancy effect, let us assume the viscoplastic material model, regarding the critical state model of the previous section as the static model representing material response at vanishing strain rate.

Consider now the viscoplastic model for which the static yield condition is specified by (6.8) Assume the dynamic yield surface in the form

$$F_d = \left[\left(\sigma_n - c_d \right)^2 + \frac{\tau_n^2}{m^2} \right]^{\frac{1}{2}} - a_d = 0 \qquad (6.19)$$

This surface has a similar shape to the static surface (6.8) but its centre is now at $(c_d, 0)$ and its major semiaxis equals a_d. Assuming the similarity mapping (6.9) with constant k, we have $c_d = k a_d$, so a_d can be assumed as the size parameter of the dynamic yield surface. Assume the viscoplastic flow rule in the form

$$\mu \dot{\varepsilon}_n = \frac{\partial F_d}{\partial \sigma_n} = \frac{\sigma_n - c_d}{a_d} \left(\frac{a_d}{a} - 1 \right)^n, \quad \mu \dot{\gamma}_n = \frac{\partial F_d}{\partial \tau_n} = \frac{\tau_n}{m^2 a_d} \left(\frac{a_d}{a} - 1 \right)^n, \qquad (6.20)$$

where $\mu = \mu h$ and μ, n are material parameters. Note that $\dot{\varepsilon}_n = \dot{\gamma}_n = 0$ when $a_d = a$, and we pass to the static case. From (6.20) it follows that

$$\frac{a_d}{a} = 1 + \tilde{\mu}^{\frac{1}{n}} \left(\dot{\varepsilon}_n^2 + m^2 \dot{\gamma}_n^2 \right)^{\frac{1}{2n}} = 1 + \tilde{\mu}^{\frac{1}{n}} \dot{\varepsilon}_e^{\frac{1}{n}} \qquad (6.21)$$

Thus $\dot{\varepsilon}_e = \left(\dot{\varepsilon}_n^2 + m^2 \dot{\gamma}_n^2 \right)^{\frac{1}{2}}$ vanishes when $a_d = a$,

6.3.1   Shear test at constant h and specified $\dot{\gamma}_n$

Consider first the simple shear program for which $\dot{\varepsilon}_n = 0$. This test corresponds to a fixed thickness h of the shear layer when the shear strain is imposed, and the normal stress σ_n is measured. From (6.20) and (6.21), we have

$$\sigma_n^c = c_d = k a_d = ka(\rho) \left[1 + \mu^{\frac{1}{n}} \dot{\varepsilon}_e^{\frac{1}{n}} \right] = ka(\rho) \left[1 + \left(\mu m \dot{\gamma}_n \right)^{\frac{1}{n}} \right]$$

$$\tau_n^c = m a_d = ma(\rho) \left[1 + \left(\mu m \dot{\gamma}_n \right)^{\frac{1}{n}} \right]$$
(6.22)

and the stress state is represented by the static critical state line

$$\tau_n^c = \frac{m}{k}\sigma_n^c \tag{6.23}$$

In other words, the constant volume shear test is represented by the critical state line but the values of both normal and shear stresses increase with the shearing strain rate. To specify the exponent n, the steady shear and normal stresses should be plotted against the shear rate, Fig. 19.

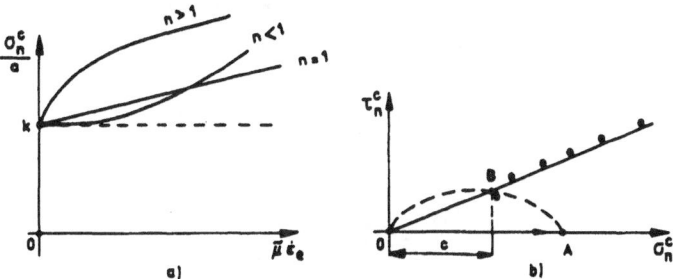

Figure. 19 a) Dependence of the critical normal stress on strain rate, b) experimental verification of the critical state line. Data taken from Savage and Mc Keown [84].

The tests data carried out by Savage and Mc Keown [84] in a rotating shear apparatus for constant height of shear layer provide us the values of n, namely, for glass $n=0.50$, beads for polystyrene beads $n=0.55$. The measured values of shear and normal stress at the steady state provide nearly linear relationship between τ_n^c and σ_n^c, Fig. 19.b).

6.3.2. Shear test at constant σ_n and specified $\dot{\gamma}_n$

Consider now the shear test program carried out at a specified value of $\sigma_n = const.$ but with the varying height of the shear layer. From (6.20), we now obtain

$$\frac{\dot{\varepsilon}_n}{m^2\dot{\gamma}_n} = \frac{\sigma_n - c_d}{\tau_n}, \quad \frac{\tau_n}{m^2} = \mu\dot{\gamma}^n a_d\left(\frac{a_d}{a}-1\right)^{-n} \tag{6.24}$$

The first equation (6.24) specifies the dilatancy relation. When $\sigma_n > c_d$, the shear deformation is accompanied by compaction, $\dot{\varepsilon}_n > 0$. However, for larger strain rates, there is $\sigma_n < c_d$ and $\dot{\varepsilon}_n < 0$. The dynamic dilatancy effect increases with the strain rate and induces the subsequent material softening. Consider, for instance, the material of density ρ_0 for which the static value of a_0 is specified by (6.13) and the corresponding static normal stress equals $\sigma_{n0} = c + a = (k+1)a_0$. From (6.13), we have

$$\rho_o = \rho_{min}\left[1+(\beta-1)\left(\frac{a_0}{a_m}\right)^{\frac{1}{p}}\right] = \rho_{min}\left[1+(\beta-1)\left(\frac{\sigma_{n0}}{\sigma_{nm}}\right)^{\frac{1}{p}}\right] \tag{6.25}$$

where $\beta = p_{\max} / p_{\min}, \sigma_{nm} = (k+1)a_m$.

Assume now that the stress is reduced to the critical value $\sigma_n^c = c_0 = ka_0$ and the granular layer is subjected to shear with the specified strain rate. The respective value of a_d is then specified by (6.21) and $c_d = ka_d$. The variation of τ_n and $\dot{\varepsilon}_n$ can be determined from (6.24). Fig. 20 a) presents typical shear curves at $\sigma_n = c$ obtained for different shear strain rates $\dot{\gamma}_n$.

Due to dynamic dilatancy effect, the shear stress curves exhibit softening and asymptotically tend to critical state. Fig.20 b) presents the respective curves for shear programs carried out at normal stress $\sigma_n > c_0$. For small strain rates the material consolidates, similarly as for quasistatic shear. However, for larger strain rate values the dynamic dilatancy and softening effect occurs.

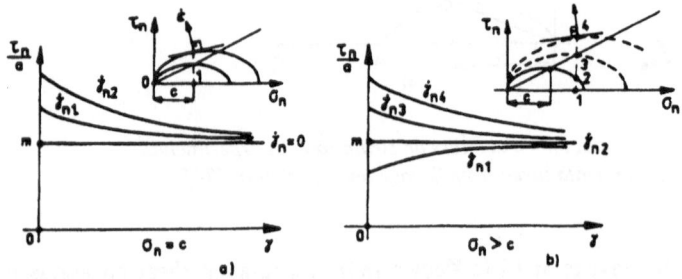

Figure 20. Shear curves for varying strain rates: a) for the case $\sigma_n = c$, b) for the case $\sigma_n > c$

Since the steady state is again represented by the critical state line, we can determine the variation of initial height of the shear layer due to dynamic dilatancy. Since at the critical state there is $\sigma_n = \sigma_n^c = ka_d$, $\dot{\varepsilon}_n = 0$, we can write

$$\frac{\sigma_n}{k} = a_d = a\left[1+\left(\mu m \dot{\gamma}_n\right)^{\frac{1}{s}}\right], \quad ka = \frac{\sigma_n}{1+\left(\mu m \dot{\gamma}_n\right)^{\frac{1}{s}}} \tag{6.26}$$

so the corresponding material density equals

$$\frac{\rho}{\rho_{\min}} = \left[1+(\beta-1)\left(\frac{a}{a_m}\right)^{\frac{1}{s}}\right] = 1+(\beta-1)\frac{\left(\frac{\sigma_n}{ka_m}\right)^{\frac{1}{s}}}{\left[1+\left(\mu m \dot{\gamma}_n\right)^{\frac{1}{s}}\right]^{\frac{1}{s}}} \tag{6.27}$$

For the initial density ρ_0 we have

$$\frac{\rho_0}{\rho_{\min}}\left[1+(\beta-1)\left(\frac{a_0}{a_m}\right)^{\frac{1}{s}}\right] \tag{6.28}$$

In view of (6.27) and (6.28), the ratio of steady state and initial shear layer thickness can be presented in the form

$$\frac{h}{h_o} = \frac{p_o}{p} = \frac{1+(\beta-1)\left(\dfrac{a_0}{a_m}\right)^{\frac{1}{p}}}{1+(\beta-1)\left(\dfrac{a}{a_m}\right)^{\frac{1}{p}}} = \left[1+(\mu m\dot\gamma_n)^{\frac{1}{n}}\right]^{\frac{1}{p}} \frac{1+(\beta-1)\left(\dfrac{a_0}{a_m}\right)^{\frac{1}{p}}}{\left[1+(\mu m\gamma_n)^{\frac{1}{n}}\right]^{\frac{1}{p}}+(\beta-1)\left(\dfrac{\sigma_n}{ka_m}\right)^{\frac{1}{p}}} \qquad (6.29)$$

Equation (6.29) provides us the analytical expression for h/h_o which can be used to verify the model predictions or to identify some material parameters. The measurement should be conducted for varying σ_n and $\dot\gamma_n$ which are now the control variables.

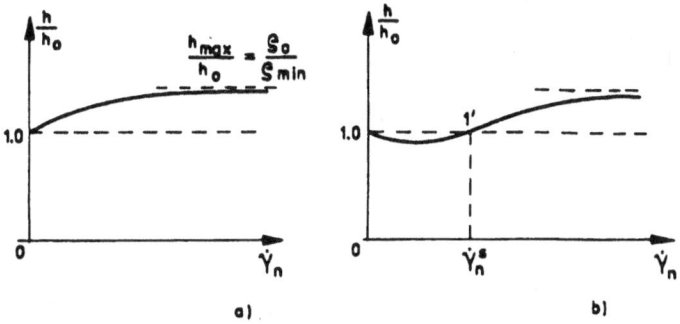

Figure 21.   Layer thickness evolution with increasing strain rate $\dot\lambda_n$ for steady states: a) for $\sigma_n\langle ka_0$,   b) for $\sigma_n\rangle ka_0$.

Fig.21 presents the typical dependence of h/h_o on the rate of shear $\dot\gamma_n$. When $\sigma_n \le c_0 = ka_0$, the layer shearing is accompanied by the continuing dilatancy and h/h_o increases monotonically to its asymptotic value $h_{max}/h_o = \rho_{max}/\rho_{min}$, cf. Fig 21 a). On the other hand, when $\sigma_n 0\rangle ka_0$, then for small rates of shear the layer consolidation occurs and for large values of $\dot\gamma_n$, the material dilates, Fig.21 b). The switching point 1' from consolidation to dilatancy corresponds to the strain rate

$$\mu m\dot\gamma_n^s = \left(\frac{\sigma_n}{ka_0}-1\right)^n \qquad (6.30)$$

The constant normal stress experiments of shear of suspensions of 2 mm glass spheres in water/glycerin mixture for three values of normal stress were carried out by Kyotomaa and Prasad [76]. Fig.5 presents the solid fraction variation for increasing rates of shear. It is seen that the solid fraction first increases and for a sufficiently large strain rate value, it starts to decrease. The increasing normal pressure shifts the diagrams upwards, thus generating increasing solid fraction values. These experimental data correspond qualitatively to the present model predictions, though the full identification of model parameters is difficult in view of incomplete empirical data.

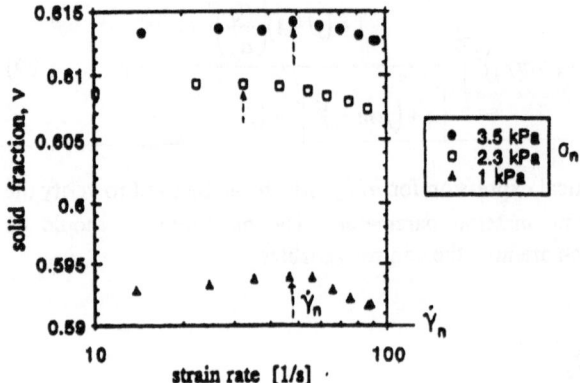

Figure 22. Variation of solid fraction n with increasing rates of shear at constant σ_n (after Kyotomaa and Prasad [76]).

6.3.3. Composite shear layer, $h = h_0 + h_s = const$

Consider now a more complex case when the granular shear layer interacts with an elastic layer of specified normal compliance K_n. The thickness of the granular layer is denoted by h_m and that of elastic layer by h_s. The initial state is specified by the normal stress σ_{n0} and the thickness $h_0 = h_{m0} + h_{s0}$. The subsequent shearing process occurs at constant $h = h_0$, so we have

$$\Delta h_m + \Delta h_s = 0 \tag{6.31}$$

where Δ denote the thickness increment.
Denoting the normal stress during shearing process by σ_{nd}, we have

$$\frac{\Delta h_s}{h_{s0}} = \frac{\Delta h}{h_{m0}} \frac{h_{m0}}{h_{s0}} = \frac{\sigma_{nd} - \sigma_{n0}}{K_n} \tag{6.32}$$

or

$$\frac{\Delta h_s}{h_{m0}} = \frac{\sigma_{nd} - \sigma_{n0}}{\overline{K}_n}, \qquad \overline{K}_n = K_n \frac{h_{m0}}{h_{s0}} \tag{6.33}$$

Using the relations

$$\frac{\rho}{\rho_{min}} = 1 + (\beta - 1)\left(\frac{a}{a_m}\right)^{\frac{1}{p}}, \qquad \frac{\rho_0}{\rho_{min}} = 1 + (\beta - 1)\left(\frac{a_0}{a_m}\right)^{\frac{1}{p}} \tag{6.34}$$

and (6.31), we obtain

$$\frac{h_m}{h_{om}} = \frac{p_0}{p} = \frac{1+(\beta-1)\left(\dfrac{a_0}{a_m}\right)^{\frac{1}{p}}}{1+(\beta-1)\left(\dfrac{a}{a_m}\right)^{\frac{1}{p}}} = 1+\frac{\Delta h_m}{h_{m0}} = 1-\frac{\sigma_{nd}-\sigma_{ns}}{\overline{K}} \tag{6.35}$$

Equation (6.35) specifies the normal stress σ_{nd} during the shearing process. Let us note that now $\sigma_{nd} = \sigma_{nd}\left(\sigma_{n0},p_0,\dot{\gamma}_n,\overline{K}\right)$. Equation (6.35) can be written in a more explicit form

$$\left[1+(\mu m\dot{\gamma}_n)^{\frac{1}{n}}\right]^{\frac{1}{p}}\frac{1+(\beta-1)\left(\dfrac{a_0}{a_m}\right)^{\frac{1}{p}}}{\left[1+(\mu m\dot{\gamma}_n)^{\frac{1}{n}}\right]^{\frac{1}{p}}+(\beta-1)\left(\dfrac{\sigma_{nd}}{ka_m}\right)^{\frac{1}{p}}}-1-\frac{\sigma_{nd}-\sigma_{ns}}{\overline{K}} \tag{6.36}$$

7. Anisotropic hardening models

In this section we shall discuss two particular forms of anisotropic hardening rules, applicable to clays and sands.

7.1 Density hardening anisotropic model.

Let us first discuss the anisotropic hardening model used usually for clays. Then, the initial consolidation pressure specifies the void ratio and the plastic void ration can be used as the hardening variable. Details of this model are discussed in [35,37], and in the review article [34].

Consider first the case of a triaxial test when the two principal effective stresses are equal, $\sigma_2 = \sigma_3$. The commonly used stress and strain components then are

$$p = \tfrac{1}{3}(\sigma_1+2\sigma_2), \quad q = \sigma_1-\sigma_2, \quad \varepsilon_v = (\varepsilon_1+2\varepsilon_2), \quad \varepsilon_q = \tfrac{2}{3}(\varepsilon_1-\varepsilon_2) \tag{7.1}$$

Assume that the degree of consolidation of soil is represented by the consolidation surface $F_c = 0$. For clays, this surface is constituted by the initial consolidation process either in natural deposit or in laboratory and it preserves the memory of maximal prestress from the loading history. The yield surface $f_o = 0$ encloses the elastic domain enclosed by the consolidation surface $F_c = 0$. Usually the elastic domain is very small and it may be assumed that this domain shrinks to a point.

In this case, the plastic strain increment occurs for any stress increment, both inside and on the consolidation surface. In view of this assumption, the distinction should be made between the *consolidation process* with the stress point remaining on the consolidation surface $F_c = 0$ and unloading or reloading process within the domain enclosed by $F_c = 0$. Let us discuss consecutively the description of these two types of processes.

Consider first the consolidation process. The stress path is then directed into the exterior of the domain $F_c \leq 0$ and the stress point remains on the consolidation surface. Assume that this surface is represented in the p, q-plane by the portions of two ellipses rotated with respect to the p, q-axes, thus

$$F_c = p^2 + \frac{2(\zeta-1)}{m_c}pq + \frac{(1-2\zeta)^2}{m_c^2}q^2 - 2a_cp - \frac{4(\zeta-1)}{m_c}2a_cq = 0, \quad q > 0,$$

and   (7.2)

$$F_s = p^2 + \frac{2(\zeta-1)}{m_s}pq + \frac{(1-2\zeta)^2}{m_c^2}q^2 - 2a_cp - \frac{4(\zeta-1)}{m_s}2a_cq = 0, \quad q < 0,$$

where $\zeta = a_c/c$, $m_c = \tan\omega_c$, $m_s = \tan\omega_s$ and ω_c, ω_s, are the angles of inclination

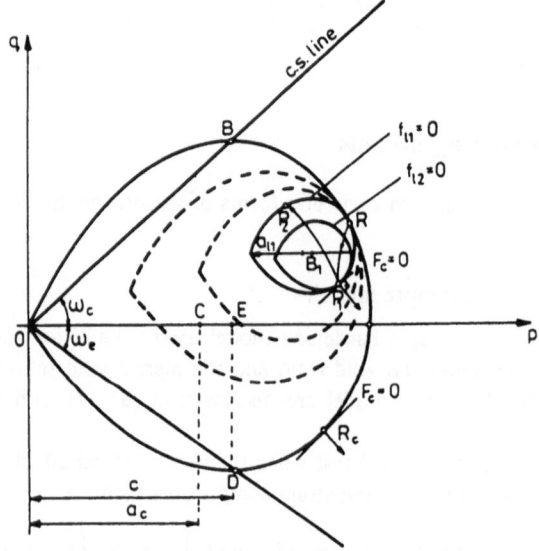

Figure 23. Consolidation and loading surfaces in the p, q-plane for clay

of the critical state lines to the p-axis. The angles ω_c and ω_s satisfy the relations

$$m_c = \frac{q_R}{c} = \frac{6\sin\varphi}{3-\sin\varphi}, \qquad m_s = \frac{q_D}{c} = \frac{6\sin\varphi}{3+\sin\varphi} \qquad (7.3)$$

following from the Coulomb yield condition, where φ denotes the angle of internal friction at failure, Figure 23. Let us note that for $\zeta = 1$ there is $a_c = c$ and the principal axis of each ellipse coincides with the p-axis, thus

$$F_c = (p - a_c)^2 + \frac{q^2}{m_c^2} - a_c^2 = 0 \qquad (7.4)$$

The value of ζ will be identified from the K_o-consolidation test as it was shown in [32] that the use of (7.4) does not provide accurate values of lateral pressures in the uniaxial consolidation. Further, it will be assumed that the consolidation surface represents the isotropic hardening of the material and the surface diameter varies with the irreversible void ratio, thus $F_c(p,q,e^s) = 0$. The parameters $m_c, m_s,$ and ζ will be assumed as constant and $a_c = a_c(e^p)$.

Writing the associated flow rule

$$d\varepsilon_v^p = \frac{1}{K_p} n_p (d\sigma \cdot \mathbf{n}), \qquad d\varepsilon_q^p = \frac{1}{K_p} n_q (d\sigma \cdot \mathbf{n}) \tag{7.5}$$

the hardening modulus K_p is expressed as follows

$$K_p = H_c \left[\left(\frac{\partial E_c}{\partial p} \right)^2 + \left(\frac{\partial F_c}{\partial q} \right)^2 \right]^{1/2},$$

$$H_c = (1+e)\frac{\partial F_c}{\partial e^p} \frac{\partial F_c}{\partial p} = -4(1+e)\frac{da_c}{de^p} \left(p + \frac{2(\zeta-1)}{m_c} \right) \left(p + \frac{\zeta-1}{m_c} q - a_c \right) \tag{7.6}$$

Now, let us discuss the reverse active loading process occurring within the domain $F_c \leq 0$. Consider the situation presented in Figure 23 where the initial consolidation process is terminated at the point R with the subsequent stress path RP₁ directed into the domain $F_c < 0$. When elastic domain vanishes, the reverse plastic strain occurs along the whole path RP₁ otherwise the yield surface $f_o = 0$ translates with the point P_1.

Let us introduce the concepts of *active loading* and *stress reversal surfaces* that will enable us to make distinction between particular loading events. Assume that after reaching the maximal prestress point R all reverse loading surfaces are tangential at R to the consolidation surface. For the loading programme RP₁, the stress point always remains on the active loading surface $f_n = 0$ translating and expanding so that it always remains tangential to $F_c = 0$ at R. The plastic response at P_1 is described by the flow rule (7.5) applied to $f_n = 0$ with the plastic modulus K_p governed, for instance, by the relation

$$K_p = K_{pc} + \left(K_y - K_{pc} \right) R_1^r \tag{7.7}$$

where

$$R_1 = \frac{a_c - a_n}{a_c - a_o} \tag{7.8}$$

where a_c, a_n, a_o are size parameters of the consolidation, active loading and yield surface, K_y is the initial modulus when the stress point reaches the yield surface, that is $a_n = a_o$ and K_{pc} is the value of the hardening modulus at the associated point R_c' on the consolidation surface for which the normal vector has the same direction as that at P_1. When $a_n = a_c$, that is the stress point reaches the consolidation surface, there is $K_p = K_{pc}$. Further, setting $a_o = 0$ in (7.8), we reduce the yield surface to a point, that is neglect the elastic domain.

The first reverse loading (or unloading) program continues when a_{11} increases, thus $a_{11} > 0$. However, when at P_1 the stress path reverses and is directed into the interior of $f_{11} = 0$, then the second reverse loading (or reloading) event commences and the active loading surface $f_{12} = 0$ for this program is tangential at P_1 to $f_{11} = 0$ and passes through the stress reversal point. The previously active loading surface $f_{11} = 0$ now becomes the *stress reversal surface* and preserves the memory of the stress reversal point P_1. Denoting the size parameter of $f_{12} = 0$ by a_{12}, the rule (7.7) for the hardening modulus variation is still used with R_1 replaced by R_2, where

$$R_2 = \frac{a_c - a_{12}}{a_c - a_o} \tag{7.9}$$

The second reverse loading process continues provided $\dot{a}_{12} > 0$ and $a_{12} < a_{11}$. When at P_2 the surface $f_{12} = 0$ and $f_{11} = 0$ coincide and the stress path is directed in the exterior of $f_{12} = 0$, the second loading event and the stress reversal point P_1 are erased from the material memory. If the stress path $P_1 P_2$ continues and moves beyond the domain enclosed by the consolidation surface, the consolidation process starts again. The memory of R is erased and a new maximal prestress point is created.

The yield and active loading surface are similar to the consolidation surface and are composed of portions of two ellipses

$$f_{1c} = \left(p - \alpha_p\right)^2 + \frac{2(\zeta - 1)}{m_c}\left(q - \alpha_q\right)\left(p - \alpha_p\right) + \frac{(1 - 2\zeta)^2}{m_c^2}\left(p - \alpha_q\right)^2 + 2a_1\left(p - \alpha_p\right)\frac{1 - \zeta}{\zeta}$$

$$+ \frac{2(\zeta - 1)(1 - 2\zeta)}{m_c \zeta}a_1\left(q - \alpha_q\right) + a_1^2\frac{1 - 2\zeta}{\zeta^2} = 0; \qquad q - \alpha_q > 0 \tag{7.10}$$

$$f_{1e} = \left(p - \alpha_p\right)^2 + \frac{2(\zeta - 1)}{m_e}\left(q - \alpha_q\right)\left(p - \alpha_p\right) + \frac{(1 - 2\zeta)^2}{m_e^2}\left(p - \alpha_q\right)^2 + 2a_1\left(p - \alpha_p\right)\frac{1 - \zeta}{\zeta}$$

$$+ \frac{2(\zeta - 1)(1 - 2\zeta)}{m_e \zeta}a_1\left(q - \alpha_q\right) + a_1^2\frac{1 - 2\zeta}{\zeta^2} = 0; \qquad q - \alpha_q < 0 \tag{7.11}$$

where α_p, α_q define the position of the centre point B_1, and the parameters ζ, m_c, m_e are the same as those for the consolidation surface. The function $a_c = a_c(e^p)$ is assumed in the usual exponential form

$$a = a_1 \exp\left(\frac{e_i^p - e^p}{\lambda - k}\right) \tag{7.12}$$

and the curves of isotropic consolidation and unloading are specified as follows

$$p = p_i \exp\left(\frac{e_i - e}{\lambda}\right), \qquad p = p_o \exp\left(\frac{e_o^* - e^*}{k}\right) \tag{7.13}$$

where p_i, e_i are the initial values of p and e in the consolidation process whereas p_o, e_o are the initial values for the unloading process; k and λ are the material parameters. The elastic strain increments are specified by the relations

$$d\varepsilon_v^p = \frac{dp}{K}, \qquad d\varepsilon_q^p = \frac{dq}{3G} \tag{7.14}$$

For any position P_1 of the stress point on the active loading surface, there exists an associated point R_c on the consolidation surface having the same direction of the normal vector. Thus, we have

$$p_c - c = \frac{a_c}{a_{l1}}(p_p - \alpha_p), \qquad q_c = \frac{a_c}{a_{l1}}(q_p - \alpha_q) \tag{7.15}$$

where p_c, q_c denote the stress components at the associated point R_c and p_c, q_p denote the components of the loading point P.

Let us discuss first the K_o-*consolidation process* from which we want to identify the parameter ζ. When $\varepsilon_2 = \varepsilon_3 = 0$ and ε_1 is the only non-vanishing strain component, then $d\varepsilon_q = \frac{2}{3}d\varepsilon_v$. Neglecting elastic strains, the uniaxial consolidation is described by the relations

$$\frac{\partial F_c}{\partial q} - \frac{2}{3}\frac{\partial F_c}{\partial p} = 0 \tag{7.16}$$

and

$$S_{ko} = \frac{q}{p}$$

$$= \frac{2m_c^2}{\left\{\left[4(\zeta-1)(3\zeta-m_c)+3\right]^2 + 4m_c\left[(1-2\zeta)^2(m_c+3\zeta-3)-4m_c(\zeta-1)^2\right]\right\}^{1/2} + 4(\zeta-1)(3\zeta-m_c)+3} \tag{7.17}$$

The value of K_o is now expressed as follows

$$K_o = \frac{\sigma_2}{\sigma_1} = \frac{3-S_{ko}}{3+2S_{ko}} \tag{7.18}$$

Thus, for a rigid plastic material the consolidation path will be a straight line in the p, q-plane, Figure 24. Equation (7.17) can be used to identify the parameter ζ by comparing the predicted and measured K_o values. Thus, for the Weald clay, the value $\zeta = 0.85$ is selected to simulate the consolidation path obtained experimentally by Skempton and Sowa [53]. Further examples of application of this model to study undrained compression and extension after anisotropic consolidation can be found in [37,38].

The derivation of constitutive equations for a general stress state can be performed by introducing three invariants $J_m, \bar{\sigma}$, and J_3 of the „translated" effective stress $\sigma_{ij} - \alpha_{ij}$, defined as follows

$$J_m = \tfrac{1}{3}(\sigma_{ii} - \alpha_{ii}), \qquad \bar{\sigma} = \left[\tfrac{1}{2}(s_{ij} - \bar{\sigma}_{ij})(s_{ij} - \bar{\alpha}_{ij})\right]^{1/2}$$

$$J_3 = \tfrac{1}{3}(s_{ij} - \bar{\alpha}_{ij})(s_{ki} - \bar{\alpha}_{ki})(s_{kj} - \bar{\alpha}_{kj}) \tag{7.19}$$

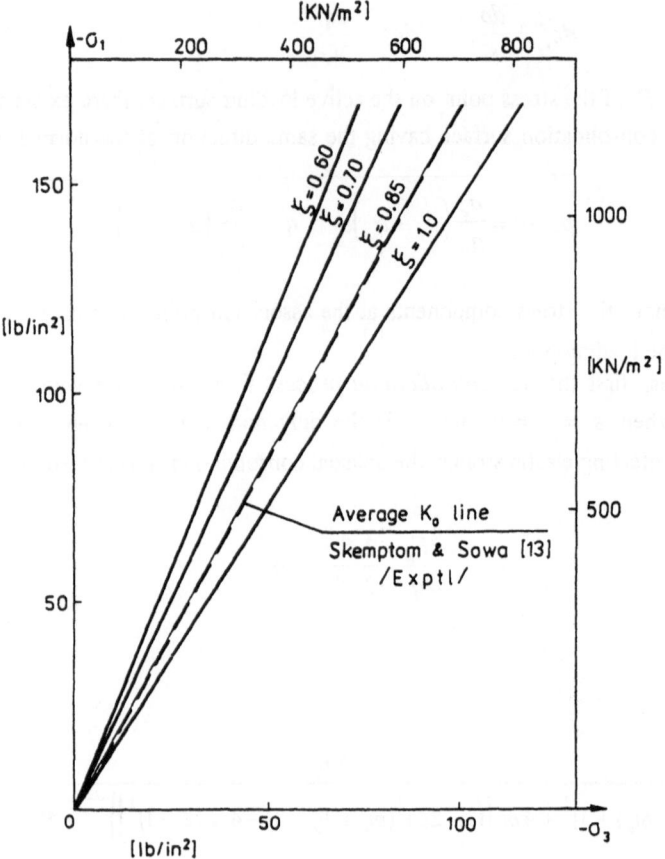

Figure 24. K_o-loading paths for the Weald clay for different values of ζ

where $\bar{\alpha}_{ij}$ and α_{ij} are the deviatoric and spherical components of α_{ij}. The angle measure of the third invariant is specified as follows

$$\theta = \tfrac{1}{3}\arcsin\left(-\frac{3\sqrt{3}}{2}\frac{J_3}{\bar{\sigma}^3}\right), \qquad -\frac{\pi}{\sigma} \le \theta \le \frac{\pi}{\sigma} \tag{7.20}$$

and this angle can be identified on the octahedral plane. The principal stresses are now expressed as follows

$$\sigma_1 = \sigma_m + \frac{2}{\sqrt{3}}\bar{\sigma}\sin\left(\theta + \tfrac{2}{3}\pi\right), \qquad \sigma_2 = \sigma_m + \frac{2}{\sqrt{3}}\bar{\sigma}\sin\theta,$$

$$\sigma_3 = \sigma_m + \frac{2}{\sqrt{3}}\sin\left(\theta + \tfrac{4}{3}\pi\right) \tag{7.21}$$

Assume now that the Π-plane section of the loading surface is described as follows

$$\bar{\sigma} = \bar{\sigma}_{+} g(\theta) \tag{7.22}$$

where $\bar{\sigma}_{+}$ denotes the values of $\bar{\sigma}$ for $\theta = \pi / 6$, $\sigma_2 = \sigma_3$ and $g(\theta)$ is assumed in the form

$$g(\theta) = \frac{2k}{1 + k - (1 - k)\sin 3\theta}, \quad k = \frac{3 - \sin\phi}{3 + \sin\phi} \tag{7.23}$$

so that $I(\pi / 6) = 1$. The equation of the loading surface (7.10, 7.11) can now be rewritten in the form

$$\begin{aligned} f_l &= 3\sin^2\phi\left(\sigma_m - \tfrac{1}{3}\alpha\right)^2 + 6\sin^2\phi\frac{1-\zeta}{\zeta}a_l\left(\sigma_m - \tfrac{1}{3}\alpha\right) \\ &\quad - \sqrt{3}\left(1-\zeta\right)\sin\phi\left(3 - \sin\phi\sin 3\theta\right)\left[\left(\sigma_m - \tfrac{1}{3}\alpha\right) - \frac{1-2\zeta}{\zeta}a_l\right]\bar{\sigma} \\ &\quad + \left[\tfrac{1}{2}\left(1 - 2\zeta\right)3 - \sin\phi\sin 3\theta\right]^2\bar{\sigma}^2 + 3a_l^2\sin^2\varphi\frac{1-2\zeta}{\zeta^2} = 0 \end{aligned} \tag{7.24}$$

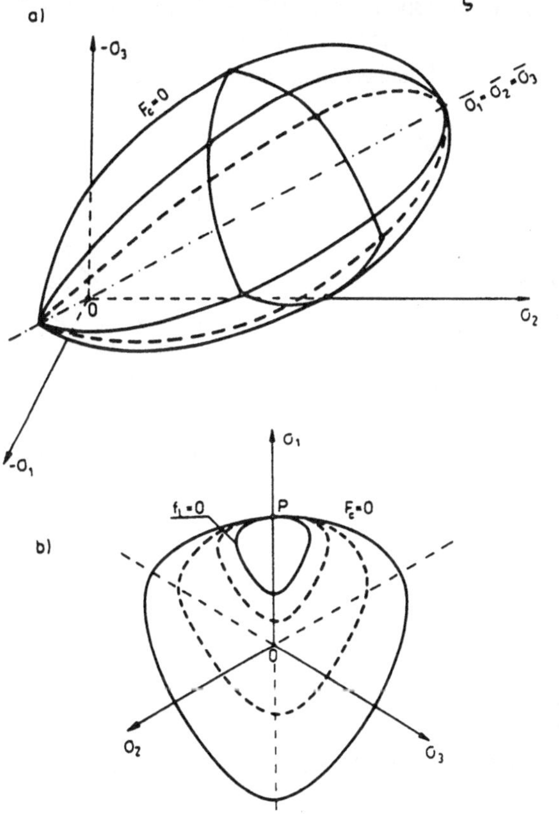

Figure 25. Consolidation and loading surfaces in the principal stress space

and is geometrically represented by a surface with similar cross-sectional shapes in all π-planes, Figure 25 Setting $\bar{\alpha}_{ij} = 0, \alpha_{ij} = -3\left(a_c / \zeta\right), a_l = a_c$, from (7.18) we obtain of the consolidation surface $F_c = 0$

7.2   Anisotropic hardening model for sands.

One of major differences between clays and sands lies in the definition of an initial consolidated state. Whereas for clays, the applied consolidation pressure p_c defines uniquely the void ratio e, for granular materials any degree of compaction may be attained by appropriate deposition of grains under gravity forces. The required consolidation pressure p_c corresponding to a given relative density or void ratio may be very high as compared to actual values of mean pressure and may not be attained at all during tests. The compressibility of sand, as well as its shear response is therefore made to depend on initial density.

   Assume that the degree of initial compaction of sand element is represented in the stress space by a configuration surface

$$F_c(\sigma,\kappa) = 0 \tag{7.25}$$

where κ is the hardening parameter defined by (4.14). Using the associated flow rule, the hardening modulus for stress states corresponding to this surface is expressed by (4.15). In actuality, the stress state corresponding to the configuration surface need not be reached during the deformation process. However, the expression (4.15)   for K_c will be used in formulating the variation rule of the hardening modulus along any stress path.

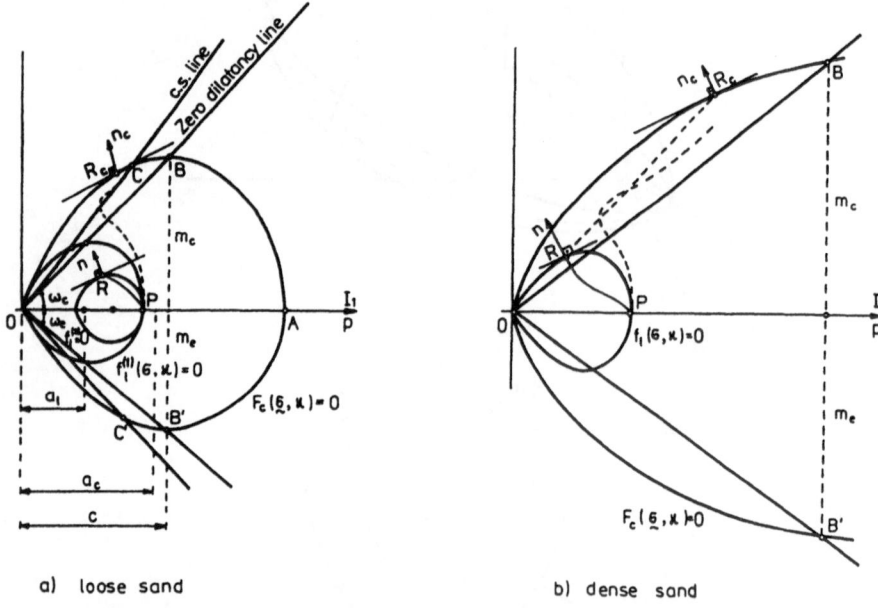

a) loose sand   b) dense sand

Figure 26. Configuration and loading surfaces for: a) loose sand, b)dense sand

Consider now the loading process for a given state of material represented by the configuration surface, Figure 26. It is assumed that the elastic domain enclosed by the yield surface does not exist and the plastic strain increment occurs for any stress increment. The concept of active loading and stress reversal surfaces discussed for clays now still applies, so the actual stress state corresponds always to the active loading surface $f_l^{(i)} = 0$ for the i-th loading event whereas the stress reversal surface $f_r^{(i-1)} = 0$ represents the maximal prestress from the past loading history. The plastic strain rate along the stress path is generated by the flow rule (7.5) associated with the active loading surface. The hardening modulus K varies between its initial high value $K = K_y$ for vanishing diameter or other size parameter a_l of the loading surface and the value $K = K_c$ corresponding to the configuration surface when $a_l = a_c$. The variation of hardening modulus can be described by (7.7) or by a similar relation

$$K_p = K_y + \left(K_{pc} - K_y\right)\left(\frac{a_l}{a_c}\right)$$

(7.26)

where K_{pc} denotes the value of K_p on the configuration surface at the associated point R_c having the same direction of exterior normal as that at the stress point R on the loading surface.

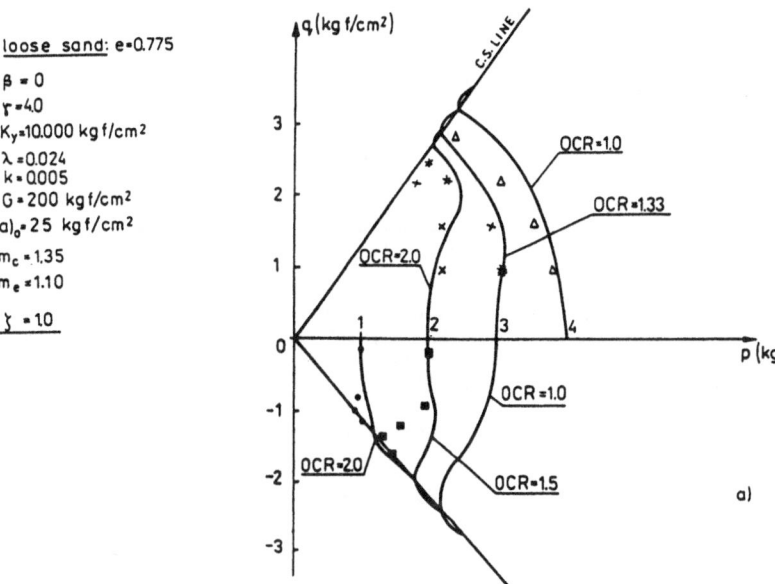

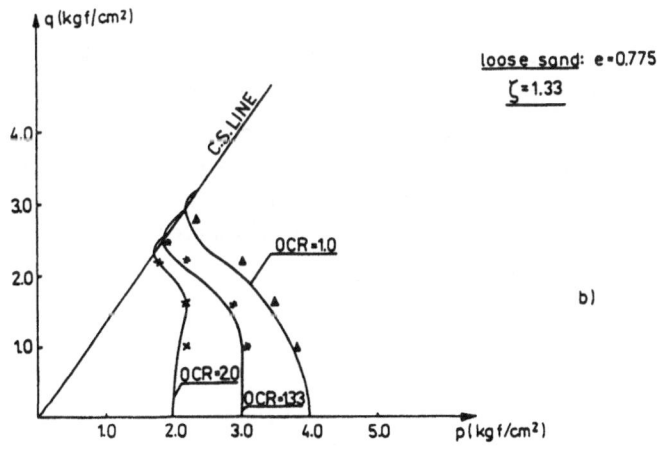

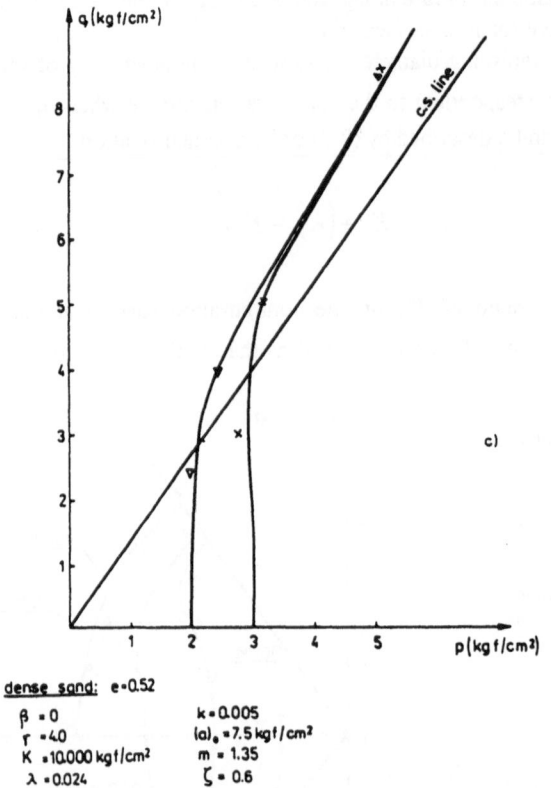

dense sand: e=0.52

$\beta = 0$ $k = 0.005$
$\gamma = 4.0$ $(\alpha)_s = 7.5\,kgf/cm^2$
$K = 10.000\,kgf/cm^2$ $m = 1.35$
$\lambda = 0.024$ $\zeta = 0.6$

Figure 27 a) Predicted and experimental undrained paths for a loose sand
for $\zeta = 1.0$, b) predicted and experimental undrained paths for a loose sand
for $\zeta = 1.3$, c) undrained paths for a dense sand ($\zeta = 0.6$)

Further development parallels closely that presented for clays. In particular, the loading and configuration surfaces are described by (7.2) and the hardening parameter κ is specified by (4.14).

The difference between initially loose and dense sand deposits lies in different size of the configuration surface and different values of ζ, Figure 26. Studying undrained paths in compression and extension is was found that the best description was obtained for $\zeta = 1.3$ for loose sand ($e_o = 0.775$) and $\zeta = 0.6$ for dense sand. Figures 27 a, b, c, illustrate the undrained paths for loose and dense sand predicted by the present model and compared with the experimental data of Ishihara and Okada [54]. Figure 28 presents the predicted and experimental curves q/p versus ε_q for sand initially precompressed and subsequently subjected to undrained compression.

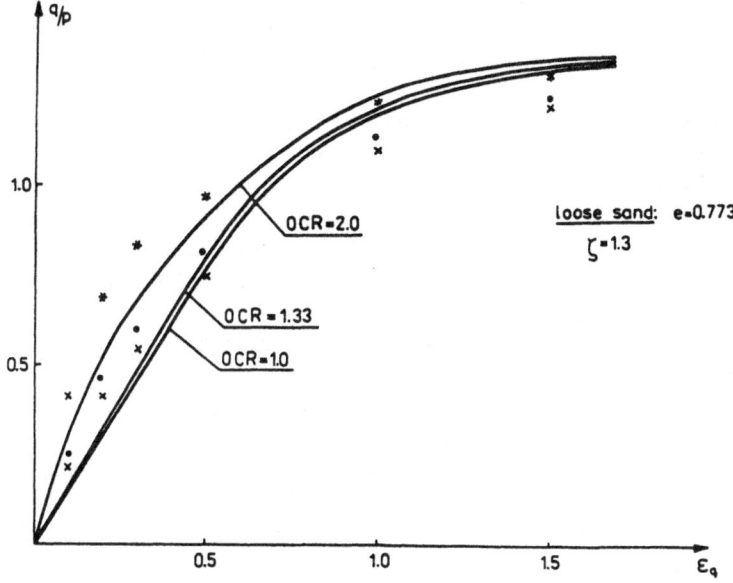

Figure 28. Predicted and experimental q / p − ε_q curves for a loose sand.

REFERENCES

1. J.H. Atkinson and P.L.Bransby , *The Mechanics of Soils. An Introduction to Critical State Soil Mechanics*, McGraw-Hill Publ. Co., 1979.
2. M.I. Baron and I.Nelson, „Application of variable moduli models to the soil behaviour", Int. J. Solids Struct., 7, 389-417, 1971.
3. Z.P.Bazant, „Endochronic inelasticity and incremental plasticity", int. J. Sol. Struct., **14**, 691-714, 1978.
4. Z.P. Bazant and S. Kim, Plastic-fracturing theory for concrete, J. Eng. Mech. Div., ASCE, **105**, No EM 3, 407-428, 1979.
5. P.M. Byrne and T.L.Eldridge, A three parameter dilatant elastic stress-strain model for sand, Proc. Int. Symp. On Numerical Models in Geomechanics, pp.73-80, Zurich, 1982.
6. F.Darve, M.Boulon, and R.Chambon, Loi rheologique incrementale des sols, J. De Mecanique, 17, 679-716, 1978.
7. C.S.Desai, A general basis for yield, failure and potential functions in plasticity, Int. J. Num. Anal. Meth. Geom., 4, 361-375, 1980.
8. C.S. Desai and H.J.Siriwardane, A concept of correction functions to account for non-associative characteristics of geologia media, Int. J. Num. Anal. Meth. In Geom., 4, 377-387, 1980.
9. E.A.Dickin and G.J.W.King, The behavoiur of hyperbolic stress-strain models in triaxial and plane-strain compression, Proc. Intern. Symp. On Numerical Models in Geomechanics, pp. 303-311, A.A.Balkema Publ., Zurich, 1982.

10. P.K.Banerjee and A.S.Stipho, Associated and non-associated constitutive relations for undrained behaviour of isotropic soft clays, Int. J. Anal. Meth. Geom., **2**, 35-56, 1978.
11. Y.F.Dafalias and E.P.Popov, Plastic internal variables formalism of cyclic plasticity, ASME J. Appl. Mech., 98, 645-650, 1976.
12. Y.F.Dafalias and L.R.Herrman, A bounding surface soil plasticity model, Proc. Int. Symp. Soils Under Cyclic and Transient Loading (Eds. G. N.Pande and O.C. Zienkiewicz), pp.335-345 A.A.Balkema Publ.,, 1980
13. Y.F.Dafalias and L.R.Herrmann, Bounding surface formulation of soil plasticity, in Soil Mechanics-Transient and Cyclic Loads (Ed. G.N. Pande and O.C.Zienkiewicz).
14. D.C.Drucker and W.Prager, Soil mechanics and plastic analysis or limit design, Quart. Appl. Math., **10**, 157-165, 1952.
15. D.C.Drucker, R.E.Gibson, and D.J.Henkel, Soil mechanics and work-hardening theories of plasticity, Trans. ASCE, **122**, 338-346, 1957.
16. R.O.Davis and G.Mullenger, A rate type constitutive model for soil with a critical state, Int. J. Num. Anal. Meth. Geom., **2**, 255-283, 1978.
17. J.M.Duncan and C.Y.Chang, Non-linear analysis of stress and strain in soils, J. Soil Mech. Found, Div. ASCE, 96, No. SM 5, 1629-1651, 1970.
18. G.Gudehus, Elastóplastische Stoffgleichung für trockener Sand, Ing. Arch., **42**, 1973.
19. K.Hashiguchi, Constitutive equations of granular media with an anisotropic hardening, Third Int. Conf. Num. Meth. Geom., Aachen, Germany, **4**, 1979.
20. K.Hashiguchi, Constitutive equations of elasto-plastic materials with elastic-plastic transition, ASME J. Appl. Mech., **47**, 266-272, 1980.
21. T.Hueckel and R.Nova, Some hysteric effects of the behaviour of geologic media, Int. J. Solids Struct., **15**, 625-642, 1979.
22. R.L.Kondner and I.S.Zelasko, Hyberbolic stress-strain response: cohesive soils, J. Soils Mech. Found. Div. ASME, **89**, No. SM 1, 115-143, 1963.
23. R.L.Kondner and I.S.Zelasko, A hyberbolic stress-strain formulation for sands, Proc. 2nd Pan-Am. Conf. Soil Mech. Found. Eng. Brazil, **1**, 289-324,1963.
24. W.D.Iwan, On a class of models for the yielding behaviour of continuous and composite systems, J. Appl. Mech. 34, 612-617, 1967.
25. R.D.Krieg, A practical two-surface plasticity theory, J.Appl. Mech. Trans. ASME, E 42, **97**, 641-646, 1975.
26. P.V.Lade and J.M.Duncan, Elastic stress-strain theory for cohesion-less sil, J. Geotechn. Eng. Div. ASCE, **101**, No. GT 10, 1037-1053, 1975.
27. P.V.Lade, Elastoplastic stress-strain theory for cohesion-less soil with curved yield surfaces, Int. J. Solids and Structures, **13**, 1019-1035, 1975.
28. Z.Mróz, On forms of constitutive laws for elastic-plastic solids, Arch. Mech. Stos., **18**, 3-35, 1966.
29. Z.Mróz, On hypoelasticity and plasticity approaches to constitutive modelling of inelastic behaviour of soils, Int. J. Num. Anal. Meth. Geom., 4, 45-55, 1980.
30. Z.Mróz, On the description of anisotropic work-hardening, J. Mech. Phys. Solids, **15**, 163-175, 1967.
31. Z.Mróz, V.A.Norris, and O.C.Zienkiewicz, An anisotropic hardening model for soils and its application to cyclic loading, Int. J. Num. Anal. Meth. Geom., **2**, 203-221, 1978.
32. Z.Mróz, V.A.Norris, and O.C.Zienkiewicz, Application of an anisotropic hardening model in the analysis of elastoplastic deformation of soils, Geotechnique, **29**, 1-34, 1979.
33. Z.Mróz, V.A.Norris, and O.C.Zienkiewicz, An anisotropic critical state model for soils subject to cyclic loading, Geotechnique, **31**, 451-469, 1981.

34. Z.Mróz and V.A.Norris, Elastoplastic and viscoplastic constitutive models for soils with application to cyclic loading, in Soil Mechanics-Transient and Cyclic Loads (Ed. G.N.Pande and O.C.Zienkiewicz), Chapter 8, pp.173-217, J.Wiley and Sons, 1982.

35. Z.Mróz and S.Pietruszczak, A constitutive model for sand with anisotropic hardening rule, Int. J. Num. Anal. Math. Geom., 7, 19-38, 1983.

36. R.Nova and D.M.Wood, A constitutive model for sand in triaxial compression, Int. J. Num. Anal. Meth. Geom., 3, 255-278, 1979.

37. S.Pietruszczak and Z.Mróz, On hardening anisotropy of K0-consolidated clays, Int. J. Num. Anal. Meth. Geom., 7, 19-38, 1983.

38. S.Pietruszczak and Z.Mróz, Description of anisotropic consolidation of clays, Proc. Symp. CRNS, Comportement Mecanique des Solides Anisotropes, pp.399-623, Noordhoff Sc. Publ., 1982.

39. S.Pietruszczak and Z.Mróz, Finite element analysis of deformation of strain softening materials, Int. J. Num. Meth. Eng., 17, 327-334, 1981.

40. J.H.Prevost, Mathematical modelling of monotonic and cyclic undrained clay behaviour, Int. J. Num. Anal. Meth. Geom., 1, 195-216, 1977.

41. J.H.Prevost, Plasticity theory for soil stress-strain behavoiur, J. Eng. Mech. Div. ASCE, 104, No. EM5, 1177-1194, 1978.

42. J.H.Prevost and K.Hoeg, Effective stress-strain strength model for soils, Proc. ASCE, 101, GT 3, 259-278, 1975.

43. G.Pande and S.Pietruszczak, A sideways look at numerical models of soils, Univ. Coll. Swansea, Civ. Eng. Dep. Report C/R/433/82, Dec. 1992.

44. M.Romano, A continuum theory for granular media with a critical state, Arch. Mech. Stos., 26, 1011-1028, 1974.

45. D.A.Kolymbas, A rate-dependent constitutive equation for soils, Mech. Res. Comm., 4, 367-372, 1977.

46. I.S.Sandler, On the uniqueness and stability of endochronic theories of material behaviour, Trans. ASME, J. Appl. Mech., 45, 263-266, 1978.

47. R.S.Rivlin, Some comments on the endochronic theory of plasticity, Int. J. Sol. Struct., 1981.

48. K.C.Valanis and H.E. Read, A new endochronic plasticity model for soils, in Soil Mechanics- Transient and Cyclic Loads (Eds. G. N. Pande and O.C.Zienkiewicz), pp.375-417, 1982, John Wiley and Sons.

49. H.Di Benedetto and F. Darve, Comparaison de lois rheologiques en cinematique rotationelle, J.de Mec. Theor. Appl. 1983.

50. A.N.Schofield and P.Wroth, Critical State Soil Mechanics, McGraw-Hill, 1968.

51. P.Wilde, Two invariant depending models of granular media, Arch. Mech. Stos., 29,199-209,1977.

52. S.Nemat-Nasser and A.Shokooh, On finite plastic flows of compressible materials with internal friction, Int. J. Sol. Struct., 16, 495-514, 1980.

53. A.Skempton and V.A.Sowa, The behaviour of saturated clays during sampling and testing, Geotechnique, 13, 269-280, 1963.

54. K.Ishihara and S.Okada, Yielding of overconsolidated sand and liquefaction model under cyclic stress, Soils and Found., 18, 57-72, 1978.

55. R.H.Boyce, A non-linear model for the elastic behaviour of granular materials under repeated loading, Proc. Int. Symp. Soils Under Cyclic and Transient Loading, Ed. G.N.Pande and O.C.Zienkiewicz, J.Wiley, 1982.

56. F.L.Di Maggio and S.Sandler, Material model for granular soils, J. Eng. Mech. Div., Proc. ASCE, 935-940, 1971.
57. H.Petersson and E.P.Popov, Constitutive relations for generalized loadings, Proc. ASCE, **103**, EM 611-627, 1977.
58. Z.Mróz and P.Rodzik, On the control of deformation process by plastic strain, Int. J. Plasticity, **11**, 827-842, 1995.
59. M.Klisiński, Z.Mróz and K.Runnesson, Structure of constitutive equations in plasticity for different choices of state and control variables, Int. J. Plasticity, **8**, 221-243, 1992.
60. A.Drescher, B.Birgisson and K. Shak, A model of water saturated loose sand , Proc. V-th Int. Symp. Num. Models Geomech., NUMOG-V, Ed. G.N.Pande and St. Pietruszczak, 109-112, Balkema.
61. A.Jarzębowski and Z.Mróz, A constitutive model for sands and its application to monotonic and cyclic loading, Proc. Int. Symp. Constitutive Equations for Granular Non-cohesive Soils, Ed. A.Saada and G.Bianchini, Balkema, 307-323, 1988.
62. D.M.Wood, K.Belkheir and D.F.Liu, Strain softening and state parameter for sand modelling, Geotechnique, **44**, 235-239, 1994.
63. J.A.Sladen and J.M.Oswell, The behaviour of very lose sand in the triaxial compression test, Canadian Geotechn. J., **26**, 103-113, 1989.
64. K.Ishihara, Liquefaction and flow failure during earthquakes , Geotechnique, **43**, 351-415, 1993.
65. P.V.Lade, Static instability and liquefaction of loose fine sandy slopes, J. Geotech. Eng., ASCE, **118**, 51-71, 1992.
66. A.Drescher and Z.Mróz, A refined constitutive model for superior sand, Proc. NUMOG-VI, Ed. St. Pietruszczak and G.N.Pande, 21-26, A.A.Balkema, 1997.
67. J.P.Bardet, A bounding surface model for sands, J. Eng. Mech. ASCE, 1198-1217, 1986.
68. C. Di Prisco, R.Nova and J.Lanier, A mixed isotropic-kinematic hardening constitutive law for sand, Modern Approaches to Plasticity, Ed. D.Kolymbas, Elsevier Sci. Publ., 83-124, 1993.
69. M.Pastor, O.C.Zienkiewicz and K.H.Leung, Simple model for transient soil loading in earthquake analysis, Int. J. Num. Meth. Geomech., **9**, 477-498, 1985.
70. B.Cambou and K.Jafari, A constitutive model for granular materials based on two plasticity mechanisms, Proc. Int. Symp. Constitutive Equations for Granular and Non-cohesive Soils, Eds. A.Saada, G.Bianchini, Balkema, 149-167, 1988.
71. A.Drescher and R.L.Michałowski, Density variation in pseudo-steady plastic flow of granular media. *Geotechnique*, **34**, 1-10,1984.
72. M.A.Goodman and S.C.Cowin, Two problems in the gravity flow of granular materials. J.*Fluid Materials*, **45**, 321-339, 1971.
73. M.A.Goodman and S.C.Cowin,.,A continuum theory for granular materials. *Arch. Rat. Mech. Anal.*, **44**, 249-266, 1972.
74. A.Goldstein and M.Shapiro, Mechanics of collisional motion of granular materials, Part I. General hydrodynamic equations, *J. Fluid Mech.*, **282**, 75-114, 1995.
75. J.T.Jenkins and S.B.Savage, A theory for the rapid flow of identical, smooth nearly elastic, circular disk. *J. Fluid Mech.*, **171**, 53-69, 1983.
76. H.K.Kytomaa. and D.Prasad, Transition from quasistatic to rate dependent shearing of concentrated suspensions. *Powders and Grounds 93, Proc. 2-nd Int. Conf. Micromechanics of Granular Media*, (Ed. C.Thorton and A.A.Balkema, 281-287, 1993.
77. C.K.K.Lun, S.B.Savage., D.J.Jeffrey. and N.Chepurny, Kinetic theories for granular flow: inelastic particles in Couette flow and sligh, J. Fluid Mech., 1984.

78. R.C.Yalamanchili, R.Gudha and K.R.Rajagopal, Flow of granular materials in a vertical channel under the action of gravity. *Powder Technology*, **81**, 65-73, 1994.

79. R.L.Michałowski, Strain localisation and periodic fluctuations in granular flow processes from hoppers. *Geotechnique*, **40**, 389-403, 1990.

80. Z.Mróz and O.C.Zienkiewicz, Uniform formulation of constitutive models for clays and sands. In *Constitutive Equations for Engineering Materials*, (Ed. C.S. Desai, and R. Gallagher). J.Wiley, 1984.

81. Z.Mróz and A.Jarzębowski, Phenomenological model of contact slip. *Acta Mech.* **102**, 59-72, 1994.

82. Z.Mróz and J.Maciejewski, Post-critical response of soils and shear band evolution. In *Localisation Phenomena in Geomaterials* (Ed. R. Chambon et al.). Balkema, 1994.

83. J.W.Rudnicki and J.R.Rice, Conditions for the localisation of deformation in pressure-sensitive dilatant-materials. *J. Mech. Phys. Solids*, **23**, 371-394, 1975.

84. S.B.Savage and S.Mc Keown, Shear stresses developed during rapid shear of concentrated suspensions of large spherical particles between concentric cylinders. *J. Fluid Mech.*, **127**, 453-472, 1983.

85. O.C.Zienkiewicz and Z.Mróz, Generalized plasticity formulation and application to geomechanics. In *Mechanics of Engineering Materials*, (Ed. C.S.Desai, and R. Gallagher). J.Wiley 655-680, 1984.

STRAIN LOCALIZATION IN GRANULAR MATERIALS

I. Vardoulakis
National Technical University of Athens, Athens, Greece

1.   Introduction
1.1   Historical Note

Contemporary localization theory [99] is a natural extension of Mohr's [50] original strength theory published in the year 1900 in a milestone paper with the title, "Welche Umstände bedingen die Elastizitätsgrenze und den Bruch eines Materials?". Mohr's question cannot be answered without resorting to experiments carefully and systematically run. Experiments, however do not give definite answers, since they are always subject to theoretical interpretation: In order to arrive to some conclusion one needs a theoretical framework within which the experiment is run and interpreted. In that sense Mohr's fundamental geometrical theory of stress analysis provided a useful tool for engineering design. Fig. 1 is taken from Mohr's paper and is usually referred to as the graphical representation of the *'Mohr-Coulomb'* failure criterion, although in Mohr's original paper no explicit reference to Coulomb's work [16] is made.

In the aforementioned paper, Mohr summarizes his observations by pointing to the following general property of *localized* deformation (p.7): "...The deformations observed in a homogeneous body after the elasticity limit [is reached] are not confined in the smallest domains of the body. They consist more or less in that, parts of the body of finite dimension displace with respect to éach other on two sets of slip bands...". Indeed one basic property of localization phenomena is some degree discontinuity of the deformation.

Today we know that prior to *localization* the governing partial differential equations of the underlying quasi-static rate-boundary-value problem are elliptic and exclude discontinuous solutions. At the onset of localization these equations are changing type and from elliptic they turn hyperbolic. *Slip-lines* and *shear-bands* are thus identified with the characteristic lines of the governing hyperbolic partial differential equations. Finally we would like to point out that we are not only interested in states of *incipient failure*. One is increasingly interested in ways to trace the deformation in the so-called *post-failure* regime. Since at *failure* the underlying mathematical problem is changing type (an elliptic-to-hyperbolic transition), post-failure we deal in general with mathematically ill-posed problems which need some degree of regularization.

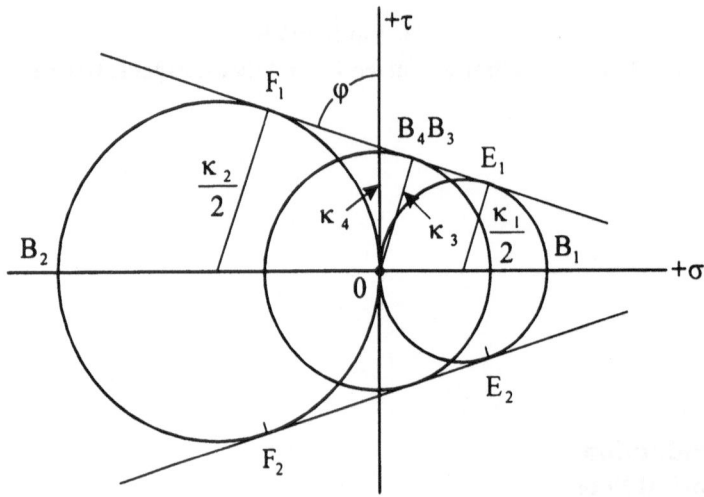

Fig. 1: Mohr's original failure criterion for a frictional or 'Coulomb' material

1.2 Granular Material Behavior

From the phenomenological point of view granular media exhibit predominantly irreversible deformations and are relatively rate-insensitive. Granular materials are ideal examples of *plastic* solids. The property however which differentiates granular materials from other plastic solids, like e.g. metals, is their pronounced *pressure sensitivity*: For pressure-sensitive materials under continued loading the stress deviator T depends on to the mean normal compressive stress p (compression negative). In form of equation one may write,

$$T = \langle f \rangle (-p) \tag{1.1}$$

The pressure-sensitivity coefficient is defined only in the compressive regime of stresses

$$\langle f \rangle = \begin{cases} 0 & \text{for} : p \geq 0 \\ c|p|^{-1} + f_C[1 + O(|p|)] & \text{for} : p < 0 \end{cases} \tag{1.2}$$

Here c is called the *cohesion* and f_C is the *Coulomb friction coefficient*. From eq. (1.2) we see that under sufficiently high stresses, geomaterials are behaving as purely frictional solids. Since sand is the ideal representative of a purely frictional material, it is dry sand which is usually selected as model material to study the basic properties of geomaterials. For example, since the mid-30's it was widely recognized that for studying the formation of geologic structures, like rock faulting, 1g- model tests with cohesionless sand are perfectly suitable [37].

Besides internal friction, geomaterials are characterized also by their plastic *dilatancy*, which is understood as a simple internal constraint between the plastic volumetric strain increment Δv^P and the plastic shear strain increment Δg^P [24]

$$\Delta v^P = d \, \Delta g^P \tag{1.3}$$

Parameter d in this relation is called *Reynolds dilatancy coefficient* [63].

Above basic assumptions constitute the frame of constitutive modeling of rate-independent granular materials. The cohesive-frictional and dilatant character of granular media is quite satisfactorily modeled within the frame of elastoplasticity theory with strain hardening/softening yield surface and non-associate flow-rule [52]. In particular one may select an appropriate plastic strain- or work-hardening parameter, and assume that both f and d are functions of it,

$$f = f(...; \psi) \qquad d = d(...; \psi) \tag{1.4}$$

Depending on the nature of strain hardening (shearing or compaction) one may assume $\psi = \int dg^P$ resp. $\psi = \int dv^P$. In case when both shearing and compaction are equally important one may use as hardening parameter the plastic work of the stress on the plastic strain, $\psi = \int dw^P$, where

$$dw^P = \sigma_{ij} d\varepsilon_{ij}^P = T dg^P + p dv^P \tag{1.5}$$

In frictional and dilatant materials the (first order) plastic work increment becomes

$$\Delta w^P = \langle -p \rangle (f - d) \Delta g^P \tag{1.6}$$

Thus from the point of view of energy dissipation a frictional and dilatant material behaves like a purely frictional material or Coulomb material with an *effective friction coefficient*

$$\bar{f} = f - d \tag{1.7}$$

The experimental evidence suggests that, $\bar{f}$ = const., and thus

$$d = f - \bar{f} \qquad\qquad (1.8)$$

This is Taylor's condition [75] for the dilatancy coefficient, and is the simplest generalization of the *normality condition* of classical flow theory which is included in it for the degenerate case of zero effective friction coefficient (f=d). It is customary, however, to call the corresponding flow-rule *non-associated*, although Taylor's rule provides a simple way to relate the normals to the plastic potential- and yield-surface by assuming that they produce always the same angle, not necessarily zero. It should be noticed also that Rowe [67,68] and de Joselin de Jong [18] have used the same concept as above; they have called the parameter $\bar{f}$ a *true angle of friction* and identified it with the inter-particle friction coefficient.

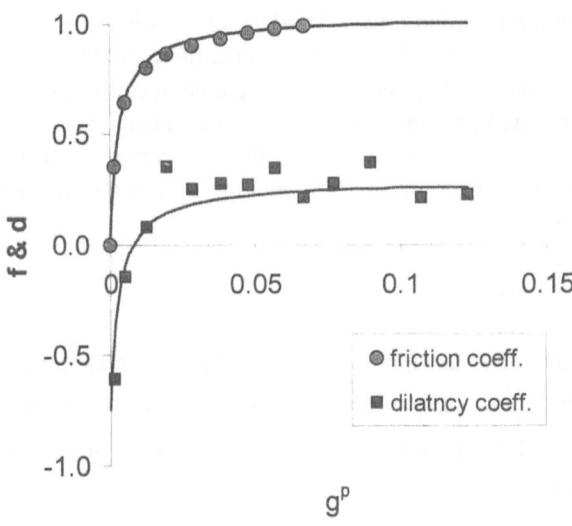

Fig. 2: Mobilized friction and dilatancy coefficients from a triaxial compression test on medium dense Karlsruhe sand [99]

1.3 Localization

In granular, cohesionless materials shear localization induces intense inter-granular slip, which in turn leads to strong dilatancy of the material inside the localized zone [23,94]. Fig. 3 shows three sequential x-ray plates of a sand specimen [96]. The first plate was taken prior to loading, at a state of isotropic compression. In the middle of the specimen a small lens of looser sand was placed to serve as a site of localized deformation. The second plate shows the same sand specimen at peak deviator, with a faint trace of localization of porosity crossing the soft sand lens. With continued deformation this localized zone extended outwards and reached eventually the boundaries of the specimen.

Strong localized material dilatancy due to grain rearrangement and grain rotation are the dominant micro-kinematical features of shear banding. Increasing porosity reduces naturally the coordination number of the granular assembly (i.e. the number of contacts per grain), yielding progressively to a weaker granular structure. On top of that and from the micro-statical point of view an important structure that appears to dominate this type of localized deformation is the formation and collapse (buckling) of grain columns [57], which in turn lead to a basic asymmetry of shear stress and to micro-polar effects [55].

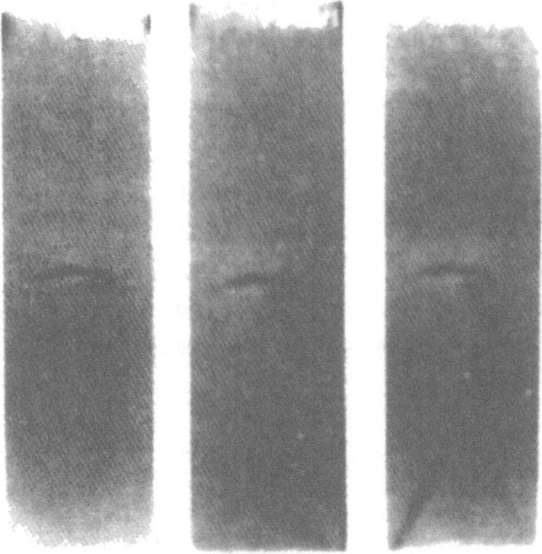

Fig. 3: Evolution of dilatancy localization in a sand specimen under biaxial compression

Reduction of coordination number and grain-column buckling lead to macroscopic *material softening* inside the localized zone. For equilibrium reasons the material outside the localized zone is unloading. Thus we perceive the *shear-band* as being separated from the rest material by a set of parallel surfaces at some distance $2d_B$, the so-called shear-band boundaries. The shear-band boundaries are modeled as elastoplastic boundaries; i.e. material surfaces which separate the elastoplatically softening shear band from its elastically unloading neighborhood.
The problem of modeling localized deformation in geomaterials is quite a challenging task, due to the mathematical difficulties which are encountered in general while dealing with non-associate and softening material behavior and boundary-value problems with moving internal elastoplastic boundaries. Thus, as first addressed by Mandel [45], questions of uniqueness and stability of solutions arise naturally within the context of shear-band analysis. It turns out that the result of such analyses depends primarily on the assumed physical non-linearities which are inherent to the underlying constitutive description and, in a lesser degree, is influenced by geometrical non-linearities.

The various drawbacks and shortcomings of the classical continuum and constitutive theories in connection with strain localization or, more generally, for problems where loss of ellipticity of the governing partial differential equations is taking place, have been discussed extensively in many recent papers (cf. [5]). The origin of this undesirable situation can be traced back to the fact that conventional constitutive models do not contain material parameters with the dimension of length, so that the shear-band thickness (i.e. the extent of the plastically softening region) is undetermined.

Vardoulakis [81] conjectured that spontaneous loss of homogeneity in the form of shear band formation is a clear indication for the existence of material length scale. Indeed, there is ample experimental evidence that shear-bands in granular materials engage a significant number of grains. Based on direct experimental observations Roscoe [66] proposed that the width of shear-bands is about 10 times the average grain diameter; see also Scarpelli and Wood [70].

In Fig. 4 X-ray radiographs of shear-bands are shown that are formed in the biaxial tests [82] reported by Vardoulakis et al. [94,96]. The first plate corresponds to a medium-grained sand from Karlsruhe, and the second plate corresponds to a fine-grained sand Dutch dune sand. Table 1 summarizes the evaluation of these plates. In this table d_{50} denotes the mean grain size of the tested sand and $2d_E$ the measured shear-band thickness. Accordingly, these experiments suggest a shear-band thickness that is about 16 times the mean grain diameter.

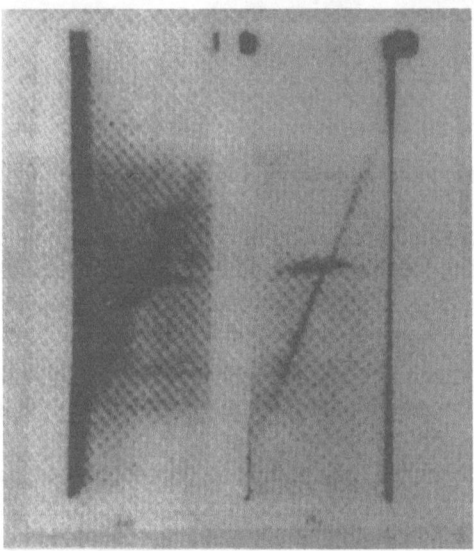

Fig. 4: Shear-band emerging out of a density inhomogeneity:(a) medium grained Karlsruhe sand ;(b) fine grained Dutch dune sand

Table 1: Measured shear-band thickness [95,96]

	d_{50} [mm]	$2d_E$ [mm]	$\dfrac{2d_E}{d_{50}}$
Fine sand (FS)	0.20	3.7	18.5
Medium sand (D)	0.33	4.3	13.0

1.3 Regularization

One could say that localization of deformation leads to a change of scale of the problem so that phenomena occurring at the scale of the grain cannot be ignored anymore in the modeling process of the macroscopic behavior of the material. Then it appears necessary to resort to continuum models with micro-structure to describe correctly localization phenomena. These generalized continua usually contain additional kinematical degrees of freedom and/or higher deformation gradients. These observations have prompted the extension of classical continuum mechanical descriptions for granular media past the softening regime by resorting to the so-called *Cosserat* [54] or *Gradient models* [92,93].

Cosserat continua and higher gradient continua belong to a general class of constitutive models which account for the material's micro-structure. The description of statics and kinematics of continuous media with microstructure has been studied systematically by many authors in the past (e.g. [29,30]). In a classical description, a continuum is a continuous distribution of particles, each of them being represented geometrically by a point $\mathbf{X}$ and characterized kinematically by a velocity $\mathbf{v}$. In a theory which takes into account the micro-structure of the material, each particle is viewed as a continuum $C(\mathbf{X})$ of small extend around the point $\mathbf{X}$. Consequently the deformation of the volume $C(\mathbf{X})$ of the particle is called micro-deformation. For example, a Cosserat continuum is a *micro-polar* medium such that the particle $C(\mathbf{X})$ moves as a rigid body; i.e. it is characterized by a particle velocity vector $\mathbf{v}$ and a particle rotation vector ω^c. The corresponding deformations, the velocity gradient $\nabla\mathbf{v}$ and the rotation gradient (curvature), $\nabla\omega^c$ are associated through the principle of virtual work with a non-symmetric stress tensor and couple-stress tensor, respectively.

Similarly, in a 2nd gradient continuum, again through the principle of virtual work, a symmetric 2nd order stress tensor and a 3rd order (double-) stress tensor are defined, which work on $\nabla\mathbf{v}$ and on $\nabla\nabla\mathbf{v}$ respectively.

Rotation gradients and higher velocity gradients introduce naturally a *material length scale* into the problem, which as already mentioned is necessary for correct modeling of localization phenomena. In this case the underlying mathematical problem is *regularized* and the governing equations remain always elliptic. This extension of the continuum description leads to robust computations and allows us to follow the evolution of the considered system in the post-bifurcation regime. The model is then capable to capture additional structural information such as shear-band thickness or to asses the effect of scale.

2. Uniqueness Theorems
2.1 Uniqueness under Dead Loading

Under dead loading conditions the virtual work equation for global continued equilibrium for a deformable body **B** with volume V and boundary ∂V in the reference configuration, reads

$$\int_V \Delta\pi_{ij}\partial_j\delta u_i dV = 0$$

$$\delta u_i = \delta_i \quad \text{on } \partial V_u \tag{2.1}$$

In these expressions $\Delta\pi_{ij}$ is the nominal (1.Piola-Kirchhoff) stress increment, referred to the current configuration [82,89,99]. Let $\left\{\Delta\pi_{ij}, \Delta u_i\right\}^{(1)}$ and $\left\{\Delta\pi_{ij}, \Delta u_i\right\}^{(2)}$ be two solutions of the incremental boundary value problem (2.1). Their difference

$$\Delta\pi_{ij} = \Delta\pi_{ij}^{(1)} - \Delta\pi_{ij}^{(2)}$$

$$\Delta u_i = \Delta u_i^{(1)} - \Delta u_i^{(2)} \tag{2.2}$$

satisfies the following homogeneous equations

$$\int_V \Delta\pi_{ij}\partial_j\Delta u_i dV = 0$$

$$\Delta u_i = 0 \quad \text{on } \partial V_u \tag{2.3}$$

The integrand of the above volume integral, denoted by the symbol

$$\Delta_2 w = \Delta\pi_{ij}\partial_j\Delta u_i \tag{2.4}$$

is called the *second-order work* of the nominal stress.

Theorem: Sufficient condition for *global uniqueness* of the incremental boundary value problem under dead loading conditions is that for *any* admissible displacement field, the global second order work of stresses,

$$\int_V \Delta_2 w \, dV \geq 0 \tag{2.5}$$

with the equal sign holding only for the trivial solution, $\Delta u_i = 0$.

For hyperelastic materials this theorem is known as *Hadamard's* linearized, dead-load stability criterion; cf. [4]. Hadamard [32] has shown that for an equilibrium configuration of a hyperelastic body under dead loads to satisfy the stability criterion (2.5) it is necessary that for all vectors g_i, n_i and everywhere in V the condition holds,

$$C_{ijkl}g_ig_kn_jn_l \geq 0 \qquad (2.6)$$

where

$$\dot{\pi}_{ij} = C_{ijkl}D_{kl} \qquad (2.7)$$

is the underlying relationship between the rate of the 1.P.-K. stress tensor and the rate of deformation tensor; Truesdell and Noll [79].
For example, for a hypoelastic materials, whose constitutive equations are derived from a hyperelasticity and are given in terms of the Jaumann rate of the relative Kirchhoff stress,

$$\tilde{T}_{ij}^t = R_{ijkl}^t D_{kl} \qquad (2.8)$$

we have

$$\dot{\pi}_{ij} = \tilde{T}_{ij}^t + W_{ik}\sigma_{kj} - \sigma_{ik}D_{kj} \qquad (2.9)$$

where σ_{ij} is the initial stress and D_{ij}, W_{ij} are the symmetric part (rate of deformation tensor) and antisymmetric part (spin tensor) of the spatial gradient of the velocity, respectively [99]. According to the above definitions the components of the material stiffness tensor obey the major symmetry conditions

$$R_{ijkl}^t = R_{klij}^t = R_{jikl}^t = R_{ijlk}^t \qquad (2.10)$$

We remark that the Cauchy/Jaumann stress increment given by a constitutive equation of the form

$$\tilde{\sigma}_{ij} = C_{ijkl}^e D_{kl} \qquad (2.11)$$

where the stiffness tensor

$$C_{ijkl}^e = R_{ijkl}^t - \sigma_{ij}\delta_{kl} \qquad (2.12)$$

does only obey minor symmetry conditions

$$C^e_{ijkl} = C^e_{jikl} = C^e_{ijlk} \tag{2.13}$$

This means that for large strain analysis, hypoelasticity laws should be expressed in terms of the Kirchhoff-Jaumann stress rate rather than in terms of the Cauchy-Jaumann one. We remark also that an elastic material which is isotropic under finite strain, develops a strain-induced anisotropy for shearing in axes parallel to the initial stretch; [10, 99].

For hyperelastic materials non-uniqueness implies instability in the sense of Hadamard. Of course, the converse does not hold; i.e. uniqueness does not imply Hadamard stability. A counter example follows from the observation that condition

$$\int_V \Delta_2 w \, dV < 0 \tag{2.14}$$

implies uniqueness.

The sufficient global uniqueness condition (2.5) is satisfied if we impose the stronger, *local sufficient* condition for uniqueness

$$\Delta_2 w > 0 \tag{2.15}$$

holding everywhere in V. Condition (2.15) requiring positiveness of the second order work of stresses, is a stronger requirement then the so-called *strong-ellipticity condition* for the stiffness tensor

$$C_{ijkl} g_i g_k n_j n_l > 0 \tag{2.16}$$

which assures that the differential equations governing the incremental boundary value problem are elliptic.

We notice that within a small velocity-gradient theory all stress rates coincide

$$\overset{\ast}{\pi}_{ij} \approx \overset{\cdot}{\sigma}_{ij} \approx \widetilde{T}^t_{ij} \approx \widetilde{\sigma}_{ij} \tag{2.17}$$

and positiveness of the second order work of stress,

$$\Delta_2 w \approx \Delta\sigma_{ij}\Delta\varepsilon_{ij} > 0 \tag{2.18}$$

implies uniqueness. This inequality is encountered in the literature as *postulate for material stability*. Materials which satisfy this postulate are called *stable materials*. If on the other hand the second order work of stress is negative then the material is called unstable. For example, unstable hyperelastic materials are characterized by a non-convex strain energy function, which is resulting to an 'unstable' stress-strain curve, which is typical for *strain-softening* material.

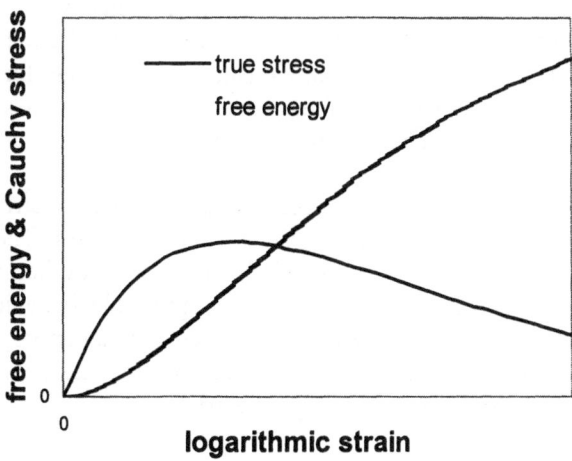

Fig. 5: Free energy and true stress for 'unstable' elastic material

2.3 Constitutive Inequalities for Geomaterials

In a small strain theory of elasto-plasticity, the total strain rate is decomposed into an elastic and a plastic component

$$\dot{\varepsilon}_{ij} = \dot{\varepsilon}_{ij}^e + \dot{\varepsilon}_{ij}^p \tag{2.19}$$

- *Elasticity*

Elastic strain-rates are linked to the stress-rate,

$$\dot{\varepsilon}_{ij}^e = M_{ijkl}^e \dot{\sigma}_{kl}^e \quad \text{or} \quad \dot{\sigma}_{ij} = C_{ijkl}^e \dot{\varepsilon}_{ij}^e \quad (\mathbf{M} = \mathbf{C}^{-1}) \tag{2.20}$$

where C_{ijkl}^e is the isotropic (Hooke) elastic stiffness tensor

$$C_{ijkl}^e = G\left(\delta_{ik}\delta_{jl} + \delta_{il}\delta_{jk} + \frac{2v}{1-2v}\delta_{ij}\delta_{kl} \right) \tag{2.21}$$

The shear modulus and the Poisson's ratio being restricted such that

$$G > 0 \quad \text{and} \quad -1 \le v \le \frac{1}{2} \tag{2.22}$$

These inequalities assure positive elastic strain energy under any circumstances.

In several finite-strain formulations of elastoplasticity we find isotropic hypoelastic constitutive equations, which relate the relative Kirchhoff/Jaumann or Cauchy/Jaumann stress rate to the rate of deformation. Constitutive equations for the Jaumann rate of the relative Kirchhoff stress with a stiffness tensor given by Hooke's law are not truly elastic, i.e. hyperelastic; [38, 47]. This is generally not considered as a major deficiency of such a theory, and in large strain analyses this assumption is often made; cf. [25]. Christoffersen [15] discussed a simple class of hyperelastic relations with isotropic rate forms which can be used in finite-strain formulations of elastoplasticity laws. These rate equations do not involve the Jaumann derivative of the Kirchhoff stress but other, suitably chosen objective stress rates and are restricted to plastic-incompressible behavior. Strictly speaking the problem remains open. This is the problem of selecting an appropriate objective stress rate such that the underlying elasticity law is isotropic for elastoplastic material with plastic volume changes such as is the case in geomaterials.

- *Flow-theory*

Plastic strain-rates are generated whenever the stress state lies on a yield surface in stress space and *loading* is taking place. Let $F = F(\sigma_{ij}, \psi)$ be the yield function, defined in terms of the stress tensor σ_{ij} and a non-decreasing plastic-hardening parameter ψ, which determines the plastic state. *Loading* is defined by the condition:

$$F = 0 \, ; \dot{F} = 0 \, ; \dot{\psi} > 0 \tag{2.23a}$$

The condition

$$F = 0 \, ; \dot{F} = 0 \, ; \dot{\psi} = 0 \tag{2.23b}$$

is called *neutral loading condition*. Thus during loading the state of stress lies and stays on the yield surface. Usually the plastic strain-rate tensor is assumed to be coaxial with the stress tensor. In particular a class of plasticity laws is based on the concept of plastic strain-rate potential: If $Q = Q(\sigma_{ij}, \psi)$ is such a plastic potential function, then the plastic strain-rates are given by the following flow-rule,

$$\dot{\varepsilon}_{ij}^{p} = \frac{\partial Q}{\partial \sigma_{ij}} \dot{\psi} \qquad \text{and} \qquad \dot{\psi} \geq 0 \tag{2.24}$$

We remark that whenever the yield surface is used as a plastic potential surface, then the flow-rule is called an *associate* flow-rule, and the plastic strain-rate vector in stress-space is normal to the yield surface. The latter is called the *normality condition* of plasticity theory.

The rate constitutive equations of small strain flow theory of plasticity take finally the following form

$$\dot{\sigma}_{ij} = C_{ijkl}^{ep}\, \dot{\varepsilon}_{kl} \qquad (2.25)$$

where the elastoplastic stiffness tensor is split in the elastic and plastic stiffness tensor

$$C_{ijkl}^{ep} = C_{ijkl}^{e} - C_{ijkl}^{p}$$

$$C_{ijkl}^{p} = \frac{<1>}{H} Q_{mn} C_{mnij}^{e} F_{pq} C_{pqkl}^{e} \qquad (2.26)$$

$$F_{ij} = \frac{\partial F}{\partial \sigma_{ij}}; \quad Q_{ij} = \frac{\partial Q}{\partial \sigma_{ij}}$$

In eq. (2.26) H is the *plastic modulus*,

$$H = H_0 + H_t \qquad (2.27)$$

where H_t is the *hardening (softening) modulus*

$$H_t = -\frac{\partial F}{\partial \psi} \qquad (2.28)$$

If $H_t > 0$, hardening is taking place, while for $H_t < 0$ softening occurs. The modulus H_0 is the *snap-back* threshold value for the softening modulus, and is usually restricted to be strictly positive in order to exclude locking behavior [55]

$$H_0 = F_{kl} C_{klmn}^{e} Q_{mn} > 0 \qquad (2.29)$$

We observe that in case of associative flow-rule $(F \equiv Q)$ this requirement always follows from the strong ellipticity condition for the elastic stiffness tensor.
In eq. (2.26) $\langle \bullet \rangle$ denote the Foeppl-Macauley brackets, which, with H>0, are defined as

$$\langle 1 \rangle = \begin{cases} 1 & \text{if} : F = 0 \text{ and } B_{kl}\dot{\varepsilon}_{kl} > 0 \\ 0 & \text{if} : F < 0 \text{ or } F = 0 \text{ and } B_{kl}\dot{\varepsilon}_{kl} \leq 0 \end{cases} ; \quad B_{ij} = F_{kl} C_{klij}^{e} \qquad (2.30)$$

Due to the Foeppl-Macauley brackets the elastoplastic stiffness tensor C_{ijkl}^{ep} is a quasi-linear operator. From eqs. (2.21) and (2.26) we see that in case of associative plasticity the elastoplastic stiffness tensor is satisfying major symmetry conditions.

For granular materials non-associativity of the flow-rule is usually restricted only for the volumetric component of the plastic strain rate. At the same time the deviatoric normality is assumed to hold [31]. This means that the tensors F_{ij} and Q_{ij} differ only in their volumetric part, i.e.

$$F_{ij} - Q_{ij} = \lambda \delta_{ij} \tag{2.31}$$

and λ is a scalar.
Finally we remark that the local dissipation inequality is connecting the entropy production to the first-order work of the stress. Thus, within the frame of a small-strain theory, for isotropically hardening, elastoplastic materials the second law of thermodynamics requires only that the *first-order plastic work* of the true stress is non-negative [52],

$$\Delta w^P = \sigma_{ij} \Delta \varepsilon_{ij}^P \geq 0 \tag{2.32}$$

For frictional-dilatant materials this inequality results in turn into a simple restriction for the underlying friction and dilatancy coefficients. For example for a Drucker-Prager model (with deviatoric but not necessarily volumetric normality) yield and plastic potential functions are defined as follows (Fig. 6)

$$F = T + f(\psi)p \quad ; \quad Q = T + d(\psi)p \tag{2.33}$$

where

$$p = \frac{1}{3}I_{1\sigma} = \frac{1}{3}\sigma_{kk}; \quad T = \sqrt{J_{2s}} = \sqrt{\frac{1}{2}s_{ij}s_{ij}} \tag{2.34}$$

are the first and second invariants of the stress and its deviator, respectively. f and d are the corresponding friction and dilatancy coefficients. In this case the dissipation inequality (2.32) requires that the dilatancy coefficient cannot exceed the friction coefficient,

$$d \leq f \tag{2.35}$$

On the other hand Drucker's stability postulate [27] restricts the *second order plastic work*. Within a small strain plasticity theory Drucker's postulate takes the following form

$$\Delta_2 w^P = \Delta \sigma_{ij} \Delta \varepsilon_{ij}^P \geq 0 \tag{2.36}$$

with the equal sign holding for elastic states or neutral loading, $\Delta \varepsilon_{ij}^P = 0$.

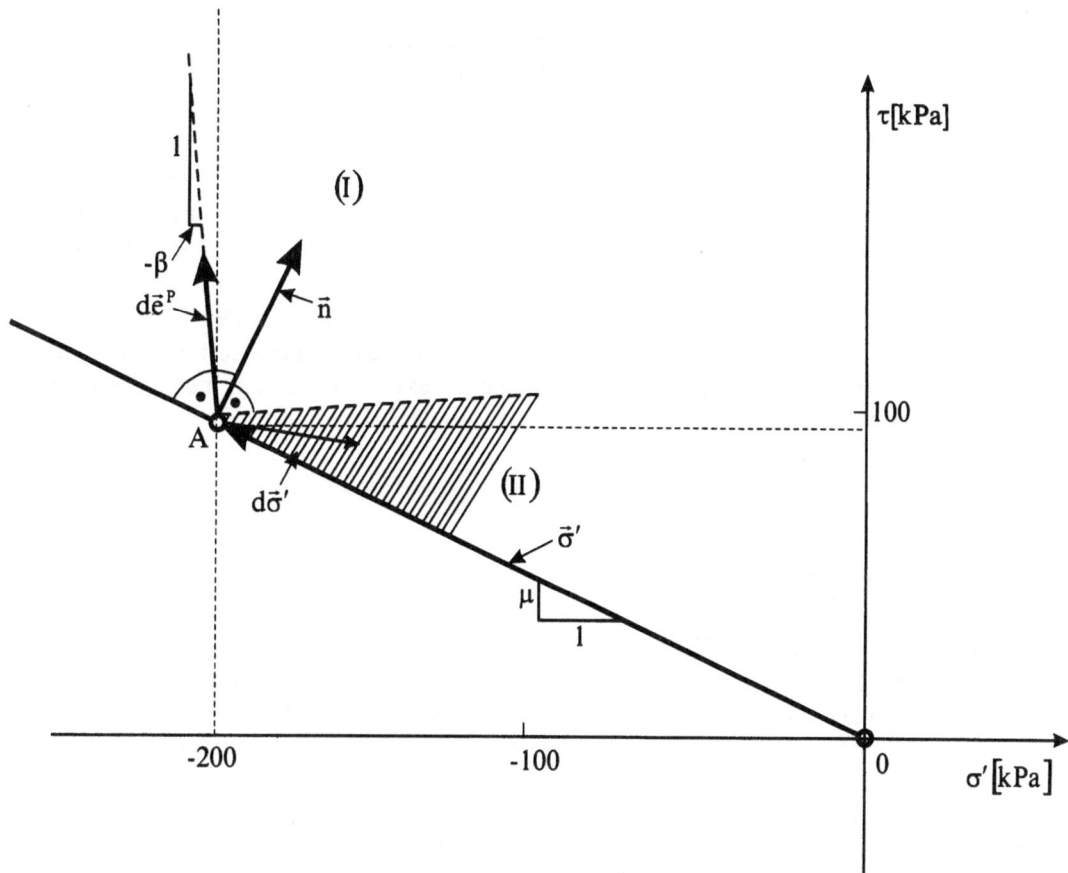

Fig. 6: Yield surface, plastic strain-rate vector and domains of positive and negative 2nd order plastic work for Coulomb material

Theorem: Drucker's stability postulate is a sufficient local criterion for uniqueness under dead loading:

$$\Delta_2 w^P \geq 0 \Rightarrow \Delta_2 w > 0 \tag{2.37}$$

Assuming that the elastic stiffness tensor is positive definite, the second-order elastic work is positive

$$\Delta_2 w^e = \Delta\sigma_{ij}\Delta\varepsilon_{ij}^e = C_{ijkl}^e \Delta\varepsilon_{ij}^e \Delta\varepsilon_{kl}^e \geq 0 \tag{2.38}$$

with the equal sign holding for $\Delta\varepsilon_{ij}^e = 0$.

Above theorem follows then directly from the additive strain-rate decomposition, in elastic and plastic parts, eq. (2.19) and inequality (2.38). Drucker's postulate is violated if material strain softening is taking place; since for continued plastic loading the stress-increment vector is pointing into the interior domain of the yield surface (F<0), whereas the plastic strain-increment is pointing into the exterior domain. It should be stressed out however that Drucker's postulate is not connected to the second law of thermodynamics and its violation does not contradict to any fundamental law of mechanics. As we will see in the last part of these lectures, the regularization of $\Delta_2 w^P$ for strain-softening materials is straightforward and in congruence with the physical problem at hand, namely that of plastic strain localization.

In case of non-associated flow-rule, inequality (2.35) is holding and Drucker's stability postulate is violated even in the hardening regime of the material behavior: For continued plastic loading the stress-increment is pointing into the exterior domain of the yield surface (F>0), and we distinguish among two domains of material behavior (Fig. 6):

- Domain (DS) with: $\Delta_2 w^P \geq 0$

- Domain (DI) with: $\Delta_2 w^P < 0$

We emphasize that domain (DI) is avoided altogether if normality (d=f) is adopted as a flow-rule. The existence of a set (DI) of stress probes with negative second order plastic work in frictional materials was first pointed out by Mandel [45], who clearly stated that Drucker's postulate is *not* a necessary condition for stability. We recall here that Mandel's idea of stability is that of linear, local stability analysis of the elastoplastic-dynamic initial value problem; i.e. it refers to the search for conditions for exponential growth of small perturbations of a given wavelength out of a given equilibrium configuration. Finally we remark that in case of severe violation of normality (d<0, e.g. contractant behavior of loose sands) and two-phase, water-saturated granular material, linear stability analysis of several bifurcation modes that characterize globally undrained condition breaks down. In this case the considered mathematical problem is ill-posedeness in the sense of Hadamard; i.e. for some characteristic directions in space there exist infinite growth of the instability for all wavelengths [85,86], this is equivalent with loss of existence of solution [90,91].

3. Localized Bifurcations
3.1 Thomas-Hill-Mandel Shear-Band Model

According to Thomas [78] shear-band is a thin material layer that is bounded by two parallel material discontinuity surfaces $D^{(\alpha)}$ ($\alpha = 1,2$) of the velocity gradient; Fig. 7. These material discontinuity surfaces are called shear-band boundaries and their distance $2d_B$ is called the *'thickness'* of the shear-band [36]. We notice that within the frame of constitutive theories without material length, the shear-band thickness is undetermined. Solutions obtained from such constitutive theories correspond to the singular limit $d_B \to 0$, of a higher grade continuum extension of the underlying classical constitutive theory, which is equipped with a material length scale [81],[54]. Based on these definitions, the so-called *Thomas-Hill-Mandel* shear band problem consists in examining whether or not a constitutive description of homogeneous deformation admits also a solution which corresponds to an equilibrium bifurcation towards a non-uniform deformation in a plane shear-band.

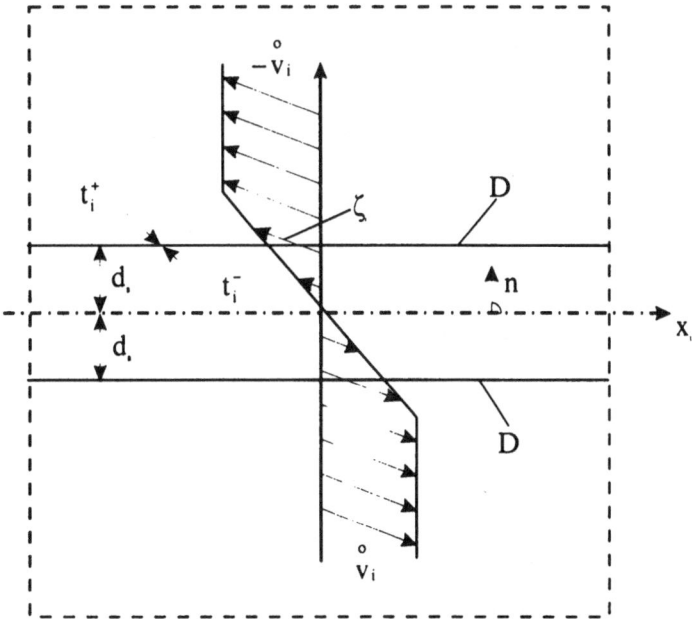

Fig. 7: The Thomas-Hill-Mandel shear-band model

3.1.1 Kinematical compatibility conditions

The velocity field outside and inside the shear-band differ drastically, since inside the band a rapid change is assumed to take place. Accordingly, in the vicinity of the shear band the exterior displacement field is varying slowly, and thus, as the shear-band thickness tends to zero, v_i is only a function of distance across the band. Let $\pm\zeta_i$ be the velocity at the shear band boundaries (Fig. 7). Notice that with the normal component $\zeta_2 \neq 0$, we do not imply separation (like e.g. in a crack) but dilatancy of the material inside the band. Under these conditions the inhomogeneous velocity field is given by the following expression

$$v_i = \begin{cases} -\zeta_i & \text{for} : d_B \leq n_k x_k \\ -\dfrac{1}{d_B}\zeta_i n_k x_k & \text{for} : -d_B \leq n_k x_k \leq d_B \\ +\zeta_i & \text{for} : n_k x_k \leq -d_B \end{cases} \qquad (3.1)$$

where n_i is the unit normal vector to the shear-band boundaries. Across these boundaries the velocity field is continuous and only the velocity gradient jumps. The corresponding kinematic compatibility conditions for the *jump* of the velocity and its gradient are [78],

$$[v_i] = v_i^+ - v_i^- = 0 \quad \text{(continuity)}$$

$$[\partial_i v_j] = g_i^{(\alpha)} n_j \quad \text{(Maxwell condition)}$$

(3.2)

In the linear approximation of the velocity field inside the shear-band, the jump vector becomes

$$g_i^{(1)} = -g_i^{(2)} = -\frac{\zeta_i}{d_B}$$

(3.3)

3.1.2 Statical compatibility conditions

Let

$$t_i = \sigma_{ji} n_j$$

(3.4)

be the traction vector at a surface element with unit normal n_i. Tractions across a material discontinuity surface D, like a shear-band boundary, must be in equilibrium. By assuming equilibrium in a given configuration C across D, (Fig. 7) we get,

$$t_i^+ = t_i^- \Rightarrow [t_i] = 0$$

(3.5)

By requiring equilibrium in the adjacent configuration $\overline{C}$ we have

$$t_i = \overline{\pi}_{ij} n_j; \quad \overline{t}_i^+ = \overline{t}_i^- \Rightarrow [\overline{t}_i] = 0$$

(3.6)

Here, $\overline{\pi}_{ij} = \sigma_{ij} + \Delta\pi_{ij}$, is the 1.P.-K. stress tensor in $\overline{C}$ referred to configuration C. To the limit $(\Delta t \to 0)$ these conditions yield a compatibility condition for the traction rate at C,

$$\dot{t}_i^+ = \dot{t}_i^- \Rightarrow [\dot{t}_i] = 0 \Rightarrow [\dot{\pi}_{ij}] n_j = 0$$

(3.7)

As already mentioned in section1, if geometric terms are negligible, then $\dot{\pi}_{ij} \approx \dot{\sigma}_{ij}$, and above compatibility conditions are expressed in terms of the Cauchy stress-rate,

$$[\dot{\sigma}_{ji}] n_j = 0$$

(3.8)

3.1.3 Bifurcation condition

The stress-rate can be expressed in terms of the strain-rate by the constitutive equations of elastoplastic materials with smooth yield and plastic potential surfaces (cf. eq. 2.25 ff). As already mentioned, elastoplastic equations are quasi-linear, and they provide different responses for loading and unloading. Thus one has to distinguish among two possibilities: (a) the constitutive behavior across the shear-band boundaries is continuous, or (b) it is discontinuous. We will consider here only continuous bifurcations. As shown by Raniecki and Bruhns [62] one can always linearize the shear-band bifurcation problem and provide upper and lower bounds to the critical load at which first shear-band bifurcation is predicted. These bounds are computed on the basis of suitable linear (comparison) solids.

Definitions:
1. Shanley's or *upper bound linear (comparison) solid* is the solid which corresponds to continuous loading; i.e. the solid which is defined through the following linear rate constitutive equations

$$\dot{\sigma}_{ij} = C^u_{ijkl} \, \dot{\varepsilon}_{kl}$$

$$C^u_{ijkl} = C^e_{ijkl} - C^{pl}_{ijkl}$$

$$C^{pl}_{ijkl} = \frac{1}{H} B^Q_{ij} B^F_{kl} \tag{3.9}$$

$$B^F_{ij} = F_{kl} C^e_{klij} \qquad B^Q_{ij} = Q_{kl} C^e_{klij}$$

2. A *lower bound linear comparison solid* is defined through the following linear rate constitutive equations with symmetric stiffness tensor

$$\dot{\sigma}_{ij} = C^\ell_{ijkl} \, \dot{\varepsilon}_{kl}$$

$$C^\ell_{ijkl} = C^e_{ijkl} - \frac{1}{4rH} (B^Q_{ij} + rB^F_{ij})(B^Q_{kl} + rB^F_{kl}); \qquad r^2 = \frac{Q_{ij} B^Q_{ij}}{F_{ij} B^F_{ij}} \tag{3.10}$$

Definitions and theorems related to the upper- and lower-bound linear comparison solids with non-associated flow-rule, are systematically presented in the aforementioned fundamental paper by Raniecki and Bruhns [62]; cf. ref. [87] for an application to granular materials. We remark also that in case of associativity the two linear comparison solids coincide, and one is recovering Hill's [34] original theory of uniqueness for elastoplastic solids. According to this theory, if normality holds, bifurcation loads predicted on the basis of Shanley's [71] comparison solid are true bifurcation loads. Concerning discontinuous bifurcations one has to examine the possibility that elastic unloading occurs outside the shear band while continued elastic-plastic loading occurs within the band. If the elastoplastic constitutive law admits a single smooth yield surface and plastic potential, Rice and Rudnicki [65] have shown that continuous bifurcation analyses on the basis of the upper-bound linear comparison solid, described here by eqs. (3.9), provide the lower limit to the range of deformations for which discontinuous bifurcations can occur. Accordingly, we restrict ourselves here to the first possibility, namely that of continuous constitutive behavior

$$[C_{ijkl}] = [C_{ijkl}^u] = 0 \tag{3.11}$$

Using the above assumption that the stiffness tensor is continuous across the shear-band boundaries the static compatibility conditions (3.8) become

$$C_{ijkl}^u [\partial_l v_k] n_i = 0 \tag{3.12}$$

By combining these compatibility conditions with kinematic compatibility conditions for the velocity gradient, eqs. (3.2), we finally obtain the *bifurcation condition*

$$\Gamma_{ik} g_k^{(\alpha)} = 0 \tag{3.13}$$

where Γ_{ik} coincides with the so-called *acoustic tensor* of the upper-bound linear comparison solid for the direction n_i,

$$\Gamma_{ik} = C_{ijkl}^u n_j n_l \tag{3.14}$$

Following these considerations we conclude to the following

Theorem: continuous shear-band bifurcations exist, if for some characteristic direction n_i the acoustic tensor of the upper-bound linear comparison solid is singular; i.e. if there is n_i such that

$$\det(\Gamma_{ik}) = 0 \tag{3.15}$$

Real n_i are then the direction cosines of the normal to the shear-band boundaries [64],[83].
In the contrary, as shown by Hill [35], the condition $\det(\Gamma_{ik}) > 0$ is sufficient for the exclusion of discontinuous solutions for the incremental displacement field.
For elastoplastic solids, the threshold to the bifurcation stress which satisfies the characteristic eq. (3.15) is usually expressed in terms of tangent hardening modulus H_t.
Because H_t is a decreasing function of the plastic hardening parameter ψ, we seek the

orientation (given by the direction cosines n_i of the normal to the discontinuity surface) for which the value of H_t is maximum [69]. For example, assuming linear isotropic elasticity, characterized by a shear modulus G and Poisson's ration ;, then the critical hardening modulus for shear band bifurcation is computed as the solution of the following constrained maximization problem [64]

$$H_t^{cr} = (2G)\max\{L(n_i)\}; \, n_i n_i = 1$$

$$L = 2F_{ik}Q_{il}n_kn_l - F_{ij}Q_{kl}n_in_jn_kn_l - F_{ij}Q_{ij} - \frac{v}{1-v}\left(F_{ij}n_in_j - F_{kk}\right)\left(Q_{ij}n_in_j - Q_{kk}\right) \tag{3.16}$$

Computational results for the critical hardening modulus and the corresponding critical orientation angles of shear-bands for various constitutive models for non-associative, frictional elastoplastic materials are given in [50]. In [59] one finds the numerical procedures for evaluating the above constrained maximization problem, whereas analytic solutions of it are given in [2] and [6,7].

3.2 Shear Band Analysis in Plane-Strain Rectilinear Deformations

As an application we restrict our demonstrations here to a 2D-constitutive model for the stress rate. We consider a plane-strain rectilinear deformation of a sand specimen, as shown in Fig. 8. Such a deformation is performed in the biaxial apparatus [94], [26].

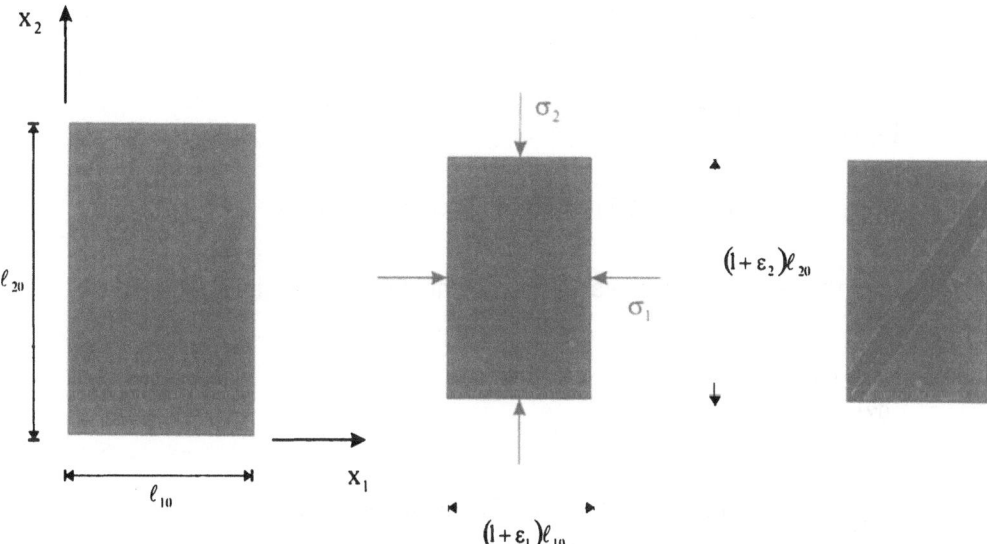

Fig. 8: Plane strain rectilinear deformation and shear-band bifurcation mode

3.2.1 Constitutive equations

The constitutive equations for the upper-bound linear comparison solid, corresponding to a rectlinear compression are conveniently derived in the coordinate system of principal axes of initial stress

$$\dot{\sigma}_{11} = L^u_{1111}\dot{\varepsilon}_{11} + L^u_{1122}\dot{\varepsilon}_{22}$$
$$\dot{\sigma}_{22} = L^u_{2211}\dot{\varepsilon}_{11} + L^u_{2222}\dot{\varepsilon}_{22} \qquad\qquad (3.18)$$
$$\dot{\sigma}_{12} = 2G\dot{\varepsilon}_{12}$$

where for the considered 2D-continuum model the components of the stiffnes tensor are given by the following expressions,

$$L^u_{1111} = G\{\kappa(1-\beta)(1-\mu) + (1+\kappa)h\}/\bar{h}$$
$$L^u_{1122} = G\{\kappa(1+\beta)(1-\mu) - (1-\kappa)h\}/\bar{h}$$
$$L^u_{2211} = G\{\kappa(1-\beta)(1+\mu) - (1-\kappa)h\}/\bar{h} \qquad \text{or} \quad L^u_{ijkl} = GL^{u*}_{ijkl} \qquad (3.19)$$
$$L^u_{2222} = G\{\kappa(1+\beta)(1+\mu) + (1+\kappa)h\}/\bar{h}$$

Here the following notation is used:

- G: elastic shear modulus

- $\kappa = \dfrac{1}{1-2\nu}$; ν : Poisson's ratio

- $\sin\phi_m = \mu(\gamma^p)$: mobilized Mohr-Coulomb friction coefficient

- $\sin\psi_m = \beta(\gamma^p)$: mobilized dilatancy coefficient after Hansen and Lundgren

- $h_t = \dfrac{d\mu}{d\gamma^p}$: tangent modulus

- $h = \dfrac{|\sigma_1 + \sigma_2|}{2G}h_t$: dimensionless tangent hardening modulus

- $\bar{h} = h + h_0$: dimensionless plastic modulus

- $h_0 = 1 - h_T$; $h_T = -\kappa\mu\beta$

3.2.2 Bifurcation analysis

By using these constitutive equations the bifurcation condition eq. (10) becomes

$$\begin{bmatrix} \Gamma_{11} & \Gamma_{12} \\ \Gamma_{21} & \Gamma_{22} \end{bmatrix} \begin{bmatrix} g_1 \\ g_2 \end{bmatrix} = \begin{bmatrix} 0 \\ 0 \end{bmatrix} \tag{3.20}$$

where the components of the acoustic tensor are given by the following equations

$$\begin{aligned}
\Gamma_{11} &= L^u_{1111} n_1^2 + G n_2^2 & \Gamma_{12} &= \left(L^u_{1122} + G\right) n_1 n_2 \\
\Gamma_{21} &= \left(L^u_{2211} + G\right) n_1 n_2 & \Gamma_{11} &= L^u_{2222} n_2^2 + G n_1^2
\end{aligned} \tag{3.21}$$

Let

$$\tan\theta = -\frac{n_1}{n_2} \tag{3.22}$$

be the shear-band inclination angle (Fig. 9). For non-trivial solutions for the jumps of the velocity gradient we obtain the following characteristic equation

$$\det(\Gamma_{ik}) = 0 \implies a\tan^4\theta + b\tan^2\theta + c = 0 \tag{3.23}$$

$$\begin{aligned}
a &= L^{u*}_{1111} \\
b &= L^{u*}_{1111} L^{u*}_{2222} - L^{u*}_{1122} L^{u*}_{2211} + L^{u*}_{2211} - L^{u*}_{1122} \\
c &= L^{u*}_{2222}
\end{aligned} \tag{3.24}$$

The condition for shear-band bifurcation is derived from the requirement that the above characteristic equation has real solutions. This condition is firstly met at a state C_B (B for bifurcation) for which the discriminant of the biquadratic equation (3.23) vanishes, and the double root is real,

$$D = b^2 - 4ac = 0; \quad \frac{b}{a} < 0 \tag{3.25}$$

For any state beyond C_B, there are four solutions for the shear-band orientation. According to the experimental evidence, however, the observed shear-bands belong to a single family of symmetric solutions.

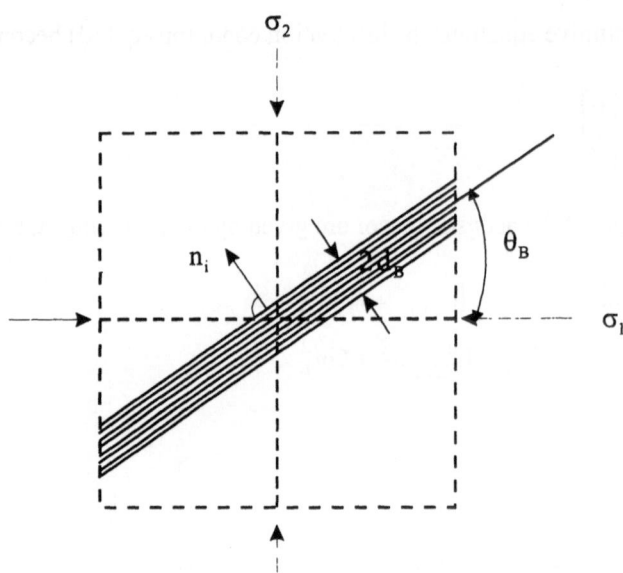

Fig. 9: Shear banding in domains under uniform state of stress, the full space solution

This observation justifies the selection of eq. (3.25) as the shear-band bifurcation condition. For the considered linear comparison solid, eq. (3.25) can be solved in terms of the critical hardening rate at the bifurcation point, resulting to the so-called Mandel solution for the critical hardening modulus [45]:

$$h = h_B; \quad h_B = \frac{(\mu_B - \beta_B)^2}{8(1-\nu)} \tag{3.26}$$

This means that cntinuous shear-band bifurcation in plane-strain deformations of non-associative frictional-dilatant materials, is predicted in the hardening regime of the mobilized friction coefficient. At C_B only two symmetric shear-band directions exist, given by the followng expressions

$$\theta_{1,2} = \theta_{3,4} = \pm\theta_B \tag{3.27a}$$

where

$$\theta_B = \arctan\left(\sqrt[4]{\frac{c}{a}}\right) = \arctan\left(\sqrt[4]{\frac{(1+\beta_B)(1+\mu_B)+(1/\kappa+1)h_B}{(1-\beta_B)(1-\mu_B)+(1/\kappa+1)h}}_B\right) \qquad (D=0; b<0) \tag{3.27b}$$

The dimensionless hardening modulus at bifurcation is a relatively small number $(h_B \ll 1)$. With the observation that usually, $(\phi_B - \psi_B) < 30°$, we obtain a simple formula for the shear-band orientation angle

$$\theta_B \approx \theta_V \qquad \theta_V = 45° + \frac{1}{2}\left(\frac{\phi_B + \psi_B}{2}\right) \qquad (3.28)$$

This formula was first proposed by Arthur et al. [1] on the basis of experimental observations and was subsequently proven theoretically and supported experimentally by Vardoulakis [84].

Let C_C (C for Coulomb) be the state of maximum stress obliquity. In this state the mobilized friction is maximum $h = h_C = 0$. Due to the Taylor rule at C_C the dilatancy angle is also maximum. It can be easily shown that at C_C two symmetric solutions for the shear-band orientations exist, namely [83]:

$$\theta_{1,2} = \pm\theta_C \qquad \theta_C = 45° + \frac{1}{2}\phi_C$$
$$\qquad\qquad\qquad\qquad\qquad\qquad\qquad\qquad (3.29)$$
$$\theta_{3,4} = \pm\theta_R \qquad \theta_R = 45° + \frac{1}{2}\psi_C$$

The first solution is the classical *Coulomb* solution whereas the second one is called the *Roscoe* solution. In case of associative plasticity C_B coincides with C_C; i.e.

$$\mu \equiv \beta \qquad \Rightarrow \theta_B = \theta_C = \theta_R = \theta_V \qquad (3.30)$$

This means that for an *associated flow-rule the Mohr-Coulomb failure criterion is derived from a bifurcation analysis*. We recall that the Mohr-Coulomb failure criterion is stating that shear-banding is occurring at the state of maximum stress obliquity and that the orientation of the shear-bands coincides with the planes across which the ratio of shear- to normal-stress is maximum

The bifurcation analysis presented above applies to a full-space domain under uniform state of stress (Fig. 9). In case however of a half-space domain under uniform stress, a shear band will intersect in general the free boundary. As pointed out by Benallal et al. [5], the effect of traction boundary conditions can be accounted for in the analysis by requiring that the same tractions exist at both sides of the shear band boundaries. This gives an additional condition [99], which together with the bifurcation condition (3.20) provides as the only solution the Roscoe solution,

$$h_B = h_C = 0 \qquad \theta_B = \theta_R = 45° + \frac{1}{2}\psi_C \qquad (3.31)$$

Since the dilatancy angle is always less than the friction angle, this theory predicts that shear bands should curve while approaching a boundary. Such a phenomenon is clearly seen in Fig. 10.

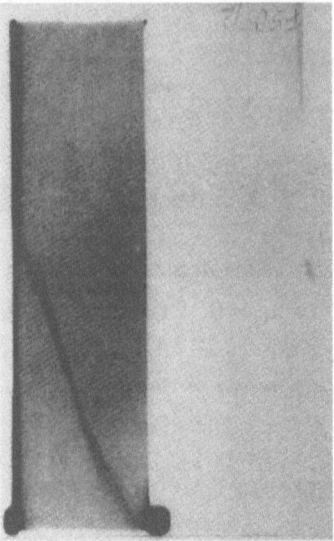

Fig. 10: Shear band emerging from the interior domain and curving as it approaches the boundaries of the specimen.

3.2.3 Bifurcation analysis of a biaxial test on sand

We consider as an example the pre-failure data from a biaxial experiment on dry, dense Dutch dune sand. As shown below the experimental results corroborate the simple Taylor stress-dilatancy rule, with

$$\bar{\mu} = \mu - \beta = \text{const.} \qquad (\bar{\mu} = 0.412)$$

Notice also that for the considered sand, tested in plane-strain compression with constant confining pressure $\sigma_c = 294$ kPa, it was found

$$G = 50. \text{ MPa}; \qquad \nu = 0.1$$

$$\mu = \frac{(\gamma^p / \gamma_{\text{ref}})}{1 + 1.393 \, (\gamma^p / \gamma_{\text{ref}})}; \quad \gamma_{\text{ref}} = 0.00179$$

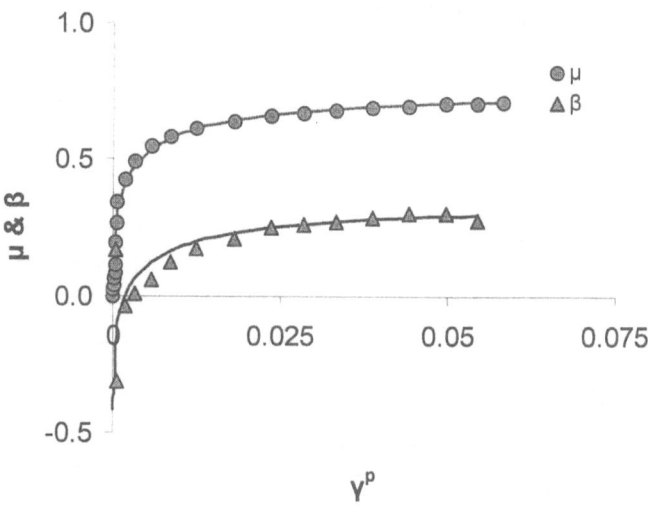

Fig. 11: Mobilized friction and dilatancy in plane-strain test on dense Dutch dune sand

Using these data on can compute the bifurcation strain γ_B, the corresponding critical hardening rate h_B and shear-band inclination angle θ_B using the formulae of the previous section. Experimental and computational results are summarized in Table 1 below. From this Table the following conclusions can be reached: (1) A typical result of flow theory as applied to granular media is that the theoretical estimate for the shear-band bifurcation strain is significantly less than a lower-bound for the shear strain at which shear-band formation is observed experimentally (here a relative error of 33% in bifurcation strain). What is not seen directly from this Table but can be easily demonstrated by parameter analysis, is that the theoretical prediction for γ_B, depends sensitively on the assumed value for the elastic shear modulus [87]. All these observations mean that nonassociative flow theory of granular media underestimates the shear-band bifurcation strain, and in that sense non-associativity severely destabilizes the constitutive response. This defect of standard flow theory for granular media is undesirable, as far as realistic constitutive modeling is concerned, and has prompted the research for further modifications of the standard plasticity model. Such modifications are the so-called *deformation* theories of plasticity [84], *yield vertex* plasticity models [69] and *non-coaxial* plasticity models [61]. (2) From Table 2 follows that the prediction in the shear band orientation angle has a relative error of about 5% and that the approximate formulae (3.26) and (3.28) give fairly good approximations of the exact theoretical results.

*Table 2.: Experimental and theoretical results on shear band formation
in a biaxial compression test on fine grained Dutch dune sand*

	γ_B	h_B	ϕ_B	ψ_B	θ_B
Experiment:					
• Lower bound	0.06	0.008	44.2°	16.6°	62.5°
• Upper bound	0.10	-	≈ 45.°	-	-
Theory:					
• Exact	0.04	0.025	43.4°	16.°	59.2°
• Approximate	-	0.024	-	-	59.8°

3.2.4 Remark on non-associate perfect-plastic laws

Starting with Hill [33], non-associated flow-rules in frictional materials have been widely discussed in the context of rigid, perfectly-plastic, or elastic, perfectly-plastic material behavior; cf. [72]. The idea was to approximate true material behavior with the relatively simple constitutive law of perfect plasticity. All these models lead to two sets of distinct characteristics, the so-called statical and kinematical characteristics; cf. eqs. (3.29). Accordingly, the domains of solution for stresses and velocities do not coincide. The existence of two sets of characteristics triggered extensive experimental investigations aiming at determination as whether or not the zero-extension lines in soils coincide with the static characteristics of the perfectly plastic solid e.g. [13].

A simple shear-band analysis for a hardening material obeying a non-associate flow rule, proves that shear banding in plane- strain occurs at positive hardening rates. Perfect plasticity presumes, however, that plastic deformation and formation of slip lines occurs at zero hardening rate. Thus, the perfectly plastic non-associate model cannot be a good approximation of the hardening non-associate plastic model. This point has been overlooked in the literature, where perfect plasticity was adopted regardless of the flow-rule. Apparently, the adoption of perfect plasticity for soils has been borrowed from metal plasticity, where the associative flow rule satisfactorily describes plastic deformation; it results directly from incompressibility, pressure insensitivity and coaxiallity. In all other materials, perfect plasticity is justified only if the flow-rule is associative.

3.3 Localization and Strain Softening Behavior

Shear-bands are not the only forms of localized behavior. In general one distinguishes among the following three types of localizations:

- *Volume instabilities* or shear-bands, wich within an acceleration waves analysis correspond to material body waves.
- *Surface instabilities* or extension boundary layers; they correspond to material surface waves.
- *Interfacial instabilities* (shear and/or compaction interface layers); they correspond to material interfacial waves.

Benallal et al. [5] have formulated necessary and sufficient conditions, which exclude the above localizations and assure the well-posedeness of the considered incremental boundary-value problem, which is defined for the upper-bound linear comparison solid. For further reading on the subject we should refer also to the comprehensive works of Bigoni and co-workers [6-9] on questions of non-uniqueness, localization and loss of strong ellipticity for elastoplastic materials with non-associative flow-rules and of Loret et al. [42] on acceleration waves and loss of hyperbolicity in elastoplastic media. It should be finally mentioned that the bulk of experimental work refers to granular, cohesionless materials or sandstones like: a) volume instabilities [84,95], [19-23], [74]; Tatsuoka et al. 1990; Ord et al. 1990). b) surface instabilities [60]. c) Interfacial instabilities [11], [12], [76] and [41].

As shown above, for a great class of tensorial fields (like plane-strain problems) continuous (localized) bifurcations in the form of shear-bands are already possible in the hardening regime of the material as a result of volumetric non-associativity [44,45], [69], [84. On the other hand for discontinuous localized bifurcations, one has to examine the possibility that elastic unloading occurs outside the shear band while continued elastic-plastic loading occurs within the band. In frictional geomaterials this solution can be justified if either, (i) the mean stress jumps across the shear band, or (ii) *strain softening* is occurring inside the shear-band.

As a general observation we may say that material softening is in most cases localized in narrow zones of material degradation; cf. the fundamental papers by Desrues et al. [22,23]. Moreover we remark that the *formation of localizations is marked by the existence of continuous or discontinuous bifurcations which in turn mark the end of classical continuum description.* In order to overcome the mathematical difficulties which arise from material strain-softening and to be able to proceed in the post-bifurcation regime with a predictive tool, one has to regularize the underlying mathematical model. This procedure is not arbitrary, since it must reflect realistically the essential physical mechanisms that govern localization phenomena: Within the frame of elastoplasticity theory, localizations are viewed as zones of high plastic strain gradients. This interpretation leads naturally to higher gradient extensions of the classical plasticity theory as will be outlined in the next sections.

4. A 2nd Gradient Plasticity Model for Granular Media
4.1 Micromechanical Considerations

In the mechanics of materials we distinguish among continuum and discrete models. Such a distinction is of course an ancient one, which can be traced back to the philosophical controversy between atomists and stoiks (see e.g. [43]). In our times this controversy still persists among those who believe that quantities which enter a continuum description should be seen as 'averages' of some other underlying 'microscopical' properties of the material, and those who do not accept this point of view. For example Truesdell and Noll in the introduction of *Non-Linear Field Theories of Mechanics* [79, sect.3], state explicitly that:"..Widespread is the misconception that those who formulate continuum theories believe matter 'really is' continuous, denying the existence of molecules. This is not so. Continuum physics presumes *nothing* regarding the structure of matter. It confines itself to relations among gross phenomena, neglecting the structure of the material on a smaller scale. Whether the continuum approach is justified, in any particular case, is a matter, not for the philosophy or methodology of science, but for the **experimental test**...". It may be so that the 'statistical' approach to continuum mechanics is not always practical, it helps however someone to become familiar with abstract concepts by thinking in terms of geometric objects like grains, atoms, their motions in space and the forces which act upon them.

Micromechanically granular media are seen as assemblies of individual grains, randomly packed and in partial contact with their neighbors. One may arrive at plausible kinematcal and statical continuum concepts by considering the motion and forces at grain level by using for example direct computer simulations [3], [17], [39], [57]. From such simulations we get that grains move more or less as **rigid bodies**, and their kinematics is very well described by grain velocity and grain spin. Grain contacts on the other hand are point-like, and transmit mainly forces and at a lesser degree couples. With such a rough micro-mechanical model in mind one may proceed towards a continuum description of the granular assembly by computing average properties, appropriately defined, so that they are representative for a small cluster, containing about 10^2 to 10^3 grains. This is by far not a trivial task and one is referred to the works of Cambou and co-workers and the literature referenced there ; see e.g. [28].

4.1.1 Granular Kinematics

Let us consider the model of two grains in contact. The relative velocity Δv_i of one grain with respect to the other may be decomposed into a component $\Delta v^{(n)} = \Delta v_i n_i$ normal to the common tangential plane with unit normal vector n_i, and into a tangential component, $\Delta v_i^{(t)} = \Delta v_i - \Delta v^{(n)} n_i$, perpendicular to it. For defining continuum point properties we have first to map on an elementary spherical surface (S) all these contacts of grain pairs, appearing in a representative volume, and compute the probability density function of contact normals. For simplicity we may assume for example that contact normals are uniformly distributed. Accordingly we introduce the following definitions:

1. The mean amplitude of the normal component of the relative velocity vector defines an average measure of the change of the distance of two grains in close proximity to each other; i.e. a local measure for rarefaction or densification.

$$\dot{\varepsilon}^P = \{\Delta v^{(n)}\} = \frac{1}{S} \int \Delta v^{(n)} dS \qquad (4.1)$$

2. The mean amplitude of the tangential component of the relative velocity vector defines an average measure of the relative slip of two grains in contact

$$\dot{\gamma}^P = \left\{ \sqrt{\Delta v_i^{(t)} \Delta v_i^{(t)}} \right\} \qquad (4.2)$$

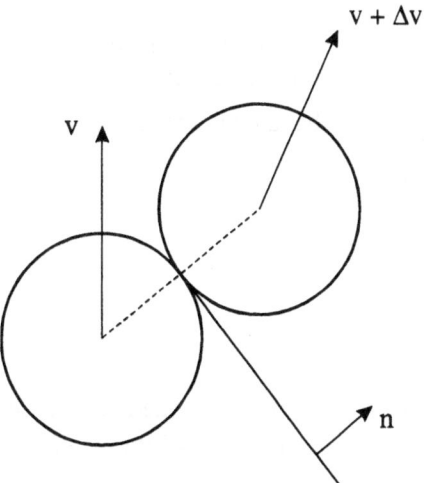

Fig. 12: Micromechanical model of granular kinematics, grain slip mode

We assume that the grains are embedded in a continuum and employ a local coordinate system which is rotating together with the neighborhood of the considered grain (i.e. the coordinate system rotates with the spin) then we can write,

$$v_i = D^p_{ij} x_j \tag{4.3}$$

where D^p_{ij} is a symmetric tensor. The superscript p (p for plastic) simply denotes the fact that in this analysis all 'elastic' grain deformation is suppressed, and the velocity field v_i corresponds to irreversible grain motion. The relative velocity field which is characteristic for two grains in contact at their periphery is computed by interpolation as

$$\Delta v_i = 2R_g D^p_{ij} n_j \tag{4.4}$$

where R_g is the radius of the grain. The normal and tangential component of the relative velocity at the contact point are given by the following expressions

$$\Delta v^{(n)} = 2R_g D^p_{ij} n_i n_j$$

$$\tag{4.5}$$

$$\Delta v^{(t)}_i = 2R_g \left(D^p_{ij} - D^p_{kl} n_k n_l \delta_{ij} \right) n_j$$

Using these expressions one can compute the averages defined above through eqs. (4.1) and (4.4). These average kinematical properties coincide with the 1st and 2nd deviatoric invariant of the irreversible part D_{ij}^{p} of the rate of deformation, respectively [99],

$$\dot{\varepsilon}^{p} = I_{1D^{p}} = D_{kk}^{p}$$

$$\dot{\gamma}^{p} = J_{2D^{p}} = \sqrt{2D_{ij}^{\prime p}D_{ij}^{\prime p}}$$

(4.6)

Above mathematical derivations can be also performed if one considers relative grain rotation as well, arriving at suitable generalizations of the J_{2}-invariant within a Cosserat (micro-polar) continuum [40], [54].

From the macroscopic point of view and for granular media under shear irreversible shear strains and irreversible volume changes are linked together. This is usually expressed by the well-known phenomenological dilatancy constraint

$$\dot{\varepsilon}^{p} = \beta\dot{\gamma}^{p}$$

(4.7)

where β is called the *mobilized dilatancy* coefficient. Thus there is great class of deformations where there is no need to treat separately irreversible volume changes from irreversible shear deformations. Moreover the constraint (4.7) allows us to choose either $\dot{\varepsilon}^{p}$ or $\dot{\gamma}^{p}$ or a linear combination of them (like e.g. the 1st order plastic work) as a measure for plastic deformation. We chose here as hardening parameter the average interparticle slip $\dot{\gamma}^{p}$, and set

$$\dot{\psi} \equiv \dot{\gamma}^{p}$$

(4.8)

In order to make this assumption more plausible we remark that for a particular deformation process equations (4.7) and (4.8) may be integrated. As is common in all rate-independent plasticity theories, time can be replaced by the hardening parameter, and we may write,

$$\gamma^{p} = \int d\psi \quad ; \quad \varepsilon^{p} = \int \beta(\psi)d\psi = D(\psi)$$

(4.9)

On the other hand, neglecting elastic deformations, plastic volume changes are directly reflected in changes of porosity,

$$\frac{dn}{1-n} = d\varepsilon^{p} \Rightarrow n = n_{0} + (1 - \exp(-D(\psi))) \approx n_{0} + D(\psi)$$

(4.10)

From the figure below we readily see that the relation between ψ and n is (for dense sand) approximately linear,

$$n \approx n_0 + c\psi \tag{4.11}$$

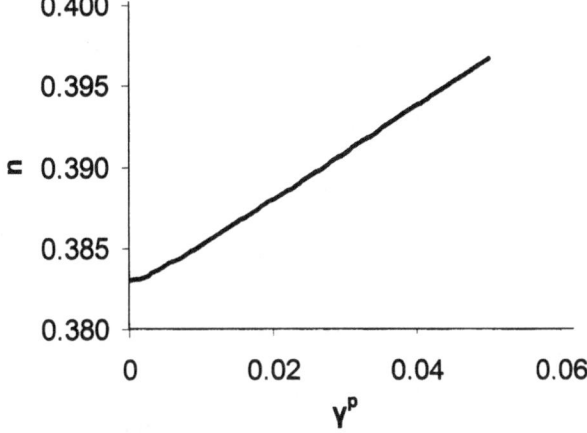

Fig. 13: Porosity evolution during a biaxial compression test on dense, Dutch dune sand

Accordingly ψ is a suitable measure for plastic deformation, and simply plays the role of 'materials own time' (it is a measure of 'endochronic' time, according to Valanis' terminology). Changes in porosity are directly linked to ψ due to the internal coupling between intergranular slip and dilatancy. Porosity is in turn a directly observable material property that accounts for irreversible structural changes of the granular assembly.

Above considerations hold also in case when the deformation is not locally homogeneous like it is the case when localization of the deformation takes place. In this case the above introduced assembly averages should be taken along the long direction of the shear band, since perpendicular to it strong changes in porosity are observed [23]. This last remark introduces the question as of who to interpret the various continuum properties in case of localized deformation. In that sense we recall below some mathematical concepts that pertain to averaging.

4.1.2 Remark on averaging

Let us for simplicity consider an 1D-case of a field $y = f(x)$, whose mean value is computed over a small but finite averaging length L around a point x

$$<y> = \frac{1}{L} \int_{-L/2}^{L/2} f(x + \xi)d\xi \tag{4.12}$$

If the field f(x) varies linearly in the considered region around point x, then it is approximated locally by a linear function, using an 1-term Taylor series expansion of the function f around point x,

$$f(x + \xi) \approx f(x) + f'(x)\xi \qquad (4.13)$$

In the trivial case when the field f is indeed constant then the first and all higher derivatives vanish and indeed the local value coincides with the average value. However, this is true in case when the field varies locally linearly. Indeed we may then identify the field with its mean value over the considered averaging length, because, following the 'trapezoidal' rule of integration, the mean value of a linearly varying field in an interval   is equal to the value of it in the *midpoint* of the sampling interval

$$y = <y> \qquad (4.14)$$

In this case the 'local' value y and the 'non-local' value <y> coincide. In a field theory where local values are identified with means according to the rule (4.14) are called *simple theories* and the corresponding continua *locally homogeneous.*
In case however where the considered field varies quadratically in the sampling region and a linear approximation is not sufficient, then we have to approximate it at least by a 2-term Taylor-series expansion around point x,

$$f(x + \xi) \approx f(x) + f'(x)\xi + \frac{1}{2}f''(x)\xi^2 \qquad (4.15)$$

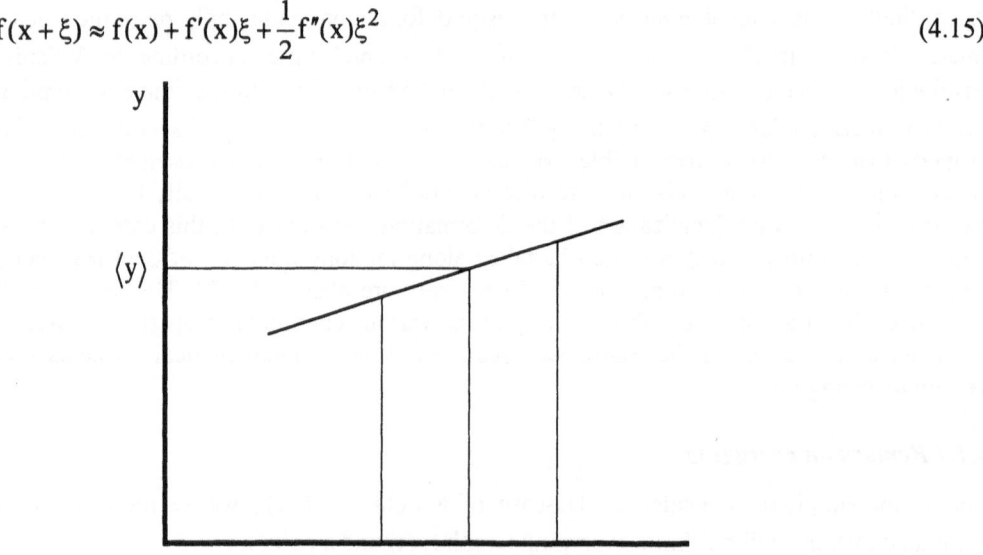

Fig. 14: The simple trapezoidal rule of averaging

We notice that in the midpoint integration rule the effect of the first derivative is null. Thus for 'quadratically' varying fields, computational rule (4.14) must be enhanced, so as to incorporate the effect of curvature,

$$y = <y> - \frac{1}{48} L^2 \left(\frac{d^2 y}{dx^2} \right)_x \qquad (4.16)$$

Field theories which are based on averaging rules that include the effect of higher gradients are called *higher gradient* theories. In particular above rule (4.16) represents a 2nd gradient rule, and can be readily generalized in 2 and 3 dimensions.

In order to illustrate the above averaging procedures, eqs. (4.14) and (4.16), we choose as an example the function y=sin(x) and select values of it at intervals $dx = \pi/6$. The corresponding values play the role of 'data'.

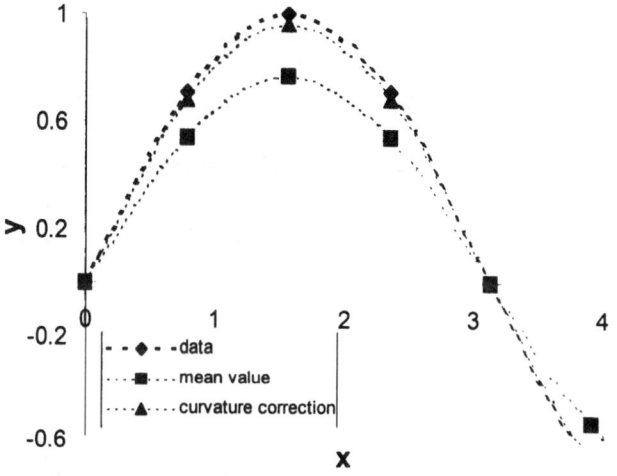

Fig. 15: The enhanced averaging rule ($y = \sin x; L = 4\Delta x$)

In the figure above 'data' are represented by black rhombic symbols. The square symbols are mean values of the data over intervals that contain 5 data points, and correspond to an evaluation according to eq. (4.14). Triangles are computed from these mean values by correcting them according to the above enhanced rule, eq. (4.16), which includes the curvature. In this example the second derivative is computed as 2nd finite difference directly from the data. (Notice that data differentiation is not a well-posed problem, and in order to obtain some reasonable results, in general, some data-smoothing is necessary prior to differentiation.). From the figure we observe that the two averaging rules reproduce very well the chosen function around its inflection points, where indeed the behavior is locally linear. At extreme points however, the behavior is at least quadratic, and the curvature correction is necessary.

Above averaging considerations find many applications in continuum mechanics. As a typical example one may refer to the *Navier-Stokes* equations for the vorticity in two-dimensional flow,

$$\frac{d\Omega}{dt} = \eta \nabla^2 \Omega$$

with η being the kinematic viscosity. According to the above considerations, the r.h.s. of the above equation can be seen as the difference between spatial average and local value,

$$\nabla^2 \Omega = \frac{1}{\ell^2}\left(<\Omega> - \Omega\right)$$

with ℓ being proportional to the averaging length. Thus in viscous fluids, spatial variability of the kinematic fields goes together with temporal variability. If conversely one assumes that vorticity is locally homogeneous $(\Omega =< \Omega >)$, then vorticity is conserved $(d\Omega/dt = 0)$. Indeed the N.-S. equations describe flows which are essentially fluctuating. These flows are called turbulent and result in multi-scale spatio-temporal fluctuations. This means in turn that even in small scale, spatial averages of the kinematic properties of the flow do not suffice for its complete description.

Another example is the hydrodynamic pressure p which is exerted on the free surface of a fluid due to *surface tension* T. This pressure is computed on the basis of a membrane analogy, and is found to be proportional to the mean surface curvature, which is computed in turn from its deflection $w = w(x,y)$ from the at rest position,

$$p = T\left(\frac{1}{R_1} + \frac{1}{R_2}\right) \approx -T\left(\frac{\partial^2 w}{\partial x^2} + \frac{\partial^2 w}{\partial y^2}\right)$$

Accordingly the effect of surface tension is only then felt when the disturbances iare relatively small, as is the case in capillary surface waves, where T changes sign and direction over small distances [73].

4.2 A Gradient Plasticity Model for Sand

As already mentioned above in section 3.3 the classical plasticity models break down at the state of localization, and must be suitably modified. Localization is manifested by intense shearing and dilatancy inside a narrow zone of few grains thick. Dilatancy and other structural changes (like e.g. grain-column buckling) lead to material strain softening that is described phenomenologically by a decreasing friction coefficient, $\mu \downarrow$: $h_t = (d\mu/d\psi) < 0$.

In order to counterbalance the destabilizing effect of material strain softening, we modify the yield function, by adopting the concept of *kinematic hardening* and we set

$$F = F(\tau_{ij}, \psi) \tag{4.17}$$

where τ_{ij} is a *reduced stress*,

$$\tau_{ij} = \sigma_{ij} - \alpha_{ij} \tag{4.18}$$

and α_{ij} is a *back stress*; cf. [51]. Here the back stress must be defined in such a way that it induces strengthening as soon as the deformation ceases to be locally homogeneous.
For simplicity we restrict our analysis into two dimensions and assume that the back stress is isotropic. Accordingly the reduced stress differs from the Cauchy stress only in its isotropic part,

$$\sigma_{ij} = s_{ij} + \sigma\delta_{ij}; \quad \alpha_{ij} = \alpha\delta_{ij} \Rightarrow \tau_{ij} = s_{ij} + (\sigma - \alpha)\delta_{ij} \tag{4.19}$$

In particular we adopt a 'modified' Mohr-Coulomb yield surface [92,93]

$$F = \tau + (\sigma - \alpha)\,\mu(\psi) \tag{4.20}$$

where μ is the mobilized friction coefficient. As in ordinary flow theory of plasticity, elastic strain-rates are defined through the elasticity law for the stress-rates, $\dot\sigma_{ij} = C^e_{ijkl}\dot\varepsilon^e_{ij}$, and plastic strain rates through a the flow rule $\dot\varepsilon^p_{ij} = \psi(\partial Q/\partial\tau_{ij})$. The plastic potential is expressed in analogy to the yield function in terms of the mobilized dilatancy coefficient

$$Q = \tau + (\sigma - \alpha)\,\beta(\psi) \tag{4.21}$$

With the decomposition of the plastic strain in deviatoric and volumetric part, $\dot\varepsilon^p_{ij} = \dot e^p_{ij} + \dot\varepsilon^p\delta_{ij}/2$, we recover that $\dot\gamma^p = \sqrt{2\dot e^p_{ij}\dot e^p_{ij}} \Rightarrow \dot\gamma^p = \psi$ and $\dot\varepsilon^p = \dot\varepsilon^p_{kk} \Rightarrow \dot\varepsilon^p = \beta\psi$.
Based on the previous remark on averaging we propose the following constitutive assumption for a back stress that counterbalances the effect of strain softening and localization:

Non-local Assumption: Back stresses depend on degree of local inhomogeneity of the granular fraction of the material. In particular we assume that back stress evolves only in regions where the local value of the solid fraction differs significantly from its local average; i.e. if $r = 1 - n$, denotes the solid volume fraction, then

$$\dot\alpha \propto (<\dot r> - \dot r) \Rightarrow \dot\alpha \approx c\nabla^2\dot r \tag{4.22}$$

This is expected to be the case inside zones of strain localization, where substantial rarefaction is taking place. The material constant c in eq. (22) has the dimension of force. If we normalize it by the elastic compression modulus (or any other material constant with dimension of stress), we obtain a new material constant with dimension of length. This constant is usually called the *material length*, and is correlated to the grain size of the granular material

$$\ell_c = \sqrt{c/K} \Rightarrow c = K\ell_c^2 \tag{4.23}$$

If we introduce in the above evolution equation (4.22) the relation between porosity and plastic volumetric strain, as well as the dilatancy constraint we obtain to some degree of approximation the following form of the evolution law for the back stress

$$\dot{\alpha} \approx -K\beta(1-n)\ell_c^2 \; \nabla^2\psi \tag{4.24}$$

The minus sign in the r.h.s. of eq. (4.23) is understood as follows: Inside the localized zone porosity varies, and assumes a maximum value somewhere at the center of it, and with that we expect that $\nabla^2\psi < 0$. This means that for dilatant material $(\beta > 0)$ we expect that $\dot{\alpha} > 0$, and that the contribution of the back stress in the yield function, eq. (4.20), will be effectively like an increase of the local confining pressure. In other words, in regions of material strain softening, a larger than local region contributes to the overall strength and non-local effects lead to overall strengthening of the material.

In the considered case, Prager's consistency condition, $F = \dot{F} = 0$, becomes

$$-\ell^2\nabla^2\psi + \psi = \frac{1}{H}B_{kl}\dot{\varepsilon}_{kl} \tag{4.25}$$

where H is the plastic modulus (cf. eqs. 2.27, 2.30 and 3.19) and ℓ is a material length. Here,

$$H = G\bar{h} > 0; \quad \bar{h} = 1 - h_T + h$$

$$h = \frac{|\sigma - \alpha|}{G}h_t < 0; \quad h_T = -\kappa\mu\beta < 0 \quad (\beta > 0); \quad \kappa = \frac{K}{G} \tag{4.26}$$

$$\ell^2 = (1-n)\frac{(-h_T)}{\bar{h}}\ell_c^2 > 0 \quad (\beta > 0)$$

The consistency condition (4.25) is a differential equation which links essentially the *micro-deformation* $\dot{\varepsilon}_{ij}^P = \psi(\partial Q/\partial\tau_{ij})$, to the *macro-deformation* $\dot{\varepsilon}_{ij}$. For homogeneous ground plastic-strain states, the term with the Laplacian on the l.h.s. of eq. (4.25) vanishes identically and the consistency condition collapses to that of the classical flow theory of plasticity; i.e. to a monomial equation which can be readily solved in terms of the plastic multiplier ψ. In general direct elimination of ψ will not be possible and one has to carry the consistency condition as an additional field equation together with the balance equations, and to treat ψ as an additional degree of freedom [53]. Here we suggest an approximate procedure, which indeed allows for elimination of ψ from the set of governing equations and simplifies the problem drastically [93]. In operator form eq. (4.25) reads,

$$(1 - \ell^2 \nabla^2)\dot{\psi} = \frac{1}{H} B_{kl}\dot{\varepsilon}_{kl} \Rightarrow \dot{\psi} \approx \{1 + \ell^2 \nabla^2 + O(\ell^4)\} \left(\frac{1}{H} B_{kl}\dot{\varepsilon}_{kl} \right)$$

or

$$\dot{\psi} \approx \frac{<1>}{H} B_{kl}\left(\dot{\varepsilon}_{kl} + \ell^2 \nabla^2 \dot{\varepsilon}_{kl} \right)$$ (4.27)

where, $\langle \bullet \rangle$ denote again the Foeppl-Macauley brackets so that $\dot{\psi} \geq 0$. Above approximation allows to derive explicitly the rate constitutive equations of *gradient flow theory of plasticity* [99]

$$\dot{\sigma}_{ij} = L_{ijkl}^{ep} \dot{\varepsilon}_{kl} - L_{ijkl}^{p} \ell^2 \nabla^2 \dot{\varepsilon}_{kl}$$ (4.28)

These constitutive equations are a straightforward *singular perturbation* of the ones of classical flow theory equations (cf. eqs. 2.24 ff).

7.4.3 Estimation of Shear-Band Thickness

The differential equations that govern continued equilibrium from a given configuration C of a soil body are the rate-equilibrium equations

$$\partial_1 \dot{\sigma}_{11} + \partial_2 \dot{\sigma}_{21} = 0$$

$$\partial_1 \dot{\sigma}_{12} + \partial_2 \dot{\sigma}_{22} = 0$$ (4.29)

We assume that at C the fields of initial stress and hardening parameter vary slowly in space. Accordingly, the components of the plastic stiffness tensor L_{ijkl}^{p} may be treated as constants. For convenience, the governing equations are written in the coordinate system of the principal axes of the stress tensor in C. Under fully loading conditions we obtain the following set of rate-constitutive equations

$$\dot{\sigma}_{11} = L^u_{1111}\dot{\varepsilon}_{11} + L^u_{1122}\dot{\varepsilon}_{22} - \ell^2 L^{pl}_{1111}\nabla^2\dot{\varepsilon}_{11} - \ell^2 L^{pl}_{1122}\nabla^2\dot{\varepsilon}_{22}$$

$$\dot{\sigma}_{22} = L^u_{2211}\dot{\varepsilon}_{11} + L^u_{2222}\dot{\varepsilon}_{22} - \ell^2 L^{pl}_{2211}\nabla^2\dot{\varepsilon}_{11} - \ell^2 L^{pl}_{2222}\nabla^2\dot{\varepsilon}_{22} \qquad (4.30)$$

$$\dot{\sigma}_{12} = 2G\dot{\varepsilon}_{12}$$

where $L^u_{ijkl} = L^e_{ijkl} + L^{pl}_{ijkli}$ are the components of the stiffness tensor of the upper-bound classical linear comparison solid, given explicitly by eqs. (3.19), and

$$L^{pl}_{1111} = G(\kappa\mu + 1)(\kappa\beta + 1)/\bar{h}$$

$$L^{pl}_{1122} = G(\kappa\mu - 1)(\kappa\beta + 1)/\bar{h}$$

$$L^{pl}_{22111} = G(\kappa\mu + 1)(\kappa\beta - 1)/\bar{h} \qquad (4.31)$$

$$L^{pl}_{2222} = G(\kappa\mu - 1)(\kappa\beta - 1)/\bar{h}$$

Introducing these expressions into the equilibrium equations results finally into the following system of partial differential equations for the components of the velocity vector

$$\ell^2 L^{pl}_{1111}\nabla^2\left(\partial^2_{11}v_1\right) + \ell^2 L^{pl}_{1122}\nabla^2\left(\partial^2_{21}v_2\right) -$$
$$L^u_{1111}\partial^2_{11}v_1 - G\partial^2_{22}v_1 - \left(L^u_{1122} + G\right)\partial^2_{12}v_2 = 0 \qquad (4.32a)$$

$$\ell^2 L^{pl}_{2211}\nabla^2\left(\partial^2_{12}v_1\right) + \ell^2 L^{pl}_{2222}\nabla^2\left(\partial^2_{22}v_2\right) -$$
$$\left(L^u_{2211} + G\right)\partial^2_{12}v_1 - G\partial^2_{11}v_2 - L^u_{2222}\partial^2_{22}v_2 = 0 \qquad (4.32b)$$

These partial differential equations constitute a *singular perturbation* of the classical ones; i.e. they reduce to the classical equations if the material length ℓ is allowed to go to zero. If one is interested in studying boundary-value or eigen-value problems which involve some geometric dimension L of a soil structure, then the coordinates x_i must be non-dimensionalized properly (by L) and the highest derivatives in eqs. (4.32) are then multiplied by the number $(\ell/L)^2 \ll 1$. In this case above equations reduce to the ones that govern deformation in a classical continuum. If, on the other hand, one is interested in determining whether or not shear-bands exist, then one has to investigate equilibrium across two adjacent planes at a distance $2d_B$ that correspond to shear-band boundaries. With the assumption that $(\ell/d_B) = O(1)$, the higher order derivatives in eqs. (4.32) become then essential. The shear band thickness $2d_B$ is called then an *internal length* of the problem, and above consideration means simply that an internal length scales with the material length per definitionem.

Accordingly the above differential eqs. (4.32) can be investigated for the special case where solutions are sought that correspond to the localization of deformation into narrow zones of intense shear, the so-called shear-bands. According to Fig. 16, the (x_1, x_2)-coordinate system is chosen in such a fashion that the x_1-axis coincides with the minor minimum principal stress σ_1 in C. Let us assume that a shear-band is forming that is inclined with respect to the x_1-axis at an angle θ. A new coordinate system is introduced with its axes parallel and normal to the shear band

$$x = x_1 n_2 + x_2 n_1 \quad ; \quad y = -x_1 n_1 + x_2 n_2 \tag{4.33}$$

where

$$n_1 = -\sin\theta \quad ; \quad n_2 = \cos\theta \tag{4.34}$$

is the unit vector that is normal to the shear band axis.

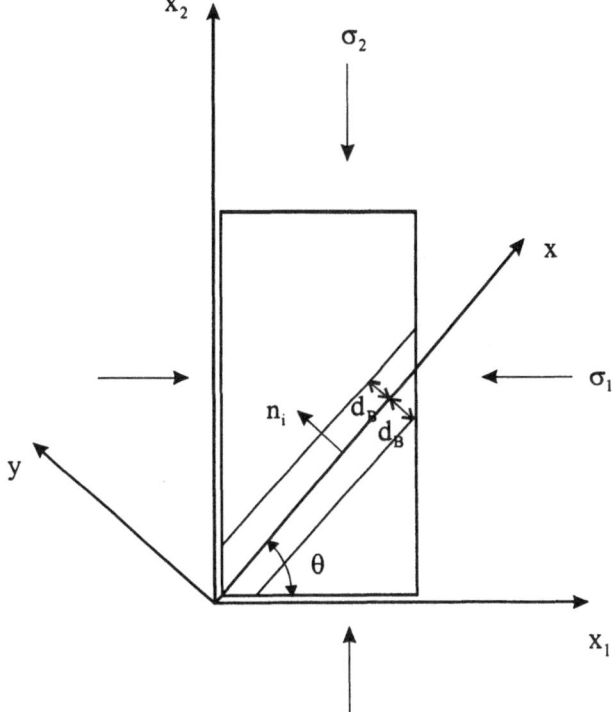

Fig. 16 : Shear band in an element test

By assuming that all field properties related to the forming shear band do not depend on the longitudinal x-coordinate and setting $(\bullet)' \equiv d/dy$, above equations reduce to the following system of ordinary differential equations:

$$\ell^2 L^{pl}_{1111} n_1^2 v_1^{(4)} + \ell^2 L^{pl}_{1122} n_1 n_2 v_2^{(4)} - \Gamma_{11} v_1'' - \Gamma_{12} v_2'' = 0$$
$$\ell^2 L^{pl}_{2211} n_1 n_1 v_1^{(4)} + \ell^2 L^{pl}_{2222} n_2^2 v_2^{(4)} - \Gamma_{21} v_1'' - \Gamma_{22} v_2'' = 0$$

(4.35)

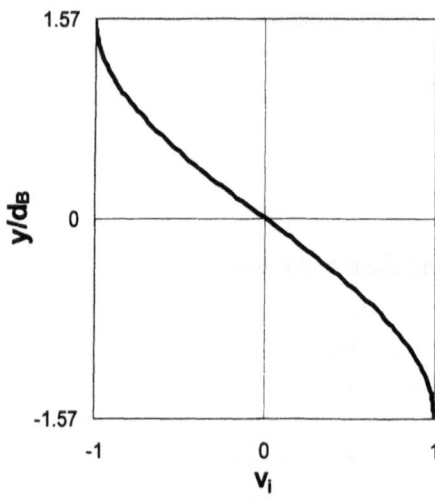

Fig. 17: Shear-band velocity field for linearized 2nd Gradient theory

We search for periodic solutions of the form

$$v_i = -\zeta_i \sin(Qy) \quad (i = 1,2)$$

(4.36)

with the *boundary condition*

$$y = \pm d_B : v_i = \mp \zeta_i \Rightarrow Q = \frac{\pi}{2d_B}$$

(4.37)

Then eqs. (4.35) yield

$$\begin{bmatrix} b_{11} & b_{12} \\ b_{21} & b_{22} \end{bmatrix} \begin{bmatrix} \zeta_1 \\ \zeta_2 \end{bmatrix} = \begin{bmatrix} 0 \\ 0 \end{bmatrix}$$

(4.38)

where

$$\begin{bmatrix} b_{ij} \end{bmatrix} = \begin{bmatrix} \ell^2 L^{pl}_{1111} n_1^2 Q^4 + \Gamma_{11} Q^2 & \ell^2 L^{pl}_{1122} n_1 n_2 Q^4 + \Gamma_{12} Q^2 \\ \ell^2 L^{pl}_{2211} n_1 n_2 Q^4 + \Gamma_{21} Q^2 & \ell^2 L^{pl}_{2222} n_2^2 Q^4 + \Gamma_{22} Q^2 \end{bmatrix}$$

(4.39)

For non-trivial solutions in terms of ζ_i the above system of equations (4.38) results to the following condition for the shear band thickness

$$a_0(n_1,n_2)(Q\ell)^2 + a_1(n_1,n_2) = 0 \tag{4.40}$$

where the coefficients of the above 'dispersion' equation are quadratic forms of the components of the unit normal vector

$$a_0 = R_{11}\Gamma_{22}n_1^2 - (R_{21}\Gamma_{12} + R_{12}\Gamma_{21})n_1 n_2 + R_{22}\Gamma_{11}n_2^2 = 0 \tag{4.41}$$

$$a_1 = \det(\Gamma_{ik})$$

It should be noticed that:
(1) continuum for any n_i the quadratic form $a_1(n_1,n_2)$ changes sign at the bifurcation point of the classical,

$$a_1(n_1,n_2) = \begin{cases} \geq 0 \text{ for}: \psi \leq \psi_B \\ < 0 \text{ for}: \psi > \psi_B \end{cases} \tag{4.42}$$

(2) Condition $a_1 = 0$ coincides with the classical bifurcation condition, eq. (3.23).
(3) The quadratic form $a_0(n_1,n_2)$ determines the type of the governing partial differential equations. It turns out that always $a_0(n_1,n_2) > 0$, which means that *the system of partial differential equations (4.32) is always elliptic*, as opposed to the classical system of governing equations which is of changing type, namely turning from elliptic to hyperbolic at the point of classical bifurcation.

Following these observations we conclude that prior to classical bifurcation there is no real solution for the shear band thickness; i.e. there exist no localization solution. At the classical bifurcation point the shear-band thickness is infinite as compared to the material length ℓ; rapidly decreasing in the post-bifurcation regime. The later is a well established qualitative result that should hold for any linear analysis that is based on a physically sound constitutive theory. In order to demonstrate this finding we extrapolated data from biaxial experiments from the hardening regime into the softening regime utilizing a continuum mixtures-theory for granular materials, described in Refs. [89] and [99]. The assumed material mobilized friction and dilatancy functions are depicted in Fig. 18. Using this as input one can estimate the shear-band thickness by evaluating the dispersion equation (4.40). The analysis was done for a shear-band angle $\theta = \theta_V = 55.9°$. In this analysis we assumed that the material length ℓ_c coincides with the mean grain size, here $\ell_c = d_{50} = 0.33\,\text{mm}$. We notice also that between ℓ and ℓ_c eq. (4.26.3) and that between Q and d_B eq. (4.37.2) are holding. The result of the analysis is shown in Fig. 19 below.

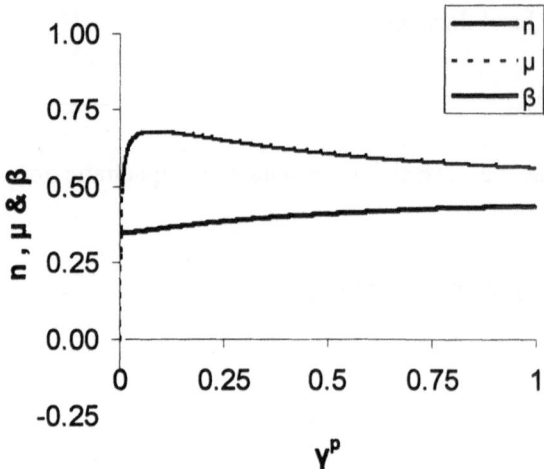

Fig. 18: Extrapolated material properties in the softening regime

X-ray radiographs of the considered sand specimens revealed a shear-band thickness that is about 13 times the mean grain diameter d_{50}, which is matched above for large plastic deformation of the shear band material. Fig. 19 demonstrates in fact that no localization is possible at the bifurcation point of the classical description and that the deformation rapidly localizes in the post-bifurcation regime.

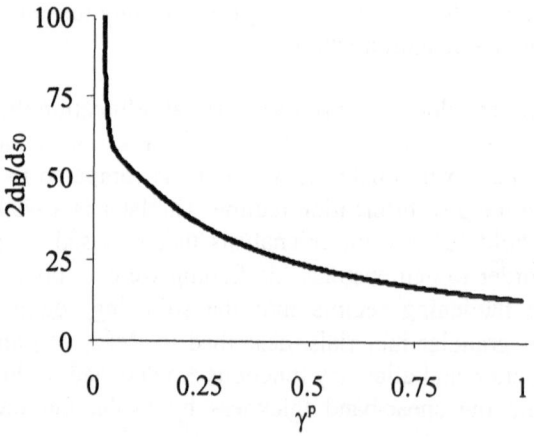

Fig. 19: Approximate shear-band thickness ($\ell_c = d_{50} = 0.33$ mm)

5. Regularization of Strain Softening Models
5.1 The Restricted MINDLIN-type Continuum

Constitutive equations (4.28) describe the behavior of a 2nd gradient elastoplastic solid which can be interpreted as follows: The true stress-rate consists of two parts,

$$\dot{\sigma}_{ij} = \dot{\sigma}_{ij}^{(0)} + \dot{\sigma}_{ij}^{(2)} \tag{5.1}$$

where $\dot{\sigma}_{ij}^{(0)}$ and $\dot{\sigma}_{ij}^{(2)}$ depend on the strain-rate and its Laplacian, respectively,

$$\dot{\sigma}_{ij}^{(0)} = L_{ijkl}^{ep}\dot{\varepsilon}_{kl}$$
$$\dot{\sigma}_{ij}^{(2)} = -L_{ijkl}^{p}\ell^2\nabla^2\dot{\varepsilon}_{kl} \tag{5.2}$$

The stress-rate tensor $\dot{\sigma}_{ij}^{(0)}$ coincides with the constitutive stress-rate of the classical flow theory of plasticity. Following Mindlin's [48] terminology, this stress is called the rate of the 'Cauchy' stress.

We define a self-equilibrating double-stress $\dot{m}_{ijjk}$ such that $\dot{\sigma}_{ij}^{(2)}$ is in equilibrium with it,

$$\dot{\sigma}_{ij}^{(2)} + \partial_k\dot{m}_{kij} = 0 \tag{5.3}$$

$\dot{\sigma}_{ij}^{(2)}$ is interpreted as the rate of a *relative stress*. From these equations the relative stress-rate of the considered Mindlin-type continuum can be eliminated, yielding

$$\dot{\sigma}_{ij} = \dot{\sigma}_{ij}^{(0)} + \partial_k\dot{m}_{kij} \tag{5.4}$$

Finally if one assumes that the double stress-rate $\dot{m}_{kij}$ is proportional to the gradient of the strain-rate,

$$\dot{m}_{kij} = L_{jkmn}^{p}\ell^2\dot{c}_{imn}\; ; \quad \dot{c}_{ijk} = \partial_i\dot{\varepsilon}_{jk} \tag{5.5}$$

and neglects non-linear gradient terms, condition (5.4) implies eqs. (5.1) and (5.2). Accordingly, the constitutive eqs. (5.1) and (5.2) of gradient dependent flow-theory with yield surface and plastic potential surface describe the behavior of a restricted Mindlin-type continuum; i.e. a micro-homogeneous continuum for which the macroscopic strain-rate coincides with the micro-deformation rate. This leads to vanishing relative deformation rate, and to a rate of micro-deformation gradient which coincides with the strain-rate gradient.

We remark finally that the Mindlin-type continuum extension of flow-theory involves only one new material parameter, the material length ℓ. This is unlike Cosserat-continuum extension of plasticity theory, where plastic solids with the microstructure of a micropolar continuum are considered, and, in addition to the material length, a shear modulus for the antisymmetric shear stress-rates has to be introduced [54].

- *Virtual Work Equation*

The weak formulation of the balance law of linear momentum together with the appropriate set of boundary conditions for the considered 2nd gradient elasto-plastic solid is achieved through the principle of virtual work. The later takes the following form [29,30] and [92],

$$\int_V \{\dot{\sigma}_{ij}^{(0)}\delta\dot{\varepsilon}_{ij} + \dot{m}_{ijk}\delta\dot{c}_{ijk} - (\dot{f}_i - \rho\partial_{tt}v_i)\delta v_i\}dV = \int_{\partial V} (\dot{t}_i\delta v_i + \dot{m}_i D\delta v_i)\,dS$$

(5.6)

$$\delta v_i = 0; \quad D\delta v_i = 0 \quad \text{on}: \partial V_v$$

In the above virtual work equation the following notation is used:
- δv_i: virtual velocity
- $\delta\dot{\varepsilon}_{ij} = (\partial_i\delta v_j + \partial_j\delta v_i)/2$
- $\dot{f}_i$: body force (rate)
- ρ: mass density of material
- $\dot{t}_i$: surface tractions (rates)
- $\dot{m}_i$: surface double-tractions (rates)
- $D = n_k\partial_k$: the normal derivative to the boundary ∂V_u and n_i is the (unique) unit outward normal on this boundary.

We notice that even with the velocity given on a boundary point, its derivative at this point is unrestricted. This means that at a boundary point the velocity and/or its normal derivative may be prescribed [99].

In order to discuss the implications of the extra gradient terms in uniqueness questions, we will consider here the simplest 1-dimensional realization of the 2nd gradient elastoplastic model. In this case we have the following reductions [98]: The strain and the strain gradient are expressed in terms of the first and second velocity gradient,

$$\dot{\varepsilon}_{ij} \rightarrow \dot{\gamma} = \nabla v$$

(5.7a)

$$\dot{c}_{ijk} \rightarrow \dot{c} = \nabla\dot{\gamma} = \nabla^2 v$$

The equilibrium stress-rate is expressed in terms of the Cauchy stress-rate and the gradient of the double stress

$$\dot{\sigma}_{ij}^{(0)} \rightarrow \dot{\tau}^{(0)}$$

$$\dot{m}_{ijk} \rightarrow \dot{m} \qquad (5.7b)$$

$$\dot{\tau} = \dot{\tau}^{(0)} - \nabla \dot{m}$$

The stiffness tensor reduces practically in an elastic and a plastic modulus,

$$L_{ijkl}^{ep} \rightarrow G^{ep} = G - G^{p}$$

$$\qquad (5.7c)$$

$$L_{ijkl}^{p} \rightarrow G^{p} = \frac{<1>}{1+h} G$$

where G is the elastic shear modulus, and

$$h = -h_s < 0; \qquad 0 < h_s << 1 \qquad (5.7d)$$

is the (dimensionless) *softening modulus*. With that the constitutive relations become

$$\dot{\tau}^{(0)} = G^{ep} \nabla v$$

$$\qquad (5.7e)$$

$$\dot{m} = G^p \ell^2 \nabla^2 v$$

We notice finally that from eqs. (5.7b) and (5.7e) we obtain the following constitutive relation for the total (equilibrum) stress

$$\dot{\tau} = G^{ep} \nabla v - G^p \ell^2 \nabla^3 v \qquad (5.8a)$$

or more explicitly

$$\dot{\tau} = \begin{cases} G\nabla v & \text{in} \quad (E) \\ \\ \frac{G}{1+h}\left(h\nabla v - \ell^2 \nabla^3 v\right) & \text{in} \quad (P) \end{cases} \qquad (5.8b)$$

where (E) signifies the *elastic domain* and (P) the *plastic-softening domain*.
With these expressions one can compute the total virtual work of internal forces for a finite volume, which in the considered one-dimensional case is a line interval [a,b].

$$\delta_2 W^{(i)} = \int_a^b (\dot{\tau}^{(0)} \nabla \delta v + \dot{m} \nabla^2 \delta v) dy \tag{5.9}$$

At the boundary points y=a and y=b kinematic or static constraints are prescribed. In classical continua these are constraints on the velocity and the traction-rate. Since in the constitutive description of 2nd grade solids, second gradients of strain-rates appear, additional kinematic constraints must be prescribed at boundary points. Even with the velocity given on a boundary point, its derivative (normal to the boundary) at this point is unrestricted . This means that at a boundary point the velocity and/or its derivative may be prescribed

$$v = V; \quad \nabla v = R \tag{5.10}$$

For the computation of the virtual second order work of external forces we have to consider the rate of surface tractions and body forces and to postulate double surface forces. By neglecting, for simplicity, volume and inertia forces the second order total virtual work of external forces becomes

$$\delta_2 W^{(e)} = [T\delta v + M\nabla\delta v]_a^b \tag{5.11}$$

where T and M are the rates of boundary traction and boundary double force, respectively. The virtual work equation becomes

$$\delta_2 W^{(i)} = \delta_2 W^{(e)} \quad \Rightarrow \quad \int_a^b (\dot{\tau}^{(0)} \nabla \delta v + \dot{m} \nabla^2 \delta v) dy = [\dot{t}\delta v + \dot{M}\nabla\delta v]_a^b \tag{5.12}$$

The integrant in the l.h.s. of eq. (5.11) is recognized as the 2nd order virtual power of the stresses

$$\delta_2 w = \dot{\tau}^{(0)} \nabla\delta v + \dot{m} \nabla^2 \delta v$$
$$= \nabla[(\dot{\tau}^{(0)} - \nabla\dot{m})\delta v] - \nabla(\dot{\tau}^{(0)} - \nabla\dot{m})\delta v + \nabla(\dot{m}\nabla\delta v) \tag{5.13}$$

From the integral equation (5.12) one recovers thecondition for continued equilibrium for the stress-rate in [a,b]

$$0 = \nabla(\dot{\tau}^{(0)} - \nabla\dot{m}) = \nabla\dot{t} \tag{5.14}$$

On the other hand, the boundary values lead to the following set of admissible static boundary conditions at y=a and/or y=b,

$$\dot{\tau} = T; \quad m = M \tag{5.15}$$

The constraint (5.15/1) takes the place of the classical boundary condition for surface tractions. However, due to the constitutive equation (5.8b), in the plastic domain this results to a boundary condition which is a linear combination of the first and third derivative of the velocity. On the other hand, due to the constitutive equation (5.7e) for the double stress rate, the constraint (5.15/2) means that at a boundary point the Laplacian of the velocity might be given,

$$-\ell^2 \nabla^3 v + h\nabla v = (1+h)\frac{T}{G}$$

$$\tag{5.16}$$

$$\ell\nabla^2 v = (1+h)\frac{M}{G\ell}$$

Parameters R and M which appear in the boundary conditions (5.10) and (5.16) are seen here as possible measures of the *'roughness'* of a structure interfacing with a material surface.

5.2 Second Order Work

Let us consider the problem od a shear band, as shown in Fig. 20 below. We distinguish betwenn two domains: (a) domain (E), external to the shear band, where elastic unloading is taking place, and (b) domain (P), internal to the shear band, where elastoplastic strain softening is ocuring. For the considered 1D-problem, and under dead loading conditions, the 2nd gradient virtual work equation (5.12) becomes

$$\int_{(P)} \delta_2 w \, dy = 0 \tag{5.17}$$

where

$$\delta_2 w = \frac{G}{1+h}\left\{ h(\nabla v)^2 + \ell^2 \left(\nabla^2 v\right)^2 \right\} \tag{5.18}$$

As outlined in section 2, the integrand $\delta_2 w$ of the above integral in eq. (5.17), has the character of second order work, and plays a central role in the formulation of uniqueness theorems. Inside the localized zone the strain-rate term in the r.h.s. of eq. (5.18) is non-positive, due to *material strain softening*, whereas the strain-rate gradient term is non-negative. Thus in spite of the gradient term, locally the second order work may become negative (e.g. in the middle of the localized zone), and the local sufficient criterion for uniqueness, inequality (2.15), breaks down. However, under some restrictions global uniqueness is restored as shown below.

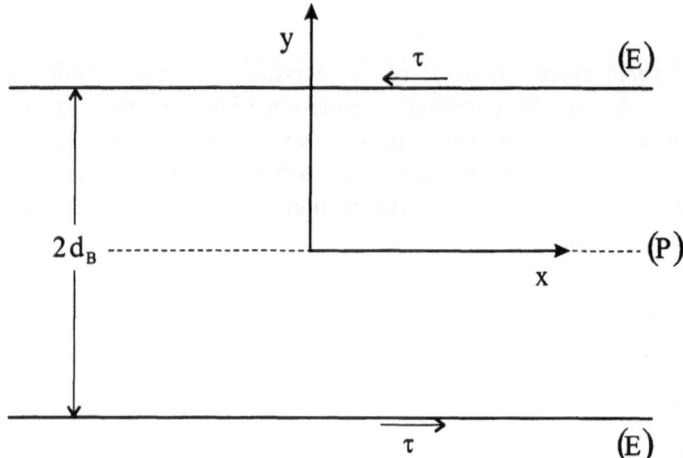

Fig. 20: The elastoplastic shear band for strain softening material

- *Rescalling*

By appropriate *rescaling* of coordinates we observe that the gradient-term may dominate, provided that the rate of material strain softening is not very large. In order to demonstrate that, let

$$\xi = \frac{y}{\ell} \tag{5.19}$$

be the dimensionless coordinate perpendicular to the shear band axis, with the origin on the shear band axis.

We assume that the extend of the localized zone must scale with the internal length; i.e. $2d_B = O(\ell)$. Whithin gradient-dependent elastoplasticity theory the shear-band thickness is determined, and the original elastoplastic problem of strain-softening material becomes a problem with internal boundaries, which separate the elastically unloading exterior domain (E) from the plastic localized zone of strain softening material (P). Thus coming back to the *global uniqueness* requirement (2.5)

$$I = \int_V \delta_2 w \, dV \geq 0 \tag{5.20}$$

we observe that this integral can be split into two parts: One computed over the elastic-unloading domain zone, outside the shear-band and the other computed over the elastoplastic-softening localized zone

$$I = I_E + I_P \tag{5.21}$$

We observe also that in the elastic-unloading zone the second-order work is always non-negative, and from eqs. (5.20) and (5.21) we obtain a requirement for I_P

$$I_E \geq 0 \Rightarrow I_P > 0 \tag{5.22}$$

The integral inside the localized zone becomes

$$I_P = \int_{-\xi_B}^{\xi_B} \frac{G}{1+h} \left\{ h \left(\frac{dv}{d\xi} \right)^2 + \left(\frac{d^2 v}{d\xi^2} \right)^2 \right\} d\xi; \quad \xi_B = \frac{d_B}{\ell} \tag{5.23}$$

The shear-band boundary is an elastoplastic boundary (EP) which separates the elastic domain from the plastic domain where plastic strains are generated. We remark that in the elastic domain (E) plastic strain generation is nil and that from this point of view (EP) belongs to (E).

Theorem: On the elastoplastic shear-band boundary (EP) plastic strain-rates are zero; i.e.

$$\dot{\gamma}^P = 0 \quad \text{on (EP)} \tag{5.24}$$

This theorem is justified taking into consideration that in the considered simple 1-D case and according to eq. (4.36) the fundamental shear-band solution is sinusoidal, and

$$\dot{\gamma} = \nabla v = \dot{\gamma}_0 \cos(q\xi); \quad q = \frac{\pi}{2\xi_B} \tag{5.25}$$

Since

$$\dot{\gamma}^P = \dot{\gamma} + \ell^2 \nabla^2 \dot{\gamma} \tag{5.26}$$

we deduce for $\xi = \pm \xi_B$ the requirement (5.24).
The integral requirement (5.22) with the estimate for the shear-band strain field, eq. (5.25), becomes

$$I_P \approx G \dot{\gamma}_0^2 \left(-h_s + q^2 \right) \xi_B > 0 \Rightarrow h_s < h_{s,max} \tag{5.27}$$

where

$$h_{s,max} = \left(\frac{\pi}{2} \frac{\ell}{d_B} \right)^2 \tag{5.28}$$

is the *maximum allowed value for the softening rate*. From this uniqueness requirement, we obtain that the shear-band thickness is bounded

$$2d_B < 2d_{B,max} \qquad (5.29)$$

where

$$2d_{B,max} = \frac{\pi}{\sqrt{h_s}}\ell \qquad (5.30)$$

is the *maximum allowed shear-band thickness* for given softening rate.
One should keep in mind that the shear-band thickness is a directly observable physical property, whereas the softening rate must be determined at best through inverse analysis on the basis of that thickness.

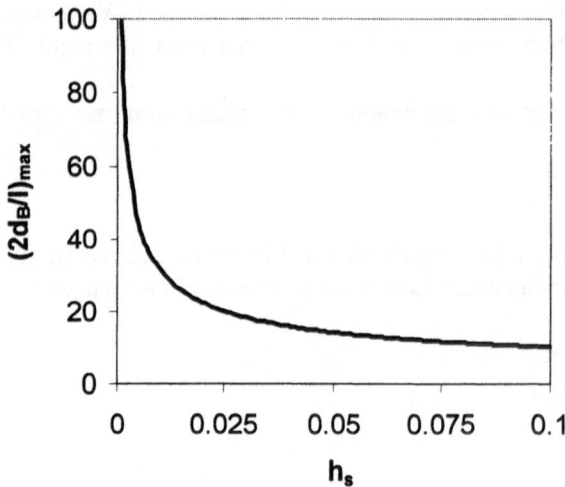

Fig. 21: Maximum shear-band thickness as function of softening rate

This means that from the constitutive modeling point of view one has to observe constitutive inequalities, like

$$h_s < \left(\frac{\pi}{2}\frac{1}{(d_B/\ell)}\right)^2 \qquad (5.31)$$

in order to assure that the gradient regularization leads to unique solution for the shear-band problem. In other words one cannot choose independently both the material length scale ℓ and rate of strain softening h_s. Finally we may conclude with the following,

Proposition: Within the frame of elastoplasticity theory, the mathematically ill-posed incremental, boundary value problem of strain-softening material is regularized by consideration of at least 1st Gradient terms in the expression of the second order work as soon as the chosen material length and the softening rate do not lead to violation of the global uniqueness requirement $I_P > 0$, inside the band.

5.3 Boundary Localizations

When a granular material is sheared against a *rough* boundary, zones of localized deformation are observed at the interface. This happens for example along the walls of silos and at the interface between soil and pile driven into it. Figure **5.1** below shows an x-ray plate of a shear-interface test performed in the biaxial apparatus [76,77].

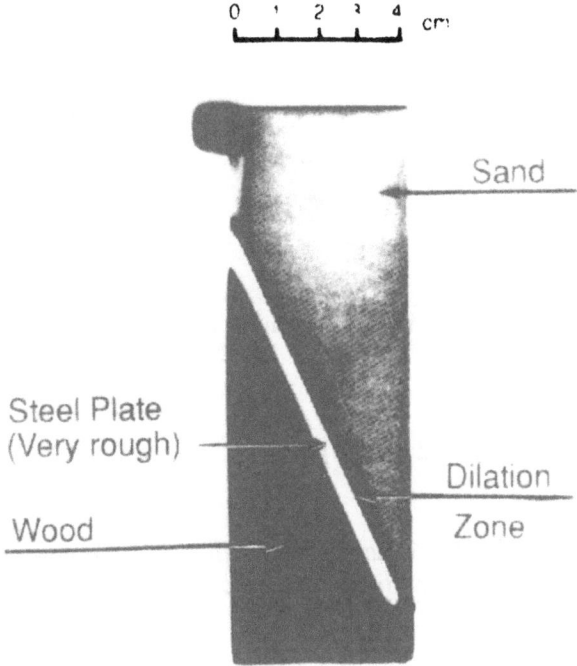

Figure 22: Interface softening dilatancy localization for a rough steel plate interfacing with sand in the biaxial apparatus [77]

In this configuration we distinguish among a softening interface layer (P) of extend d, adjacent to the rough boundary $(-d < y \leq 0)$ and an elastic layer (E) below $(-L \leq y < -d)$, bounded below by the remote boundary. Both layers are separated by the elastoplastic boundary (E/P) $(y = -d)$. Continued equilibrium results in

$$\frac{d\dot{\tau}}{dy} = 0 \Rightarrow \dot{\tau} = \text{const.} \tag{5.32}$$

everywhere in the considered body. In particular in (E) we retrieve the classical elastic solution with linearly varying velocity,

$$\frac{\dot{\tau}}{G} = \dot{\gamma} = C_0 ; \quad \dot{\gamma} = \nabla v \Rightarrow v = C_0(y + L) \quad (E : -L \leq y \leq -d) \tag{5.33}$$

Notice that due to the assumed material strain-softening, $\dot{\tau} < 0$, and with that from eq. (5.33) $C_0 < 0$.

In (P) the full gradient constitutive equations hold (5.7e) and (5.8) for the stress-rate and the double stress-rate,

$$\frac{\dot{\tau}}{G} = \frac{1}{1+h}\left(h\dot{\gamma} - \ell^2 \nabla^2 \dot{\gamma}\right) \tag{5.34}$$

$$\frac{\dot{m}}{G} = \frac{\ell^2}{1+h} \nabla \dot{\gamma} \tag{5.35}$$

From eqs. (5.32) and (5.34) we obtain the governing differential equation for the velocity field in (P),

$$-\ell^2 \frac{d^3 v}{dy^3} + h\frac{dv}{dy} = (1+h)C_0 \quad (P : -d < z \leq 0) \tag{5.36}$$

In order to integrate this differential equation, compatibility conditions at the elastoplastic boundary (EP) as well as boundary conditions at the surface must be specified.

• *Compatibility Conditions:*
Let

$$[Z] = Z^+ - Z^- \tag{5.37}$$

be the jump of a quantity $Z(y)$ across (EP), with

Following the works by Bogdanova-Bontcheva and Lippmann [11] and Unterreiner and Vardoulakis [80], controlled interface shear tests on granular materials were also performed recently by Lerat, in a newly developed Ring-Shear Apparatus [41].

Modeling Considerations
As already oulined above in section 5.1, the appearance of strain-rate gradients into the constitutive equations raises naturally the problem of higher order boundary conditions. Since strain-gradients become important in phenomena of strain localization, one realizes that the simplest class of boundary value problems which involve strain gradients and the effect of higher-order boundary conditions is that of *interface localizations*. An interesting discussion of the question of uniquness within the frame of 2nd order gradient continuum theories can be found in a arecent paper by Chambon et al. [14]. Here we will restrict our analysis to a model problem that was studied earlier by Vardoulakis et al. [98].

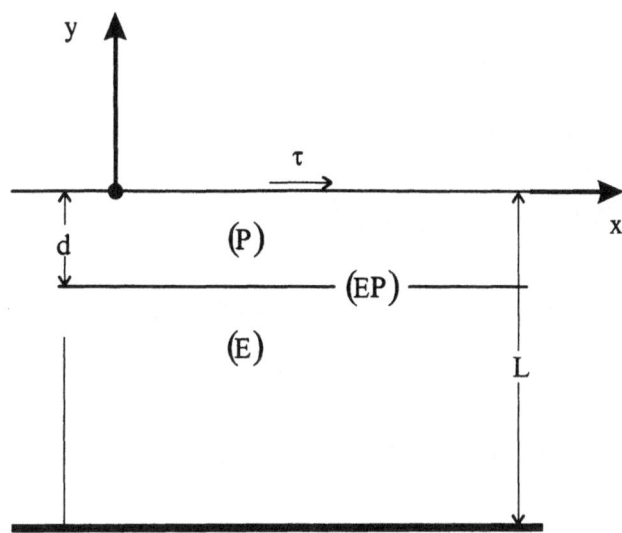

Figure 23: Long strip interfacing with a 'rough' structure under constant normal stress and controlled shear displacement [98].

We consider a long material layer with thickness L. We introduce a vertical coordinate axis z, with its origin at the surface of the layer. The only significant stress component is the shear stress τ, acting on planes parallel to the surface. The corresponding shear velocity is v. Static and kinematic constraints are prescribed at the boundaries $y = 0$ and $y = -L$. Boundary conditions will be discussed in detail below. At this point we differentiate verbally among the two boundaries by calling the upper boundary $(y = 0)$ the *rough interface* and the lower boundary $(y = -L)$ the *remote boundary*.

$$Z^+ = \lim_{y \to -d+} Z; \quad Z^- = \lim_{y \to -d-} Z \qquad\qquad (5.38)$$

being the one-side limes as (EP) is approached from the interior of (P) and (E) respectively.

At (EP) we impose the following requirements,

1. *Continuity of velocity, stress-rate and double stress-rate*

$$\left[v\right] = 0 \Rightarrow v^+ = v^-$$

$$\left[\dot{t}\right] = 0 \Rightarrow h[\dot{\gamma}] - \dot{\gamma}^- - \ell^2 \nabla\dot{\gamma}^+ = 0 \qquad\qquad (5.39)$$

$$\left[\dot{m}\right] = 0 \Rightarrow \left[\nabla\dot{\gamma}\right] = 0 \Rightarrow \nabla\dot{\gamma}^+ = 0$$

2. *Vanishing plastic strain- rates*

$$\dot{\gamma}^{p+} = 0 \Rightarrow \dot{\gamma}^+ + \ell^2\nabla^2\dot{\gamma}^+ = 0 \Rightarrow \nabla^2\dot{\gamma}^+ = -\frac{1}{\ell^2}\dot{\gamma}^+ \qquad\qquad (5.40)$$

Combining eqs. (5.39/2) and (5.40) results in continuity of strain-rate,

$$\left[\dot{\gamma}\right] = 0 \Rightarrow \dot{\gamma}^+ = \dot{\gamma}^- = C_0 \qquad\qquad (5.41)$$

Summarizing the above results in terms of velocity we have the following conditions at (EP) as it is approached from (P),

$$v^+ = C_0(L-d)$$

$$\left[\dot{\gamma}\right] = 0 \Rightarrow \left(\frac{dv}{dy}\right)^+ = C_0 \qquad\qquad (5.42)$$

$$\nabla\dot{\gamma}^+ = 0 \Rightarrow \left(\frac{d^2v}{dy^2}\right)^+ = 0$$

*   *Boundary Condition*
At the free surface we specify first the velocity

$$v(0) = v_0 > 0 \qquad\qquad (5.43)$$

The 3rd order differential equation (5.36) is introducing three integration constants, say C_1, C_2, C_3. To these unknowns one has to add the integration constant C_0, appearing in the solution eq. (5.33) for the velocity in the elastic domain (E) as well as the thickness $d = C_4$ of the softening interface layer (P). For the determination of these five unknowns we have available the three compatibility equations (5.39) and the boundary condition (5.43) for the velocity itself. This means that we need one additional boundary condition at the surface in order to define a well-posed boundary-value problem. The specification of this is additional boundary condition constitutes the central problem of any higher gradient theory. We recall as an example the famous Stokes non-slip boundary condition for the Navier-Stokes equations that has been controversial for many years.

In the above cited paper by Vardoulakis et al. [98] the additional boundary conditions where derived from the virtual work equation as outlined above in section 5.1. According to the order of the imposed boundary constraint one may distinguish among the following two problems:

1. *Dirichlet Problem*: The first derivative (normal to the surface) of the velocity is prescribed,

$$\left(\frac{dv}{dy}\right)_{y=0} = R_0 \qquad\qquad (5.44a)$$

2. *Neumann Problem*: The double stress-rate is prescribed on the surface, $\dot{m}(0) = M$, i.e.

$$\left(\frac{d^2v}{dy^2}\right)_{y=0} = \frac{M_0}{G\ell^2}(1+h) \qquad\qquad (5.44b)$$

The parameters R_0 and M_0 which appear in the boundary conditions (5.44) are seen here as possible measures of the *roughness* of the structure interfacing with the deferrable material surface. As pointed out in the aforementioned paper by Vardoulakis et al.[98] these boundary data must fulfill some restrictions that involve the softening rate, similarly to the ones derived in the previous section for the shear band.

In Fig. 24 below a typical profile for the velocity and the plastic strain-rate with the distance from the surface are shown. These results correspond to the Neumann problem. One can see also from these plots that an elastoplastic softening interface layer bears some similarity with the boundary layer of fluid mechanics as far as boundary conditions are concerned. This can be seen directly from the governing differential equation (5.36) where the leading term is multiplied by the 'small' parameter ℓ^2, and its importance is restricted only close to rough interface, where also the extra boundary conditions (5.44) are prescribed (see e.g. [56]). The main difference from the boundary layer is that the *internal* and *external* solutions are not matched in an asymptotic sense but they are made compatible in an exact manner up to the second derivative along the (generally moving) elastoplastic boundary.

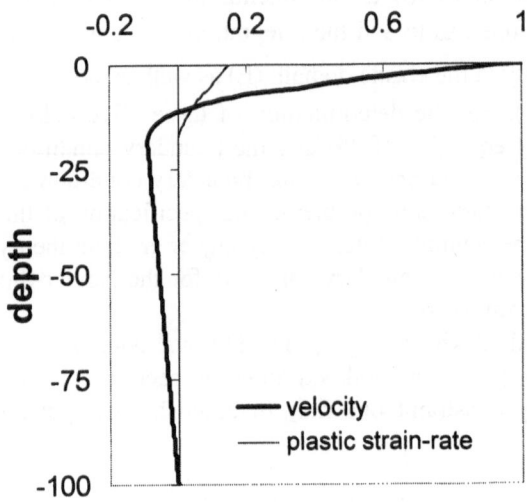

Figure 24: Interface layer solution for Neumann boundary condition

6. References

1. Arthur, J.R.F., Dunstan, T.,Al-Ani, Q.A.J. and Assadi, A. : Plastic deformation and failure of granular media. Géotechnique, 27 (1977), 53-74.
2. Bardet, J.P.:Orientation of shear bands in frictional soils. J. Eng. Mech. ASCE, 117, (1991), 1466-1484.
3. Bardet, J.-P., Proubet, J.: A numerical investigation of the structure of persistent shear bands in granular media. Géotechnique, 41 (1991), 599-613.
4. Beatty, M.F.: Some static and dynamic Implications of the general theory of elastic stability. Arch. Rat. Mech. Anal., 10 (1966), 167-186.
5. Benallal, A., Bilardon, R. and Geymonat G.: Conditions de bifurcation à l'interieur et aux frontières pour une classe de matériaux non standards. Acad. Sci., Paris, 308, série II (1989), 893-898.
6. Bigoni, D. and Hueckel, T.: A note on strain localization for a class of non-associative plasticity ru;es. Ingenieur Archiv, 60 (1990), 491-499.
7. Bigoni, D. and Hueckel, T.: Uniqueness and localization-I. Associative and non-associative plasticity. Int. J. Soilds Structures, 28 (1991), 197-213.
8. Bigoni, D. and Zaccaria, D.: Loss of strong ellipticity in non-associativive elastoplasticity. J. Mech. Phys. Solids, 40 (1992), 1313-1331.
9. Bigoni, D. and Zaccaria, D.: Strong ellipticity of comparison solids in elastoplasticity with volumetric non-associativity. Int. J. Soilds Structures, 29 (1992), 2123-2136.
10. Biot, M.A.: Mechanics of Incremental Deformations, Wiley, 1965.
11. Bogdanova-Boncheva, N. and Lipmann, H.: Rotationsymmetrisches ebenes Fliessen eines granularen Modellmaterials. Acta Mechanica, 21 (1975),93-113.

12. Boulon, M.: Basic features of soil-structure interface bahavior. Computers and Geotechnics, 7 (1989).

13. Bransby, P.L. and Milligan G.W.E.: Soil deformations near cantilever sheet pile walls, Géotechnique, 25 (1975), 175-195.

14. Chambon, R., Caillerie, D. and El Hassan N.: Etude de la localisation unidimensionelle à l'aide d'un modèle de second gradient. C.R. Acad. Sci. Paris, Série II b (1996), 231-238.

15. Christoffersen, J.: Hyperelastic relations with isotropic forms appropriate for elastoplasticity. Eur. J. Mech. A/Solids., 10 (1991), 91-99.

16. Coulomb Charles-Augustin: Essai Sur une application des règles de Maximis & Minimis à quelques Problèmes de Statique, relatifs à l'Architecture, 1773. In: J. Heyman, Coulomb's Memoir on Statics, Cambridge University Press, 1972.

17. Cundall, P.A. and Strack, O.D.L.: A discrete numerical model for granular assemblies. Géotechnique, 29 (1979), 47-65.

18. De Josselin de Jong: Rowe's stress-dilatancy relation based on friction. Géotechnique, 26 (1976), 527-534.

19. Desrues, J.: La Localization de la Déformation dans les Matériaux Granulaires. Thése de Doctorat es Science, USMG & INPG, Grenoble, 1984.

20. Desrues J.: Shear band initiation in granular materials: Experimentation and theory. In: Geomaterials: Constitutive Equations and Modelling (Ed. F. Darve) 1987.

21. Desrues, J. and Hammad, W.: Etude expérimentale de la localisation de la déformation sur sable:Influence de la contrainte moyenne. 12th I.C.S.M.F.E., **1/9** (1989), 31-32.

22. Desrues,J., Mokni, M. and Mazerolle, F.: Tomodensitométrie et la localisation sur les sables. 10th E.C.S.M.F.E., (1991), 61-64.

23. Desrues, J., Chambon, R., Mokni, M. and Mazerolle, F.: Void ratio evolution insideshear bands in triaxial sand specimens studied by computed tomography. Géotechnique, **46** (1996), 529-546.

24. Dietrich, Th.: Der Psammische Stoff als mechanisches Modell des Sandes. Dissertation Universität Karlsruhe 1976.

25. Doris, J.F. and Nemat-Nasser, S.: Instability of a layer on a half space. J. Appl. Mech., 102 (1980), 304-312.

26. Drescher, A., Vardoulakis, I. and Chunhua Han : A Biaxial Apparatus for testing Soils. Geotechnical Testing Journal, GTJODJ, 13 (1990), 226-234.

27. Drucker, D.C.: A more fundamental approach to stress-strain relations. Proc. U.S. National Congress of Applied Mechanics, ASME (1951), 487-491.

28. Emeriault, F. and Cambou, B.: Micromechanical modelling of anisotropic non-linear elasticity of granular medium. Int. J. Solids Structures, 33 (1996), 2591-2607.

29. Germain P.: La méthode des puissances vituelles en mécanique des milieux continus. Part I , J. de Mécanique, 12 (1973), 235-274.

30. Germain P.: The Method of virtual power in continuum mechanics. Part 2: Microstructure. SIAM J. Appl. Math, 25 (1973), 556-575.

31. Gudehus, G.: Elastic-plastic constitutive equations for dry sand. Arch. Mech. Stosowanej, 24 (1972), 395-402.

32. Hadamard, J.: Leçons sur la propagation des ondes at les équations de l'hydrodynamique. Paris:Hermann 1903. Reprinted New York: Chelsea Publishing Co. 1949.

33. Hill, R.: Mathematical Theory of Plasticity, Clarendon Press, Oxford, 1950.

34. Hill, R.: Eigenmodal deformation in elastic/plastic continua. J. Mech. Phys. Solids, 15, (1958), 371-386.
35. Hill R.: Acceleration waves in solids. J. Mech. Phys. Solids, 10 (1962), 1-16
36. Hill, R. and Hutchinson, J.W.: Bifurcation phenomena in the plane tension test. J. Mech. Phys. Solids, 23 (1975), 239-264.
37. Hubbert, M.K.: Mechanics of deformation of crustal rocks: Historical development. In: Mechanical Behavior of Crustal Rocks-The Handin Volume (Ed. N.L. Carter et al.) Americal Geophysical Union, 1981, 1-9.
38. Hutchinson, J.W.: Finite strain analysis of elastic-plastic solids and structures.In: Numerical Solution of Nonlinear Structural Problems (Ed. R.F. Hartung) ASME, 1973.
39. Jean, M.: Frictional contact in collections of rigid or deformable bodies: a numerical simulation of geomaterial motion. In: Mechanics of Geomaterials Interfaces, (Ed. A.P.S. Selvadurai) 1995, Elsevier .
40. Ken-Ichi Kanatani: A micropolar continuum theory for the flow of granular matereals. Int. J. Engng. Sci., 17 (1979), 419-432.
41. Lerat, P., Schlosser, F., and Vardoulakis, I.: Nouvel appareil de ciseillement pour l'étude des interfaces matériau granulaire-structure. 14th ICSMFE, (1997, in print.
42. Loret, B., Prevost, J.H. and Harireche, O.: Loss of hyperbolicity in elastic-plastic solids with deviatoric associativity. Eur. J. Mech.,A/Solids, 9 (1990), 225-231.
43. Lloyd, G.E.R.: Early Greek Science, Norton, 1970.
44. Mandel, J.: Ondes plastiques dans un milieu indéfini à trois dimensions. Journal de Méchanique, 1 (1962), 3-30.
45. Mandel, J.: Conditions de stabilité et postulat de Drucker. In: Rheology and Soil Mechanics, Springer, 58-67, 1966.
46. Mandel, J.: Propagation des surfaces de discontinuite dans un milieu elasto-plastique. In: Stress Waves in Anelastic Solids, Springer, Berlin, 1964, 331-341.
47. McMeeking, R. M. and Rice, J.R.: Finite-element formulations for problems of large elastic-plastic deformation. Int. J. Solids Struct., 11 (1975), 601-616
48. Mindlin, R.D.: Micro-structure in linear elasticity. Arch. Rat. Mech. Anal., 10 (1964), 51-77.
49. Mohr, O. Welche Umstände bedingen die Elastizitätsgrenze und den Bruch eines Materials? Zeitschrift des Vereines deutscher Ingenieure, 44 (1900), 1-12.
50. Molenkamp, F.: Comparisn of frictional material models with resoect to shear band initiation. Géotechnique, 35 (1985),127-143.
51. Mroz, Z. (1963). Non-associate flow laws in plasticity. J. de Mécanique, 2, 21-42.
52. Mroz, Z. Mathematical Models of Inelastic Behavior. University of Waterloo Press, 1973
53. Mühlhaus, H.-B. and Aifantis, E.C.: A variational principle for gradient plasticity. Int. J. Solids Structures, 28 (1991), 845-857.
54. Mühlhaus, H.-B. and Vardoulakis, I.: The thickness of shear bands in granular materials. Géotechnique, 37 (1987), 271-283.
55. Nguyen, Q.S. and Bui H.D.: Sur les matériaux élastoplastiques à écrouissage positif ou négatif. Journal de Méchanique,3 (1974), 322-432.
56. Ockendon, H. and Ockendon, J.R.: Viscous Flow. Cambridge University Press, 1995.
57. Oda, M., Kazama, H. and Konishi, J.: Effects of induced anisotropy on the development of shear bands in granular materials. Mechanics of Materials, 25 (1997), in print.

58. Ord, A., Vardoulakis, I. and Kajewski, R.: Shear band formation in Gosford Sandstone. Int. J. Rock Mech. Min. Sci. & Geomech. Abstr., 28 (1991), 397-409.
59. Ortiz, M., Leroy, Y. and Needleman, A.: A finite element method for localized failure analysis. Comput.Meth. Appl. Mech. Eng., 61 (1987), 189-194.
60. Papamichos, E. Labuz, J.F. and Vardoulakis, I.: A surface instability detection apparatus. Rock Mech. And Rock Eng., 27 (1994), 37-56.
61. Papamichos, E. and Vardoulakis, I.: Effect of confining pressure in shear band formation in sand.Géotechnique, 45 (1994), 649-661.
62. Raniecki, B. and Bruhns, O.T.: Bounds to bifurcation stress in solids with non-associated plastic flow law at finite strain. J. Mech. Phys. Solids, 29 (1981), 153-172.
63. Reynolds, O.: On the dilatancy of media composed of rigid particles in contact. With experimental illustrations. Phil. Mag. (2) 20 (1885), 469-481. Also: Truesdell, C. and Noll, W.: The Non-Linear Field Theories of Mechanics, Handbuch der Physik Band III/3, section 119, Springer 1965.
64. Rice, J.R.: The localization of plastic deformation. Theoretcal and Applied Mechanics (Ed. by W.T. Koiter), Proc. 14th IUTAM Congress, Delft, 1977, 207-221.
65. Rice, J.R. and Rudnicki, J.: A note on some features of the theory of localization of deformation. Int. J. Solids Structures, 16 (1980), 597-605.
66. Roscoe, K.H.: The influence of strains in Soil Mechanics. Géotechnique, 20 (1970), 129-170.
67. Rowe, P.W.: The stress-dilatancy relation for static equilibrium of an assembly of particles in contact. Proc. Roy. Soc., 269 (1962),500-527.
68. Rowe, P.W.: Theoretical meaning and observed values of deformation parametrs for soil. Proc. Roscoe Mem. Symp., Cambridge, (1971), 143-194.
69. Rudnicki, J. and Rice, J.R.: Conditions for the localization of deformation in pressure-sensitive dilatant materials. J. Mech. Phys. Solids, 23 (1975), 371-394.
70. Scarpelli, G. and Wood, D.M.: Experimental observations of shear band patterns in direct shear tests. In: Deformation and Failure of Granular Materials, Balkema, (1982), 473-484.
71. Shanley, R.F.: Inelastic column theory. J. Aeronaut., 4 (1947), 101-104.
72. Shield, R.T.: Mixed boundary value problems in soil mechanics, Q. Appl. Math., 11 (1953), 61-75.
73. Sommerfeld , A.: Mechanik der deformierbaren Medien. Verlag Harri Deutsch (1948) 1992.
74. Tatsuoka F., Nakamura S., Huang C.-C. and Tani K.: Strength anisotropy and shear band direction in plane strain tests on sands. Soils and Foundations, 30 (1990), 35-54.
75. Taylor, D.W.: Fundamentals of Soil Mechanics, John Wiley, 1948.
76. Tejchman, J.: Scherzonenbildung und Verspanungseffekte in Granulaten unter Berücksichtigung von Korndrehung. Dissertation Univesität Krlsruhe, 1990.
77. Tejhman , J. and Wei Wu: Experimental and numerical study of sand-stell interfaces. Int. J. Num. Anal. Meth. Geomech., 19 (1995), 513-536.
78. Thomas T.Y.: Plastic Flow and Fracture, Vol.2., Academic Press, 1961.
79. Truesdell C. and Noll W.: Nonlinear Field Theories of Mechanics, Handbuch der Physik, Vol. III/3, Sections 68,68bis,69,99,100, Springer 1965.
80. Unterreiner, F. and Vardoulakis, I.: Interfacial localisation in granular media. Computer Methods in Geomechanics (Ed. Siriwardane & Zaamn), Balkema, (1994), 1711-1715

81. Vardoulakis I.: Berechnungsverfahren für Erdkörper mit plastischer Ver- und Entfestigung: Entstehung und Ausbreitung von Scherfugen. DFG Report GU 103/16, 1974.
82. Vardoulakis I.: Die lineare Näherung des Prinzipes der virtuellen Verschiebungen. Int. Conf. on Num. Meth. in Soil and Rock Mech., Karlsruhe 1975, Universität Karlsruhe, (Ed. G. Borm and H. Meissner) (1975), 39-46.
83. Vardoulakis I.: Equilibrium theory of the shear bands in plastic bodies. Mech. Res. Comm., 3 (1976), 209-214.
84. Vardoulakis I.: Shear band inclination and shear modulus of sand in biaxial test. Int. J. Num. Anal. Meth. in Geomechanics, 4 (1980), 103-119.
85. Vardoulakis, I.: Stability and bifurcation of undrained plane rectilinear deformations on water-saturated granular soils. Int. J. Num. Anal. Meth. Geomechanics, 9 (1985), 399-414.
86. Vardoulakis, I.: Dynamic stability of undrained simple shear on water saturated granular soils. Int. J. Num. Anal. Meth. Geomechanics, 10 (1986), 177-190.
87. Vardoulakis I.: Theoretical and experimental bounds for shear-band bifurcation strain in biaxial tests on dry sand. Res Mechanica, 23 (1988), 239-259.
88. Vardoulakis, I.: Shear-banding and liquefaction in granular materials on the basis of a Cosserat continuum theory. Ingenieur Archiv, 59 (1989), 106-113.
89. Vardoulakis, I.: Potentials and limitations of softening models in Geomechanics (the role of second order work). Eur. J. Mech. A/Solids, 13 (1994), 195-226.
90. Vardoulakis, I.: Deformation of water saturated sand: I. Uniform undrained deformation and shear banding. Géotechnique, 46 (1995), 441-456.
91. Vardoulakis, I.: Deformation of water saturated sand: II. The effect of pore-water flow and shear banding. Géotechnique, 46 (1995),457-472.
92. Vardoulakis, I. and Aifantis, E.C.: A gradient flow theory of plasticity for granular materials. Acta Mechanica, 87 (1991), 197-217.
93. Vardoulakis, I. and Frantziskonis, G.: Micro-structure in kinematic-hardening plasticity. Eur. J. Mech./Solids, 11 (1992), 467-486.
94. Vardoulakis I. and Goldscheider M.: A biaxial apparatus for testing shear bands in soils. 10th Int. Conf. Soil Mech. Found. Engineering, Stockholm 1981, Vol. 4/61, (1981), 819-824, A.A. Balkema
95. Vardoulakis, I. and Graf B.: Imperfection sensitivity of the biaxial test on dry sand. IUTAM Conf. on Deformation and Failure of Granular Materials, Delft 1982, 485-491, A.A. Balkema.
96. Vardoulakis, I. and Graf, B.: Calibration of constitutive models for granular materials using data from biaxial experiments, Géotechnique, 35 (1985), 299-317.
97. Vardoulakis I., Graf B. and Hettler A.: Shear-band formation in a fine-grained sand. 5th Int. Conf. Num. Methods in Geomechanics, Nagoya 1985, 517-522, A.A. Balkema.
98. Vardoulakis, I., Shah, K.R. and Papanastasiou P.: Modelling of tool-rock interfaces using gradient-dependent flow theory of plasticity. Int. J. Rock Mech. Min. Sci. & Geomech. Abstr., 29 (1992), 573-582.
99. Vardoulakis I. and Sulem J.: Bifurcation Analysis in Geomechanics, Blackie Academic and Professional, 1995.